To

Dr. HAMILTON GRAY
Professor of Soil Mechanics and
Chairman of the Deptt. of
Civil Engineering,
Ohio State University,
Columbus, Ohio (U.S.A.),
in
appreciation of the valuable
guidance the author received
during his doctorate work under
him, from 1962 to 1965.

PREFACE TO THIRD EDITION

The author has received several requests from colleagues across the country and from abroad to bring out another edition of this book. Consequently, a third edition has been brought out by the author acting as a publisher also. The present edition is in a slightly revised form only. It is believed that the readers would like the present form as their predecessors and colleagues like the second edition.

S. B. Sehgal

PREFACE TO SECOND EDITION (ABBREVIATED)

The author takes the pleasure of introducing this volume on Soil Mechanics, to the students in Civil Engineering and the field engineers. Soil Mechanics emerged as a separate branch of Civil Engineering only in recent years. Its importance at the University level in India was realized only after 1950, when it was introduced as a full course for the B.Sc. Engineering Degree at most of the Institutions. In the later half of the last decade, this subject was introduced for specialization at the M.Sc. level in some selected Universities and Institutions.

The author is aware that a number of books by well renowned authors are available in this important field. However, this volume has some unique features.

* The book is written for the B.Sc. Engineering curriculum being followed at the Indian Universities, Colleges and higher Technological Institutes. Since it is meant for the first engineering degree, the subject-matter has been presented in a concise but precise manner keeping in view the time at the disposal of a student who has usually to follow a heavy schedule. However, no important detail has been omitted.

* The subject matter is divided into 16 chapters. Each chapter is followed by a list of relevant references and University questions.

* While dealing with the subject matter, the needs of both the average student as well as the more ambitious student have been kept in mind. Whenever an important statement has been made in the text without a detailed discussion, a reference number within the brackets is quoted and this reference number corresponds to the serial number of the relevant reference in the list at the end of the chapter. This is aimed to help a more ambitious student to look up

the details in the relevant reference without loss of time. Any student who later on may wish to undertake higher studies in the subject would be very much benefited by this system of making references. The average student may not be interested in details that have been byepassed.

* Chapters on Soil Moisture and Permeability, Consolidation Characteristics of Soils, Shear Strength of Soils and Bearing Capacity of Soils have been dealt in greater detail, incorporating such modern developments on these topics, as are not available in many text books on the subject.
* In order to explain the applications of the principles dealt with in the text, a number of examples have been solved in each chapter. The metric system has been used all along.
* The book contains over 150 illustrations.
* Subject index provides convenience to the readers.
* The symbols used are the ones commonly used in the literature in this area and mostly conform to the ones suggested in the Third International Conference on Soil Mechanics & Foundation Engineering.

To write this volume, reference has been made to the works of many authors and organisations. Soil Indentification Tests mentioned on page 32 under article 3.3 and table 4·8 in chapter 4 have been reproduced from IS: 1498-1959. Figure of thin wall tube sample in Chapter 16 and the related material have reproduced from IS : 2132-1963. Also table III in the Appendix belongs to IS : 1904-1961.* For fuller details the readers may refer to these references. The permission of the ISI for reproduction is gratefully acknowledged.

M/S. John Wiley and Sons of New York have also been kind to give permission to reproduce some figures in chapters 7, 11 and 12 from the classical works of Terzaghi, Peck and Taylor. The permission is gratefully acknowledged. Any other specific information referred to from the other sources and by omission not incorporated here is gratefully acknowledged.

The author has spent many long hours in preparing this manuscript. He wishes to place on record his sincere appreciation of the patience shown by his wife Mrs. Vinod Sehgal and the encourage ment received by him when the book was under preparation.

The author wishes to thank Professor Kulbhushan of the Punjab Engineering College, Chandigarh for helping him in reading some of the proofs when the manuscript was under print.

Constructive criticism from colleagues in the various Institutions and the student-engineers will be highly appreciated.

S. B. Sehgal

---

* It is desirable that for more complete details reference may be made to the latest version of these standards available from Indian Standards Institution, Manak Bhavan, New Delhi.

CONTENTS

Chapter 1
INTRODUCTION

Chapter 2
BASIC DEFINITIONS, WEIGHT AND VOLUME RELATIONSHIP, ETC.

Chapter 3
SIMPLE PHYSICAL PROPERTIES OF SOIL AND CLASSIFICATION TESTS

Chapter 10

SHEARING CHARACTERISTICS OF SOIL

Chapter 11

EARTH PRESSURE, EARTH PRESSURE THEORIES AND SIMILAR PROBLEMS

Chapter 12

BEARING CAPACITY OF SOILS

(v)

Chapter 13

STABILITY OF SLOPES

Chapter 14

COMPACTION OF SOILS

1

Introduction

1.1 Soils (Definition)

The word *Soil* is used in many professional fields and depending upon the context in which it is used, it has different meanings. Chamber's dictionary defines it as, "the mould on the surface of the earth which nourishes plants". Webster defines it as, "the loose surface material of the earth in which plants grow, in most cases consisting of disintegrated rock with an admixture of organic matter". Both these definitions are akin to the agricultural Soil Science, and may safely be accepted as the ones that describe the agriculturist's concept of the soils. A geologist on the other hand uses this term in a restricted sense to define the material which is produced as a result of disintegration of rocks and which has not been transported from its original position. A pedologist will define it quite differently. For example, pedologist Joffe has defined soil as follows : "The soil is a natural body, differentiated in horizons of mineral and organic constituents, usually unconsoildated, of variable depth, which differs from the parent material below in morphology, physical properties and composition, and biological characteristics."

As such, each field of specialization defines this term to incorporate the material that concerns the respective field.

The main job of a *Civil Engineer* is the design and construction of buildings, highways, airports and in fact any structure that rests on the ground. In executing his job, he comes in contact with all the material on the earth's crust and as such in the broadest sense this word *i.e.* soil in the civil engineering field would define all the loose unconsoildated, disintegrated material, transported or untransported from its original position, lying on top of hard rock below. The solid rock underneath is defined as *bed rock.*

Thus an agronomist is concerned with only the top superficial material, perhaps a couple of metres deep in which the plants grow, a geologist with the disintegrated material on top of original rock, but a civil engineer has to go to greater depths, sometimes even to bedrock or even deeper to rest his structure and all this material that he comes across, may be defined as *Soil* for him.

It may be mentioned here that when we term a material as soil, besides the mineral aggregate, it might contain some water or some decomposed organic matter as its constituent part and the general definition will embrace these materials too.

Till very recently, in the Civil Engineering field, no distinction was made between what a geologist would define as "soil" and as "rock" and the term soil would include the soild rock also when used in relation to foundations. However, the profession is now beginning to distinguish between the top loose material and the bottom solid rock and two different sciences viz. "Soil Mechanics" and "Rock Mechanics" have emerged, giving an insight into the properties of each of the two materials.

1·2. Soil Mechanics

Till the beginning of the present century, the behaviour of "Soils" as a material was never studied scientifically, although it is as important a material as any other engineering material like steel, cement or surkhi, etc. Perhaps the impact of its importance was not felt in the earlier eras of history, although one would imagine that foundation problems must have existed when pyramids of Egypt or aqueducts and roads of the Roman Empire were built. It is only after the turn of the last century that soils received the due attention and the investigations of their behaviour on a scientfic basis was taken up, possibly after the famous land slides of railway tracks in Sweden occured and the difficulties felt in the construction of Keil Canal (in Germany) and the famous Panama Canal.

Credit for developing ways and means of studying soil properties, as we do to-day, goes to Dr. Karl Von Terzaghi who is rightly called the father of modern Soil Mechanics, which term broadly includes all the principles and techniques by which the soil properties are studied scientifically. It is Terzaghi's concepts which have been developed further by his students and successors and in this process of development, altogether new concepts and theories have emerged and these concepts and theories have specifically assumed an important and separate entity in the field of civil engineering under the name Soil Mechanics.

1·3. Principles of Mechanics Applied to Soils

Principles of mechanics of solids involving stress-strain relationships are applied to all solid engineering materials such as steel, concrete, etc. ; to study their fundamental behaviour and to evolve ways and means to design the structural members made from these materials. Laboratory procedures are developed to study such fundamentals as stress, strain and strength and design procedures are drawn out to see that the strength exceeds the stress by a reasonable margin of safety. Theoretical studies, specially in materials, which are elastic in nature, are very important. In some other materials, theoretical studies regarding strees-strain behaviour coupled with experimeneal observations help in understanding their basic properties. Sometimes the material is such that the engineer has to depend upon purely experimental data.

Application of the stress-strain relationship studies to *Soils* is very recent. However, some of the important practical problems such as stress-distribution under foundations and foundation settlement analyses are based upon the principles of *Mechanics* and the theories developed on these principles. Furtheron, shear failure of foundations is akin to the shear failure in any other material and here again the principles of mechanics of solids are applied to evolve the shear stress due to loads and shear strength of the soils underneath.

Apart from the mechanics of solids, some of the investigations of fluid mechanics are akin to the considerations in seepage through soils. Similarly, recent studies of technology of materials have shown that some of the problems in "Soil mechanics" also may have a direct link with the rehological approach of the study of materials. Clays have, of late, been thought to possess visco-elastic properties and attempts are being made to study them from a point of view of the principles of visco-elasticity.

It may, however, be mentioned here that simple elastic and plastic properties are not enough to understand the behaviour of soils, because in no way any soil can be compared to a purely elastic or plastic material. Number of parameters needed to comprehend the behaviour of soils is much more compared to any other material. Moreover, soil variation occurs from point to point on the ground and at the same spot it varies from layer to layer as you dig deep into the ground.

As a conclusion, an understanding of the principles of mechanics and their application, is very essential to the study of soil mechanics but this wholly does not constitute the basis for the study of soil mechanics, because some applications of other basic sciences such as chemistry and physics have also contributed a good deal to the understanding of soil action. Moreover, some laboratory investigations and the conclusions evolved from these studies have added a good deal to the understanding of soils. However, the term "Soil Mechanics" will embrace all such studies as would include principle of mechanics, chemistry or physics that help in the study of soils.

1.4. Importance of Soil Mechanics

As stated earlier a study of the properties of soils escaped the notice of civil engineer till the beginning of the present century. Most of the foundation engineering work for the important historical building seems to have been done on the basis of experience in the area and the knowledge inherited from the forefathers by their children. No evidence of any scientific approach to the layout and construction of the foundations of these important buildings exists. However, some failures in important construction projects diverted the attention of the civil engineers to this important field.

The life of any structure naturally depends upon its foundations and since the foundations have either to rest directly on the soil surface or to come in contact with the soil in one way or the other, a knowledge of the properties of soils, in general, is very important. Some of the soils such as peat and organic silts are so compressible that they cannot be used as foundation materials, whereas others like sand and gravel are excellent foundation materials for most of the construction projects. Moreover, soil properties vary from point to point on a construction site. As such, a differentiation of soils and understanding of the properties of the soils existing at the place of construction is very important.

It may be emphazised here that a civil engineer may take a chance to guess the soil properties at a site for a small-sized project, but for a large project such as a multi-storeyed building, a highway or an air-port, merely guessing the soil properties can lead to foundation failures and ultimate economic loss. For such projects, a comprehensive soil engineering study is very essential in order to get fairly complete information with regard to the foundation design.

1.5. Origin and Formation of Soils and Soil Categories

Rocks are the parent material from which the *soils* are formed. When a rock surface gets exposed to the atmosphere for an appreciable time, considerable changes take place in its structure, such changes lead invariably to its disintegration and with further action of weathering agencies[1], the particles change their size and shape by further disintegration. The particle size may vary from a large-sized boulder to a small crystal of clay. The process of change resulting from an atmospheric exposure, is called *weathering of rocks*. Besides such rock weathering, sometimes internal changes take place in the rock due to presence of water, carbon-di-oxide or some organic matter which bring a change in the chemical composition of the mineral constituting the rock. Such a process is called *alteration*. Temperature of the atmosphere is also an important factor that brings in changes. So a soil derived from the same parent rock may or may not have the same minerological composition, as the original rock has. Furtheron the process of *erosion*[2] which results in the transportation of the weathered product from its original position occurs

---

1. Weathering agencies are many and it seems difficult in this text to describe the effect of these agencies on rocks. The student may consult any book on Engineering Geology.
2. Erosion can be by wind or by water.

simultaneously with the rock-weathering process and it is difficult sometimes to separate out the effort of weathering from that of erosion. Besides these complex factors that go to form the soils, rocks are also of three major types—*Igneous, Sedimentory* and *Metamorphic*—and depending upon the rock type, from which the soil originates ; the soil type will also vary, although the process of weathering and alteration may be the same. Observations have also been made indicating that the nature of weathering and rate of weathering depend upon the rock. Harder minerals will naturally be more resistant to weathering action. Also more chemically stable minerals will retain their identity when chemical changes tend to occur. For example, the latter is true in free quartz, which is chemically quite stable and hence resistant to changes. Some chemical actions that take place span over a long duration. Prolonged chemical action sometimes effects even some stable minerals that have been formed as a result of rock weathering and lead to the formation of secondary products of weathering. That is how the clay[1] minerals are formed from silicates, when moisture ac s for a long time.

The soils formed by weathering may remain in position on the parent rock, in which case they are termed as *residual soils* or they may get transported by the various transporting agencies such as wind, water, ice or even gravity, etc. when they are termed as *transported soils*. During the process of transportation, soil of one origin may get mixed with a soil of another origin. After mixing, they may undergo further changes and form a completely different soil. These transported soils finally get deposited, as the velocity of the transporting medium reduces. The coarser particles get settled first, and with further reduction in speed the finer particles will also settle down eventually. At the place of settlement, further weathering of the material can take place.

The transported materials can be further divided into many types. Nomenclature for these types is presented in Table 1.1 below, depending upon their transporting medium and the place of deposition.

---

1. Clay minerology is a very complicated and highly intricate subject and research is underway to study some of basic properties of clay that seem to have a bearing on the difficulties encountered in dealing with them on actual construction projects.

TABLE 1·1

Nomenclature for Soils

1,	Transported by running water e.g. rivers and streams	Alluvial
2.	Transported by wind	Aeoline
3.	Transported by Glaciers	Glacial
4.	Deposited in Lakes	Lacustrine
5.	Deposited in sea	Marine

The broad classifications given above are important to understand the general nature of soils in a geological region, e.g. the Great Plains of India mostly have alluvial soils. For a Soils Engineer, however, the soils can be classified into two broad but distinct classes :

    1.   Coarse grained soils.

    2.   Fine grained soils.

These two classes of soils, have different physical properties and in general exhibit different behaviour as foundation materials and therefore it is important to broadly categorise soils as such. The physical distinction between the two categories is made on the basis of grain analysis, details of which will be taken up in Chapter 3. However, two typical grain-size distribution curves are shown in Figure 1.1.

On the horizontal axis is plotted the grain-size of the soil in millimeters on a logarithmic scale and on the vertical axis is plotted the percentage finer than the sizes shown.
    The distinction between the *coarse-grained* and *fine-grained* categories is associated with 0 075 millimeter grain-size, as shown. A further classification of these two categories into different sizes is also shown in accordance with the Indian Standard specifications mentioned in the I.S.I. 1498-1959, except for the boulder size. Here, it has been assumed that any material size greater than 60 m.m. will be called boulder.

    It may be categorically stated and emphasized here that just categorization of a soil into these two categories does not give all the necessary information to predict their behaviour. For example a soil with a very high percentage of inert clay will be much less troublesome than a soil with far less percentage of very active clay, although the grain size analysis would indicate the former soil as inferior material. However, as stated earlier, they do

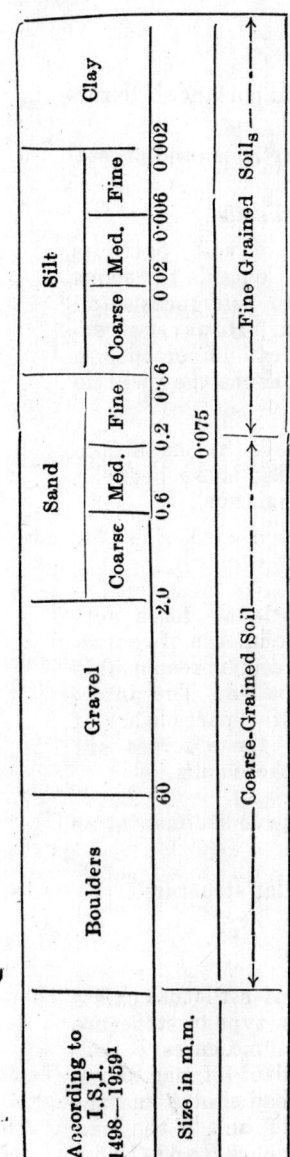

According to I.S.I. 1498—1959[1]	Boulders	Gravel	Sand			Silt			Clay
			Coarse	Med.	Fine	Coarse	Med.	Fine	
Size in m.m.		60	2.0	0.6	0.2 0.6	0.02	0.006	0.002	

Coarse-Grained Soil — · — · —   Fine Grained Soil₅ — · — · —

0.075

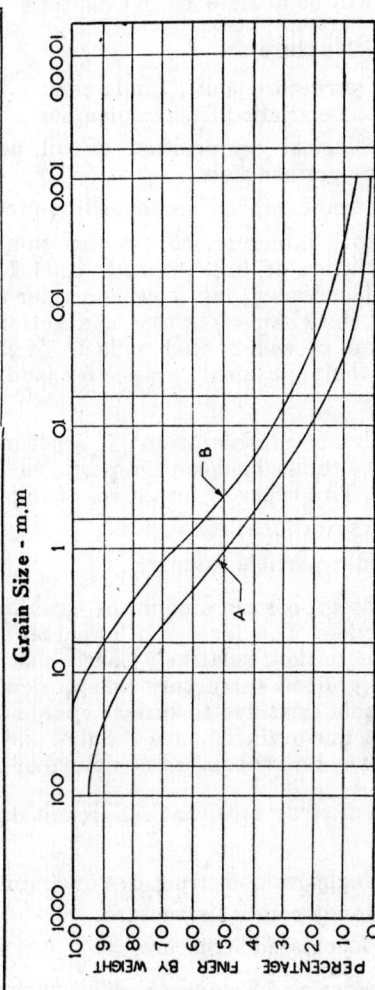

Fig. 1.1. Typical Grain-Size Distribution Curve

1, Except for the Boulders

exhibit different physical properties and general conclusions regarding their behaviour can be arrived at just by categorizing soils as such.

The details of general properties of boulders, gravels, sands, silts and clays will be dealt with in Chapter 4.

1.6. Soil Structure

Study of structure is of fundamental importance. Structure of a soil can be studied from two angles :

1. Minerological composition of soil particles and crystal formation in fine grained soils.

2. Structural composition of sedimented soils.

Study of the minerological composition of soil particles and crystal formation of fine grained soils is outside the scope of this book. The reader may, however, refer to references 4 and 7 (at the end of this chapter) in this connection. However structural composition of sedimented soils is of great importance in understanding their physical properties and hence they will be described here.

Most important consideration of stucture is the magnitude and nature of the forces between the particles and those between soil and water. Two types of forces are of importance :

     1. Gravitational force.

     2. Inter-particle force.

Due to gravity, certain amount of gravitational force acts on each soil particle. This force is of importance in case of coarse-grained soils due to their relatively large size and is responsible for their single grained structure when deposited. The inter-particle forces that exist due to surface charge of a particle are of importance in fine-grained soils only and these forces are responsible for the honey-combed structure of these soils.

The types of structures that can result due to sedimentation of the soil are :

     1. Single-grained structures or granular structure.

     2. Honey-combed structure.

     3. Flocculated structure.

1. **Single-grained Structure.** This type of structure exists only in *granular soils*. An idealization of this type of structure is presented in Fig. 1. (a) and (b). The figure indicates a two-dimensioned view of the placement of a handful of lead shots. In Fig, 1.2. (a), the lead shot A rests on top of lead shot B and the lead shot C rests on top of lead shot D and so on. If the lead shots are assumed as exact spheres, then A touches B and C at a

point and D touches B and C again at a point only. There is a big gap E between these four lead shots. All of them are only acted upon by the force of gravity and there is no inter-particle force among them. Now, if this arrangement is disturbed, then they will come into stable positions again as shown in Fig. 1.2. (b). The gap E which was formerly surrounded by the four lead shots A, B, C and D is narrowed out and C has been pushed out of position and the gap E is surrounded by only A, B and D. The first configuration shown in Fig. 1.2 (a) may be designated as loose packing and the second in Fig. 1.2 (b) as dense packing of lead shots.

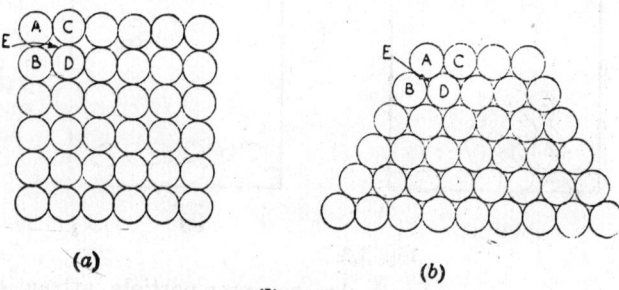

(a) (b)

Fig. 1.2.

The type of structure indicated here is found in cohesionless soils, which do not have any cohesion between the particles.

Theoretical analysis to find out the porosity[1] and void ratio[2] of a set of perfect and equal spheres has been done (8) and it has been found that porosity ranges from 26 to 48 per cent and the void ratio ranges from 0.35 to 0.91.

Cohesionless soils are not exactly an accummulation of equal spheres, however, the limiting values of porosity and void ratio are identical with the theoretical results obtained above for spheres. Granular soils usually exhibit porosities between 23 and 50% and their void-ratios lie between 0.3 and 1.0. The remarkable resemblance of these values with the ones obtained for spheres can be seen.

2. Honey-combed Structure. *Fine grained* soils do always have certain amount of inter-particle forces and consequent cohesion. *Honey-combed* structures occur in such soils only.

The inter-particle forces result from the intermolecular attraction which is predominant in fine-grained soils due to their relatively small size. In these soils the force of gravity is of secondary importance. Consider the conditions of sedimentation in a lake, as shown in Fig. 1.3. The first layer of particles is shown at the bottom. Designate one of the particles of this layer

---

1 & 2 For difinitions see Chapter 2.

as *A*. Let another particle of soil travel down to the bottom and let it be assumed that it will firstly touch particle *A* of the layer. If it is a fine-grained soil in which the inter-particle attraction is predominant, then this particle *B* on touching particle *A* will remain in position due to this attraction, as shown in Fig. 1.3 (*a*) If, however, it is a coarse-grained soil, then due to the predominant effect of the force of gravity, on touching *A*, *B* will assume a more convenient position as shown in Fig. 1.3.(*b*).

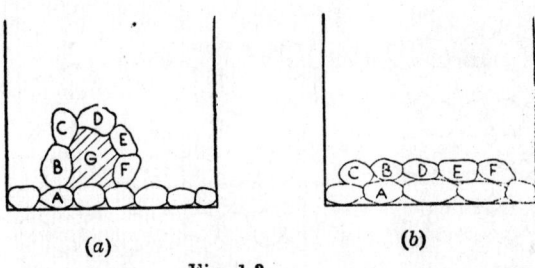

(*a*)   (*b*)

Fig. 1.3.

So in the fine-grained soils due to inter-particle attraction, the particles will assume positions as shown in Fig. 1.3 (*a*) and leave a big gap *G*, shown hatched, whereas in coarse-grained soils the positions assumed may be as shown in Fig. 1.3 (*b*) Evidently, the structure formed by fine-grained soils will be more porous and this relatively more porous structure containing large voids is called a *honey-combed structure*. This type of structure is mostly found in fine silts and clays.

3. Flocculant Structure. *Flocculant structure* occurs only in very fine grained soils. Both the force of gravity and the molecular forces play their roles in its formation. When a solid particle is suspended in water, it will start settling down under gravity and the rate of settlement or the velocity of settlement wli be proportional to the square of the diameter. So, for every fine particle of the order of 0.5×10^{-4} cms in diameter, the rate of settlement under gravity is almost nil. Technically, a suspension of such particles is called a *colloidal suspension* and the liquid is given the name *continuous phase* and the solids, the *disperse phase*.

The molecules of the continuous phase of such a suspension are continuously vibrating and strike against the particles of disperse phase. If the particles are large compared to the molecules, the impact forces on the particles will balance each other, due to the law of probability. On the other hand, if the particles approach the size of molecules, these impact forces will not be balanced and the vibrating molecules will set irregular motion of the particles of the soil. This phenomenon is called *Brownian movement* and can be seen under a microscope.

The particles of soil in suspension do not collide with one another due to the same surface charge that they carry, as like poles repel each other. The liquid always carries the opposite charge If, now, the mutual repulsion of the particles is removed by adding some chemical, the particles will collide with one another and, due to cohesion in the soil, these particles will form a bigger particle called a *floc*. The process of forma-tion of flocs is called floc culation and these flocs are large enough to stop the Brownian movement, by balancing the impact forces. Consequently, the particles will start settling, too. However, the flocs are small enough and only settle down in a honey combed structure. The grains group around voids larger than grain size of the soil to form flocs and these flocs group around voids larger than the flocs themselves, as shown in Fig 1.4.

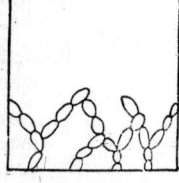

Fig. 1.4.

It may be stated here that mixed structure incorporating the various features of this individual structure discussed above is also possible and this mixed structure is typical of the marine clays. However, details of such a structure will not be discussed here.

1·7. Application of Properties of Soils

Study of the properties of soil is very important as they find their application in almost all projects that a civil engineer has to tackle. Some typical applications are enumerated below only to illustrate this point.

1. Permeability characteristics of the soil are important in all hydraulic structures resting on soils. Seepage problems can-not be tackled without the knowledge of the permeability of the soil encountered.

2. Compressibility and consolidation properties are impor-tant in the settlement analysis of foundations of all large or small structures.

3. Shear strength properties are applicable in the stability and bearing capacity analyses of foundations.

4, Certain strength properties such as *modulus of subgrade reaction* find their applications in the design of foundations.

5. Chemical properties of the soil are helpful in understand-ing the stabilization procedures and techniques.

All the above properties and many more will be discussed thoroughly in the following chapters.

CHAPTER BIBLIOGRAPHY

1. Bowles, E. "Foundation Analysis and Design, "McGraw Hills Book Co. New York, N.Y. 1968.

2. Chamber's Twentieth Century Dictionary.

3. Dawson, Raymond F, Laboratory Manual in Soil Machanics, second Edition, New York, London, Pitman Publishing Corporation.

4, Grim, R.E., Clay Mineralogy, New York, McGraw Hill Book Co. inc 1953.

5. Indian Standard Institution, Classification and Indentification of soil for General Engineering purposes ; I.S. 1498-1959, New Delhi.

6. Joffe J. S. Pedology, New Brunswick, New Jerssey-Rutgers, University Press, 1936.

7. Lambe, T.W., "Structure of Organic Soil" proceedings, American Society of Civil Engineers, Volume 79, New York-1953.

8. Muskat, M., The Flow of Homogeneous Fluids through Porous Media : New York, McGraw Hills-1937.

9. Peck, R.B., Hanson, W.E., and Thornburn, T.H., Foundation Engineering, J. Wiley & Sons, Inc., N.Y., 5th prining, 1957.

10. Scott, R.F., "Principles of Soil Mechanics", Addison Wesley Publishing Co., Inc., Massachusetts, 1963.

11. Webster's International Dictionary of the English Language, second Edition Springfield, Mass., G. and C. Merrian Company, Publishers, 1856.

QUESTIONS

1. Define Soil.

2. Why is Soil Mechanics important for a Civil Engineer ?

3. What are the two distinct categories of soils that are important to a Civil Engineer.

4. Describe the features of the various forms of structure found in sedimentary soils ?

5. List the important problems associated with soil in enginering practice (Mosore Univ., 1966).

A TEXT BOOK OF SOIL MECHANICS

When the voids contain only air as in Fig. 2·1 (b), the voids are usually referred to as air voids.

2

Basic definitions, weight and Volume Relationships, etc.

2·1 Introduction

Before presenting any detailed theories of "Soil Mechanics", it is essential that some basic properties are presented in order to give a clear picture to the student of what follows hereafter. This chapter, is being devoted completely to the basic definitions, weight-volume relationships and the like material.

2·2 Phase Diagrams

If a mass of soil is examined. it is found to be a skeleton of the type shown in Fig. 2·1 (a) and consists of :

1. Solid soil particles enclosing voids which will contain air only, if the soil is completely dry

or 2. Solid soil particles enclosing voids which will contain air and water, if the soil is partially dry and partially saturated with water.

or 3. Solid soil particles enclosing voids which will contain water alone, if the soil is fully saturated with water.

These three states of the soil can be represented diagrammatically as shown in Fig. 2·1 (b), (c) and (d). Each figure is in two dimensions.

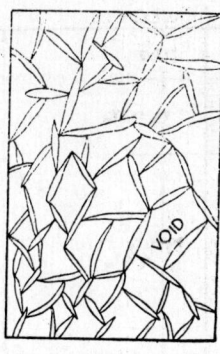

(a)

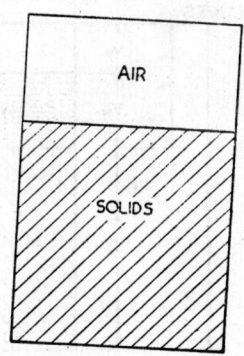

COMPLETELY DRY SOIL
(b)

Fig. 2.1.

When the voids contain only air as in Fig. 2·1 (b), the voids are usually referred to as *air voids*.

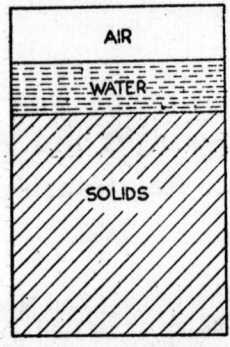
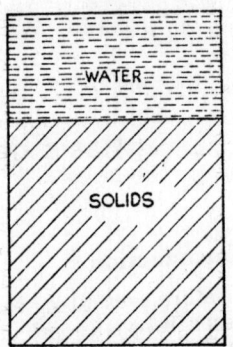

PARTIALLY SATURATED SOIL   FULLY SATURATED SOIL
(c)   Fig 2·1   (d)

A diagram of this type which represents the components of the soil is called a *phase-diagram*. Since Fig. 2·1 (b) and (d) contain only two materials *i.e.*, solids and air, and solids and water respectively, they are called *two-phase diagrams* and the Fig. 2·1 (c) which represents three materials *i.e.* solids, air and water may be called a *three-phase diagram*. The solids, air and water may be called the components of soil mass.

These diagrams are helpful in developing the weight volume relationships of the dry, partially saturated or fully saturated soils. However, it may be noted that they are just diagramatic representations, as it is not physically possible to place solids in one space and voids in the other.

2·3 Basic Definitions

Refer to the phase diagram in Fig. 2·2. The total volume

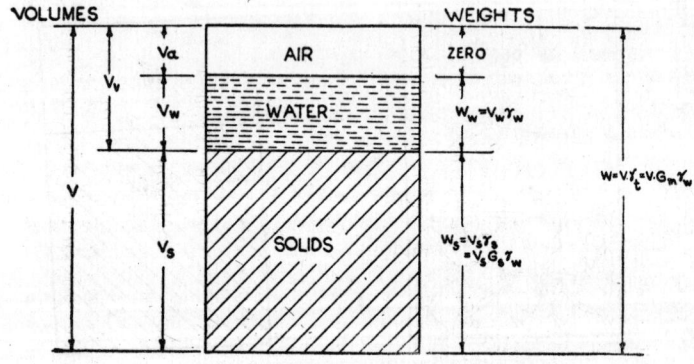

Fig 2·2.

of the mass of soil is represented as V, the volume of solids as V_s volume of water as V_w and the volume of the air as V_a.

Volume ratios which are widely used in soil mechanics are, the *porosity, void ratio* and *degree of saturation*. They aae defined as under.

Porosity (Defn.) *Porosity* of of a mass of soil is defined as the ratio of the *volume of total voids* (including air voids and voids containing water), to the *total volume of the mass of soil*. It is denoted by the letter η and is usually represented as a percentage.

∴
$$\eta = \frac{V_v}{V} \times 100 \qquad \ldots(2\cdot1)$$

where $\qquad\qquad V_v = V_a + V_w$ and $V = V_s + V_a + V_w.$

Void Ratio (Defn) *Void ratio* of a mass of soil is defined as the ratio of the *volume of total voids* (including air voids and voids containing water), to the *volume of the solids* in the mass of soil. It is denoted by the letter e and is usually expressed as a fraction.

∴
$$e = \frac{V_v}{V_s} \qquad \ldots(2.2)$$

where $\qquad\qquad V_v = V_a + V_w$

Degree of Saturation (Defn) *Degree of saturation* of a mass of soil is defined as the ratio of the *volume of voids actually filled with water* to the *total volume of voids*. It is denoted by the letter S and is usually expressed as a percentage[1].

$$S = \frac{V_w}{V_v} \times 100 \qquad \ldots(2\cdot3)$$

where $\qquad\qquad V_v = V_a + V_w$

When the degree of saturation S of a soil is zero percent, the soil is in a dry state and when the degree of saturation is 100 percent the soil is called fully saturated. In both these states, it represents a two phase system. When S lies between zero percent and 100 percent, the soil is moist or partially saturated and represents a three-phase system.

Void ratio is a quantity which is more often used than porosity in soil mechanics because of the fact that the solids are in general assumed incompressible and hence the denominator V_w of the void ratio remains constant and change under pressure can only occur in the quantity in the numerator which indicates the volume of voids. Because of the variation both in the numerator

---

[1] It may be stated here that in most of the numerical work, the degree of saturation S which occurs frequently is taken as a fraction *i.e.*, ratio of (V_v/V_w) and not as percentage.

and denominater of the expression for porosity, this term is not often used.

The total weight of the mass of soil is represented by W. It constitutes of the weight of solids represented as W_s and weight of water represented as W_w. The weight of air is considered negligible and hence is always taken as zero.

$$\therefore \qquad W = W_s + W_w \qquad \ldots(2.4)$$

A very important and frequently occuring term called *water content* has relation to the weights of soil components.

Weter content (Defn.) *Water content* of a soil mass is defined as the ratio of the *weignt of water* to the *dry weight of solid mass.* It is denoted by letter w. The water content is usually expressed as a percentage.

$$w = \frac{W_w}{W_s} \times 100$$

This definition of water content is different from the one used in geology where it is defined as a ratio of the weight of water to the weight of total soil mass. Hence, the difference may be noted.

Water in clayey soils exits in two different forms. In the first place, the water exists in the voids or pores of soil mass. This water is usually called pore water or free water. Secondly, water also exists in the soil mass as a part of the mineral grains. This water is called adsorbed water or bound water.

So when the above difinitions of porosity, void ratio, degree of satuiation and water content of soil are used, it is essential to know which water is being referred to in these definitions. It is an ideal state of affairs to assume that water and solids are two different phases as represented in the phase diagram. In actual effect, due to the presence of bound or adsorbed water, the above assumption is not correct.

For the purpose of the definitions, it will be assumed, that water is that part of the soil mass which evaporates when the wet soil is put in an oven for drying at a temperature 105°—110°C, till it assumes a constant weight, and that solids is the left over mass. The time for drying in laboratory practice is usually 24 hours for most of the soils. In practice, it is assumed that drying the soil under above conditions will just drive off the pore-wa er ; however, even if some adsorbed water gets driven off under above conditions, the quantity will be negligible.

It is, however, noteworthy, that a departure from the above temperature will change the quantity of water that will evaporate. If higher temperatures are used for drying, the adsorbed water may also get evaporated and the water content so obtained will be different for each temperature for the same soil.

For pure sands, however, changes in temperature of drying may not have much effect on the water content values; so long as the soil mass attains a constant weight.

2.4. Unit Weights and Specific Gravities

In developing the weight-volume relationships, terms like the unit weights and specific gravities of soil mass, solids and water are applied.

Mass Unit Weight (Defn). *Mass unit weight* also called the *bulk unit weight* or *sometimes as bulk density* of a soil is the weight per unit volume of the soil mass. It is denoted by the letter γ_t.

$$\therefore \qquad \gamma_t = \frac{W}{V} \qquad (\ldots 2.6)$$

where $\qquad W = W_s + W_w$

and $\qquad V = V_s + V_w + V_a$

Unit Weight of Solids (Defn). *Unit weight of solids* also called *absolute density of a soil* is the weight of solids per unit volume of solids alone. It is denoted by the letter γ_s.

$$\therefore \qquad \gamma_s = \frac{W_s}{V_s} \qquad \ldots(2.7)$$

Unit Weight of Water (Defn). *Unit weight of water* is the weight of water per unit volume of water at a specific temperature.

The unit weight of pure water is taken as one gm./c.c. at a temperature of 4° C. If any other temperature is used in an experimental work, corresponding value of the unit weight of water may be used (See Table 3.1). Unit weight of water at any temperature is denoted by γ_w but for this standard temperature of 4°C, it is denoted by γ_0.

$$\therefore \qquad \gamma_w = \frac{W_w}{V_w} \qquad \ldots(2.8)$$

Dry Unit Weight of Soil (Defn).

In the Soil Mechanics and foundation Engineering field, "*dry unit weight of a Soil*" is a very common term. It can be defined as the weight of solids per unit volume of the soil mass. It is usually denoted by the letter γ_{dry}.

$$\therefore \qquad \gamma_{dry} = \frac{W_s}{V} \qquad \ldots(2.9)$$

Specific Gravity (Defn). *Specific gravity* of a material is the ratio of unit weight of the material and the unit weight of some reference material. The reference material usually taken is pure water at 4°C. With this general definition, the *mass specific gravity* also called *bulk* or *apparent specific gravity ; specific gravity of solids* also called *absolute specific gravity and the specific gravity of water* can be written as,

$$\text{Mass Specific Gravity} \quad = G_m = \frac{\gamma_t}{\gamma_0} \qquad \qquad ...(2.10)$$

$$\text{Specific Gravity of solids} = G_s = \frac{\gamma_s}{\gamma_0} \qquad \qquad ...(2.11)$$

$$\text{Specific Gravity of water} = G_w = \frac{\gamma_w}{\gamma_0} \qquad \qquad ...(2.12)$$

where γ_0 is the unit weight of water at 4°C.

The unit weight of water under normal conditions ranges from 1·00 to 0.995 gms./c.c. and as such ordinarily, no differentiation between γ_0 and γ_w is made and invariably γ_0 in the above three equations is replaced by γ_w.

Density (Defn). *Density* of a substance is defined as its *mass per unit volume* Mass is equal to the weight divided by the acceleration due to gravity (g).

$$\therefore \quad \text{Density of Soil Mass} = \frac{W}{g} \times \frac{1}{V}$$

$$\text{or} \qquad \text{Density} = \frac{1}{g}\left(\frac{W}{V}\right) = \frac{\gamma_t}{g} \qquad \qquad ...(2.13)$$

Units. The unit weights of the materials can be expressed as lbs./c. ft. in the F.P.S. system and as gms./c.c. in the M.K.S. (metric) system.

Specific gravity is a ratio of the unit weights of two materials and hence is a dimensionless quantity.

Many times in soil mechanics literature, the student will find that the subscript "t" with γ_t and subscript s with G_s are dropped and the mass unit weight of soil and the specific gravity of solids are represented as γ and G respectively. In this book, henceforth, the specific gravity of solids will represented by the symbol G and the bulk density as γ[1].

---

1. This is in keeping with the recommendations published in Vol. III of the Proceedings of the Fifth International Conference on Soil Mechanics and Foundation Engineering (*Paris 1961*).

2·5. Some Derivations using Basic Definitions, Unit Weights and Specific Gravities.

(a) Derive : $\qquad e=\dfrac{w.G.}{S}$ $\qquad$ or $\qquad S.e=w.G.$

By definition $\qquad e=\dfrac{V_v}{V_s}=\dfrac{V_v}{V_s}\times\dfrac{V_w}{V_w}$

or $\qquad e=\dfrac{V_v}{V_w}\times\dfrac{V_w}{V_s}=\dfrac{1}{\dfrac{V_m}{V_v}}\times\dfrac{V_w}{V_s}$

But $\qquad V_w=\dfrac{W_w}{\gamma_w}$ $\qquad\qquad$ (By defn.)

and $\qquad V_s=\dfrac{W_s}{\gamma_s}=\dfrac{W_s}{G.\gamma_w}$ $\qquad$ (By defn.)

and $\qquad \dfrac{V_w}{V_v}=S$ $\qquad\qquad$ (By defn.)

Substitute in the R.H.S.

∴ $\qquad e=\dfrac{1}{S}\cdot\left(\dfrac{W_w}{\gamma_w}\right)\left(\dfrac{G.\gamma_w}{W_s}\right)$

or $\qquad e=\dfrac{1}{S}\left(\dfrac{W_x}{W_s}\right)G$

But $\qquad \dfrac{W_w}{W_s}=w$ $\qquad\qquad$ (By defn.)

∴ $\qquad e=\dfrac{1}{S}\cdot w.G=\dfrac{G.w}{S}$

or $\qquad S.e=w.G$ $\qquad\qquad$ (2·14)

(b) Derive : $\qquad \gamma=\dfrac{1+w}{1+e}\cdot G.\gamma_w$

By definition $\qquad \gamma=\dfrac{W}{V}=\dfrac{W_s+W_w}{V_s+V_v}$

or $\qquad \gamma=\dfrac{W_s\left(1+\dfrac{W_w}{W_s}\right)}{V_s\left(1+\dfrac{V_v}{V_s}\right)}$ $\qquad$ ...(2·15)

But $\qquad \dfrac{W_w}{W_s}=$ water content $=w.$ $\qquad$ (By defn.)

and $\qquad \dfrac{V_v}{V_s}$ = void ratio = e. $\qquad\qquad$ (By defn.)

Substitute these values in equation (2·14).

$$\therefore \qquad \gamma = \frac{W_s(1+w)}{V_s(1+e)}$$

Again $\dfrac{W_s}{V_s} = \gamma_s = G.\gamma_w \qquad$ [From equation (2·7) and (2·10)]

$$\therefore \qquad \gamma = \frac{1+w}{1+e} G.\gamma_w \qquad\qquad (...2\cdot16)$$

This soil has been assumed as partially saturated and this weight γ is the unit weight of the partially saturated soil. Unit weight of partially saturated soil is sometimes denoted by γ_{sat}, although γ_{sat} is normally used for the completely staurated soil for which S is unity.

$$\gamma = \gamma_{sat} \text{ (Partial saturation)} = \left(\frac{1+w}{1+e}\right) G.\gamma_w \ ...(2\cdot17)$$

(c) Derive : $\gamma = \dfrac{G+Se}{1+e}\gamma_w$

From equation (2·15)

$$\gamma = \frac{1+w}{1+e} G.\gamma_w = \left(\frac{G+w.G}{1+e}\right)\gamma_w$$

But $\quad w.G = S.e \qquad\qquad$ [From equation...(2·14]

$$\therefore \qquad \gamma = \frac{G+Se}{1+e}.\gamma_w \qquad\qquad ...(2\cdot18)$$

When the degree of saturation is 100 per cent, S in the above equation is unity.

Dry Unit Weight : When the soil is dry $w = o$ and $S = o$. Substituting these values in (2·16) and (2·17) the dry unit weight of the soil designated as γ_d is given by :

$$\gamma_d = \frac{G.\gamma_w}{1+e}. \qquad\qquad ...(2\cdot19)$$

2·6. Submerged Unit Weight

When the soil is submerged in water, in addition to the weight of soil acting downwards, the effect of bouyancy acts upwards. So the net unit weight of the submerged soil denoted by γ_{sub} will be given by :

$$\gamma_{sub} = \gamma - \text{(effect of bouyancy)}$$

or $\quad \gamma_{sub}=\gamma-\gamma_w$[1]

because the effect of bouyancy is equal to the unit weight of water.

Therefore $\quad \gamma_{sub}=\dfrac{G+Se}{1+e}\gamma_w-\gamma_w$

Usually the submerged soils are in a state of complete saturation, so that the value of $S=1$.

Substituting this value in the above equation :

$$\gamma_{sub}=\dfrac{G+e}{1+e}\gamma_w-\gamma_w$$

or

$$\gamma_{sub}=\left(\dfrac{G+e}{1+e}-1\right)\gamma_w$$

or

$$\gamma_{sub}=\dfrac{G-1}{1+e}\gamma_w \qquad (2.20)$$

The sub-merged saturated unit weight is usually designed as $\overline{\gamma_b}$:

It may be noted here that a saturated soil is different from a sub-merged soil. A saturated soil need not essentially be submerged but a submerged soil is invariably completely saturated. The difference can be visualised by comparing equations (2.17) and (2.19). The numerical value of the unit weight of completely saturated soil is higher than the numerical value of the unit weight of submerged soil by the unit weight of water. This is verified as under :—

When the soil is completely saturated $S=1$.
Therefore from equation (2.17)

$$\gamma_{sat}=\dfrac{G+e}{1+e}\gamma_w$$

In the case of submerged soil

$$\gamma_b=\dfrac{G-1}{1+e}\gamma_w$$

Difference of the above two expressions is γ_w i. e.

$$\gamma_{sat}-\gamma_b=\gamma_w \qquad (2.21)$$

2.7. Examples

Example 1. Using a *phase diagram*, find each of the following relationships in terms of given quantities. Assume that the unit weight of water γ_w is always given.

---

1. Archimedes principle says that the loss of weight in water of a substance is equal to the weight of water displaced by it. This ... in weight is due the bouyancy in water, hence the value γ_w.

(a) Given the porosity η find out the void ratio e ?

(b) Given specific gravity of solids G, and water content w. find out void ratio for a fully saturated soil ?

(c) Given specific gravity of solids G, water content w and degree of saturation S, find out mass unit weight γ. [See also Q. 7(a) at the end of this Chapter].

(d) Given specific gravity of solids G· and water content w, find out the submerged unit weight γ_b of a fully saturated soil ?
Solution. (a) Refer to Fig. 2.3(a).

Assume the total volume of the soil as one unit, say 1 c.c.

$$\text{Porosity} = \eta \qquad\qquad \text{(Given)}$$

∴

$$V_v = \eta$$

$$V_s = (1 - \eta) ;$$

both shown on the phase diagram in Fig. 2,3(a).

∴ Void Ratio

$$e = \frac{V_v}{V_s} = \frac{\eta}{1 - \eta}$$

Ans.

Note. Value of η will be given by

$$\eta = \frac{e}{1 + e}$$

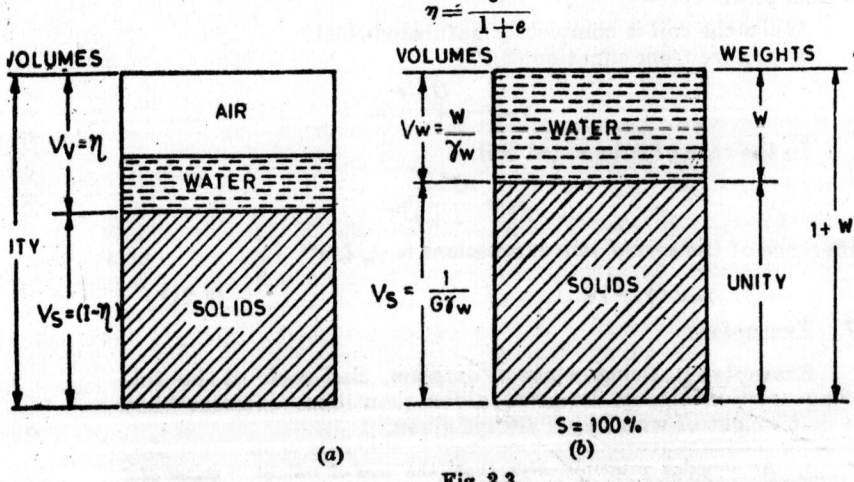

Fig. 2 3.

(b) Since the soil is fully saturated, a two-phase diagram is applicable and is shown in Fig. 2.3(b).

Assume the weight of solids W_s of the soil as unity, say 1 gm. Water content $= w$

Now
$$\frac{W_w}{W_s} = w \qquad \text{(Given)}$$

(By defn.)

$\therefore \qquad\qquad W_w = w,$ as W_s is unity.

$$V_w = \frac{W_w}{\gamma_w} = \frac{w}{\gamma_w} = V_v;$$

since the soil is fully saturated,

and
$$V_s = \frac{1}{G\gamma_w};$$

both the above quantities are shown in the phase diagram 2.3(b).

$\therefore$ Void Ratio $\qquad e = \dfrac{V_v}{V_s} = \dfrac{w}{\gamma_w} \quad \dfrac{1}{G\gamma_w} = G.w.$

Ans.

Another approach. Equation (2.13) gives $S.\, e. = G.w.$ If the degree of saturation is 100 per cent then, $S = I$ in this equation.

$\therefore$ Void ratio $\quad e = G.w.$

(c) Refer to Fig. 2.3(c). Assume the volume of solids, as unity say 1 c.c.

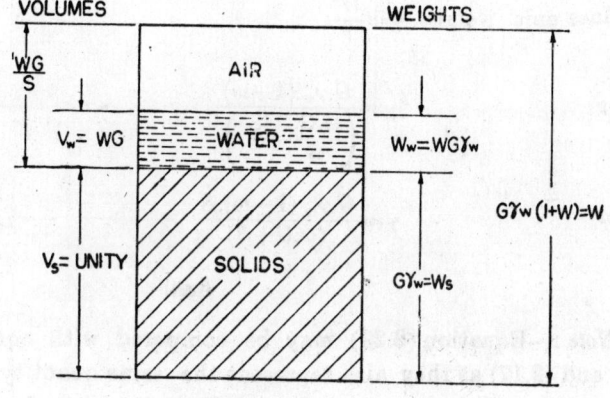

(c)
Fig. 2.3.

Then $\qquad\qquad V_s = 1$ c.c,
$$W_s = G.\gamma_w.V_s$$
$$= G.\gamma_w. \text{ gms.}$$

Denote water content as w

$\therefore$ $$W_w = w.W_s = w.G.\gamma_w$$

$\therefore$ Total weight of soil $W = W_s + W_w$

or $$W = G.\gamma_w.(1+w) \qquad ...(2.22)$$

Volume of water $V_w = \dfrac{W_w}{\gamma_w}$

or $$V_w = \dfrac{w.G\gamma_w}{\gamma_w} = w.G$$

Denote degree of saturation as S

$\therefore$ $$\dfrac{V_w}{V_v} = S \qquad \text{(By defn).}$$

or $$V_v = \dfrac{V_w}{S} = \dfrac{w.G.}{S}$$

$\therefore$ Total volume of soil $V = \left(\dfrac{w.G}{S} + 1 \right) \qquad ...(2.23)$

From equation (2.21) and (2.22).

Mass unit weight $\gamma = \dfrac{W}{V}$

or $$\gamma = \dfrac{G.\gamma_w.(1+w)}{\left(\dfrac{w.G.}{S} + 1 \right)}$$

or $$\gamma = \dfrac{G.\gamma_w (1+w).S}{(w.G+S)} \qquad ...(2.24)$$

Ans.

Note :—Equation (2.23) may be compared with equations (2.16) and (2.17) as they also represent the same quantity. However, equation (2.16) expresses the mass unit weight of a soil in terms of G, w. e and γ_w, equation (2.17) expresses it in terms of G, S, e and γ_w and equation (2.23) expresses it in terms of G, S, w and γ_w.

(d) Since the soil is fully saturated, a two-phase diagram is applicable and is shown in Fig. 2.3(d).

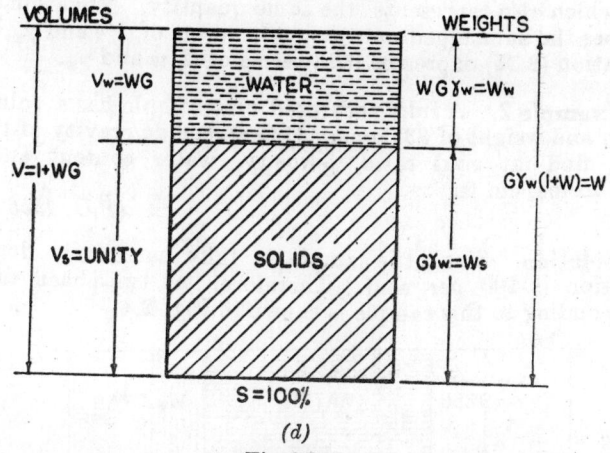

(d)

Fig. 2.3.

Assume the volume of solids as unity, say 1 c.c.

$$\therefore \qquad V_s = 1 \text{ c.c,}$$

Weight of solids $W_s = G.\gamma_w. \ V_s = G.\gamma_w,$

$\therefore$ Water content $= w$ (given)

Weight of water $W_w = w.W_s = w.G\gamma_w$

$\therefore$ Total weight of soil $W = W_s + W_w$

or $\qquad W = G\gamma_w (1+w)$

Volume of water $V_w = W_w/\gamma_w = w.G.$

$\therefore$ Total volume of the soil $V = V_s + V_w$

or $\qquad V = 1 + w.G.$

$\therefore$ Saturated unit weight of the soil

$$\gamma_{sat} = \frac{W}{V} = \frac{G \gamma_w (1+w)}{1 + w.G.}$$

By defn. submerged unit weight

$$\gamma_b = \gamma_{sat} - \gamma_w$$

$$\therefore \qquad \gamma_b = \frac{G.\gamma_w. (1+w)}{1 + w.G.} - \gamma_w$$

or $\qquad \gamma_b = \gamma_w \left[\frac{G(1+w)}{1 + w.G.} - 1 \right]$

or $\qquad \gamma_b = \frac{G-1}{1 + wG}. \gamma_w.$ $\qquad$ ...(2.25)

Ans.

Note :—Equation (2.24) may be compared with equation (2.19) which also represents the same quantity. Equation (2.19) expresses the submerged unit weight in terms of G, e and γ_w, whereas equation (2.24) expresses it in terms of G, w and γ_w.

Example 2. A fully saturated clay sample has a volume of 185 c.c. and weight of 331 gms. If the specific gravity of the soil is 2.67, find out void ratio, porosity, water content and unit weight in lbs. cu. ft.

(P.U. 1956 Supp.)

Solution. Since the sample is fully saturated, degree of saturation is 100 per cent. Therefore, a two-phase diagram corresponding to this sample is shown in Fig. 2.4.

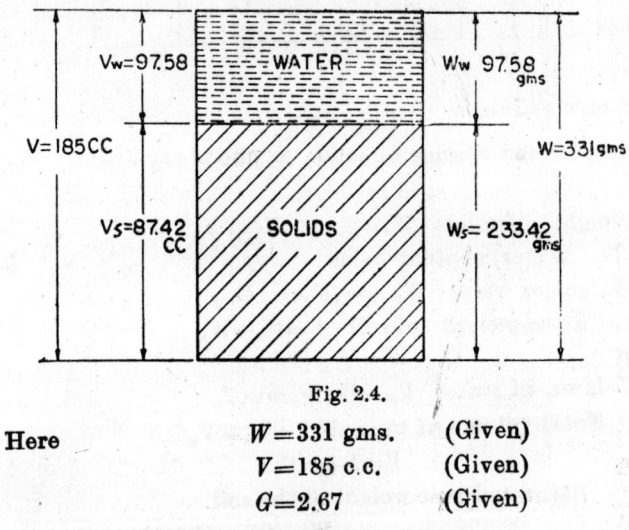

Fig. 2.4.

Here

$$W = 331 \text{ gms.} \quad \text{(Given)}$$
$$V = 185 \text{ c.c.} \quad \text{(Given)}$$
$$G = 2.67 \quad \text{(Given)}$$

Let volume of solids be $= V_s$

$\therefore$ $W_s = V_s \times G \times \gamma_w$

where $\gamma_w = 1 \text{ gm./c.c.}$

or $W_s = 2.67 \times V_s$

$\therefore$ Weight of water $= W_w = W - W_s$
$$= (331 - 2.67 \, V_s) \text{ gms.}$$

$\therefore$ Volume of water $= V_w = (331 - 2.67 \, V_s) \times 1$ c.c.

$\therefore$ Total volume $= V = 331 - 2.67 \, V_s + V_s$

This must equal the given total volume of 185 c.c.

$\therefore$ $331 - 2.67 \, V_s + V_s = 185$

$$\therefore \qquad 1.67 \ V_s = 146$$

$$V_s = \frac{146}{1.67} = 87.42 \ \text{c.c.}$$

$$\therefore \qquad V_w = 185 - 87.42 = 97.58 \ \text{c.c.}$$
$$W_s = 87.42 \times 2.67 = 233.42 \ \text{gms.}$$

and $\qquad W_w = 331 - 233.42 = 97.58 \ \text{gms.}$

$\therefore$ Void Ratio $\quad e = \dfrac{V_v}{V_s} = \dfrac{V_w}{V_s} = \dfrac{97.58}{87.42} = 1.12$

and Porosity $\quad \eta = \dfrac{V_v}{V} = \dfrac{V_w}{V} = \dfrac{97.58}{185} 0.528$

and water content $w = \dfrac{W_w}{W_s} \times 100 = \dfrac{97.58}{233.42} \times 100 = 41.6\%.$ $\Bigg\}$ **Ans.**

Mass unit weight $\quad \gamma = \dfrac{W}{V} = \dfrac{331}{185} = 1.79 \ \text{gms./c}$

In terms of lbs./cubic ft., mass unit weight

$$\gamma = \frac{1.79}{453.6} \times \cfrac{1}{\cfrac{1}{(2.54)^3 \times 12^3}}$$

$$= 105.5 \ \text{lbs./cu. ft.}$$

Ans.

Example 3. A soil is at a void ratio of 0·9 with the specific gravity of the solid particles of 2·70.

(a) Can the water content be determined from this information? If so, find it. If not, why not?

(b) If the water content cannot be determined, can the upper and lower limits be determined? If so find them.

Solution. (a) The water content cannot be determined since the degree of saturation is not known.

(b) The upper and lower limits can be determined by assuming the lower limit of degree of saturation as zero and the upper limit as 1.

From equation (2·13)

$$S \times e = w \times G$$

If $\qquad S = 0, \ w = 0\%.$

Ans.

If $\qquad S = 1, w = \dfrac{0.9}{2.7} \times 100 = 33 \cdot 3\%.$

Ans.

Example 4. A moist soil sample has a volume of 464 c.c. in natural state and a weight of 793 gms. The dry weight is 735 gms. specific gravity of soil grains is 2.68. Determine the void ratio, the porosity, water content and the degree of saturation?

(*P. U. 1960*)

Solution. Refer to Fig. 2·5 for the phase diagram.

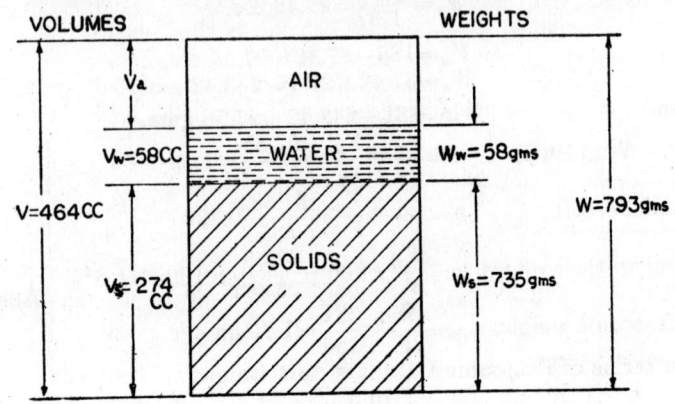

Fig. 2·5.

$$W = 793 \text{ gms.} \quad \text{(Given)}$$
$$W_s = 735 \text{ gms.} \quad \text{(Given)}$$
$$W_w = 793 - 735 = 58 \text{ gms.}$$
$$G = 2·68 \quad \text{(Given)}$$

∴ Volume of solids $= V_s = \dfrac{735}{2·68} = 274 \text{ c.c.}$

Volume of water $= V_w = 58 \text{ c.c.}$, Since $\gamma_w = 1 \text{ gm./c.c.}$

∴ Volume of voids $= V_u = V - V_s = 464 - 274 = 190 \text{ c.c.}$

∴ Void Radio $\qquad e = \dfrac{V_v}{V_s} = \dfrac{190}{274} = 0·693$

and porosity $\qquad \eta = \dfrac{V_v}{V} = \dfrac{190}{464} = 0·410$

and water content $\qquad w = \dfrac{W_w}{W_s} \times 100 = \dfrac{58}{735} \times 100 = 7·9\%$

and degree of saturation $S = \dfrac{V_w}{V_v} \times 100 = \dfrac{58}{190} \times 100 = 30·5\%$

$\left. \right\}$ **Ans.**

Example 5. A mass of soil coated with a thin layer of paraffin weighs 485 gms. When immersed in water it displaces 320 c.c. of water. After the paraffin is pealed off, it is found to weigh 18 gms. If the specific gravity of the soil grains is 2.70 and that of paraffin is 0·9,

(a) What is the void ratio of the soil, if it is assumed dry ?

(b) What is the void ratio of the soil, if it has a water content of 10 per cent ?

(c) What is the degree of saturation in (b) ?

Solution. (a) Since the soil is dry, the phase diagrams when the soil is coated with praffin will constitute of the solids, paraffin and air and when the paraffin is pealed off, the phase diagram will constitute of solids and air alone. Refer to Fig. 2·6 (a) for these phase diagrams[1].

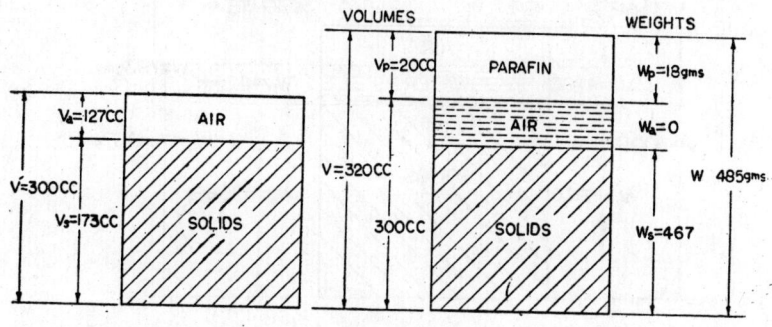

(a)

Fig. 2·6.

Weight of paraffin $=W_p=18$ gms.

Specific gravity $=0·9$

∴ Volume of paraffin $=V_p=\dfrac{18}{0·9\times1}=20$ c.c.

∴ Total volume=volume of water displaced $=V=320$ c.c.

∴ Volume of solids+air $=V'=V_s+V_a=320-20=300$ c.c.

Weight of solids $=W_s=485-18=467$ gms.

∴ Volume of solids $=V_s=\dfrac{467}{2·70\times1}=173$ c.c.

and volume of air $=300-173=127$ c.c.

∴ Void ratio $e=\dfrac{V_v}{V_s}=\dfrac{127}{173}=0·733.$

Ans.

(b) Refer to the phase diagram in Fig. 2·6 (b).

Water content $=10\%$ (Given)

Weight of solids+water=485−18=467 gms.

---

1. These phase diagrams are not drawn to any scale.

$$\therefore \qquad W_s = \frac{467}{(1+w)} = \frac{467}{1+0\cdot1} \approx 425$$

$$W_w \approx 467 - 425 \approx 42 \text{ gms.}$$

$$V_s = \frac{425}{2\cdot7 \times 1} = 158 \text{ c.c.}$$

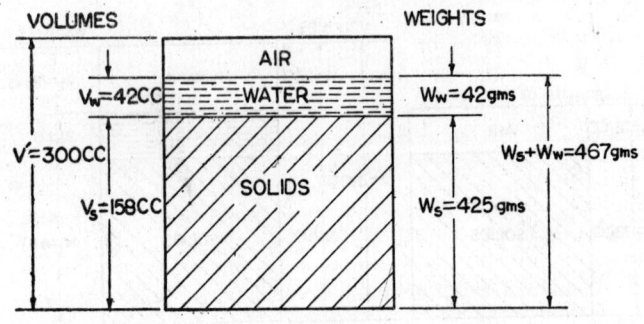

(b)

Fig.2 6.

and $\qquad V_w = 42\cdot00$ c.c.. since $\gamma_w =$ unity.

Now total volume of soil, water and air from $(a) = 300$ c.c.

$\therefore$ Volume of voids in the sample $= V_v = 300 - 158 = 142$ c.c.

$\therefore$ Void Ratio $\qquad e = \dfrac{V_v}{V_s} = \dfrac{142}{158} = 0\cdot90$

Ans.

(c) Degree of saturation in $(b) = S = \dfrac{V_w}{V_v} \times 100$

or $\qquad S = \dfrac{42\cdot0}{142} \times 100 = 29\cdot5\%,$

Ans.

CHAPTER-BIBLIOGRAPHY

1. Capper P.L., Cassie W.F., Geddes J.D., Problens in Engineeris soils, E and F.N. Spon Ltd, London, 1966.

2. Reynolds, Henry R and Protopapadakis P.. Practical Problems in Soil Mechanics, Frederick Unger Publishing Co, New fork, 1959.

3. Sehgal S B ; Practical Problems in Soil Mechanics, Foundatsul Engineering and Pavement Desigs, International Educational Publishers, Consutants and Distributors Inc, Sterling Heights Michigan (1974-in press).

QUESTIONS

1. (a) Define the terms void-ratio, mass specific gravity, degree of saturation and dry unit weight. Develop a relation for the void-ratio in terms of specific gravity of a soil and water content, for a saturated soil.

(b) A sample of most silty soil has a volume of 14·88 c.c. and weighs 28·81 gms. After complete drying out in oven its weight is 24·83 gms. The unit weight of solid constituents is 2·7 gms./c.c. Calculate the void-ratio and degree of saturation of the sample. (P.U. 1959 Annual)

2. Define the terms porosity, mass unit weight, water content and degree of saturation. Develop a relationship for submerged unit weight in terms of specific gravity of soil, void-ratio and unit weight of water. (P.U. 1959 Supp).

3. Define the terms void-ratio, dry density, submerged unit weight. Deduce an expression for partially saturated density of a soil in terms of its water content, void-ratio, specific gravity of soil grains and the unit weight of water. (P.U. 1960)

4. A 50 c.c. volume of moist soil weighs 95 gms. Its dry weight is 75 gms. and the specific gravity of solids is 2.68 Compute the water content, void-ratio, porosity and weight per c. ft. (P.U. 1961)

5. In a three-phase soil system, derive an expression for

(i) Void ratio as a function of porosity.

(ii) Porsosity as a function of void ratio (P.U. 1962)

6. In an oil-well drilling project, a heavy viscous liquid (drilling mud) was used to keep the drilling hole open. It consists of a suspension in water of the following proportions per litre of volume :

380 gms. of clay of $G = 2·82$

82 gms. of sand of $G = 2-68$

300 gms. of iron fillings of $G = 7·13$

Assuming that the unit weight of water $= 1·00$ gms, per c.c. and a uniformly mixed suspension, what is the unit weight of the suspension ? (P.U. 1964)

7. (a) By the three-phase soil system, show that the degree of saturation "S_r" (as ratio) in terms of mass unit weight γ_m, water content ratio w and specific gravity of soil G_s and unit weight water γ_w is given by the expression.

$$S_r = \frac{W}{\dfrac{\gamma_w}{\gamma_m}(1+w) - \dfrac{1}{G_s}}$$

*If 10% of W_s is taken, it will come to 42·5 However, the difference is immaterial in the numerical work presented herein.

(b) A pyconometer weighing 620 gms. was used in the following measurements on samples '*A*' and '*B*' of the same ƚsoil. Sample '*A*' was oven dried and '*B*' was completely saturated. Weight of pyconometer when filled with water only was 1495 gms.

	A	B
Weight of sample only (gms.)	980	1,020

Weight of pyconometer full of soil sample and water 2,112   2,030
Find (*i*) the specific gravity of the soil, (*ii*) the water content and void-ratio of sample *B*.   (*P.U. Nov 1964*)

8. (a) Define water content, void ratio. porosity, degree of saturation and dry unit weight.

    (b) Establish relationship between

        (*i*) Void ratio and porosity.

        (*ii*) Void ratio, water content, degree of saturation and unit weight of solids.

(c) A saturated soil sample has water content of 30 per cent. The specific gravity of solids is 2·70. Determine its void ratio and dry unit weight.   (*P.U. Nov. 1965*)

9. A sample of sand has a porosity of 40% and the specific gravity of its solids is 2·75. Find out the dry unit weight. saturated unit weight, when fully saturated, submerged unit weight and the saturated unit weight, when the degree of saturation is 50%.

(10) The mass specific gravity of a soil is 1·82 and the specific gravity of solids is 2·67. Water content in the wet soil is 30%, Find out its void ratio, porosity and degree of saturation ?

11. Differentiate between

    (a) Absolute density and bulk density of soils.

    (b) Absolute specific gravity and bulk specific gravity of soil.

12. Prove that

    (a) $\gamma_{dry}=(1-\eta)\,G.\gamma_w$, where γ_{dry} is the dry unit weight, η is the porosity, G is the specific gravity of soilds and γ_w is the unit weight of water.

    (b) Volume of air in a unit volume of soil is given by $V_a=(1-S)\,\eta$ where V_a is the air content, S the degree of saturation and η the porosity ?

13. The wet density of a soil is 2 gms./c.c. The specific gravity of solids is 2·70 and the moisture content of the soil is 15%. Calculate :

    (a) Dry density ;

    (b) Porosity and void ratio ;

    (c) Degree of saturation ;

    (d) Percentage of air voids.

14. A soil sample in its natural state a volume of ·310 c.c. and a weight of 528 gms. On even drying, its weight was 490 gms. Determine its void ratio porosity, water content and degree of saturation. Assume the specfic gravity of soil grains to be 2·67.   (*P.U. Nov. 1966*)

15. A subgrade soil of $G=2.67$ and $\gamma_d=(r_{dry})=95$ PCF is given. Mix with this soil an aggrgate of same specilic gravity in a proportion of 75% of aggagrete and 25% of soil by weight. The mixtu⸱e is then compacted to a $\gamma_d=115$ PCF. Whenthe aggregate has a moisture

content of 3% it is completely saturated. What is the saturation moisture content for the soil in the compacted mixture ? (*Mysore 1967*)

16. A sample of soil has a water content of 30 percent and a specific gravits of solids 2·70. Its unit weight is 1·50 kg/c.c. Determine void ratio, porosity and degre of saturation. (*Warangal 1968*)

17. A fully saturated clay sample has a volume of 185 c.c. and weighs 331 gms. If the specific gravity of solids is 2·67, find out void ratio, porosity, water content and unit dry weght. (*Baroda 1966*)

18. A fully saturated soil sample has a volume of 1 c. ft and weighs 130 lbs. The specific gravity of soil particles is 2·79. Determine the water content and void ratio.

(*i*) When the pore fluid is pure water having the unit weaght of 1 gm/c.c. and.

(*ii*) When the pore fluid is "Salt water" having a specific gravity of 1·025. (*Baroda 1965*)

Simple Physical Properties of soil and Classification Tests

3·1. Introduction

To physically identify the soils in the field as also in the laboratory and to determine some of their simple physical properties such as water content, specific gravity, void-ratio, plasticity indices and gradation, etc., some procedures have been developed. These procedures are not based upon any complicated theories but involve very simple calculations. However, determination of these physical features is very important in the field of soil mechanics. Many agencies such as the A.S.T.M., B.S.I. and I.S.I. have laid down set procedures standardised to ensure uniformity in practice. Detailed presentation of the procedures laid down by each individual agency is not possible here. However, fundamental concepts involved will be presented, and where thought necessary, a reference to the appropriate standard procedure will be made and the student may look up the relevant standard. Along with these procedures, some physical properties of soils, will also be dealt with side by side, so as to give an integrated account.

3·2. Colour of Soils

When one looks at a soil, colour is the first feature observed. Soils can have varieties of colours ranging from white through gray to red and even black. In a dry soil sample, the colour of mineral grains, the quantity of organic matter and the amount of oxidising compounds mainly affect its colour. Iron compounds under favourable conditions of moisture and temperature get oxidised and hydrated to yellow, brown and red compounds. Highly decayed organic matter and presence of manganese compounds lend black colour to the soil. Larger is the quantity of such materials in a soil, darker will be the colour. Green and blue colours in soils are rarely found but if they are present, they are due to the presence of ferrous compounds. Gray colour is a result of either quartz, kaolonite or small quantities of decayed organic matter.

Presence of moisture lends a darker colour to the same soil which is lighter when dry. This characteristic of the soil can be very well seen if an aerial photograph of a region is examined. The darker patches will invariably represent clayey soils which have the moisture-retention properties and the lighter patches on the photograph will show granular soils which do not retain moisture.

A wet dark soil will turn lighter in colour when dried in an oven.

3·3. Field Identification of Soils

Principal terms used by civil engineers to designate soils are *gravel, sand silt* and *clay*. Only artificially manufactured soil will contain one of the above varieties alone. For example, kaolin most widely used in ceramic industry is a pure clay. Most natural soils found in the field, are mixtures of two or more of the above varieties and may contain as an admixture same organic material in a partially or fully decomposed state. The mixture is then named after the major component and the minor component is added as an adjective. For example, mixture in which the major portion is sand and the minor portion is silt is called silty-sand. The properties of silty-sand are predominantly that of sand but it also exibits some characteristics of silt.

A soil which contains sand, silt and clay is given a special name, loam.

As already indicated in chapter I, gravel and sand and possibly coarse silt will fall into the coarse grained variety of soils and the silt and clay will fall in the fine-grained category. These two categories *viz.* coarse-grained and fine-grained soils, are distinguished accordingly as the individual soil particle can be seen with naked eye or not. Individual coarse grained particles can be seen visually, but the particles of silt can not be seen with the naked eye.

The distinction between gravel and sand is usually based upon the grain-size whereas this technique cannot be used to distinguish silt from clay and *vice versa*. Field identification can be done as under.

Gravel Versus Sand. From grain size considerations, any individual particle of soil larger than 2.0 m.m. is called *gravel.* The size lying between 2·0 m.m. and 0·06 m.m. is called *sand.* So roughly gravel is larger than the size of lead in an ordinary pencil and if individual particles in a sample can be seen with the naked eye and are less then the size of lead in an ordinary pencil, they may be classified as sand.

The above limits[1] on the sizes have been placed arbitrarily of course, bnt have been accepted by almost all agencies dealing with soil mechanics problems.

---

1. The Indian Standard I.S. : 1498-1959 also places the above limits on grain sizes of gravels and sands.

Field identification of gravel and sand should, if possible also include identification of the mineralogical composition and shape. The shape may be described as angular, sub-angular, rounded or sub-rounded. Shape of the grains has an important bearing on the suitability of the material for a construction project or its stability as a foundation material.

Sand Versus Silt. It many times becomes very difficult to distinguish *sand* from *silt* specially when the sand is very fine as a simple visual examination of their grain sizes will not indicate anything. Both the materials when dry look like dust excepting that silt may be slightly darker in colour than sand. However, they can be differentiated by conducting the *Dispersion Test.*

The dispersion test consists of pouring a spoonful of soil sample in a jar of water. If the material is sand, it will settle down within a minute and in case it is silt, it will take anywhere from 15 minutes to an hour to settle down. In both cases nothing will be left in suspension after the lapse of the respective timings.

Silt Versus Clay. Visual examination of the grain sizes of silts and clays is not possible and only a microscopic examination of such particles is possible in the laboratory. However, there are four simple tests which can fairly distinguish between the two varieties.

(a) *Shaking Test.* Prepare a pat of the material in water and put it in the hand. With the palm open, give this pat a shaking motion. If silt is present, due to higher *permeability*[1] of this material, the water will come to the surface and will give the soil a shining appearance. If the pat is now kneaded, the shining water at the surface will vanish and get absorbed in the soil. If it is clay, due to the lower permeability of clay, on shaking, the water will not come to the surface and the surface will still be dull.

In a mixture of silt and clay, the presence of relative amount of silt can be seen by observing the time taken to get the shining surface. If the reaction is rapid, the material is predominantly silt ; if it is slow, the quantity is relatively less and if it non-existant, then the silt is absent.

(b) *Strenght Test.* Prepare a small briquette of the material and dry it in the oven or in the sun. After drying, try to break the briquette. In case the material is silt, the briquette will break easily and if it is clay, it will need considerable effort to do it. Also, in case of silt, one can dust the material off the briquette just by rubbing a finger and in case of clay, dusting off is not possible.

---

1. For definition of *permeabilit* see Chapter V, Art.

(c) *Rolling Test.* Clay has the property of plasticity and cohesion. A soil, which is predominantly clay at certain moisture contents can be moulded in hand in the form of threads about 3 mm in diameter without disintegration. A thread of this size made out of clay and about 30 cm. long can support its own weight, when held in hands from the two edges, whereas it is not possible to make out a thread of this size from pure silt without crumbling or disintegration, short of its being able to stand any tensile stress due to its own weight.

(d) *Dispersion Test.* Pour about a spoonful of the soil in a jar of water. In case the soil is mainly silt, the particles will settle down in about 15 minutes to an hour, In case the soil is predominantly clay, it will form a suspension which will remain as such for hours and may be for days, unless the material gets flocculated into flocs and the flocs settle down.

It may, however, be indicated that considerable amount of experience is needed in the field for identifiation of the soils. A knowledge of the geology of the area can help in understanding the types of soils encountered in the region. An examination of any stratifications or in case a test pit has been dug out, an examination of the walls of the test pit can add to the results.

The above identification tests have a bearing on preliminary classification of the soils as presented in the next Chapter.

In their standard I.S. 1498—1959,. *Classification and Identification of soils for General Engineering Purposes*, the Indian Standard Institution has laid down their own procedure to conduct the field identification of the soils. .With the permission of the Institution, the tests as per their version are reproduced below from their standard[1] quoted above.

"Field identification[2] of soils is carried out by

(a) Visual Examination.

(b) Wet and Manipulated strength.

(c) Thread test.

(d) Dilatancy test.

(e) Dry strength.

The last three tests are only applicable to the fine-grained soils.

Visual Examination. The visual examination shall be carried out by naked eye with reference to size, angularity. touch and grading characteristics.

---

1. The I.S. 1498—1959, is under revision. It is only appropriate that the student refers to the new version of this standard when made available.

2. Reproduced by permission of the Indian Standards Institution.

Wet and Manipulated Strength Test. Take a small quantity of the soil specimen in hand, moisten it if needed, and work it with fingers and feel it. If the soil is clayey, a soapy touch is felt; if the soil is sandy, a feeling of roughness is experienced and in the case of silts when the soil is squeezed in between fingers, the moisture comes out. Also clay sticks to the fingers and dries slowly, but silt dries fairly quickly and can be dusted off the fingers leaving only a stain. The test helps to distinguish the predominant soil characteristic, that is whether it is clayey, sandy or silty.

Thread Test. Take a specimen of soil about one centimeter cube in size, moisten if needed, and roll it between the palms of the hands or on a flat, smooth surface into a thread of about 3 mm., in diameter. If crumbling does not occur, fold the thread, knead and re-roll as before. Repeat the process until the moisture content of the soil has been reduced, by drying manipulation, to the plastic limit, which is indicated by the crumbling which occurs as the soil is being rolled. The characteristic of the thread as it approaches the plastic limit affords the means of indentifying the soil in the manner indicated in Table[3] 1.

Dilatancy Test. After removing the particles retained on I.S. Sieve 40, prepare a pat of soil of a size 2 cm. cube. Add enough water, if necessary, to make the soil soft but not sticky. Place the pat in the open palm of one hand and shake horizontally, striking vigorously against the other hand several times. A positive reaction consists of the appearance of water on the surface of the pat which changes to a livery consistency and becomes glossy. When the sample is squeezed between the fingers, the water and glass disappear from the surface, the pat stiffens and finally it cracks or crumbles. The rapidity of appearance of water during shaking and of its deappearance during squeezing assist in identifying the character of the fines in a soil. Very fine clean sands give the quickest and most distinct reaction whereas a plastic clay has no reaction. Inorganic silts, such as as a typical rock flour, show a moderately quick reaction.

Dry Strength Test. After removing particles retained on I.S. Sieve 40, mould a pat of soil to the consistency of putty, adding water if necessary. Allow the pat to dry completely by oven, sun or air-drying, and then test its strength by breaking and crumbling between the figures. This strength is a measure of the character and quantity of the colloidal fraction contained in the soil. The dry strength is characteristic for clays of CH group. A typical inorganic silt possesses only very slight dry strength. Silty fine sands and silts have about the same dry strength, but can be distinguished by the feel when powdering the dry specimen. Fine sand feels gritty whereas a typical silt has the smooth feel of flour.

---

3. This table has been reproduced in Chapter 4, as table 4.8 and the student may refer to this table.

3 4. Determination of moisture content

Strength and stability of soils in almost all cases, depend upon its moisture content and hence it is invariably required to find out the moisture contents of the soils. The moisture content in a soil may vary from near saturation stage to a very small quantity acquired by the soil when in contact with the atmosphere. The quantity of moisture acquired from atmosphere is usually termed as *hygroscopic moisture*.

Moisture content determinations invariably form a part of all the elaborate laboratory tests in soil-mechanics. The Procedure normally adopted is as under.

The weight[1] of the wet soil sample normally taken for moisture content determination is nearly 50 gms. The sample is placed in a weighed container and the weight of wet soil and the container taken. The container is then placed in the oven pre-set at a temperature of $105°—110°C$ for about 24 hours when it will attain a constant weight. The following calculations are then made.

Weight of the container $= W_c$

Weight of wet soil + container $= W_1 = W_c + W_s + W_w$;

where W_s and W_w denote the weight of dry soil and weight of water respectively.

Weight of wet soil $= (W_1 - W_c)$

Weight of oven dried soil + container $= W_2 = W_c + W_s$

Weight of oven dried soil $W_s = (W_2 - W_c)$ ...(3.1)

Weight of water evaporated $W_w = (W_1 - W_2)$ ...(3.2)

From (3.1) and (3.2) by definition ;

Percentage moisture content of soil

$$= \frac{W_w}{W_s} \times 100 = \frac{(W_1 - W_2)}{W_2 - W_c} \times 100\% \qquad ...(8.3)$$

3·5. Determination of specific gravity

Specific gravity of the soil particles is needed in determining the void-ratio, porosity and degree of saturation, particle size distribution ; etc. of the soil. It is also used in the studies of *critical hydraudic gradient*[1] when *quick sand*[1] conditions are implied and in the zero-air void estimation in the compaction theory. When the soil particles are of the same size, determina-

---

1. For ISI specifications see IS : 2720 (Part II)—1964, *Methods of Tests for soils—Part II Determination of Moisture Content.* (Reference 11 at the end of this Chapter).

tion of the specific gravity is no problem. However, if the soil constitutes of a mixture of gravel, sand and clay, the average specific gravity of the mixture may be quite different from the specific gravities of coarse aggregates and fine-aggregates determined separately. Unless otherwise required, an average specific gravity is determined for practical purposes. The test[2] may be carried out as per procedure under A.S. T.M. Designation D 854— 58. The following is the general procedure.

The method recommended for determination is called the pycnometer method. About 100 to 200 gms. of dry soil are taken in a 500 c.c. pycnometer and the rest of the volume upto 500 c.c. mark of the pycnometer is filled with distilled water. All the air in the soil-water mixture is expelled by either applying a vacuum or by heating the mixture gently. Both while evacuating or boiling, the pycnometer is continuously shaken to agitate the mixture, which process helps in expelling air as well as in keeping the whole mixture at a uniform temperature.

How much inaccuracy is involved, by not expelling air completely out of the voids can be imagined by going through example 3.1. It is, therefore, very essential that the air is completely expelled from the sample. Application of vacuum or heating gently for about 30 minutes normally expels all air out. However, this limit may be taken only as a guide. The measurements needed are.

Dry weight of the soil$=W_s$

Weight of pycnometer+soil+water$=W_1$

Temperature of the mixture$=T°C$

The pycnometer is then emptied and after cleaning it, is filled with distilled water upto the 500 c.c. mark. Then, the pycnometer filled with water is weighed. Let

Weight of pycnometer+distilled water $=W_2$

Following calculations, from the above observations, will give the specific gravity of solids. A phase diagram shown in Fig. 3.1 helps in the calculations.

Note that the volume in both weight measurements W_1 and W_2 is the same.

Now, from the figure, $W_1-W_2=$Weight of solids—weight of an equal volume of water. If γ_w is the unit weight of water at the temperature $T°C$ of the test, then

$$W_1-W_2=W_s-V_s\gamma_w,\text{ where } V_s \text{ is the volume of solids.}$$

or $$V_s\gamma._w=W_s-W_1+W_2 \qquad\qquad ...(3.4)$$

---

1. For definitions see Chapter VI.

2. For ISI Specifications see IS : 2720 (Part III)—1964, *Methods of Test for soils Part III, Determination of Specific Gravity.* (Reference 12 at the end of this Chapter).

If G_r is the specific gravity of water at the temperature $T°C$, then $\gamma_w = G_T\gamma_o$ where γ_o is the unit weight of water at 4°C.

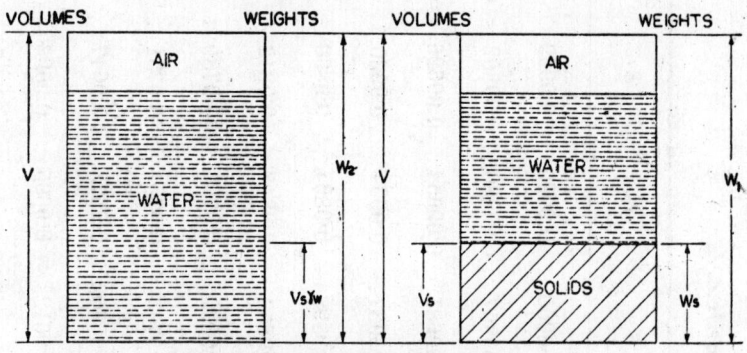

DETERMINATION OF SPECIFIC GRAVITY OF SOLIDS

Fig. 3.1.

Substituting this value of γ_w in equation (3.4)

$$V_s.G_T\gamma_v = W_s - W_1 + W_2 \qquad \qquad ...(3.5)$$

If G is the specific gravity of solids, then

$$W_s = G\gamma_o. \ V_s$$

$$\therefore \qquad G = \frac{W_s}{\gamma_o . V_s} \qquad \qquad ...(3·6)$$

Substituting $\gamma_o . V_s$ in equation (3.6) from equation (3.5)

$$G = \frac{W_s . G_T}{W_s - W_1 + W_2} \qquad \qquad ...(3·7)$$

which is the required specific gravity of solids.

The specific gravities of distilled water at different temperatures is given in Table 3.1 on page 42.

It can be seen from Table 3.1 that the specific gravity of water over the range of temperatures indicated varies from 0·96 approximately to 1·0 at the most. Therefore, if the temperature correction is neglected and the value of G_T be taken as unity, *i.e.*, the unit weight of water taken as 1 gm./c.c., then equation 3·7 will reduce to

$$G = \frac{W_s}{W_s - W_1 + W_2} \qquad \qquad ...(3·7a)$$

If the denominator of equation (3·6) or (3·7) is critically examined, it is nothing but the weight of a volume of water displaced by the soil. This is quite clear from the phase diagram in Fig. 3·1.

TABLE 3·1

Specific Gravities of Distilled Water in gms./c.c.

Temp. °C	0	1	2	3	4	5	6	7	8	9
0	0·9999	0·9999	1·0000	1·0000	1·0000	1·0000	1·0000	0·9999	0·9999	0·9998
10	0·99973	0·99963	0·99952	0·99940	0·99927	0·99913	0·99897	0·99880	0·99862	0·99843
20	0·99823	0·99802	0·99780	0·99757	0·99733	0·99707	0·99681	0·99654	0·99626	0·99597
30	0·99568	0·9954	0·9951	0·9947	0·9944	0·9941	0·9937	0·9934	0·9930	0·9926
40	0·9922	0·9919	0·9915	0·9911	0·9907	0·9902	0·9898	0·9894	0·9890	0·9885
50	0·9881	0·9876	0·9872	0·9867	0·9862	0·9857	0·9852	0·9848	0·9842	0·9838
60	0·9832	0·9827	0·9822	0·9817	0·9811	0·9806	0·9800	0·9795	0·9789	0·9784
70	0·9778	0·9772	0·9767	0·9761	0·9755	0·9749	0·9743	0·9737	0·9731	0·9724
80	0·9718	0·9712	0·9706	0·9699	0·9693	0·9686	0·9680	0·9673	0·9647	0·9660
90	0·9653	0·9647	0·9640	0·9633	0·9626	0·9619	0·9612	0·9605	0·9598	0·9591

Where the identity of the soil has been established beyond doubt, the Engineer may not like to do too much of laboratory work and as a guide for him specific gravity values of some typical soils are given in Tables 3.2 below.

TABLE 3.2

Specific gravity values of some typical soils

Serial No.	Soil Types*	Specific Gravity G.
1.	Quartz Sand	2·64—2·65
2.	Quartzite	2·65
3.	Silt	2·68—2·72
4.	Silt with organic matter mixed in it	2·40—2·50
5.	Loess	2·65—2·75
6.	Lime	2·70
7.	Clay	2·44—2·92
8.	Kaolin	2·47—2·58
9.	Chalk	2·63—2·81
10.	Bentonite clay	2·34
11.	Peat	1·26—1·80
12.	Humus	1·37

*For a description of some of these soil types, see Chapter IV.

Example 3·1 Specific gravity for a soil was obtained in a laboratory test. Following measurements were made :

$$W_s = 100·0 \text{ gms.}$$

$$W_1 = 608·0 \text{ gms.}$$

$$W_2 = 550·0 \text{ gms.,}$$

where the various symbols have the meanings given in article 3.5 By oversight, 2 c.c. of air remained entrapped in the suspension when the weight W_1 was taken.

(a) will the value of G be lower or higher than the true value ?

(b) what is the percentage error ?

Neglect the effect due to temperature.

Solution. (a) The specific gravity of solids, neglectnig the effect of the temperature is given by

$$G = \frac{W_s}{W_s - W_1 + W_2} \qquad \text{(From eqn. 8·7)}$$

W_1 occurs in this equation in the denominator with a negative sign. So, if 2 c.c. of air was entrapped, it just replaced an equal quantity of water which would have otherwise occupied this space. As such, W_1 is smaller by 2 gms. As such the value of denominator in the case would be more compared to its correct value. So the value of G would be lower than the true value. This can be seen by calculations.

$$\text{Measured } G = \frac{100}{100-608+550}$$

$$= \frac{100}{42} = 2\cdot38$$

$$\text{Correct } G = \frac{100}{100-(608+2)+550}$$

$$= \frac{100}{40} = 2\cdot50.$$

Hence measured value is lower than the correct value.

(b) Percentage $\text{error} = \dfrac{(2\cdot50-2\cdot38)}{2\cdot50} \times 100$

$$= \frac{0\cdot12 \times 100}{2\cdot5} = 4\cdot8\%$$

Ans.

3·6. Determination of void ratio

Void ratio characterizes the natural state of density of a soil. This parameter is used in the soil mechanics field very often and hence it is important to determine it. The computations for void ratio are based upon the volume V of the soil, the dry weight of the sample W_s and the specific gravity G of the solids. These observations are made with great accuracy because a little inaccuracy can affect the value of void ratio considerably.

By definition from equation (2.2)

$$\text{Void ratio } e = \frac{V_v}{V_s} = \frac{V-V_s}{V_s}$$

$$= \frac{V}{V_s} - 1$$

But $\qquad V_s = \dfrac{W_s}{G.\gamma_w}$

$$e = \frac{V}{W_s/G.\gamma_w} - 1$$

$$= \frac{V.\gamma_w G}{W_s} - 1 \qquad\qquad \dots(3\cdot8)$$

It may be stated here it is only the volume measurement of the soil, which presents the real problem[1] in the void ratio determination and the observer has to be really very careful in determining this parameter.

Table 3·3 gives the void ratio, porosity[2] and dry unit weight; etc. of some typical soils in their natural state. These values are of only qualitative significance and for a particular, soil, numerical values must be determined actually in the laboratory.

TABLE 3.3

Values of void Ratio, Porosity and Unit weight, etc. for some typical soils in a natural state

S. No.	Description of soil	Void Radio. e	Porosity η	Water content in saturated state, w	Unit weights	
					γ_d gms./c.c.	γ_{sat} gms./c.c.
1.	Dense uniform sand	0.50	0.333	19	1.75	2.10
2.	Loose uniform sand	0.85	0. 46	32	1.43	1.90
3.	Dense Mixed-grained sand	0.42	0.295	16	1.86	2.16
4.	Loose Mixed-grained sand	0.66	0.395	25	1.59	2.00
5.	Soft clay (Glacial)	1.20	0.55	45	—	1.77
6.	Soft Slightly organic clay	1.90	0.66	70	—	1.58
7.	Soft very organic clay	3. 0	0.75	110	—	1.43
8.	Soft Bentonite	5. 2	0.84	195	—	1.27

1. All other parameters are easy to determine.
2. Already defined in Article 2·3 Chapter 2.

3.7 Index properties of soils

Field identification methods for the soils have been described in Section 3.3. Identification of a soil in the field is always rough. By laboratory analysis, soils are grouped into classes that exhibit similar properties. Method of classifying the soils after laboratory testing is called classification of the soils. Soil classification systems are described in Chapter 4. In this article some testing procedures used in classification systems are only given. Numerical results, which are obtained from such tests are called the *Index Properties* of the soils. Index properties of a soil can be studied under two distinct headings—properties concerning the individual grains studied from disturbed soil samples and the ones concerning the soil as an aggregate, studied from the mode, history of deposition and the structure of the soil.

The first analysis carried out in the laboratory for soil classification is the grain-size analysis. Grain-size analysis of a soil is carried out in two stages :

(1) Sieve analysis for the coarse grained particles.

(2) Hydrometer analysis or the Pipette analysis for the fine-grained particles.

Line of division between coarse-grained and fine-grained soils is the B.S., A.S.T.M. or A.A.S.H.O. Sieve No. 200 or I.S. Sieve designation 75 micron corresponding to 0.75 m.m. aperture width. This size corresponds to the lower limit for sands in the I.S. and M.I.T. Soil Classification Systems. Each of the above two procedures will be described separately in the following sections.

Besides the grain-size analysis, the Atterberg Limits of the soil are determined which give some numerical values that are required in certain classifications systems. Also these values help in making a preliminary estimate of the behaviour of the soil. as a construction material or as a foundation material. In the highway Engineering field, certain pavement design procedures are also based upon the sieve analysis and the Atterberg limits, results.

3·8. Sieve analysis of soils

Agencies entrusted with the standardisation work for the soil mechanics field have standardised certain sieve sizes for sieve analysis of soils. The important agencies whose standards may be referred to, for sieve sizes, are the Indian Standards Institutions (I.S. 460—1962 Revised), the American Society of Testing Materials (Designation E 11·58 T and E 11-62), British Standards Institution (B. S. 410—1962) and the International Standards Organisation. A comparison of the sieve sizes and their designations by some of these agencies is given in Table 3.4, for ready reference.

TABLE 3-4
Important Sieve Sizes and their Designations[1]

Sieve Size m. m.	Old I. S. Designations	New I. S. Designations	B. S. I. Designations	A. S. T. M. Designations
5.66	570	—	—	3½
5.60	—	5.60m.m.	—	—
4.76	480	—	4	4
4.75	—	4.75	—	—
4.00	400	4.00	—	5
3.35	340	3.35	5	6
2.818	280	—	6	7
2.800	—	2.80	—	—
2.411	240	—	7	8
2.360	—	2.36	—	—
2.032	200	—	8	10
1.70	—	1.70	—	—
1.676	170	—	10	12
1.405	140	—	—	—
1.40	—	1.40	—	—
1.204	120	—	14	16
1.18	—	1.18	—	—
1.00	100	1.00	16	18
0.853	85	—	18	20

0.850	—	850micron	—	—
0.710	—	710	—	—
0.708	70	—	22	25
0.600	—	600	—	—
0.599	60	—	25	30
0.500	50	500	30	35
0.425	—	425	—	—
0.422	40	—	36	40
0.355	—	355	—	—
0.351	35	—	44	45
0.300	—	300	—	—
0.295	30	—	52	50
0.251	25	—	60	60
0.250	—	250	—	—
0.212	—	212	—	—
0.211	20	—	72	70
0.180	—	180	—	—
0.177	18	—	85	80
0.152	15	—	100	100
0.150	—	150	—	—
0.075	8	75	200	—
0.074	—	—	—	200

1. An effort has been made in this table to relate the various sieve sizes as best as was possible. The student may find some marginal adjustments.

For details of the various standards, the reader may refer to I. S. 460—1962 (old and revised), B. S. 410—1962 and A.S,T.M. E 11—62.

The test procedure for seive analysis has also been standardised by some of these agencies, in their relevent standards. The readers may refer to the Indian Standards Institution IS:2720 (Part IV)—1965, or A.S.T.M. Designation 422—54 T or B. S. 1377 (1948) Test No. 8, or the A.A.S.H.O.[1] Designation T 88—54, for details as mentioned by these agensies. The following is the procedure that is generally employed.

A known weight of the soil, usually 500 gms., is taken. It is ensured before taking this sample, that the sample is representative of the whole lot of soil and that no soil nodules exist in the sample, i. e., the soil has been ground to individual grains; otherwise the results will be erroneous.

It is also necessary to see that the soil is dry. In case it is moist, it is dried out in an oven at a temperature of 105° C to 110° C for twentyfour hours, so that it attains constant weight. If, however, the soil contains considerable amount of organic matter which might get burnt in the oven, air-drying is resorted to.

A set of sieves is arranged in such a way that the sieve with the largest opening is at the top and the sieve with the smallest opening is at the bottom of the set. The selection of sieves for a given test depends upon the soil to be tested, e. g., the coarser the soil, the larger the top sieve. Sieve sizes usually adopted are the ones that correspond to B. S. designations, 4, 8, 14, 25, 52, 100 and 200. The student may use the appropriate set available to him by comparing these sizes with the ones he has, from Table 3.4.

The weighed quantity of the soil is placed in the top-most sieve and the set is placed in a sieve-shaker and shaken for 10 to 15 minutes. Soil gets graded according to the individual grain sizes. Some of the soil will pass through the smallest sieve size corresponding to 0·075 m.m. width of opening. This portion of the soil will need further particle size analysis of its grains by the hydrometer or the pipette analysis procedures.

The 0·075 m.m. width of the sieve is so small that sometimes the material does not pass through this sieve, although individual grains are smaller than this size. In that case, then, the soil grains have to be washed through this sieve. In case of many clays, it is not possible to break nodules into individual grains mechanically and in such soils invariably, the whole of the analysis is done by washing the material through the various sieves and getting their weights after evaporation of water in an oven. In any case, whether the dry sieve analysis or the wet sieve analysis is carried out, weight retained on each sieve and the weight passing through No. 200 B. S. sieves are noted.

---

1. American Association of State Highway Officials.

From the dry soil weights retain d on eacn sieve, the weights passing through each sieve are calculated. A cumulative percentage of the dry soil weight passing a sieves is obtained by dividing the weight passing each sieve by the total dry soil weight originally taken and multiplying it by 100. The results of this analysis are represented on a semi-logarithmic plot showing cumulative percentage finer by weight on the ordinate on an arithmetic scale and the corresponding particle diameter sizes or the sieve sizes on the abscissa on a logarithmic scale. Figure 3.2 shows such curves. It may be noted here that from the sieve analysis results, the shape of the curve can be drawn upto 0.075 m.m. size only as shown in the figure and for the material finer than this size, the results are taken from the hydrometer or pipette-analysis, described in Art. 3.10 and Art. 3.11 respectively.

Curves of the type shown in Fig. 3.2 are called particle-size distribution[1] curves. A logarithmic scale for the particle diameters gives the best representation of size distributions because a wide range of particle diameters between the coarsest and

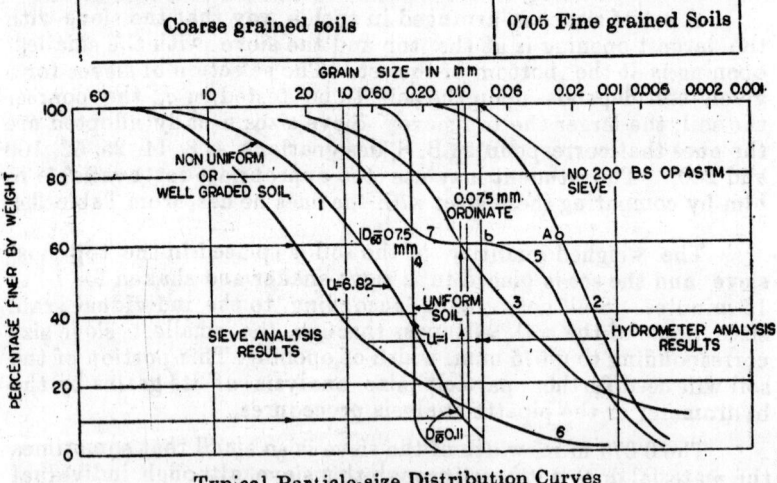

Typical Particle-size Distribution Curves
Fig. 3·2

finest sizes can be well presented on such a scale. Also the results of the finer fraction, *i.e.* minus 0·075 m.m. size can be plotted with a great degree of accuracy without choosing smallar scale lengths on the abscissa.

---

1. Sometimes these curves are designated as particle accumulation curves. If a histogram is drawn from the fracrion retained between successive sizes and then a smooth curve is drawn through this histogram, this smooth curve is someitmes designated as particle-size distribution curve. However, at the undergraduate level, the distinction is not important and the curves shown in Fig. 3.2 may be called particle-size distribution curves.

The characteristics of these curves are discussed in Art.3·12

3.9. Stoke's Law.

The limiting width of individual opening of wire cloth for the sieves is 0·075 m.m. Materials finer than this size cannot therefore, be sieved. Their particle size distribution is found out by hydrometer or pipette analysis, which is based upon Stoke's Law for the velocity of falling spheres in a liquid medium. The principle involved is that spheres of different sizes fall through a l·quid at different velocities. If a single sphere is allowed to fall in an infinite liquid medium without any interference, its velocity first increases under the action of gravity, but in a matter of seconds, the sphere attains a constant velocity, which it maintains indefinitely, unless the boundary conditions or other experimental conditions are changed. This constant velocity which the sphere attains is called terminal velocity and is given by the equation

$$v = \frac{\gamma_s - \gamma_w}{18\mu} D^2 \qquad ...(3.9)$$

where
$v =$ Terminal velocity in cms/sec.
$\gamma_s =$ Unit weight of a sphere in gms/c.c.
$\gamma_w =$ Unit weight of water. gms/c.c.
$\mu =$ Viscosity of water in gms sec/cm^2.
$D =$ Diameter of the sphere in cms.

At a temperature of 20° C and specific gravity γ_s of the sphere as 2·67 and for the sphere falling in water of rough average viscosity of 10^{-5} gms. sec/cm., the above equation reduces to.

$$v = 9000 \, D^2 \qquad ...(3·10)$$

where v is in cms./sec. and D is in cms.

The unit weights of pure water for different temperatures are given in Table 3.1 and the viscosities of pure water for different temperatures are given in Table 3·5 on page 53.

Stoke's Law is applicable to spheres between about 0·2 m.m and 0·0002 m.m. in diameters, falling freely through water and having specific gravity close to that of soil grains.

Application of the Stoke's Law to determine the grain-size distribution of soils is not wholly justified unless some assumptions are made. The reasons are as under :

(1) Soil particles are never perfect spheres. They are either flake like or needle like. So to overcome this shortcoming, the concept of equivalent diameters has been evolved. A particle is said to have an equivalent diameter D, if it has the same velocity of fall as a sphere of same unit weight and of diameter D.

(2) Stoke's Law is applicable to a sphere which falls freely without any interference, in an infinite liquid medium. The hydrometer or pipette test is conducted in a 1000 c.c. cylinder. So the medium of water does not extend indefinitely. Also

among the falling particles there is always interference from each other, due to limited space and hence the fall is not free. Moreover, the glass walls of the jar interfere with the free fall of particles near it.

However, it is assumed that if upto 50 gms. of soil per litre of water are used in a 1000 c.c. container, the interference is negligible.

(3) All soil grains do not have the same specific gravity due to variation in sizes. The specific gravity may have a range of values.

It is assumed, however, that the variation in specific gravities of various particles for minus 0.075 m.m. size of a soil may not be of any significance and an average specific gravity will be a fair approximation for all particles in soil mechanics work.

(4) Soil particles of minus 0·075 m.m. size being clayey carry surface electric charge and sometimes tend to form flocs. So, unless this tendency of formation of flocs is checked, the diameters measured will be those of flocs and not of soil particles.

This difficulty is overcome by adding a diflocculating agent to the soil. Many deflocculating agents are available, most commonly used ones are, sodium silicate. sodium oxalate and sodium hexa-metaphosphate.

3·10 The Hydrometer Analysis

Hydrometer is a device (see Fig. 3·3) which measures the specific gravity of a liquid. In the case of a soil supension, since the particles start settling down from the time of preparation of the suspension, the unit weight of suspension varies from top to the bottom of the suspension. However, the principle of measurment of specific gravity by a hydrometer can be extended to soil suspension of variable density, thus making it possible to determine its grain size distribution.

Firstly it must be established, however, that measurement of unit weight of the suspension at any known depth at an elapsed time from the time the suspension is made, can give a point on the grain size distribution curve.

Let W = weight of the soil mixed in water
V = volume of suspension.

As soon as the suspension is prepared, in a unit volume of the suspension, the weight of solids is $\left(\dfrac{W}{V}\right)$ and the volume of solids is $\left[\dfrac{W}{V} \times \dfrac{1}{G\gamma_w}\right]$

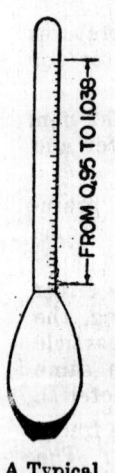

A Typical Soil Hydrometer
Fig. 33

FROM 0.95 TO 1.038

TABLE 3·5

Viscosity of Water at Different Temperatures in Millipoises.

C°	0	1	2	3	4	5	6	7	8	9
0	17·94	17·32	16·74	16·19	15·68	15·19	14·73	14·29	13·87	13·48
10	13·10	12·74	12·39	12·06	11·75	11·45	11·16	10·88	10·60	10·34
20	10·09	9·84	9·61	9·38	9·16	8·95	8·75	8·55	8·36	8·18
30	8·00	7·83	7·67	7·51	7·36	7·21	7·06	6·92	6·79	6·66
40	6·54	6·42	6·30	6·18	6·08	5·97	5·87	5·77	5·68	5·58
50	5·49	5·40	5·32	5·25	5·15	5·07	4·99	4·92	4·84	4·77
60	4·70	4·63	4·56	4·50	4·43	4·37	4·31	4·24	4·19	4·13
70	4·07	4·02	3·96	3·91	3·86	3·81	3·76	3·71	3·66	3·62
80	3·57	3·53	3·48	3·44	3·40	3·36	3·32	3·28	3·24	3·20
90	3·17	3·13	3·10	3·06	3·03	2·99	2·96	2·93	2·90	2·87
100	2·84	2·82	2·79	2·76	2·73	2·70	2·67	2·64	2·62	2·59

1 dyne sec. per sq. cm. = 1 poise
1 gram sec. per sq. cm. = 980·7 poises
1 pound sec. per sq. ft. = 478·69 poises
1 poise = 1000 millipoises.

Since this one unit is composed of solids and water,

$$\text{volume of water} = 1 - \frac{W}{V.G.\gamma_w} \text{ and}$$

$$\text{weight of water} = \gamma_w\left[1 - \frac{W}{V.G.\gamma_w}\right] = \gamma_w - \frac{W}{V.G}.$$

Initial weight of one unit volume of suspension=weight of solids +weight of water.

$$= \frac{W}{V} + \left[\gamma_w - \frac{W}{V.G}\right]$$

Let this initial weight be designated as γ_i.

$$\therefore \qquad \gamma_i = \frac{W}{V} + \left[\gamma_w - \frac{W}{V.G}\right]$$

$$\text{or} \qquad \gamma_i = \gamma_w + \left[\frac{G-1}{G}\right]\frac{W}{V} \qquad \ldots(3\cdot11)$$

Now consider a point A at a depth of z from the surface of suspension (See Fig. 3·4) and let t be the time elapsed since the particles started settling.

Particles of diameter D which have just fallen from the surface to the level A in time t will be given by :

$$v = \frac{\gamma_s - \gamma_w}{18\mu}D^2 = \frac{z}{t}$$

$$\text{or } D = \sqrt{\frac{18\mu}{\gamma_s - \gamma_w} \cdot \frac{z}{t}} \quad \ldots(3\cdot12)$$

where symbols have the significance already mentioned.

Above the level of A, all the particles of soil greater in diameter than D, do not exist. They have already crossed this point. Now if an element of depth dz is chosen at the level of A, as shown, the solid particles are uniformly dispersed in this element because initially the suspension was uniform and all particles of same size have settled to the same depth in the same time interval.

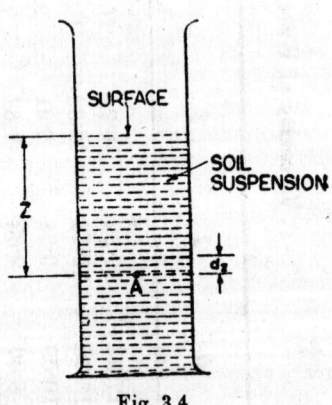

Fig. 3.4.

With in this element now, therefore, exist particles of diameter smaller than D. Let the ratio between the weights of particles smaller then D in the element dz and the original weight of the soil particles be n.

Per unit volume of suspension at a depth z, weight of solids $= n.\dfrac{W}{V}$ and the unit weight of suspension by similar reasoning as per equation (3·11) is :

$$\gamma_z = \gamma_w + \left(\frac{G-1}{G}\right)\cdot n\cdot\frac{W}{V}$$

$$n = \frac{G}{G-1}\ (\gamma_z - \gamma_w)\cdot\frac{V}{W} \qquad\qquad ...(3·13)$$

n in equation (3·13) represents the fraction of soil smaller than diameter D represented by equation (3·12).

Hence, at the known depth z, after an elapsed time t, if the suspension unit weight γ_z could be found, the percent of fines smaller than diameter D can be established from equation (3·13).

This unit weight of the suspension is found out with the help of the hydrometer, since as already stated in the beginning, a hydrometer gives the specific gravity of a liquid.

The fact that the unit weight of the suspension of soil varies from top to bottom of the suspension, causes slight difficulties. However, Casagrande has demonstrated that in case this variation in density is assumed to be linear, then the hydrometer reading say R gives the specific gravity of the soil suspension at the level of the centre of immersed volume. This assumption causes an error of 3% with respect to the actual conditions and as such is considered negligible.

I.S : 2790 (Part IV)—1965 or A.S.T.M. Designation $D432$—$54T$ or B.S. 1377 (1948) or A.A.S.H.O. Designation B—88—54 may be consulted for their details on this test. The general proceduce is as under :

About 50 gms. of the soil passing No. 200 B.S. or ASTM sieve or their equivalent is mixed with 250 c.c, of distilled water and an appropriate quantity[1] of a deflocculating agent is added. The mixture is thoroughly agitated using a mechanical mixer, so as to ensure proper dispersion of the particles.

The mixture is then transferred to a 1000 c.c. jar and distilled water added to it to bring the suspension to the level of the

---

1. There is no single figure for the amount of deflocculating agent to be added. Different standards quote different type of deflocculating agents and different amounts of the same. If sodium hexa-meta-phosphate is used, add 20 gms. of it to the suspension.

1000 c.c. mark. This mixture is then agitated by turning the jar upside down to ensure that the density of the whole mixture is uniform. The cylinder is then placed on a level platform and stop-watch started. A stream lined hydrometer is then suspended very gently into the suspension and hydrometer readings taken at intervals of 30 seconds, one minute and two minutes and then the hydrometer is taken out of the suspension and placed in another jar full of distilled water alone.

Again the readings are taken at elapsed intervals of 5, 10. 20, 40 and 100 minutes and so on at regular intervals till such time that the hydrometer reading has decreased to a value of 1·008. For all these readings after the 2 minutes reading the hydrometer is placed very gently in the suspension just before taking the reading, and after taking the reading, it is immediately removed. This is done to ensure that there is no error caused due to the presence of the hydrometer itself. The temperature of the suspension is recorded throughout the test and it is ensured that the jar is away from direct sunlight or any local source of heat such as an oven. This avoids the setting in of any convection currents and non-uniform temperatures.

In order to get good results, it should be ensured that the temperature should not vary from an average temperature of the test by more than $2°C$.

Hydrometer readings need certain corrections before the data can be processed. The corrections needed are as under.

Hydrometer Corrections : There are three corrections which need be mentioned.

(a) *Meniscus Correction.* Hydrometers are designed so that the readings are taken at the lower level of the meniscus. Since the soil suspension is opaque, readings can only be taken at the upper level of the meniscus. Hence a meniscus correction is needed. The difference between the lower and upper levels of meniscus is noted by suspending the hydrometer in clean distilled water. The readings on the stem of a hydrometer go on increasing from top to bottom. So if the reading in suspension is taken at the upper level of the meniscus, this reading will be less than the actual reading. Hence the meniscus correction is additive.

(b) *Temperature Correction.* Hyrometers are calibrated at a temperature of $20°C$. In case the test temperature is different, the temperature correction is needed. For this purpose, the hydrometer is placed in clean distilled water at different temperatures and required correction noted. Then these corrections are plotted against the range of temperatures studied and the corrections applied accordingly.

Casagrande's nomographic Chart given in the Appendix automatically takes care of this correction.

(c) *Correction for the use of deflocculating agent.* The calculations assume that the soil is dispersed in clean distilled water.

However, some deflocculating agent is always used for this test. The density of the solution of deflocculating agent in water is always more than that of pure water for any temperature. Hence, a correction is needed. The correction is done by noting the difference between the hydrometer reading in water and the hydrometer reading in the solution of the deflocculating agent used and then this correction is applied to the readings.

A composite correction for all the above three corrections can be made by noting the readings of hydrometer in a solution of the deflocculating agent at different temperatures.

Besides these above corrections needed, the *hydrometer immersion correction* will be needed for all the readings after the 2 minute reading.

(d) *Hydrometer Immersion Correction.* Refer to Fig. 3·5 (a) and (b). Consider a point P in the suspension at the section BB

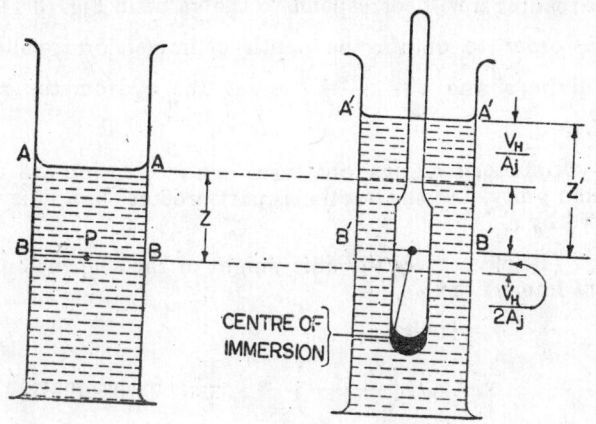

(a)   (b)
Hydrometer Immersion Correction
Fig. 3·5.

in figure 3·5 (a) before the hydrometer is suspended. In time t after the particles started settling, all particles larger than diameter D have crossed this point. Now, the hydrometer is inserted in the suspension to find its specific gravity as shown in Fig. 3·5 (b). The depth of immersion as indicated in Fig. 3·5 (b) is z'.

However, a comparison of Fig 3·5 (a) and 3·5 (b) shows that due to immersion of the hydrometer, the level of section BB in Fig. 3·5 (a) jumped to $B'B'$ in Fig. (3·5) (b) and the level of surface AA in Fig. 3·5 (a). Jumped to $A'A'$ in Fig. 3.5 (b).

If V_H is the volume of the hydrometer and A_j is the area of the jar in which suspension is put, then the rise of $A'A' = \dfrac{V_H}{A_j}$, since below the level of $A'A'$ the hydrometer is fully in water and rise $B'B'$ is $\dfrac{V_H}{2A_j}$, because below the level of $B'B'$, only half of the hydrometer is in water.

In fact the particles had fallen through the depth z in time t as shown in Fig. 3·5 (a) and not through z' as indicatad in Fig. 3·5 (b).

From the figure $z = z' - \dfrac{V_H}{A_j} + \dfrac{V_H}{2A_j}$,

or $$z = z' - \frac{V_H}{2A_j} \qquad \qquad ...(3·14)$$

In the equation (3·12), for the first few readings upto the 2 minute reading, the value of z is the same as z' but for the subsequent reading z will correspond to the value in Fig. (3·14).

In order to obtain the depth of immersion z, calibration curves giving z' and $\left(z' - \dfrac{V_H}{2A_j} \right)$ versus the hydrometer reading R are drawn.

Hydrometer readings obtained in a suspension are usually more than unity. Let the fractional part greater than unity be represented by r.

∴ $(1+r)\gamma_w$ gives the unit weight of the suspension at the centre of immersion, i.e.

$$\gamma_z = (1+r)\gamma_w \qquad \qquad ...(3·15)$$

Also $\quad \gamma_z = \gamma_w + \left(\dfrac{G-1}{G} \right) . \; n \; . \; \dfrac{W}{V}$, from equation (3·13)

∴ $\quad \gamma_w + \left(\dfrac{G-1}{G} \right) . \; n \; \dfrac{W}{V} = \gamma_w + r\gamma_w$

or $\quad n = r . \; \gamma_w . \dfrac{G}{G-1} . \dfrac{V}{W} \qquad \qquad ...(3·16)$

If the unit weight of water is taken as 1 gm./c.c., the equn. reduces to, $\qquad \qquad (3·16)$

$$n = r . \left(\frac{G}{G-1} \right) \frac{V}{W} .$$

Percentage of particles smaller than D is given by multiplying by 100.

∴ Percentage of particles smaller than equivalent diameter

$$D = r \cdot \left(\frac{G}{G-1} \right) \cdot \frac{V}{W} \times 100 \qquad \qquad ...(3\cdot17)$$

Equations (3.12) and (3.17) give the co-ordinates of a point on the particle distribution curve. Such co-ordinates are computed for all the readings taken in the analysis. These points are then plotted on the particle size distribution sheet on the same scale as the points from sieve analysis, as shown in Fig. 3·2. Then a smooth curve is drawn through all these points. Examples 3·2 and 3·3 illustrate these concepts very well.

Examples 3.2. Fifty gms. of a soil of specific gravity 2.70 are mixed with 1000 c.c., of distilled water for a hydrometer analysis. Appropriate quantity of the deflocculating agent is also added. Reading of the hydrometer taken after 4 minutes is 1·020. The volume of the hydrometer is obtained and found as 60 c.c The internal area of the jar containing the suspension is 60 cm.^{2} The depth of suspension z' below this reading is found as 14 cms. Assuming the values of $\gamma_w = 1$ gm / c.c and $\mu = 10.00 \times 10^{-6}$ gms. sec./cm.^{2}, find out a point on the particle size distribution curve corresponding to this reading.

Solution. As per nomenclature in the preceding article,

$$V_H = 60 \text{ c.c.}$$

$$A_j = 60 \text{ cm.}$$

and $\qquad\qquad z' = 14 \text{ cm.}$

Actual depth of fall of the particles for the 4 minute reading

$$= z = z' - \frac{V_H}{z A_j}$$

$$= 14 - \frac{60}{2 \times 60} = 11 \text{ cm.}$$

Also $\qquad \mu = 10\cdot00 \times 10^{-6}$ gms. sec./cm^2

$\qquad\qquad \gamma_s = \gamma =$ unit weight of particles

$\qquad\qquad = 2\cdot70 \times 1 = 2\cdot70$ gms./c.c.

$\qquad\qquad t = 4$ minutes $= 240$ seconds.

From equation (3.12),

$$D = \sqrt{\frac{18\mu}{(\gamma_s - \gamma_w)} \times \frac{z.}{t}}$$

$$= \sqrt{\frac{18 \times 10 \times 10^{-6}}{(2 \cdot 70 - 1)} \times \frac{11}{240}}$$

or

$$D = 10^{-3} \sqrt{\frac{18 \times 10}{1.7} \times \frac{11}{240}}$$

$$= 10^{-3} \sqrt{4.85} \text{cm.}$$

or $$D = 0 \cdot 22 \text{ m.m.}$$

For the equation (3.17)

$$r = 1 \cdot 020 - 1 = 0.020$$

$$G = 2 \cdot 70$$

$$W = 50 \text{ gms.}$$

and $$V = 1000 \text{ c.c.}$$

∴ Percentage of particles smaller than 0.022 m.m. diameter obtained above

$$= r \left(\frac{G}{G-1} \right) \cdot \frac{V}{W} \times 100$$

$$= 0.02 \left(\frac{2 \cdot 70}{2 \cdot 70 - 1} \right) \times \frac{1000}{50} \times 100$$

$$= \frac{0 \cdot 02 \times 2 \cdot 70}{1.70} \times 2000$$

$$= 63.5\%.$$

The diameter 0·022 m.m. forms the abscissa and the precentage 63·5% forms the ordinate of the point on the particle size distribution curve, of this soil. This point is shown as A on curve 9 in Fig. 3.2.

Example 3.3. Particles of 5 different sizes are mixed in the proportions shown below and enough water is added to make 1000 c.c. of the suspension. The temperature of the suspension is 20°C.

Particle Size	Weight
0·050 m.m.	6gms.
0·020 m.m.	20 gms.
0·010 m.m.	15 gms.
0·005 m.m.	5 gms.
0·001 m.m.	4 gms.
Total ...	50 gms.

It is ensured that the suspension is thoroughly mixed so as to have a uniform distribution of the particles. All particles have a specific gravity of 2.70. Assume $\gamma_w = 1$ gm./c.c.

(a) What is the largest particle size present at a depth of 6 cms. after 5 minutes of start of sedimentation.

(b) What is the specific gravity of the suspension at a depth of 6 cms. after 5 minutes of start of sedimentation.

(c) How long should the sedimentation be allowed so that all the particles have settled below 6 cms.

Solution. (a) From table 3·5, viscosity of water at 20°C=10.09 millipoises or 0·0109 poises or $\dfrac{0.0109}{980\cdot7}$gm. sec./cm.

As per nomenclature in the preceding article.

$$z = 6 \text{ cms.}$$

$$t = 5 \times 60 = 300 \text{ seconds.}$$

$$\gamma_s = 2\cdot70.$$

From equation (3·12)

$$D = \sqrt{\frac{18\mu}{(\gamma_s - \gamma_w)} \quad \frac{z}{t}}$$

$$= \sqrt{\frac{18}{(2\cdot70 - 1)} \left(\frac{0\cdot0109}{980\cdot7}\right)\left(\frac{6}{300}\right)}$$

$$= 1\cdot47 \times 10^{-3} \text{ cms.}$$

$$= 0\cdot0147 \text{ m.m.}$$

So this is the largest size of the particle that can be present at 6 cms, after 5 minutes. The size that fits in the material introduced is 0·01 m.m. **Ans.**

(b) All particles of same size of the soil settle down at the same rate. Hence, the specific gravity of the suspension at 6 cms. after 5 minutes of sedimentation will be same as if the particle size <0·01 m.m. were suspended in 1000 c.c. of suspension.

Weight of all particles <0·01 m.m.=W_s=15+5+4=24 gms.

$$\therefore \quad \text{Volume of the soil} = \frac{24}{2\cdot70 \times 1} = 8\cdot9 \text{ c.c.}$$

Volume of water=100—8·9=991·1 c.c.

Weight of water=991·1 gms.

$\therefore$ Total weight of suspension=991.1+24=1015·1 gms.

Unit weight of suspension $= \dfrac{1015.1}{1000} = 1\cdot015$ gms. /c.c.

$\therefore$ Specific gravity $= 1\cdot015$. **Ans.**

(c) $$v = \frac{\gamma_s - \gamma_w}{18\mu} D^2$$

But $\qquad v = \dfrac{z}{t}$ also

$\therefore \qquad \dfrac{z}{t} = \dfrac{\gamma_s - \gamma_w}{18\mu} \cdot D^2$

Now, smallest size of the particle is $0\cdot001$ m.m. or $0\cdot0001$ cms. For all particles to settle below 6 cms.,

$z = 6$ cms., $\gamma_s = 2\cdot70$, $\gamma_w = 1$ gm./c.c.

$D = 0\cdot0001$ cm. and $\mu = \dfrac{0\cdot0109}{980\cdot7}$ gm. sec. /cm$_\frac{a}{2}$.

Substituting in the above equation.

$\therefore \qquad \dfrac{6}{t} = \dfrac{2\cdot70 - 1}{18 \times \dfrac{0\cdot0109}{980\cdot7}} \cdot (0\cdot0001)^2$

or $\qquad t = \left(\dfrac{18 \times 0\cdot0109}{1.70 \times 980\cdot7 \times (0.0001)^2} \times 6 \right)$ secs.

or $t = 18$ hours, after proper conversion.

So all particles well settle down after 18 hours. **Ans.**

3·11. Pipette analysis

Another method to find out the particle size distribution of the minus $0\cdot075$ soil size is by the sampling technique in which a pipette is used to take out soil samples from a soil-water-suspension. The suspension is prepared from a properly dispersed soil sample in water. Use is made of a deflocculating agent, if the material is suspected to contain colloidal particles.

As for the hydrometer analysis, this suspension is also placed usually in a 1000 c.c. jar. The suspension is thoroughly mixed so that there is a uniform distribution of all particle sizes throughout the jar. Thorough mixing and uniform distribution is achieved by covering the mouth of the jar with the palm of a hand and turning it upside down at least eight-ten times, by holding it in hand. As soon as the jar is set down for settlement of the particles, a stop watch is started. With the help of a pipette, soil-water suspension samples are taken out from some known depth of the jar at regular intervals of time. Usually a 10 or 20 c.c. pipette is employed. Data regarding the depth from which suspension sample is taken, corressponding time after start of settlement and the exact volume in the pipette is recorded, as each set of such readings furnishes one point on the particle size distribution curve.

A suspension sample so taken is put in an oven and the dry weight of the soil contained in it is found out. It may be emphasied, here that the density of the soil suspension should be sufficiently high so that a pipette sample has enough solids to ensure accurate weighings, but it need not be that there is mutual nterference in the settlement of particles as individual units.

Principle involved in the calculations, again, is the Stoke's law. Since the soil particles are dispersed uniformly throughout the suspension and according to Stoke's Law particles of same size settle at the same rate, particles of a given size wherever they exist, have the same degree of concentration as at the beginning. As such, particles smaller than a given size in a pipette sample, will be present in the same degree of concentration as at the beginning. As such, particles smaller than a given size in a pipette sample, will be present in the same degree of concentration as at the beginning and particles larger than this size would have settled already below the depth at which the sample has been obtained and are, therefore, not present in the pipette sample. Calculations can be carried out as under.

Let W_s be the weight of the soil used in the suspension of volume V and W_D be the weight of soil particles smaller than diameter size D in the entire suspension. Further, let W_p be the weight of solids in the pipette sample of volume V_p. Then, by the argument presented in the preceding paragraph, the following relationship follows :

$$\frac{W_D}{V}=\frac{W_p}{V_p}$$

Or

$$W_D=\frac{W_p}{V_p}V=W_p\frac{V}{V_p}$$

The percentage by weight smaller than the diameter size D is given by ;

$$\text{Percentage } W_D=\frac{W_D}{W_s}\times100=\frac{W_p}{W_s}.\ \frac{V}{V_p}\times100...(3.18)$$

The diameter D of the largest particle present in the pipette sample is calculated by equation (3.12) and in the same manner[1] as indicated in example 3.2. So, against this particle diameter is plotted the percentage value calculated by equation (3.18). This gives one point on the particle size distribution curve. Similar points are calculated from other sets of readings.

---

1. Velocity of fall in this analysis is given by the depth at which the sample is taken divided by the time after start of settlement.

It may be mentioned here, that it obligatory in this procedure to take a pipette sample from a depth not below the depth at which the preceding samples have been taken.

Example 3·4. is an illustration of this method.

Example. 3·4 An observation was recorded during a pipette analysis for determining the particle size distribution. The data obtained were :

Weight of weighing bottle+dry soil = 22·90 gms.

Weight of empty bottle = 22·61 gms.

Time of sampling after the start

of settlement =8 minutes,

Depth of sampling =10 cms.

Capacity of pipette =10 c.c.

The volume of soil suspension at the start of the test was 500 c.c. and it contained 50 gms. of dry soil. Assume the specific gravity of solids as 2.70 and the temperature of test as 20° C. Take the unit weight of water as unity, *i.e.*, neglect the temperature effect on the unit weight of water. Determine the co-ordinates of a point on the particle size distribution curve from this data.

Solution. From the data given above and the nomenclature given in the preceding article.

$$W_p = 22·90 - 22·61 = 0·29 \text{ gms.}$$
$$V = 500 \text{ c.c.}$$
$$V_p = 10 \text{ c.c.}$$

From equation (3·18), percentage $W_D = \dfrac{0·29}{50} \times \dfrac{500}{10} \times 100 = 29\%$

From Table 3·5, μ for 20° C = 10·09 millipoises

$$= \frac{0·01009}{980·7} \text{gms. sec./cm}^2$$

Depth from which the sample has been taken=10 cms.

Time after start of the settlement=8 minutes=480 seconds.

∴   Velocity $v = \dfrac{10}{480}$ cms./sec.

γ_s =unit weight of soil[1]

=2·70 gms./c.c.

---

1. Unit weight of a soil in the C.G.S. units equals the specific gravity of the soil, when γ_w=unity.

From Stoke's Law,

$$D = \sqrt{\frac{18\mu \cdot v}{\gamma_s - \gamma_w}}$$

$$= \sqrt{\frac{18}{(2\cdot 70 - 1)}\left(\frac{0\cdot 0109}{980\cdot 7}\right)\left(\frac{10}{480}\right)}$$

$$= 1\cdot 565 \times 10^{-3} \text{ cms.} = 0\cdot 01565 \text{ m.m.}$$

So (0·01565 m.m., 29%) forms the co-ordinates of a point on the particle size distribution curve. **Ans.**

3·12. Characteristics of particles size distribution curves

Particle size distribution curves of soils indicate many important features of the soil. In the *first instance,* a curve clearly indicates whether the soil is predominantly coarse-grained or fine grained. Curve 1 of figure[1] 3·2 shows that the soil is coarse grained ; it only contains about 4 percent of the fine-grained material ; whereas curve 2 shows that soil is fine grained and it contains only 6 percent of coarse grained material. *Secondly,* a curve shows whether the soil is uniformly graded over the entire range of particle sizes, *i.e.* whether it is *well graded* or contains a single particle size.

Curve 3 is an example of all sizes. whereas curve 4 is an example of a soil that contains a single particle size. This latter curve shows that the soil is medium sand only and about 90% of the soil is of size 0.30 m.m. diameter. Curve 3 shows that the soil contains equal amounts of coarse and fine-grained soil well distributed over the range of the particle sizes.

Thirdly, a curve clearly indicates whether the soil is well-grained over the entire range of particle sizes, or it is well graded over the range of coarse-grained sizes only or it is well-graded over the range of fine-grained sizes alone. This can be seen by comparing curves 3, 1 and 2 which show respective characteristics.

Fourthly, if a soil is deficient in particular particle size of the gradation, the curve becomes flat at this stage as can be seen in curve 5 of the figure. In this soil, the fine-sand gradation is almost missing.

Lastly, a particle size distribution curve of a soil also indicates sometimes the type of the soil and mode of its deposition. For *example* curve 6 in the figure would indicate young

---

1. Curve 4 also indicates that the soil is coarse-grained.

residual soils and as the time passes, the particle sizes due to weathering will become smaller and the curve will look like 3, *i.e.*, it will become more smooth. Absence of some medium and completely fine varieties of sand in curve 5 is commonly an indication of a mixture of sand and gravel deposited by swiftly flowing water in a river or stream, carrying large quantities of the material. They may be termed poorly graded soils. A break like "*ab*" of curve 5 can, however, be also an indication of a material deposited by two agencies *e.g.* one fraction might have been washed in a glacial lake by a stream and the other fraction by the floathing ice when it melts.

Sometimes a *well-graded soil* is termed as *non-uniform soil* and a soil containing single-sized particles is termed as *uniform soil.*

In the case of granular soils or coarse-grained soils, degree of uniformity is of special significance. Whether a granular soil contains particles of all sizes or it contains single size particles *i.e.* its degree of uniformity, is very well expressed by the *coefficient of uniformity*. The more non-uniform a soil is, the flatter is its curve and the more uniform a soil is, steeper is its curve, as can be seen from curves 1 and 4.

Because the particle-size distribution curve indicates the uniformity or otherwise of a soil, sometimes this curve is called *uniformity curve.*

From the uniformity curve of a soil, the coefficient of uniformity is found in the form of a number and this number gives the degree of uniformity of the soil. Coefficient of uniformity is defined as the ratio of the diameter of the particle which has 60 percent of the soil finer than the size shown, to the diameter of the particle which has 10 percent of the soil finer than this size, *viz.*

$$U = \frac{D_{60}}{D_{10}} \qquad ..(3\cdot19)$$

where
U = Coefficient of uniformity
D_{60} = 60% diameter of soil particle
D_{10} = 10% diameter of soil particle.

The 60% and 10% diameter sizes are directly taken from the particle-distribution curve.

The 10% diameter size *i.e,* D_{10} is also called the effective size of the particle or the effective diameter of the soil sample.

This special name of 'effective diameter' to this size of the particles has been given, because it had been found that this diameter in actual spheres has the same effect as a given soil.

For the soil represented by curve 1 of Fig. 3·2, *for example,*

$$D_{60} = 0·75$$

and

$$D_{10} = 0·11$$

$$U = \frac{0·75}{0·11} = 6·82.$$

When the soil is very uniform $U < 5$; when the soil is of medium uniformity, U lies between 5 and 15, and when the soil is very non uniform $U > 15$.

A uniformity coefficient of *one* represents a soil with practically the same particle size, as can be seen for the soil represented by curve 4 in Fig 3·2.

3·13. Grain Shape

Shape of grains is an important property of soil grains but obviously cannot be given a numerical value. There is also no standard universally recognised way to describe the shape of soil grains.

In the case of the granular soils, the shape of a grain may be described as *angular, sub-angular, rounded, sub-rounded* or *well rounded.* However, the shapes of the clayey particles cannot be seen with the naked eyes. Microscopic studies of such soils have revealed that the grain shape of fine soils are either flaky or needle-like. These shapes are found in the Montmorillonite, Illite and Kaolinite group of minerals that go to constitute the clays.

3·14. Atterberg Limits.

Water as a component of a soil plays an important role in molding its phvsical behaviour. Properties of fine-grained soils *specially,* are effected by the amount of moisture present in them. A given clay suspension can pass from a liquid state to that of a plastic material and to a semi-solid state as the amount of moisture content in it is reduced. Through a series of tests proposed by Atterberg in 1911, it has been possible to demarcate the boundaries of the liquid, plastic and semi-solid states of the solids. The demarcation is in the form of numerical limits on the moisture contents present in the soil in each state when remolded[1]. These numerical limits are not the same for all the soils but vary from soil to soil. depending upon their physical properties, in the remolded state. So, determination of these

---

1. A soil is said to be remolded when its natural structure which it attains on deposition is destroyed. Remolding can occur due.to disturbance to the soil or can be accomplished by manipulation in hand or by some instrument such as a pastula.

numerical limits on moisture contents furnishes data to predict the physical behaviour of a soil and hence these limits are important index properties of a soil.

After the name of Atterberg who proposed these limits, they are called *Atterberg Limits* (these limits are also called *consistency limits*) of the soil, since they convey the same meaning as the words *very soft, soft, medium* and *hard* used to describe the consistency of soils. *See* (Art. 3·15) and the tests which are conducted to determine these limits are called Atterberg-limit tests.

A soil containing water is said to be in a liquid state, when it offers no resistance to deformation. The shear strength of the soil in this state is nil. As the water content of the soil is reduced, it becomes stiffer and stiffer and at some stage starts offering resistance to deformation. Here the soil is said to be passing from a liquid to a plastic state. The moisture content which marks the boundary of the liquid and plastic states of the soil is called its *liquid limit*, usually denoted by L_w. So, as a definition, *liquid limit* (L.L.) of a soil is the percentage moisture content above which the soil is in a liquid state and below which the soil passes into the plastic state. At this moisture content, the soil has a very small value of shear strength.

As the moisture content is reduced further, the soil passes from a plastic state to a semi-solid state. In its plastic state the soil does not show any sign of formation of cracks, when remolded or deformed. However, as soon as it passes into the semi-solid state, cracks start forming, when the soil is remolded in hand. The percentage moisture content at which the soil changes from a plastic to a semi-solid state is called the *Plastic limit* (P.L.) of the soil and is usually designated as P_w.

The difference between the numerical values of liquid limit and plastic limit of a soil is called the *Plasticity Index* (P.I.) of the soil and is donated by I_w and this indicates the range of moisture contents over which the soil remains in plastic state. Thus,

$$P.I. = I_w = L.L. - P.L. = L_w - P_w \qquad \ldots(3.20)$$

On further evaporation of water plastic limit is crossed and the soil starts behaving as a semi-solid. Upto the end of semi-solid state the volume of soil-water system goes on decreasing as the moisture content is reduced. But further reduction of the moisture content does not bring in any change in the volume. At the moisture content, where the volume does not change on further reduction of water of the system, the soil is said to have reached the *Shrinkage limit*. So as a definition, *Shrinkage limit* may be defined as the percentage moisture content below which the volume of the soil does not change, when the soil-water system is further dried. At this moisture content, the soil is assumed to have changed from semi-solid to a solid state. Shirnkage limit may be denoted by S_w.

During the liquid, plastic and semi-solid state of a soil, the soil is in a fully saturated condition, i.e., all the voids in the soil are filled with water, but below the shrinkage limit, the soil while still wet passes into a partially saturated state and some of its voids contain air or gases, while others contain water.

A further reduction of the moisture content ultimately makes the soil dry. All the above limits on the moisture content are chosen on the basis of systematic laboratory tests described in the next article. A further discussion on their utility and the indices derived from them is given in Chapter 4.

The Atterberg limits of a soil may be classified as the soil aggregate properties.

3·15. Determination of Atterberg Limits

(a) **Liquid limit.** There are two methods of determining the liquid limit L_w of a soil, one is known as the hand method and the other called the mechanical method. Hand method of determination of liquid limit is very rough and is in a way more or less guess work. This method will not be discussed here.

The mechanical method employs an apparatus called the liquid limit device, shown in Fig. 3·6. Fig. 3·6 (a) is the front view of the liquid limit device cup and Fig. 3·6 (b) is the side elevation of the apparatus. The cup is mounted on a hinge. A handle with a cam arrangement is provided and when the handle is rotated, this cam, engages itself with the piece attached to the cup and lifts the cup about the hinge and above the base. When the instrument is in adjustment, this lift of the lowest point of the cup above the rubber base is exactly 1 cm. As the lowest point of the cup reaches this height, the cam automatically gets disengaged and the cup falls down on the rubber base a distance

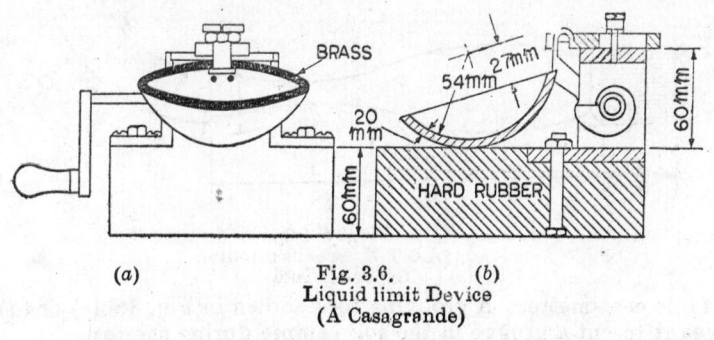

(a)   Fig. 3.6.   (b)
Liquid limit Device
(A Casagrande)

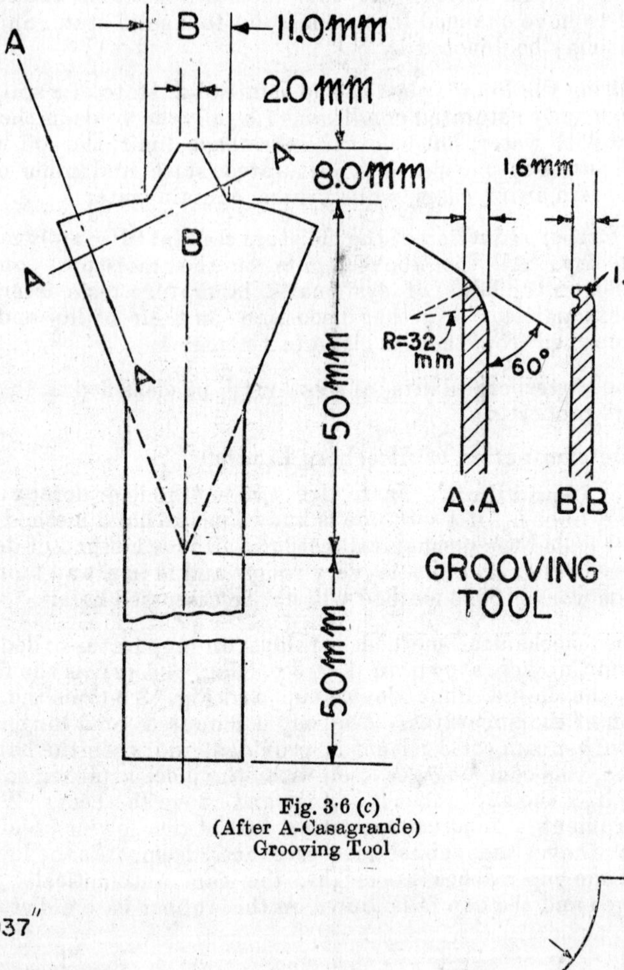

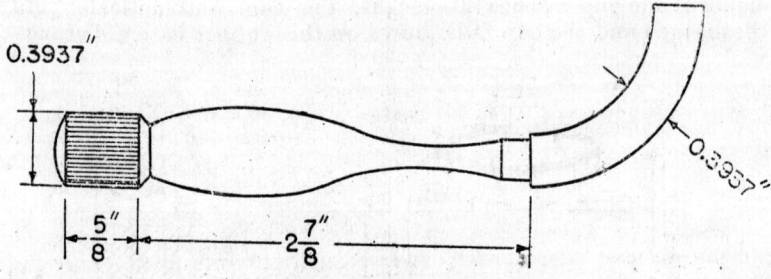

Fig. 3·6 (c)
(After A-Casagrande)
Grooving Tool

Fig. 3·6 (d)
(A.S.T.M. Specification)
Grooving Tool

of one centimetre. A grooving tool shown in Fig. 3·6 (c) or (d) is
meant to cut a groove in the soil sample during the test.

Air dried sample of the soil passing through No. 40 A.S.T.M. or No. 36 B.S. Sieve is mixed with water and the mixture is thoroughly manipulated to ensure uniformity of the paste. Some investigators put water in the soil, however, others put the soil in water. It is felt that putting the soil in water gives better uniformity to the mixture since the material can be manipulated easily.

Some agencies (10) suggest mixing time of not less than 5 minutes and not more than 10 minutes to ensure better consistancy. Also it is suggested, therein, that the prepared paste be cured for at least 30 minutes and not more than 90 minutes in a humidifier.

Drying a sample in an oven before finding out its *L.L.* usually lowers the *L.L.* of the soil[1]. However, data exist that show that in some soils. the *L.L.* increases after drying. As a general rule, however, air-dried sample is taken for *L.L.* determinations, as indicated above.

The paste of the soil is placed in the liquid limit cup and with the help of a spatula, it is levelled off. Then with the aid of the grooving tool shown in Fig. 3·6 (c) or (d), a groove is cut in the soil and the soil paste after cutting the groove has a section

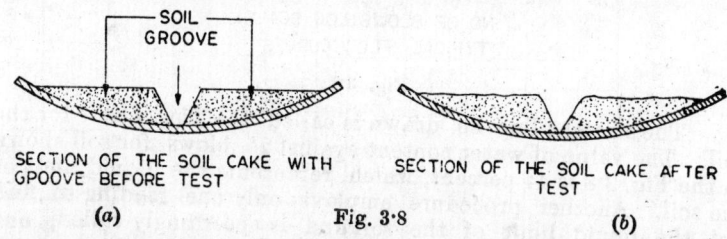

SECTION OF THE SOIL CAKE WITH GROOVE BEFORE TEST

SECTION OF THE SOIL CAKE AFTER TEST

(a)   Fig. 3·8   (b)

indicated in Fig. 3·7 (a). The handle is then rotated at the rate of 2 revolutions per second. The cup rises a distance of one centimeter above the base and then falls. This procedure is continued till the soil paste parted in two portions as in Fig. 3·7 (a) joins at the base as shown in Fig. 3·7 (b) for a length of 1·25 cms. ($\frac{1}{2}''$). The number of blows after which the material flows like this is noted. A soil sample is then taken near the point where the soil has flown to determine the water content.

The above procedure is repeated at least four times all over again by adding more water to the soil and mixing the paste thoroughly.

Then points are plotted on a semi-log scale between the percentage water content on an arithmetic scale along the ordinate

---

1. This phenomenon is very true in the case of soils containing organic matter.

and the number of blows on a log scale along the abscissa, as shown in Fig. 3·8. A straight line is drawn through these points. The water content against 25 blows is taken as the liquid limit of the soil.

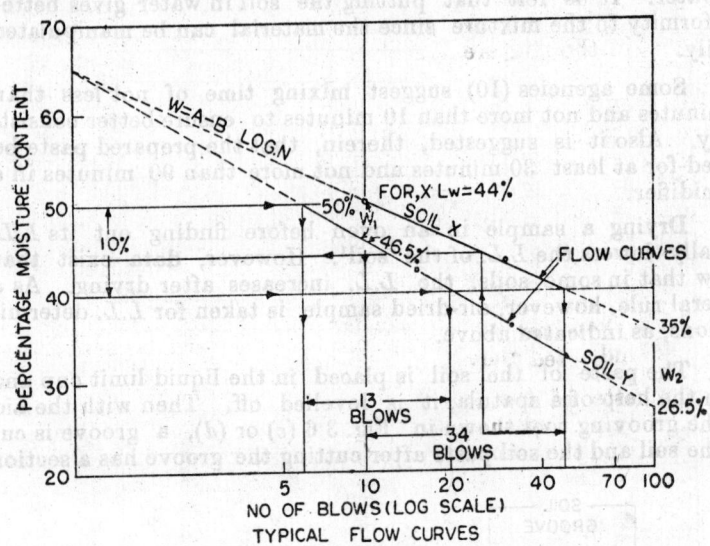

Fig. 3·8

The straight line so drawn is called the *Flow Curve* for the soil. The value of water content against 25 blows for soil shown in the Fig. 3·8 is 44 percent, which represents the liquid limit of the soil. Another procedure employs only one reading to find out the liquid limit of the soil and is accordingly called, **one point method of liquid limit determination.**

The one point method (4) of L.L. determination simplifies the laboratory work and there is no need to draw the flow curve. If the number of blows required to close the groove as shown above lies between 22 and 28, there is no need to repeat the procedure by adding more water. In that case, just this one observation is enough. The liquid limit in that case is determined from the equation.

$$L_w = W_n \left(\frac{N}{25}\right)^n \qquad \ldots(3·20)$$

where

L_w = Required liquid limit

w_n = Water content at N blows

N = No. of blows used

n = An index

Different numerical values for the index n of have been proposed by different agencies. The Waterways Experiment Station of Vicksburg, U.S.A. proposes a value of $n=0\cdot121$ and the Road Research Laboratory at Harmmondsworth, England, proposes value of $n=0\cdot092$. The method is based on the assumption the the slope of the plot of logarithmic of the water content w_n and logarithmic of the corresponding number of blows N is a constant quantity, as can be verified by taking the logs of both sides of equation (3·21).

The liquid limit test may also be done with the help of Dr. Uppal's device[1], invented at the Central Road Reserved Institute. For want of space, it is not being described here.

(b) **Plastic limit.** The plastic limit P_w of the soil is founded by rolling soil samples prepared in the same way as for the liquid limit test, but at lowest moisture contents. The soil samples are rolled into threads by hand on a smooth glass plate. The water content at which a thread of soil 1/8 inch (approx. 0·32 cms.) diameter shows signs of crumbling is taken as the plastic limit of the soil. The water content is determined from the crumbled soil itself, by putting it in the oven.

(c) **Shrinkage Limit.** Two procedures can be employed to find out the shrinkage limit of a soil—one involves the use of the specific gravity of the soil in the calculations, and the other does not involve it. Shrinkage limit for a soil can be found for the undistributed as well as remolded soil sample. A remolded soil sample for the shrinkage limit is prepared with a moisture content above the liquid limit add then it is allowed to dry downin an oven.

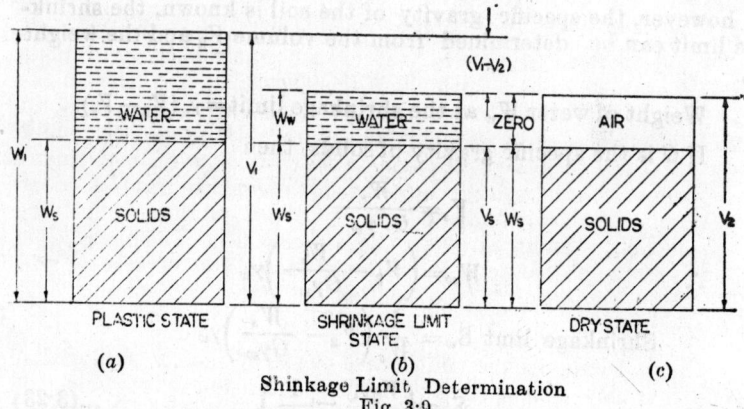

PLASTIC STATE   SHRINKAGE LIMIT STATE   DRY STATE

(a)   (b)   (c)

Shrinkage Limit Determination
Fig. 3·9

Fig. 3·9 is a phase diagram for determining the shrinkage limit of the soil. Upto the shrinkage limit of the soil, the soil is

1. For details, see Indian Road Congress Research Bulletin 4, December 1957.

fully saturated. Fig. 3·9 (a) represents a soil in a plastic state. Its volume in this state is V_1 and its corresponding weight is W_1 out of which W_s is the weight of solids. This volume V_1 of the soil is known from the volume of a container in which the soil is placed and levelled and the weight W_1 is found out by deducting from the total weight of the soil and container, the weight of empty container.

The soil, in this container is dried slowly so as not to form any cracks. After drying, the soil weight is found. This weight is nothing but the weight W_s of the solids. The volume V_2 of this dry soil pat is found by immersing the pat in a known volume of mercury. The mercury which gets spilled out is weighed and this weight divided by the unit weight of mercury of 13·6 gms/c.c. gives the volume V_2 of the soil.

This volume V_2, in the dry state, represented in Fig 3·9 (c) is the same as the volume of saturated soil at the shrinkage limit, in Fig. 3·9 (b), since the volume of the soil after this shrinkage limit does not change.

$\gamma_w(V_1 - V_2)$ gives the weight of water that got evaporated between the plastic state and the shrinkage limit state. Weight of water at the plastic state is given by $(W_1 - W_s)$.

$\therefore$ Weight of water at the shrinkage limit.

$$W_w = (W_1 - W_s) - \gamma_w(V_1 - V_2)$$

and weight of solids is W_s.

$\therefore$ Shrinkage limit $S_w = \dfrac{(W - W_s) - \gamma_w(V_1 - V_2)}{W_s}$ ...(3·22)

If, however, the specific gravity of the soil is known, the shrinkage limit can be determined from the volume V_2 and the weight W_s.

Weight of water W_w at the shrinkage limit $= \gamma_w(V_2 - V_s)$.

If G is the specific gravity of solids, then

$$V_s = \frac{W_s}{G \cdot \gamma_w}$$

$\therefore$

$$W_w = \left(V_2 - \frac{W_s}{G\gamma_w}\right)\gamma_w$$

Shrinkage limit $S_w = \dfrac{1}{W_s}\left(V_2 - \dfrac{W_s}{G\gamma_w}\right)\gamma_w$

or $\qquad S_w = \left(\dfrac{V_2\gamma_w}{W_s} - \dfrac{1}{G}\right)$ ...(3·23)

Sometimes it is necessary to find out the *shrinkage ratio* of a soil. By definition, shrinkage ratio of a soil is the ratio between a given volume soil in the specimen expressed as a percentage of dry volume of the specimen and the corresponding change in the water content above the shrinkage limit, expressed as a percentage of oven dry weight of the soil specimen.

From Fig. (3·9) the change in volume of the specimen above the shrinkage limit is $(V_1 - V_2)$ and the corresponding change in the weight of water is $\gamma_w(V_1 - V_2)$. So from definition.

$$\text{Shrinkage ratio} = \frac{\dfrac{V_1 - V_2}{V_2} \times 100}{\dfrac{\gamma_w(V_1 - V_2)}{W_s} \times 100} = \frac{W_s}{\gamma_w . V_2}$$

$$= \frac{1}{\gamma_w}\left(\frac{W_s}{V_2}\right)$$

$$= \text{Apparent specific gravity of the specimen, by definition.}$$

So shrinkage ratio is nothing, but the apparent specific gravity of the soil specimen.

3·16. Simple soil properties and their determination

Four more simple soil aggregate properties, namely, the *relative density, unconfined compressive strength*, the *sensitivity* and *thixotropy* need be mentioned here, since they are as important for some soils as the Atterburg limits.

Relative density. Relative density is a term which is used in connection with granular soils only, specially in connection with sands and it does not concern silts or clays. The concept of relative density has been developed to describe the relative compactness of these soils.

Relative density is, in fact, a function of the void ratio or porosity of a soil. So, if D is the relative density of a soil, then,

$$. \quad D = f(e \text{ or } \eta) \qquad \qquad ...(3·24)$$

where e is the void ratio and η is the porosity of the soil.

In a given container having a specified volume, one can pour the sand in such a manner that volume remaining constant the solid particles are minimum and voids are maximum or alternatively the solid particles are maximum and voids are minimum. In the former case, the sand is in its loosest state and in the latter case it is in its densest state ; so that

$$e_{max} \frac{m}{,,} = \frac{\text{Maximum } V_v}{\text{Minimum } V_s}$$

and

$$e_{min} \frac{m}{,,} = \frac{\text{Minimum } V_v}{\text{Maximum } V_s}$$

$e_{max} \frac{m}{,,}$ is assumed to correspond to the minimum value of relative density D i.e. $D=0$ and $e_{min} \frac{m}{,,}$ is assumed to correspond to the maximum value of relative density D i.e. $D=1$.

So, if for the soil being studied, $e_{maximum}$ is plotted againt $D=0$ and $e_{minimum}$ is plotted against $D=1$ as in Fig. 3·10, the relative density for any intermediate value of e can be written as :

$$D = \frac{e_{max} - e}{e_{max} - e_{min}} \qquad (3·25)$$

Relative density is at time presented as a percentage rather than a fraction. In that case, right hand side of equation 3·25 is multiplied by 100.

Since dry density or dry unit weight of a soil is a more common parameter used in practice, it becomes convenient to covert equation (3·25) in terms of dry densities at respective void ratios. Recalling equation (2·19) for dry density as,

$$\gamma_d = \frac{G\gamma_w}{1+e} \qquad ...(3·26)$$

one may notice that e appears in the denominator. Assuming that G and γ_w for the soilds and water are constants, γ_d is inversly proportional to the void ratio.

∴ At e_{min}, γ_d will be maximum and

At e_{min}, γ_d will be minimum.

If subscript d is dropped and the maximum and minimum dry densities are expressed as γ_{max} and γ_{min}, then

$$\gamma_{max} = \frac{G\gamma_w}{1+e_{min}} \qquad ...(3·27)$$

$$\gamma_{min} = \frac{G\gamma_w}{1+e_{max}} \qquad ...(3·28)$$

Equation (3·25) can then be expressed as

$$D = \left(\frac{1}{\gamma_{min}} - \frac{1}{\gamma} \right) \left(\frac{1}{\gamma_{min}} - \frac{1}{\gamma_{max}} \right) \qquad ...(3·29)$$

wherein γ represents the dry density at a void ratio e. The student may make substitutions from equations (3·26), (3·27) and (3·28) into equation (3·25) and cancel G and γ_w in the numerator and denominator and he will find that the result will be equation (3·29).

In actual practice, the relative density or relative compactness of a soil is measured by a penetration test, described in Chapter 16.

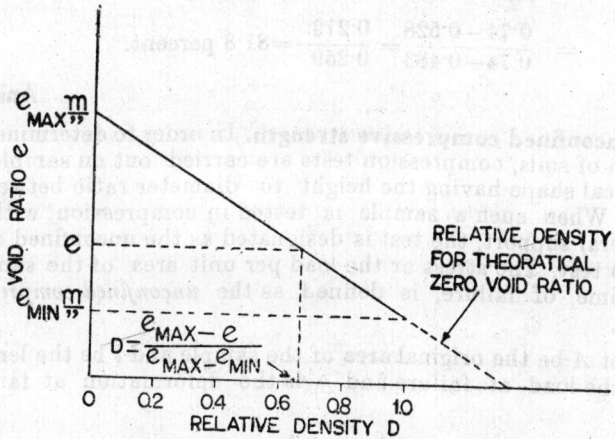

Fig. 3·10.   Relative Density D vs. Void Ratio e.

Example 3·5

A soil has an ordinary range of void ratios from 0·30 to 1·00 Calculate the maximum and minimum dry unit weights in kg/cm³, assuming that the specific gravity of solids is 2·70. If the maximum density attainable in the field is 1·90 kg/cm³, find out the corresponding relative density of the soil.

Solution   The unit weight of a soil is given by

$$r=\frac{G+Se\gamma_w}{1+e} \qquad \qquad ...(3.30)$$

with the usual meanings of the various symbols.

If the degree of saturation is zero i.e $S=0$ because the soil is dry, then

$$\gamma_{dry}=\frac{G\gamma_w}{1+e} \qquad \qquad ...(3.31)$$

Using equation (3·31) maximum γ_{dry} will be obtained when e is minimum and minimum γ_{dry} will be obtained when e is maximum.

$$\therefore \ \ \text{Max } \gamma_{dry}=\frac{2.70}{1.03}\times1=\ 2.07 \text{ kg/cm}^3$$

$$\therefore \text{ Minimum } \gamma_{dry} = \frac{2 \cdot 70}{1 + 1 \cdot 0} \times 1 \cdot 0 = 1 \cdot 35 \text{ kg./cm}^3.$$

Drop the subscripts and use $\gamma = 1 \cdot 90$ in equation (3·29).

$$\therefore \quad D = \frac{\dfrac{1}{\gamma_{min}} - \dfrac{1}{\gamma}}{\dfrac{1}{\gamma_{min}} - \dfrac{1}{\gamma_{max}}} = \frac{\dfrac{1}{1 \cdot 35} - \dfrac{1}{1 \cdot 90}}{\dfrac{1}{1 \cdot 35} - \dfrac{1}{2 \cdot 07}}$$

$$= \frac{0 \cdot 74 - 0 \cdot 528}{0 \cdot 74 - 0 \cdot 483} = \frac{0 \cdot 212}{0 \cdot 259} = 81 \cdot 8 \text{ percent.}$$

Ans.

Unconfined compressive strength. In order to determine the strength of soils, compression tests are carried out on samples of cylindrical shape having the height to diameter ratio between 2 and 3. When such a sample is tested in compression, without any lateral support, the test is designated as the unconfined compression test. The stress or the load per unit area of the sample, at the time of failure, is defined as the *unconfined compressive strength*.

Let A be the original area of the sample and l be the length. If P is the load at failure and $\triangle$ is the deformation at failure, then,

$$\text{Strain at failure} = \varepsilon = \frac{\triangle}{l}.$$

If the sample even at failure is assumed as a cylinder, then it can be shown[1] that the area A_f at failure, is given by,

$$A_f = \frac{A}{(1 - \varepsilon)}$$

$\therefore$ Compressive strength at failure

$$= \frac{P}{A_f} \tag{3.30}$$

Unconfined compressive strength of the soil measures the consistency of a cohesive soil. *Consistency* of a cohesive soil, by definition, is that property of the material because of which it shows resistance to flow. In fact the *Soil Consistency* is a term which shows how much cohesion the soil possesses. Consistency is usually described by terms such as *very soft, soft, medium* and *hard* etc. Method of representing the soil consistencies is through their unconfined compressive strength. Table 3·5 gives the consistencies of soil against their unconfined compressive strength, usually denoted as q_u.

---

1. *See* Chapter 10 for *Area Correction*.

TABLE 3.6

Consistencies of soils in terms of unconfined compressive strength.

Consistency of soil	Unconfined compressive strength q_u in kg/cm^2
Very soft	<0.25
Soft	0.25—0.50
Medium	0.50—1.00
Stiff	1.00—2.00
Very stiff	2.00—4.00
Hard	>4.00

Sensitivity. A cohesive soil in a natural state of deposition has a certain structure and consequently certain amount of strength. When the soil is disturbed or manipulated in hand, its structure gets destroyed. Also the orderly arrangement of molecules in the absorbed layers gets deranged, temporarily. The soil in this disturbed state is called remoulded. The remoulded strength of a cohesive soil at a moisture content equal to its natural moisture content is not the same as its strength when it is in a natural state, before remoulding.

The ratio of the unconfined compressive strength of a cohesive soil in an undisturbed state to its strength in a remoulded state at equal moisture contents is called the *sensivity* of the soil. The degree of sensitivity of a cohesive soil may be described by a ratio given below:

$$\text{Degree of sensitivity} = S = \frac{q_u \text{ (undisturbed or natural state)}}{q_u \text{ (remoulded state)}}$$

where q_u is the confined compressive strength of the soil.

The degree of sensitivity or simply sensitivity for different clays is different. For the same clay, the sensitivity at different moisture contents may be different. If the natural water content w_n of a clay is less than its liquid limit Lw, it is likely to be less sensitive than when its natural water content w_n is close to its liquid limit. A very sensitive clay when remoulded turns almost into a slurry that can flow.

The values of sensitivity S for different clays is as given in Table 3.7 below.

TABLE 3.7
Values of Sensitivity

Type of clay	Sensitivity
Low sensitivity	Between 2 and 4
Medium sensitivity	Between 4 and 8
Extra sensitivity	Between 8 and 15
Quick	Above 15

It may be stated here, that a change in the consistency brought out in a sensitive soil by disturbance is accompanied by a change in its permeability too.

Thixotropy. It has been indicated above that in the remoulded state the soil loses its strength compared to that in its natural state at identical moisture contents. This loss is partly due to the permanent destruction of the structure which the soil attains during deposition and partly to the derangement of molecules in the absorbed layer. If this remoulded soil is allowed to stand without loss of moisture, it regains some strength over a period of time. This phenomena of regaining of strength is termed *thixotropy*. This regain of strength is due to the rehabilitation of the molecular structure of the soil. However, the strength loss due to the destruction of the soil structure is permanent, and cannot be regained over a period of time.

CHAPTER—BIBLIOGRAPHY

1. A.S.T.M. Standards (for testing of materials), 1958.

2. A.S.T.M., "Annual Book of ASTM Standards", ASTM, Philadelphia Pa. 1972.

3. A Atterberg Die Plastizalaet and Bindigkeit Liefernde Bestandtile der Tone, Int. Mitteil Bodenkunde, Volume 3, 1913.

4. Bowls, E., "Foundation Analysis and Design", McGraw Hill Book Co., New York, N. Y., 1968.

5. Casagrande, A., Research on the Atterberg limits of Soil, Public Roads Volume 133, 1948.

6. Dawson, R. F., Laboratory Manual in Soil Mechanics, Pitman Publishing Corporation, Second Edition, 1959.

7. Dept. of Public Works, State of California, Materials Manual, Testing and Central Procedures, Vol. 1, 1963.

8. Hazen, A., Some physical properties of sands and gravels with special reference to their use in filteration, 24th report of the State Board of Health of Massachusettes, 1892, Public document number 88, Boston, Wright and Potter Co., 1893.

9. Indian Standards Institution, Methods of Test of Soils, Part II, Determination of Moisture Contents, I. S. 2720 (Part III)—1994 New Delhi.

10. Indian Standards Institution, Methods of Test of Soils, Part III—Determination of Specific Gravity; I. S. 2720 (Part III)—1964 New Delhi.

11. Norman, L. E., The one point method of determining the value of limit of soil, Geotechnique Volume 9, 1959.

12. Rutledge, P. C., Description and Identification of Soil types, Proceeding of Purdue Conference on Soil Mechanics, 1940.

13. Terzaghi, K. and Peck, R. B., Soil Mechanics in Engineering Practice, New York, John Wiley and Sons, Inc., 1968.

14. Tschebotarioff, G. P., "Soil Mechanics, Foundation, and Earth Structures", McGraw Hill Book Co., Inc., N. Y., 1951.

15. Waterways Experiment Station, Vickberg, U. S. A. Simplification of liquid limit procedure, Tech Memo number 3-286.

QUESTIONS

1. (a) What do you understand by any four of the following :
(i) Consistency (ii) Thixotropy (iii) Base Exchange (iv) Soil Structure (v) Uniformity Coefficient (vi) Optimum Moisture Content.

(b) What do you understand by Atterberg's limits ? Explain how these limits are helpful to engineers in indicating general properties and classification of soils.
(P. U. 1959—Supp.)

2. What is Stoke's law ? Describe briefly the pipette method of particle size analysis of soils ?

In a test 10 gms of fine gained soil of specific gravity 2.71 was dispersed in 500 c. c. of water (viscosity 0.01 c. g. s. units). A sample of volume 10 c. c. was taken by means of a pipette at a depth of 10 cms, 46 minutes after the commencement of sedimention. The sample was found to contain 0.026 gms. of solids. Calculate

(1) The largest size of particle remaining in suspension at 10 cm. depth.

(2) The percentage of particles finer than this size in the original soil. (1 gm. sec. per sq. cm.=981 poises) *(P. U. 1960 Supp.)*

3. Define "Stoke's Law". *(P. U. 1962)*

4. Describe briefly the complete procedure of performing the hydrometer analysis for obtaining particle size distribution of a clayey soil. How are hydrometer readings related to the particle sizes and their distribution ? *(P. U. 1963)*

5. An undisturbed saturated specimen of clay has a volume of 19.57 c.c. and a weight of 30.2 gms. On oven drying, the weight reduces to 18·0 gms. The volume of dry specimen as determined by displacement of mercury is 9.86 c.c. Find out the shrinkage limit of the soil. *(P. U. 1963)*

6. (a) The following observations were recorded during the pipette analysis for determining the particle size distribution of a soil:—

Observation No.	Sampling Time	Weight of weighing bottle+dry soil	Wt. of empty bottle
	Minutes	Gms.	Gms.
1.	0.5	22.85	22.02
2.	1	22.33	21.80
3.	4	23.47	23.13
4.	8	22.89	22.60
5.	15	23.25	23.02
6.	30	23.05	22.86
7.	60	23.99	23.82

The capacity of the pipette is 10 ml. and the depth of the sampling for each observation is 10 cms. The soil suspension at the start of the test was 500 ml. in volume and contained 50 gms. of dry soil and 1.5 gms. of dispersing agent. Assume the soil specific gravity as 2.70, viscosity of water 0.01 dyne sec/cm^3. Determine the percentage finer and particle size corresponding to the above sampling times.

(b) Define shrinkage limit, shrinkage ratio, liquid and meniscus correction for hydrometer analysis. *(P. U. 1963 Supp.)*

7. Define Liquid Limit, Plasticity Index and Shrinkage Limit of a soil. How is shrinkage limit determined in the laboratory ? *(P. U. Nov., 1965)*

8. From equation (3 21) for the shrinkage limit of a soil, derive an expression for the specific gravity of solids .

9. Calculate the coefficients of uniformity of only the coarse grained soils, the particle size distribution curves of which are shown in Fig. 3.2 and describe the uniformity of these soils.

10. Why are the semi-logarithmic plots more suitable for the particle-size distribution of soils than plots on arithmatic scale ? Dis-

cuss the characteristics of the semi-logarithmic plots of particles size distribution, for soils.

11. Two samples of dry soil each weighing 500 gms. are subjected to mechanical analysis. The test results are given below :—

(a) Plot the grain size distribution curves for both the soils. Use semi-logarithmic scale, showing equivalent grain diameter *versus* percentage finer by weight.

(b) Discuss the distinguishing characteristics of each curve. Can you suggest a mode of deposition of the soils that will result in these gradations ?

DATA

Sieve No. A.S.T.M. or B.S. Designation	Diameter (mm)	Soil A weight Retained in gms.	Soil	
			Weight Retained in gms.	Hydrometer analysis Results, % age Finer by wt.
3/8''	9.52	—	21.0	—
4	4.76	—	11.5	—
8	2.38	—	22.5	—
16	1.19	—	35.0	—
40	0.420	6.0	97.0	—
60	0.250	39.0	—	—
80	0.177	55.5	108.2	—
120	0.125	114.5	—	—
140	0.105	133.0	—	—
200	0.075	127.0	92.5	—
—	0.050	—	—	73.6
—	0.020	—	—	48.3
—	0.010	—	—	34.1
—	0.005	—	—	26.8
—	0.002	—	—	17.1

12. (a) Describe how a grain size analysis of a soil is performed. List the Indian Standard Sieves in such a test. Mention the opening size of sieves also.

(b) Grain size analysis of a sandy soil yielded the following data :

Diameter m.m.	Percent Finer
0.5	95
0.3	80
0.2	60
0.1	5

Plot the grain size curve on the semi-log graph and determine its uniformity coefficient. *(P. U. April. 1966)*

13. (a) Define Liquid limit, Plastic limit and Shrinkage limit of a soil.

(b) Explain how shrinkage limit is determined in the laboratory. Derive the expression also. *(P. U. April, 1966)*

14. Briefly describe the apparatus and explain the procedure for the determination of liquid limit of a given soil. *(P. U. Nov., 1966)*

15. Following results have been obtained in a liquid limit test on a soil having plastic limit=17.0

Numbers of biows	10	15	27	35	44
Moisture content percent	41.60	41.0	39.70	40.20	40.10

Determine the liquid limit and plasticty index of the Soil. *(Baroda 1966)*

16. Calculate the shrinkage limit of a soil given the following data. A pat of dry sample 5 cms(2'') dia and 1.25 cms. (1/2'') high weighs 50 gms. The specific gravity of solids is 2.65. *(Mysore 1968)*

17. An undisturbed saturated specimen of clay has a volume of 23.4 cm^3 (1.44 inch^3) and a weight of 42.5 gms (0.094 lbs). On oven drying the weight reduces to 29.6 gms (0.065 lbs). The volume of dry soil as determined from displacement of mercury is 18.1 cm^3 (1.1 inch^3). Determining (1) Shrinkage limit (2) Specific gravity (3) Shrinkage Ratio (4) Volumetric shrinkage. *(Mysore Oct. 1968 Sch. S & F. E.)*

4

Soil Classification

4·1 Introdcution

If an area is studied from soil-engineering point of view, it may contain several soil deposits of different types that in general exibit different soil characteristics. Natural soil deposits are never homogeneous in their nature. A soil is said to be *homogeneous*, if it has alike properties at all points in the given mass and it is impossible to find a natural deposit that has alike properties at all points within the deposit. Within a deposit, the soil varies from point to point. A deposit may constitute of a mixture of soils of many types and sizes. Soil properties may vary with the depth of the deposit as well as its extent in the horizontal direction. Such variations in soil properties in a deposit may be termed as *local variations*. In spite of such local variations, the deposit as a whole may have average properties that might be the same for all portions of the deposit.

Deposits that exhibit, in general, similar average properties can be grouped together, as a class. Classification of soils is essential because, by classifying a soil, a fairly accurate idea of its average properties, can be made and an estimation of these average properties just by classification is helpful in *ordinary* soil engineering projects. The opinion, however on the applicability of classification of soils differs among soil-engineers. The difference of opinion arises mainly due to the fact that many classification systems exist. Some systems take into consideration only the grain-sizes of the soils, while others take into consideration the plastic properties, while quite another classification exists, which defines soils in terms of their origin and general features. So, a soil classified into a group according to one system may fall into quite another group of another system. Also in the same system there is a possibility that a soil may form a border-line case, in which case, it could be classified under two groups.

Classification of a soil is a *mere pointer* to its general properties and behaviour of a soil cannot be solely predicted from its classification alone. Certain other important tests and parameters would, of course, be necessary before the conclusions about its behaviour as a construction or foundation material could be deduced.

In spite of the inadequacies pointed out above, it is essential that a soil for an engineering project is classified and then further tests made to predict its suitability or otherwise for the project.

Soil classification by origin has been referred to in Article 1·5 and for want of space shall not be discussed in this chapter any further. Classification of soil by mineral types, forms, is more a part of geology than of soil-mechanics, although the types of minerals in soil have an important bearing on soil behaviour. A reference to this system of classification can be made from a text book on geology.

Following systems of classification will be discussed in this chapter.

1. Descriptive classification of soils based upon the properties of undisturbed soils.

2. Grain-size classification based upon the mechanical analysis or grain-size analysis of soils. Under this heading the following systems will appear :

   (a) U.S. Bureau of Soils Classification System,

   (b) M.I.T. Classification System,

   (c) U.S. Bureau of Soils Triangular Classification System.

3. Soil classification according to a A. Casagrande's Plasticity Chart.

4. Highway Research Board Soil classification System.

5. United Soil Classification System.

6. Indian Standard Soil Classification System.

After the presentation of the first two classification systems, a brief account of the utility of the consistency limits and the indices derived from them will be given, as these index properties form a part of the last four systems of classification.

4·2. Descriptive classification of soils

Under this system of classification, soils are designated as boulders, graval, sand and silt ; etc. This nomenclature is universal and each name is associated with a soil that has certain general physical features. As such, it is only desirable to familiarise oneself with these soil types and their general physical features.

Boulders, gravel and sand. These are coarse grained soils and soil particles that belong to this category possess no cohesion among themselves. The main distinguishing physical feature among boulders, gravels and sands is the size of their particles. Boulders are larger than gravel particles and gravel particles larger than sand particles. Various systems based upon mechanical analysis place different limits on the sizes of these three categories. Most of the systems recognise any material larger than 7·62 cm (*3 inches*) size as a boulder and any material between 2 m.m. and 0·06 m.m. as sand. The Shape of all of these can be described as angular, sub-angular, well rounded, rounded and sub-rounded. Field identification of this material is done as described in Section 3·3.

The gravel and sand fractions are further sub-divided, based upon their sizes, into coarse, medium and fine varieties.

Silt. Silt is a fine grained soil and occurs in two varieties— the non-plastic variety including what is called *rock flour* and the plastic variety containing finely divided particles of organic matter. The two varieties are also designated as inorganic silt and organic silts. The inorganic variety of silt usually has a size larger than 0·002 m.m, except the variety which is termed as rock flour, which contains paricles finer than 0.002 m.m too. The organic silt particles range between 0·06 and 0·002 m.m.

The inorganic silt may constitute or bulky grains of plate-like grains. The former type is incompressible whereas the latter type is relatively compressible. The compressible variety may show little plasticity when wet, although in almost all cases, inorganic silt has little or no cohesion when dry and has a gritty feeling. It has sufficiently high permeability to allow the pore-water to move.

The organic silt is usually dark to brown in colour. They are usually highly compressible soils with sufficient cohesion or plasticity. They have relatively low permeability. Because of high compressibility and poor permeabilty, they form very weak foundation materials. As against gritty touch of inorganic silts when dry, the organic variety has soft touch when wet. They also have a peculiar odour due to the decomposition of organic matter.

Clay. Soil particles smaller than 0·002 m.m. size and capable of showing plasticity when wet are called clays. The property of plasticity of clays is exhibited over a large range of moisture-contents.

Clays are believed to be formed by certain secondary products of weathering. Prolonged action of water on silicate minerals lead to these secondary products of weathering. These minerals exist in a colloidal state. The three important groups of these mineral are *montmorillonite, illite* and *kaolinite.*

Dry specimens of clay are very hard due to the cohesion when wet. Unlike silts, it is not possible to rub any powder from the dry clay nodule. The *permeability*[1] of these soilds is very very low.

When clay is mixed with some amount of silt, it is termed as *lean clay*, whereas fine clays of high plasticity are termed as fat clays.

In its wet condition the clay sticks to the hand. When it is partially wet, it sticks to the hands so hard that it becomes difficult to remove it. Stiff clays in the field are very hard and cannot be excavated without a pick-axe, Such varieties are sometimes useful foundation materials.

The organic variety of clay contains finely divided organic matter. These soils are highly compressible when saturated and from very poor foundation materials. Highly organic clays give stinking ordour and are dark in colour.

Peat. Peat is a partly carbonized organic matter and is fibrous in nature. Its specific gravity ranges from 0.5 to 0.80 and it can be seen that it is very light compared to other soils. Due to its fibrous nature it is extremely compressible and is the poorest foundation material.

Loess. Loess is a fine-grained yellow coloured soil. It is an airtrausported material of uniform size and occurs in thick beds rather than startifications. It has high void ratio and when dry carries cementitious properties. Due to the cementitious nature, it can stand in vertical cuts. The material is largely silt-sized. When partially or fully saturated, it loses its cohesive nature.

Hardpan. Hardpan is a term used for soil-strata which remain hard when wet. Boulder clays or glacial tills are sometimes given this name. It is very difficult to drill in such starta with ordinary tools.

Bentonite. Bentonite is the name given to decomposed volcanic ash. This material contains high percentage of montmorillonite. When wet this material shows high swelling characteristics and on drying it shows considerable shrinkage.

Topsoil. Disintegrated surface materials that support plant life are called topsoil.

Varved clay. When fat clay of glacial origin and silt gets deposited alternately in layers, *varved clay* is formed. This is essentially a *lacustrine* deposit and it is believed that the high

---

1. For definition, see Chapter 5.

flood periods of summer and spring deposit the silt layers from melting ice and the clay particles settle down in winter. Thin layers are called varves and hence the name *varved clay*. The thickness of each layer is only a fraction of a centimeter. As foundation material, the general character of this soil resembles that of silt.

Black-cotton soil. This soil occurs in the central and western parts of India. It contains high percentage of montmorillonitic clay, and because of the high percentage of this clay, it has high swelling and shrinkage properties, during wetting and drying. When dry, the soil is very hard but on getting wet, it loses all its hardness. As such it has very low bearing capacity when wet and because of this low bearing capacity and high swelling and shrinkage characteristics, it forms a very poor foundation material. Black soils of Russia and *adobe* of California (U. S. A.) can be compared to the black cotton soils of India.

As the name suggests, the soil is black in colour. A typical black cotton soil has a liquid limit of 90% and plastic limit of 43%. Forty to sixty percent of the soil has a size less than 0.001 mm. At the liquid limit the volume change is of the order of 200 to 300 percent and results in swelling pressures of as high as 8 to 10 kgs/cm^2.

Moorum. Moorum is nothing but ironstone gravels mixed with red clay, resulting from the disintegration of rocks.

Laterites. Laterite is a material which may be described as having a structure which is perforated and cellular. The material is found in some parts of Rajasthan. It is deep brown in colour and is very soft to cut from the quarries but on exposure to air for a few months, it becomes hard due to the formation of hydrated iron oxides.

4.3. Grain-size classification of soils

Behaviour of clays to a large extent depends upon their grain sizes. It is grain sizes of individual grains that lead to the types of structures indicated in Article 1.6. On the basis of grain size analysis, many agencies have proposed grain-size classification. A few of these classification systems are given below.

(a) **U. S. Bureau of Soil Classification System.** The Table 4.1 below gives the type of the soil and the corresponding range of sizes according to this classification system.

TABLE 4.1

U. S. Bureau of Soil Classification based upon Grain-size

S. No.	Type of Soil	Range of Sizes in m.m.
1.	Fine Gravel	2.00 —1.00
2.	Coarse Sand	1.00 —0.500
3.	Sand	0.500—0.25
4.	Fine Sand	0.25 —0.20
5.	Very Fine Sand	0.100—0.05
6.	Silt	0.050—0.005
7.	Clay	Below 0.005

(b) **M. I. T. and I. S. I. Soil Classification Systems.** Two other systems namely the M. I. T. grain size classification system and the I. S. I. grain size classification system give similarly the type of soils corresponding to their grain sizes. These two systems are fairly identical. Table 4.2 gives the type of the soil and the corresponding range of grain sizes according to these two systems.

The I. S. I. Code IS: 1498-1959 also lays down particle size limits for gravels i. e. any material between 2 mm. and 60 mm. diameter size may be called *gravel.*

The M.I.T. system further classifies the clay into *coarse clay* ranging from 0.002 mm. to 0.0006 mm., *medium clay* from 0.0006 mm. to 0.0002 mm. and *fine or colloidal clay* below 0.0002 mm. size.

As can be seen from Table 4.2, it is easy to remember the M.I.T. and I. S. I. classification systems. All that is needed is to remember two numbers 2 and 6 and one boundary between grain sizes. Thus, a lower limit of 2.00 mm. may be assigned to gravels, below this size from 2.00 to 0.6 mm. is coarse sand; from 0.6 mm. to 0.2 mm. is medium sand and from 0.2 to 0.06 mm. is fine sand and so on; in the same manner one can write the sizes of silts and clays, splitting them into coarse, medium and fine varieties. This is an excellent example of the use of metric system.

TABLE 4.2

M. I. T.[1] and I. S. I.[2] Systems of Grain Size Classification.

S. No.	Type of Soil	Range of sizes in m. m.
1.	Coarse Sand	2.00 —0.60
2.	Medium Sand	0.60 – 0.20
3.	Fine Sand	0.20 —0.06
4.	Coarse Silt	0.06 —0.02
5.	Medium Silt	0.02 —0.006
6.	Fine Silt	0.006—0.002
7.	Clay	Below 0.002

(c) **U. S. Bureau of Soils Textural Classification System.** Another system of Grain-size classification of soils developed by the U. S. Bureau of Soils, known as the U. S. Bureau of Soils Textural Classification system, uses a chart given in figure 4.1. This system assumes that a soil constitutes of the fractions of sand, silt and clay sizes and the coarser sizes above 2·00 mm. do not exist and any of the fractions may, of coarse, be missing, in a particular variety. The chart assumes that the sizes of sand, slit and clay correspond to the U. S. Bureau of Soils Classification given in Table 4.1. Different names as shown are given to different soil mixtures. If the percentage of sand, silt and clay are known in a mixture, the soil can be located in the chart by going along the arrows shown and appropriate name given to it.

As an example, a soil containing 30% sand, 30% silt and 40% clay is located by starting from these points on the respective scales for sand, silt and clay and going along the arrows and ending in point A which lies in the clay category.

The percentage ranges of sand, silt and clay in each type of soil represented in the chart of Fig. 4.1 are given in Table 4.3.

It can be seen that in any of the soil classification systems given above, there is no mention of the plasticity characteristics

1. M. I. T. is abbreviation for the Massachusettes Institute of Technology, located at Boston, Massachusettes, U. S. A.

2. I. S. I. is abbreviation for the Indian Standards Institution located at New Delhi, India.

of a soil. This is a shortcoming in these systems, since even if a soil falls by grain size analysis in the category of clays, it may not possess properties of plasticity to justify its classification as a clay and hence by grain size classfication procedure this soil would be wrongly classified. A concrete example of such a misclassification by these procedures is fine *rock-flour*, having particle sizes lying between 0.001 and 0.002 m.m.

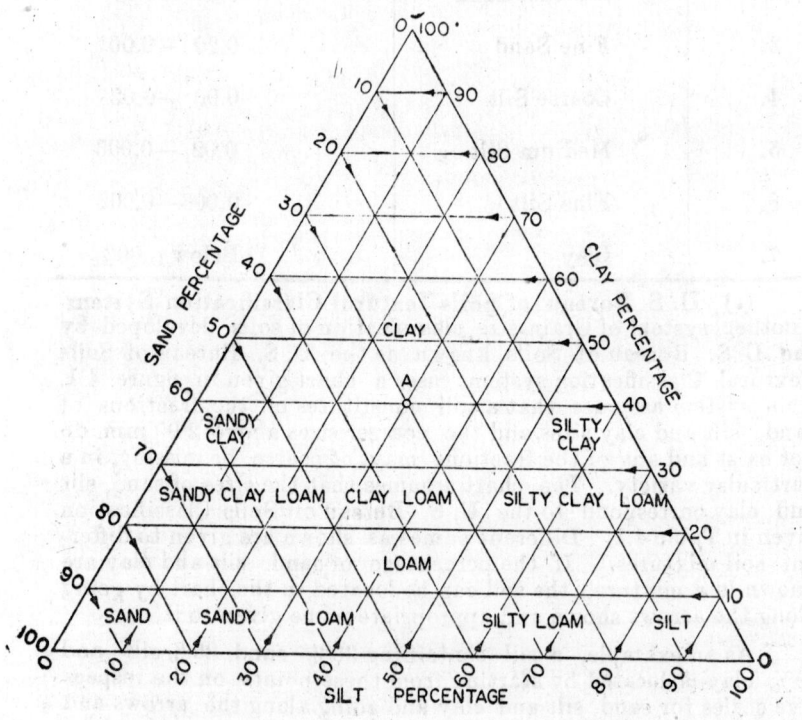

U. S. Bureau of Soils Textural Classification Chart.
Fig. 4·1.

TABLE 4.3

Percentage ranges of sand, silt and clay present in a soil

(*U. S. Bureau of Soils Textural Classification System*)

S. No.	Soil Description	Percentage of Soil Separates Present		
		Sand	Silt	Clay
1.	Sand	80—100	0—20	0—20
2.	Sandy loam	50—80	0—50	0—20
3.	Sandy clay loam	50—80	0—30	20—30
4.	Sandy clay	50—70	0—20	30—50
5.	Clay	0—50	0—50	30—100
6.	Silty clay	0—20	50—70	30—50
7.	Silty clay loam	0—30	50—80	20—30
8.	Silty loam	0—50	50—80	0—20
9.	Silt	0—20	80—100	0—20
10.	Clay loam	20—50	20—50	30—20
11.	Loam	30—50	30—50	0—20

It may be noted that the grain-size limits in all the above classification systems have been placed arbitrarily.

4·4. Atterberg Limits and Indices

Methods of determining the Atterberg limits have been given in Article 3·14. These Atterberg limits define the physical states of fine grained soils at different moisture contents. The various states in which soil can exist are *liquid, plastic, semi-solid* and *solid* and the numerical values of liquid, plastic and shrinkage limits define the boundaries of these states. Above the liquid limit, the soil is considered to be in a liquid state, between the liquid and plastic limits the soil is considered to be in a plastic

state, between the plastic and shrinkage limits, the soil is considered to be in a semi-solid state and below the shrinkage limit,
a soil approaches a solid state. Upto the shrinkage limit
the soil is in a fully saturated state and below this limit it is
partially saturated. Upto the shrinkage limit, with a decrease in
the moisture content the volume of soil-water mixture decreases,
but after the shrinkage limit, the volume remains constant.

The above states of the soil corresponding to various moisture
contents may be placed on a scale termed as the *moisture scale*
shown in Fig. 4·2.

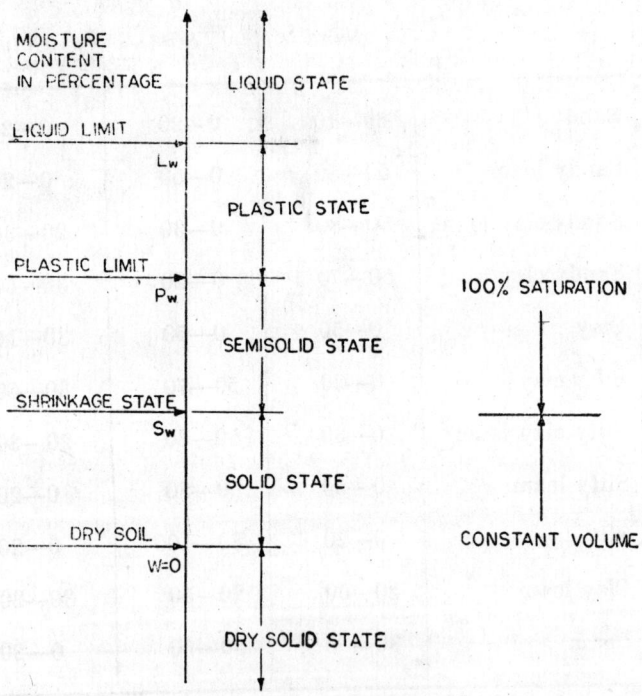

Fig. 4·2. Moisture Scale

The range of liquid limit (L.L.) for cohesionless soils is of
the order of 20 per cent and for the silty and clayey soils it may
even be more than 100 percent. An average value of liquid limit
for really cohesive soils is 60 percent. The relatively high liquid
limit in the case of cohesive soils is due to the fact that for the
same volume, the surface area of the particle is very high as
compared to the surface area of cohesionless soils, because the
clayey particles are flaky in nature. The finer and flatter the
grains of clay are, more is the surface area of grains and hence

more will be the quantity of water required to coat them. With the addition of sand or silt the clay becomes leaner and the liquid limit of a claysand or clay-silt mixture is lower. However addition of sand or silt does not bring down the plastic limit of a clay so rapidly as it brings down the liquid limit. Hence admixture of coarser particles causes a simultaneous decrease of P.I. value. Sandy soils and specially pure sands cannot be subjected to a plastic limit test and hence these soils that cannot be subjected to this test are termed as non-plastic soils (N.P) Soils containing silts, clays and particles sizes of colloids are essentially plastic in nature and consequently are cohesive. The reverse is also true. Any soil, which is cohesive and plastic. shall essentially contain silts, clays and particle sizes of colloids. Presence of organic matter in the soil increases its plasticity and results in high plastic limit values. However, organic matter does not have any significant influence on the liquid limit. So, soils with organic matter have low plasticity indices with high liquid limits.

The range of plasticity of a cohesive soil is usually defined by its plasticity index, which as stated in equation (3.19) is the difference between the liquid-iimit and plastic-limit of a soil. The state of plasticity is described by such terms as : *Non-plastic, Slight, Low, Medium, High* and *Very high*, depending upon the plasticity index, the soil possesses. Table 4·4 gives such a classification according to Burmister (3) and this scale gives fairly wide range of description of plasticity of soils. According to

TABLE 4·4

Plasticity Indices and Corresponding States of Plasticity according to Burmister

S. No.	Plasticity Index	State of Plastic
1.	0	Non-plastic
2.	1—5	Slight
3.	5—10	Low
4.	10—20	Medium
5.	20—40	High
6.	Greater than 40	Very high

Atterberg (1,2) who first proposed these limits, P.I. of sands is zero and they may be called no-plastic or non-cohesive ; P.I.

of silts is less than 7 and they may be considered to have low plasticity and are partly cohesive ; P.I. of silty clays or clayey silts is more than 7 and less than 17 and they may be considered to have medium plasicity and are cohesive in nature ; P.I. of clays is more than 17 and they are highly plastic and cohesive in nature.

Some engineers feel that there is no practical utility of the consistency limits of a soil, whereas research on these empirical parameters has revealed that they are quite useful and point out clearly the behaviour of soils in many respects.

These limits are used in the classification procedures. The three important systems worth mentioning here are the A. Casagrande's, the Unified soil classification and the I.S.I. soil classification systems described in the latter articles.

As a general rule, liquid limit gives the soil type. *For example*, soil cohesion increases with the increase in percentage of clay in a soil. Cohesion retards the flow of the material. So, liquid limit of a soil is a pointer towards the relative amount of clay in a soil. The higher is the liquid limit, greater is percentage of clay in the soil.

A. Casagrande (4) points out that the force resisting the flow of the grooved soil pat in the liquid limit device cup is a measure of the shear resistance of the soil. So, the number of blows required to close the groove for the required length indicates its shear strength as a relative measure, at the moisture content of the soil. It may be deduced from this argument that the shear resistance of all soils at their respective liquid limits must be a certain constant value.

The slope of a semi-log plot of number of blows and corresponding moisture contents is called the *flow index* of the soil and is denoted by I_f. So, for the two soils X and Y, shown in Fig. 3·8 page 72, the flow indices may be derived as under.

Soil X The equation[1] of the flow curve may written as

$$w = A - B \log N \qquad \qquad ...(4·1)$$

where w is any moisture content (as a percentage), A is the intercept on the Y-axis, B is slope of the line and N is the number of blows, corresponding to w.

Easiest method to find B is to take

---

1. Compare it with the equation of line $y = a + bx$ where a and b are constants and x, the x-coordinate and y, the y-cordinate. a in the equation is the intercept on the y-axis and b is the slope of the equation.

$N=10$ and $N=100$ and find out the corresponding moisture contents on the curves, say w_1 and w_2 and substitute in the equation (4.1)

$$w_1 = A - B \log 10$$
$$= A - B$$
$$w_2 = A - B \log 100$$
$$= A - 2B \; ; \text{ Subtraction will give}$$
$$B = w_1 - w_2 = 50 - 35$$
$$= 15 \text{ (For the flow curve of soil } X)$$

$\therefore \; I_f$ for soil $X = 15$

Soil Y By a similar procedure as for the soil X the slope B for the flow curve of soil Y is

$$= 45 \cdot 5 - 26 \cdot 5 = 19.$$

$\therefore \; I_f$ for soil $Y = 19$

This method of using water contents against blows over one cycle of the log scale is very convenient as it eliminates many calculations. These flow indices give relative shear strengths of the two soils. If *for example*, it is assumed that the P.I. for the two soils is the same, then the soil X having a lower flow index of 15 has better shear strength than soil Y having higher flow index of 19.

Another way of visualizing the same property is this : for an equal decrease in moisture contents of the two soils, say, 10% from 50% to 40%, the difference in the number of blows for soil X is 34 blows and for soil Y is 13 blows, thus indicating that the soil X has greater resistance to flow than soil Y.

It can be seen, therefore, that the flow index of a soil gives an indication about the change in shear strength of the soil with a change in the water content. The flow indices of the two soils give a relative measure of their shear strengths as a function of water content.

Shear strengths of various cohesive soils present a good deal of variation at their plastic limits. Some soils at their plastic limit are much more tougher than others. To express this property of the soils, *toughness index* is used. *Toughness index* of a soil is the ratio of its plasticity index and its flow index *i.e.*

Toughness Index $\qquad = I_t = \dfrac{I_w}{I_f} \qquad \qquad \text{...(4·2)}$

The larger is the value of the toughness index, the better is the strength of the soil at its plastic limit.

The natural water content w of a soil is no indication of the state in which the soil exists. Its position with respect to liquid limit or plastic limit value for the same soil, on the moisture scale of Fig. 4·2 gives an idea of the state in which it exists. As this moisture content approaches the plastic limit P_w, the stiffness and relative compactness of the soil increase.

The ratio of (liquid limit—natural moisture content) to the plasticity index of the soil gives the *relative consistency* of a cohesive soil. In the form of an equation it can written as :—

$$\text{Relative consistency} \quad C_r = \frac{L_w - w_{nat}}{L_w - P_w}$$

$$= \frac{L_w - w_{nat}}{I_w} \qquad \ldots(4\cdot3)$$

where w_{nat} is the natural moisture content, the soil and all the other terms have their usual meaning. If the natural or exsisting moisture content w_{nat} of the soil is equal to the liquid limit L_w, then $C_r = 0$ and if it is equal to the plastic limit P_w, then $C_r = 1$. Plot these two values of C_r on the moisture scale as shown in Fig. 4·3, point A representing $C_r = 0$ and the ordinate CB representing $C_r = $ unity.

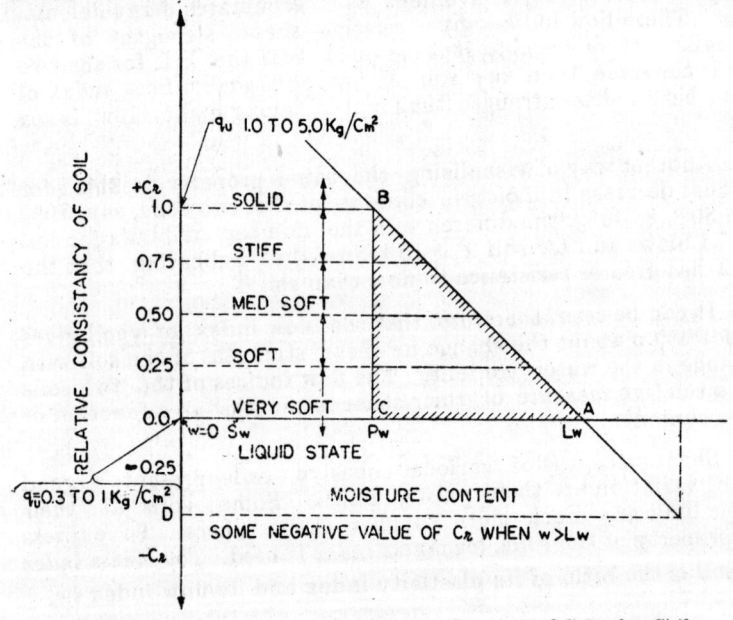

Relative Consistency *versus* Moisture Content of Cohesive Soils

Fig. 4.3.

The values of C_r give an indication of the physical behaviour of u soil, If C_r is negative corresponding to $w_{nat} < L_w$ as, for example, shown by point D, the soil at this moisture content w_{nat} becomes a viscous slurry on remoulding. If C_r is equal to or less than unity, the soil can be remoulded. If, however, C_r is more than unity, the soil cannot be remoulded.

Also the unconfined compressive strength of clays with C_r near zero usually ranges between 0.3 and 1.00 kg/cm.[2].

The relative consistency of cohesive soils may be considered analogous to the relative density of cohesionless soils as given in section 3.15.

The state of a soil at its natural moisture content is sometimes also represented by what is known as *liquidity index or consistency index or relative plasticity index*. This parameter is defined as :

Liquidity Index
$$I_L = \frac{w_{nat} - P_w}{L_w - P_w}$$

$$= \frac{w_{nat} - P_w}{I_w} \qquad \ldots(4.4)$$

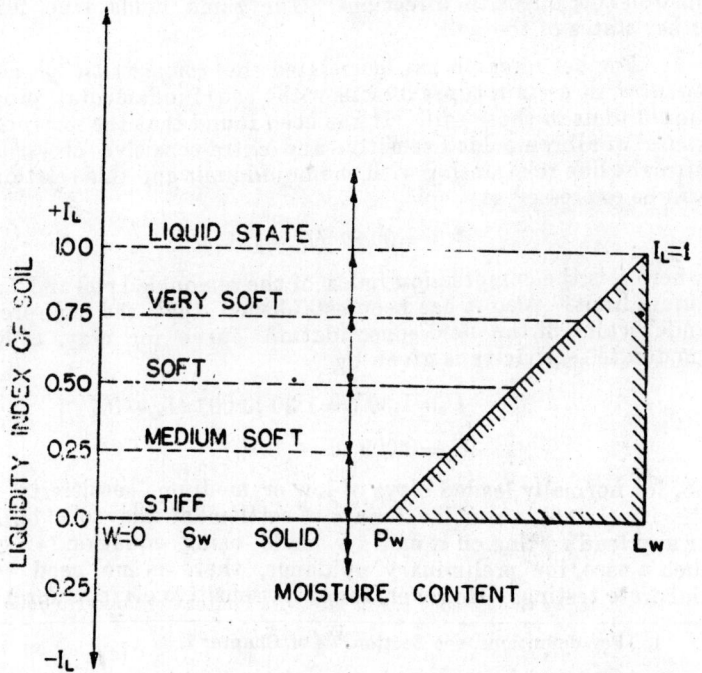

Fig. 4.4. Liquidity Index *versus* Moisture Coontent of Cohesive Soils.

Like C_r this I_L can also be plotted on the moisture scale by assigning w_{nat}, a value of L_w in equation (4.4) when the value of I_L will be unity and giving w_{nat}, a value of P_w when the value of I_L will work out to be zero. I_L is plotted on the moisture scale in Fig. 4.4.

When the liquidity index is more than unity, the soil is in a liquid state ; when it lies between 0.75 and 0·5 it has soft consistency ; between 0·5 and 0·25 it is of medium soft consistency, between 0·25 and zero, it is stiff and below zero I_L, it is a solid. These states are also shown in the figure.

Addition of equations (4·3) and (4·4) shows that

$$C_r + I_L = 1 \qquad \qquad ...(4·5)$$

As such, the relative consistency and the liquidity indices are just related by a constant only and therefore relative consistency or liquidity indices are just two ways of representing the same properties of a soil.

Therefore, various states of the soil at its natural moisture content, can be expressed by relative consistency values also, as shown in Fig. 4·3. The only difference is that liquid state, for example, on the liquidity index scale is represented for $I_L > 1$ and on the relative consistency scale is represented by $C_r < 0$, just on the opposite directions. The same holds true for the other states of the soil.

Further research has correlated the *compressibility*[1] *characteristics* of certain types of soils with the fundamental property liquid limit of those soil. It has been found that the *compression index*[1] of all remoulded sensitive and extra-sensitive clays has a straight line relationship with the liquid limit and this relationship can be expressed as

$$C = 0·007 \; (L_w - 10\%) \qquad \qquad ...(4·6)$$

where C is the compression index of the remoulded soil and L_w is its liquid limit. Also it has been established that the compression index value of the field consolidation curve for clays of low or moderate sensitivity is given by

$$C_c = 1.30 \; C = 1·30 \; [0·007 \; (L_w - 10\%)]$$

or $\qquad \qquad C_c = 0·009 \; (L_w - 10\%) \qquad \qquad ...(4·7)$

So, for normally loaded clays of low or medium sensitivity (sensitivity $S < 8$), a rough estimate of settlement due to a building or any load resting on it can be made using equation (4·7). In such a case, for preliminary guidance, there is no need to do elaborate testing. However, for extra-sensitive clays or for a clay

---

1. For definitions, see Section 7·4 of Chapter 7.

having L_w more than 100 or if the water content of the clay at a depth of more than 8 or 10 metres below the surface is more than the liquid limit or if the clay contains high percentage of organic matter, the estimates based upon equation (4·7) will be erroneous.

Research is underway to correlate the shear strength parameters found by elaborate testing procedures, with these basic Atterberg limits and it is just possible that the parameters of shear strength may one day be deduced from a study of these limits alone.

Some highway agencies placed a limit on the P.I. value of soils used close to the pavement. This way, too, these limits find their practical utility.

Concept of shrinkage limit is very useful in evaluating the behaviour of earth embankments, cuts and earthen dams, specially in estimating the possibility of development of shrinkage cracks. Swelling and shrinkage of cohesive soils is attributed to the presence of organic and colloidal size material in soils. Therefore, soils are sometimes qualitatively evaluated on the basis of what is defined as degree of shrinkage. Degree of shrinkage is given by,

$$D_s = \frac{V_1 - V_2}{V_1} \times 100 \qquad \qquad ...(4\cdot8)$$

where D_s is the degree of shrinkage of the soil, V_1 is the volume of the soil sample before drying and V_2 is the volume of the dry sample after drying, as shown in Fig. 3·9.

Scheiding (7) classifies the soils qualitatively on the basis of D_s. as given in Table 4·5 below.

TABLE 4·5

Degree of Shrinkage versus Quality of Material

S. No.	Degree of Shrinkage D_s	Quality of Material
1.	5%	Good
2.	Between 5 and 10%	Medium good
3.	Between 10 and 15%	Poor
4.	More than 15%	Very poor

The following examples will illustrate further, the use of the consistency limits

Exnmple 4·1. A sample of soil has a liquid limit of 25% and a plastic limit of 20% and ano her has a liquid limit of 50% and the plastic limit is 35%. What could be estimated about these soils as to their type and quality ?

Solution. 1st soil sample $L.L.=25\%$

$$P.L.=20\%$$
$$P.I.=5\%$$

2nd soil sample $L.L.=50\%$

$$P.L.=35\%$$
$$P.I.=15\%$$

Sample number 1 is slightly plastic as indicated on the scale given bv Burmister, in Table 4.4. Its $P.I.$ is less than 7 and hence the soil may just be silt. The low $L.L.$ of the soil confirms this conclusion. Low plastic limit value indicates the absence of any organic material. So in all probability the soil is inorganic silt. Possibility of the presence of some fine sand in the soil cannot be overruled.

Sample number 2 is of medium plasticity as indicated in the scale given by Burmister, in Table 4·4. Its $P.I.$ is more than 7 and less than 17 and the soil can either be silty clay or clayey silt, according to the $P.I.$ figures for these soils given by Atterberg. The high value of the plastic limit may mean the presence of sufficient amount of organic matter in the soil.

Example 4·2. Two soils A and B are compared by means of consistency limits. The following data were obtained in the laboratory :

Soil A		Soil B	
Liquid limit			Liquid limit
N (Blows)	w (%)	N (Blows)	w (%)
4	48	7	61
10	43	15	59
20	40	25	58
40	36	40	57
Plastic Limit	$=20\%$	Plastic Limit	$=20\%$

The natural moisture content in the field for the soils is 42% for soil A and 50% for soil B.

(a) Which soil has higher digree of plasticity ? Explain.

(b) Which clay will most probably be better, foundation material when remoulded and why ?

(c) Which soil will have better strength as a func ion of water content ?

(d) Which soil will have better strength at the plasti limit ?

(e) Do these soils seem to contain some organic material ?

Solution. Plot the flow curves for the two soils as show in Fig. 4·5.

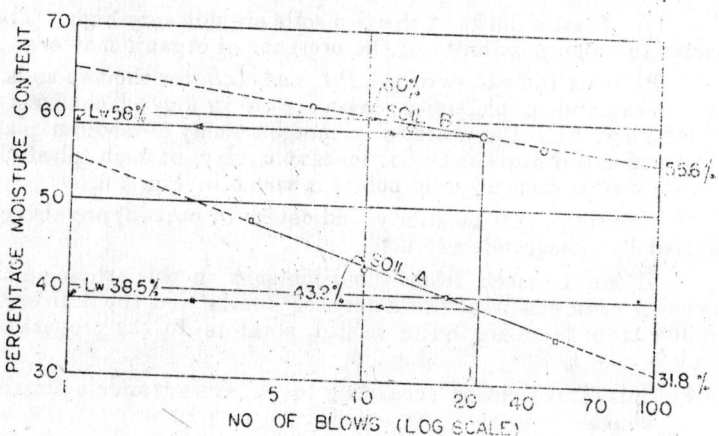

Fig. 4·5. Flow Curves for Soils A and B.

L.L. for soil $A=38.5\%$
L.L. for soil $B=58\cdot0\%$

(a) P.I. for soil $A=38\cdot5-20=18\cdot5\%$
P.I. for soil $B=58-20=38\%$

Soil B has a higher degree of plasticity.

According to Burmister, A may be said to have medium plasticity and B high plasticity.

(b) C_r for soil $A=\dfrac{L_w-W_{nat}}{L_w-P_w}=\dfrac{38\cdot5-42}{18\cdot5}=-\dfrac{3\cdot5}{18\cdot5}=-0\cdot189$

C_r for soil $B=\dfrac{L_w-W_{nat}}{L_w-P_w}=\dfrac{58-50}{38}=\dfrac{8}{38}=+0\cdot215$

Since C_r for soil A is a negative quantity, it will become a viscous slurry on remoulding, whereas the other soil B will not. Hence, soil B will behave as a better material on remoulding.

(c) From the flow curves in Fig. 4·5, considering the water contents at 10 and 100 blows, flow indices for the two soils are

Flow index of soil $B=I_f=43\cdot2-31\cdot8=11.4$
Flow index of soil $B=I_f=60\cdot0-55\cdot6=4\cdot4$

Since the flow index of soil B is less than that of soil A, soil B has better shear strength as a function of water content.

(d) From the P.I. values and the I_f values for the soils, the toughness indices work out as :—

Thoughness Index for Soil $A = \dfrac{P.I.}{I_{fA}} = \dfrac{18.5}{11.4} = 1\cdot 62$

Thoughness Index for Soil $B = \dfrac{P.I.}{I_{fB}} = \dfrac{38}{4\cdot 4} = 8\cdot 63$

$\therefore$ Soil B possesses better strength at the plastic limit.

(e) Plastic limits of the two soils are not very high. There seems to be no possibility of the presence of organic matter.

Plotting the values of the $P.I.$ and $L.L.$ for the two soils, in A. Casagrande's plasticity chart shown in Fig. 4·6 confirms the above view. Soil A appears to be inorganic clay of medium plasticity and soil B appears to be inorganic clay of high plasticity. The two soils are shown by points A and B in Fig. 4·6.

The states of plasticity indicated in part (a) are also confirmed by Casagrande's Chart.

It can be seen from the discussion in this article and the ensuing examples, that the Atterberg limits and the indices that follow from them are quite useful pointers to the properties of soil.

4·5 Soil classification according to A. Casagrande's plasticity chart.

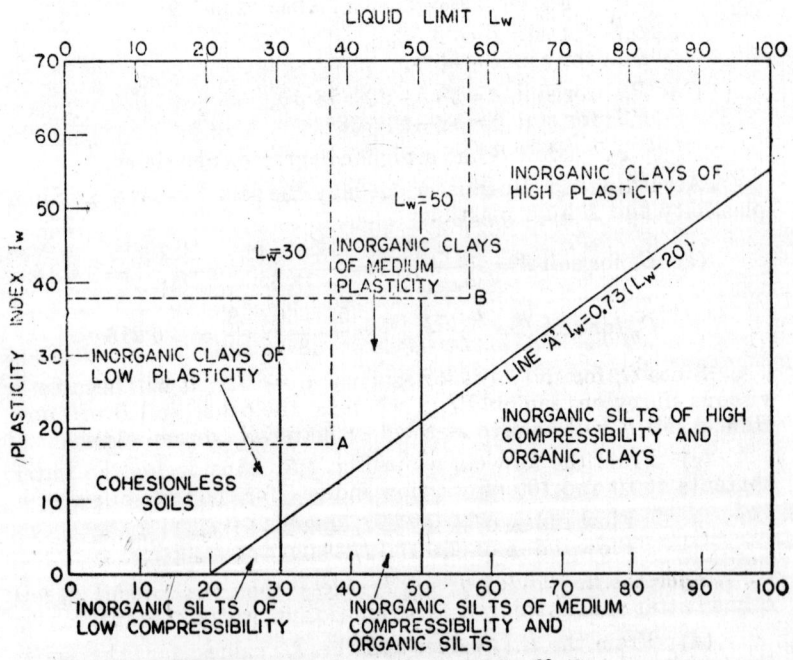

Fig. 4·6. A Casagrande's Plasticity Chart.

An important use of the Atterberg limits is found in the plasticity Chart of A. Casagrande shown in Fig. 4·6. The chart shows the values of liquid limit plotted against the plasticity index values. A line "A" and the ordinates corresponding to liquid limits of 30 and 50 divide the chart in to six regions each region defining certain types of soils, descriptions of which are also given. If the liquid limit and plasticity index of a soil are available, its group can be located on this chart, as already shown in example 4·2 above for the soils A and B, plotted as points A and B in figure 4·6.

In two of the six regions of the chart, soils of different types plot in the same region i.e., organic clays fall in the same region as the inorganic silts of high compressibility and organic slits fall in the same region as the inorganic silts of medium compressibility. The distinction between the organic and the inorganic material can be done either on the basis of odour which is emitted from the organic materials or alternatively the liquid limits of the natural air dried samples and oven dried samples be found. In case the soil contains organic material, the liquid limit value of the oven-dried sample will be less by more than 30 per cent compared to the liquid limit value of the fresh sample. The distinction between two soils one organic and the other inorganic falling in the same region can also be made from the dry strength of the soils. In case of inorganic soil, the dry strength will be less than that of the other soil, which is organic.

The Chart shows that the inorganic soils fall both above and below the line "A". Experience has shown that for the inorganic soils from different localities, having the same $L.L.$ the dry strength of the soils increases with the increasing $P.I.$ values.

4·6. Highway Research Board Soil Classification System.

The Highway Research Board Soil Classification System originates from the Public Roads Administration (P.R.A.) system (revised) of soil classification. This system takes into consideration the index properties of a soil derived both from mechanical analysis and from the consistency tests. From the sieve analysis, a parameter called the *Group Index* of the soil is found out. Some maximum and/or minimum limits are placed on the material passing No. 10, No. 40 and No. 200 A.S.T.M. sieves or their equivalents for different soil groups, along with the group index value or the maximum limit on the group index value, for each soil group. Similarly, minimum and/or maximum values of liquid limit and plasticity index are specified for each soil, these limits to be found on the material passing No. 40 A.S.T.M. sieve or its equivalent. The parameter group index (usually denoted as G.I.) is found in the following manner.

$$\text{G.I. of a soil} = 0\cdot 2a + 0.005ac + 0\cdot 01bd \qquad \ldots(4\cdot 9)$$

Table 4.6 Highway Research Board Soil Classification – 2
(Classification of Highway Subgrade Materials)

General classification	Granular materials (35% or less passing No. 200)			Silt clay materials (More than 35% passing No. 200)			
Group classification	A-1	A-3	A-2	A-4	A-5	A-6	A-7
Sieve analysis, per cent passing							
No. 10							
No. 40	50 max	51 min					
No. 200	25 max	10 max	35 max	36 min	36 min	36 min	36 min
Characteristics of fraction passing No. 40:							
Liquid limit				40 max	41 min	40 max	41 min
Plasticity index	6 max	NP		10 max	10 max	11 min	11 min
Group index			4 max	8 max	12 max	16 max	20 max
General rating as subgrade	Excellent to good			Fair to poor			

(Subgroups)

General classification	Granular materials (35% or less passing No. 200)							Silt-clay materials (More than 35% passing No. 200)			
Group classification	A-1-a	A-1-b	A-3	A-2-4	A-2-5	A-2-6	A-2-7	A-4	A-5	A-6	A-7-5, A-7-6
Sieve analysis, per cent passing											
No. 10	50 max										
No. 40	30 max	50 max	51 min								
No. 200	15 max	25 max	10 max	35 max	35 max	35 max	35 max	36 min	36 min	36 min	36 min
Characteristics of fraction passing No. 40:											
Liquid limit				40 max	41 min	40 max	41 min	40 max	41 min	40 max	41 min
Plasticity index	6 max	6 max	NP	10 max	10 max	11 min	11 min	10 max	10 max	11 min	11 min
Group index	0	0	0	0	0	4 max	4 max	8 max	12 max	16 max	20 max
Usual types of significant constituent materials	Stone fragments, gravel and sand		Fine sand	Silty or clayey gravel and sand				Silty soils		Clayey soils	
General rating as subgrade	Excellent to good							Fair to poor			

Note :- These sieve numbers in this table are according to the ASTM. Specifications.

where a = That portion of percentage passing No. 200 A.S.T.M. (or equivalent) sieve greater than 35 and not exceeding 75, expressed as a positive whole number from zero to 40.

b = That portion of percentage passing No. 200 A.S.T.M. (or equivalent) sieve greater than 15 percent and not exceeding 55 percent expressed as a positive whole number from zero to 40.

c = That portion of the numerical liquid limit greater than 40 and not exceeding 60 expressed as a positive whole number from zero to 20.

d = That portion of the numerical plasticity index greater than 10 and not exceeding 30 expressed as a positive whole number from zero to 20.

No negative values a, b, c or d are to be taken. In case any index value is less than the lower limit for it under a, b, c or d, the corressponding value of the constant is taken as zero.

The various groups into which the soils are classified under this system are given in Table 4·6. The classification procedure is as follows :—

Carry out the sieve analysis and Atterberg limits tests and find out what percentages of the soil pass through No 10, No 40 and No100 sieves. Also determining the liquid limit of the soil, its plasticity index and the Group Index. With this data in hand proceed from left to right in the table. Go on eliminating the groups in which the data does not fit in. The first group from the left into which the data fits in, is the proper soil group for the soil. For splitting up the A—7 group inito the sub groups A—7—5 and A—7—6, take the plasticity indices as under :

Plasticity index of A–7–5 $\leq (L_w - 30)$ and plasticity index of A–7–6 $> (L_w - 30)$.

General rating of the soil as a subgrade material for highways is given in the last row.

This system with suitable modifications to suit local conditions is used by most of the highway departments in U.S.A. for classification of subgrade soils for the highway pavements.

Example 4.3. Sieve analysis on a soil sample indicates that 100% of the soil passes No. 10 ASTM sieve, 65% passes No. 40 ASTM sieve. and 28% passes number 200 ASTM sieve The Atlerbeag limit tests indicate a Liquid limit of 26.3% and a plasticty index of 9·0%. Find out the group index of the soil and classify it according to the HRB classfication. system.

Solution :   Group index of the soil is giveen by

$$G.I. = 0.2a + 0.005ac + 0.01bd.$$

Refor to the descriptions of a, b,c, and d in article 4.6. According to the data obtained from this sample, these values work, out as follows.

(a)= 0 since the percentage passing No 200 sieve is only 28% and is less then 40%

(b)= 28−15=13, since percentage passing No 200 sieve is more then 15% but less then 55%

(c)= 0, since the liquid limit is only 26·3 % and is less then 40%.

(d)= 0, since the plasticity index is only 9·0% and is less than 10%.

$$\therefore G.I = 0.2\,(0) + 0.005\,(0)(0) + 0.01\,(13)\,(0) = 0$$

With the sieve analysis data and G.I. of zero, the first group into which the data fits in table 4·6 is $A-2$ and the first subgroup into which the data fits in is $A-2-4$. Hence, the soil may be classified as $A-2-4$ **Ans;**

4·7. Unified Soil Classification System :

The system presently called the Unified Soil Classification system was originally developed by A, Casagrande and adopted in 1942 by the Corps of Engineers (U.S.A.) as "Airfield Classification". The system has since been revised and in its present form (10) is not only applied to airfield soils but also to soils used under foundations for embankments and for other engineering projects.

Soils under this system are classified mainly in three categories—coarse-grained soils with material passing No.200 A.S.T.M. sieve upto 50% and fine-grained soil with material passing No. 200 sieve more than 50% and organic soils. The distinguishing characteristics between the first two categories is the degree of plasticity. The last category of soils can be identified bycolour, odour and their fibrous nature.

The fine grained soils are further divided into two subgroups on the plasticity chart shown in Fig. 4.7. The first group has a liquid limit of less than 50 and the second group has a liquid limit of more than 50.

Each major soil component in the system is assigend a symbol and two symbols can go to give the soil group symbol. The soil components are designated as :

Gravel : G ; Silt : M ; Organic O ;

Sand   : S ; Clay : C ; Peat   : Pt.

In the coarse grain variety the soil can either be well graded, when it is represented by the symbol *W* or it can be poorly

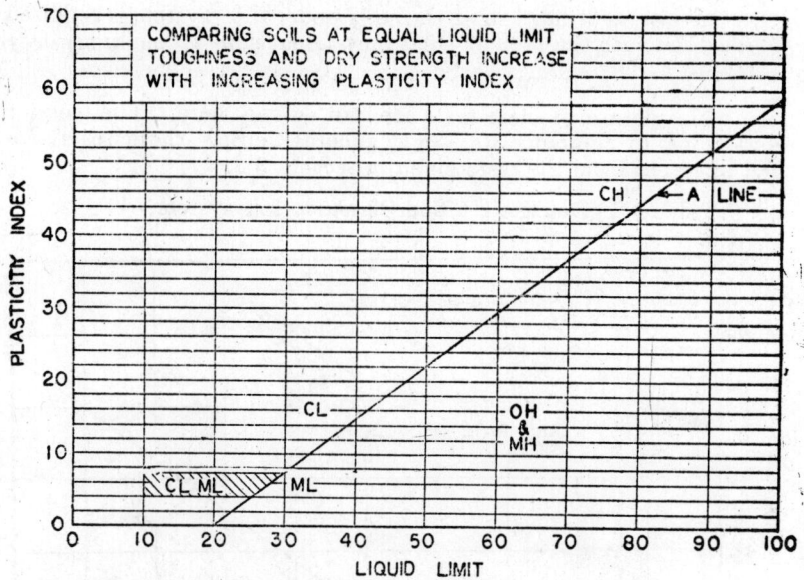

Fig. 4·7. Unified Soil Classification System Plasticity Chart

graded, when it is designated by the symbol *P*. In the fine-grained category, the soil belonging to the first sub-group *i.e.* having *L.L.* <50 are designated by the symbol *L* and the soils belonging to the second sub-group *i.e.* having *L.L.* >50 ara designated by the symbol *H*, thus signifying "Low" liquid limit, and "High" liquid limit.

The particle size distribution curve of a soil can give a fairly good idea of whether the soil is well graded or poorly graded. To further strengthen the estimate, the coefficients of uniformity (Equation 3·19) and coefficient of curvature are found and limit exist on their numerical values to define whether the soil is well graded or poorly graded. These coefficients are given by,

$$\text{Coefficient of uniformity } U = \frac{D_{60}}{D_{10}} \qquad \ldots(4\cdot10)$$

and Coefficient of Curvature $C_e = \dfrac{(D_{30})^2}{D_{10} \times D_{60}}$ ...(4·11)

where the D_{60}, D_{10}, etc., represent the particle sizes against these percentages on a particle size distribution curve of the types shown in Fig. 3.2.

For a well graded gravel U is more then 4 and for a well graded sand, it is more than 6. Again for a well graded soil (gravel, sand or a mixture), C_e lies between unity and 3.

Field identification of the soils under this system is required to be done with the help of Dilatancy (Reaction to shaking) Dry strength and Toughness tests.

The details of these tests are not given here. The reader may refer to reference 10. In a general sense, these tests are not different from the tests given in section 3·3.

Table 4·7 gives the Unified Classification system.

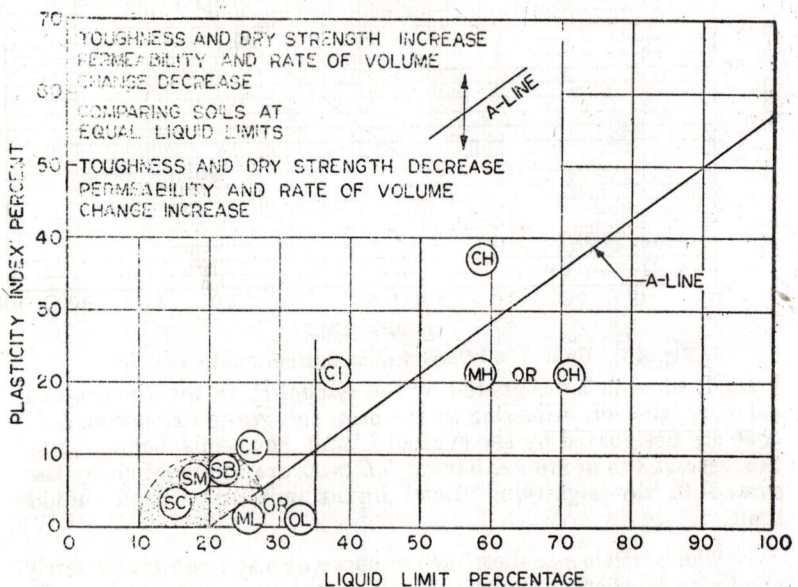

I.S.I. Soil Classification Plasticity Chart (Courtesy Indian Standards Institution)

Fig. 4·8.

Corp of Engineers of U.S.A. based upon their experience have developed a table, shown as table 4·8 in this text, which gives soil characteristics pertinent to the design of highway and airport pavements. Besides the general characteristics regarding drainage, frost action, compressibility and expansion, Columns (11) to (13) in the table give ranges of numerical values that could be assigned to the classified soils. The C.B.R. (California Bearing Ratio) and Subgrade Modulus values could be effectively used to design prvement thicknesses, where such tests are considered expensive from the total project cost point of view. An experienced designee can use these two Columns for almost all the projects, except the very large ones.

Example 4·4 Sieve analysis for a soil sample is plotted in Fig 1·1, curve B.

- (a) What is effective size for the soil
- (b) What is coefficient of uniformity for the soil
- (c) Find out percentages of sand, silt and clay

(*Mysore 1965 modified*)

Solution. (a) Effective size of the soil is D_{10}.

From Fig 1·1 $D_{10} = 0·005$ m.m.

(b) Coefficient of Uniformity $= \dfrac{D_{60}}{D_{10}} = U$

From Fig 1·1 $U = \dfrac{0·85}{0.008} = 106$

(c) Drop verticals from the sand, silt and clay sizes in Fig 1·1, on to curve B.

∴ The sample contains

Sand = 41%

Silt = 19%

clay = 6% **Ans.**

4·8 Indian Standard on Soil Classification.

The I.S. 1498—1959 describes the Indian Standard[1] on soil classification, evolved in 1959 for classification of soils for general engineering purposes. In this system, the soils are classified into four categories *viz.* :

1. Coarse grained inorganic soils, more than half of which is larger than I.S. Sieve No. 8[2].

2. Fine-grained inroganic soils, more than half of which is smaller than I.S. Sieve No. 8[2].

3. Silt and Clay with organic content.

4. Peat.

Each of the first three categories has two subdivisions.

The fine grained soils are classified on the plasticity chart shown in Fig. 4.8. Fine grained soils are divided on this chart

---

1. This classification system is under revision and it is appropriate that the reader refers to the latest standard when made available.

2. These are old sieve numbers given in I.S. 460—1958.

into three divisions on the basis of the liquid limit values, liquid limits of 35 and 50 percent being the dividing lines.

Table 4·8 gives this system of soil classification. The various letters describing the groups mentioned in the table have the following meaning :

Letter	Meaning
G.	Gravel
S.	Sand
M.	Silt
C.	Clay
O.	Organic Silts and Clays
Pt.	Peat
W	Well-graded coarse grained soils with little or no fines
B	Well graded coarse grained soils with clay binder.
P	Poorly graded coarse grained soils with little or no fines.
L	Fine grained soils with low compressibility ; L.L. varying from 0 to 35
I	Fine grained soils with medium compressibility ; L.L. varying from 35·50.
H	Fine grained. soils with high compressibility L.L. > 50.

The table is self explanatory. The field identification tests have already been described in Article 3.3 and the reader may refer to them.

4.9. Comparison of Indian and Unified Soil Classification Systems

The Indian Soil Classification system presented in Article 4·8 and the Unified Soil Classification system presented in Article 4·9 have some similarities and some points of difference. A comparison is presented in this Article.

1. Both the classification systems are based upon the physical properties inherent in a soil.

2. The ISI Classification system divides the soils in four major groups which are subdivided into eighteen sub-groups, each having different characteristics ; whereas the Unified Soil Classi-

cation system divides the soils in three major groups which are sub-divided into fifteen sub-groups, each again having different characteristics.

3. The symbols used to define the same soil in the two systems, differ.

4. The ISI system gives field identification tests, namely Visual examiation, Wet and Manipulated S rength Test, Thread Test, Dilatency Test and Dry Strength Test, whereas the Unified Soil Classification system does not make mention of the Wet and manipulated Strength Test.

5. The ISI system gives the general characteristics of the soil, such as, strength in natural state, strength when wet and remoulded, permeability and bulk density, whereas the Unified Soil Classification system does not mention these characteristics.

6. The ISI system gives the relative suitability of the soils for general engineering purposes, whereas the Unified Calssification system does not give such information.

7. To distinguish between whether a soil is well-graded or it is poorly graded, the Unified Soil Classification system places numerical limits on the Co-efficient of Uniformity and Co-efficient of Curvature, whereas the ISI does not specify any such limits.

8. For some of soils under the coarse grained group, the Unified Soil Classification system places numerical limitations on the Atterberg limits by specifying whether these limits should lie below or above the "A" line in the plasticity chart and also places limits on the P.I. values. The ISI system does not specify any such limits.

CHAPTER—BIBLIOGRAPHY

1. Atterberg A., Die Plastizitat der Tone, International Mitteilungen for Bodenkunde, Volume 1, No. 1, 1911, Verlagfiir Fachliteratur G.M.B.H. Berlin.

2. Atterberg A., Die Plastizitat and Bundigkeit liteferenden Bestaudfiel der Tone. Vol. 3. No. 4, 1913, Verlagfiir Fachliterature G.M.B.H. Berlin.

3. Bowls, E., "Foundation Analysis and Design," McGraw Hill Book Co., New York, N.Y. 1968.

4. Burmister, D.M., Principles and Techniques of Soil Identification, Proceedings of Annual Highway Research Board Meeting, Washington D.C., 1949, Vol. 29.

5. Casagrande, A., Research on the Atterberg Limits of Soils, Public Roads, Oct. 1932, Vol. 13, No. 8.

6. Casagrande, A., Applications of Soil Mechanics in Designing Building Foundation, Trans. American Society of Civil Engineers, 1944.

7. Casagrande, A., Classification and Indentification of Soils Transactions ASCE, New York, Paper No. 2351, Vol. 113.

8. Casagrande, A., Classification and Identification of Soils, proceeding ASCE, New York, Vol. 73, No. 6, Part I, 1947.

9. Hogentogler, C.A., The Sub-grade Soil, Public Roads, July 1931, Vol. 12, No. 5.

10. Scheidig, A., in Kogler F. and Scheidig A., Baugrund and Bauwerk, 5th, Berlin, Wilhelms Erust and Sohn, 1948.

11. Spangler Merlin G., Handy Richard L., "Soil Engineering," Internt Educational Publishers, N.Y., 1973.

12. Terzaghi K. and Peck R.B., Soil Mechanics in Enineering Practice, New York, John Wiley and Sons, 1968.

13. Tschebotarioff, G.P., Soil Mechanics in Foundations and Earth Structures, New York, McGraw Hill Book Co., Inc., 1951.

14. The Unified Classification System, Appendix B, Technical Memo No. 3-357, March 1953, revised 1959 U.S. Army Engineers Waterways Experiment Station.

QUESTIONS

1. What do you understand by any four of the following :

    (i) Consistency.   *(ii)* Thixotropy

  (iii) Base exchange   *(iv)* Soil structure

    (v) Uniformity coefficient   *(vi)* Optimum moisture content.

(b) What do you understand by Atterberg's limits ? Explain how these limits are helpful to engineers in indicating general properties and classification of soils.   *(P. U. 1959 Supp.)*

2. Calculate the coefficient of curvature for only the coarse-grained soils, the particle size distribution curves for which are given in Fig. 3 2. Are these soils well graded or poorly graded, according to the United Soil Classification system ?

3. A soil sample is analysed by sieve analysis. The percentages passing No. 10, No. 40 and No. 200 ASTM sieves are as under :

Sieve No.	Percentage passing
10	95
40	80
200	68

The Atterberg limits for the fraction passing No. 40 sieve are determined and found as under :

L. L.,   50 percent
P. I.,   15 percent

Find out the group index of the soil and classify it according to the P. R. A. system.

(*Hint.* For the group index $a = 68 - 35 = 33$

$b = $ Since % age, passing No. 200 exceeds 55%, this value of b is taken as 40

$c = 50 - 40 = 10$

$d = 15 - 10 = 5$

Now, find out the value of G. I. and classify the soil according to Table 4.6).

4. Write a short note on "Atterberg limits of consistency of Soils".
(*P. U. 1960*)

5. Write short note on, "Atterberg limits" and their use in Soil Mechanics.
(*P. U. 1961*)

6. Write short note on, the "Casagrande's Soil Classification" system.
(*P. U. 1963*)

7. What is the need of classifying natural soils ? Describe a system of classification of soils. How does it, suit to soils in India.
(*P. U. 1963 Supp.*)

8. (a) For the sieve analysis of a silty sand ; state the I. S. Sieves you will use ? Give the sieve designation and size of opening for each sieve.

(b) How is the I. S. Soil Classification different from the unified Soil Classification system.
(*P. U. 1965*)

9. Discuss the significance of the Atterberg limits in Civil Engineering practice.
(*P. U. Nov., 1965*)

10. Draw a qualitative diagram showing the volume change with the change in moisture content of a saturated soil specimen, along the moisture scale of the type shown in Fig. 4·2.

11. Describe the method of soil classification by the triangular classification chart. Indicate the soil type, if it contains 42% clay, 28% sand and 30% silt.
(*P. U. Nov., 1966*)

12. List the important physical properties that distinguish between granular soils and fine grained soils.
(*Mysore 1965*)

13. Describe the following soils using proper terminology, given the following data, from laboratory tests

Soil Type	Colour	Grain Size		Atterberg Limits	
A	Brown	% Passing } 100% 5%		{ 25%	22%
		Sieve No. } 4 200		{ L. L.	P. L.

B	Black	% Passing	100%	80%	80%	30%
		Sieve No.	4	200	L. L.	P. L.

(Mysore 1958)

Note :- The author believes that the Atterberg Limits and Grain Sizes given in the table above are not very consistent.

14. (a) How is soil classified according to grain size ?

(b) Describe how the "grain size distribution curve" for a given sample of soil is obtained from laboratory tests. Why is Semi-logritamic plot preferred for such a curve.   (*Warangal 1968*)

15. The mechanical analysis of a soil sample is given below. Draw a grain size curve and give approximate classification according to (a) Grain size classification (b) Textural classification (c) U.S.B.P.R. classification.

Mechanical Analysis Data
Present Passing

Sieve Size in Millimeters

26.7	18.8	9.4	4.75	2.0	0.42	0.25	.105	.074	.05	.005	.001
100	90	80	72	67	56	44	24	24	21	11	4

(*Punjabi Univ. 1967*)

Soil Moisture, Permeability and Related Characteristics

5·1 Introduction

As pointed out earlier, behaviour of a soil is influenced to a large extent by the presence of moisture in it. Therefore, it is essential to have an insight into the character of moisture in the soils and its influence on soil properties. Capillary rise of moisture in soils and consequent capillary forces have a significant influence on fine-grained soils, such as silts and clays and hence need careful attention. Furtheron, soils are continuously connected porous media and under proper hydraulic conditions any fluid can flow through the pores of the soil. Flow of petroleum oils into oil wells and flow of water into tube wells are clear examples of the flow of fluids through the soil pores. The rate of flow of a fluid through a soil depends upon its property known as **permeability**, which may be defined as that property of a soil which allows a fluid to flow through it. Permeability varies from one soil to another and the relative comparison is made on the basis of numerical values of the **coefficient of permeability**. Some soils are then designated as *permeable* or *pervious* and the others as relatively *impervious*. As will be seen later, the cohesionless soils are relatively pervious compared to the cohesive soils, which are impervious.

Study of soil moisture, permeability and related characteristics is, therefore, very important. Some examples of the influence on behaviour of soils of moisture and permeability are : shear strength of a saturated soil depends on its water content and its determination is based upon the free flow of water through soil pores ; settlement of structures founded on clay beds is dependent upon the permeability of the soil ; seepage forces under hydraulic structure are a result of the free flow of water in the soil pores ; stability of embankments, levies and earthen dams is dependent upon the shear strength of the soils, which in turn depends upon soil moisture.

5.2. Varities of soil moisture

Soil in the ground does not exist in a dry state. It always carries some moisture with it. If the smallest particle of the size of a colloid is considered, the water surrounding this particle exists in two forms. Very close to the solid particle, a moisture film exists, which is termed as the **adsorbed moisture** film as shown in Fig. 5 1 This film encasing the soil particle is, for all practical purposes, a part of the soil grain itself. As indicated in the figure, the absorbed moisture layer may by divided into two portions, the so-called *solidified water* very close to the soil grain and the surrounding highly viscous water termed as *cohesive water*. The entire film is sometimes called **bound water** or **hydrated water**. Surrounding this film is the water that fills the voids of the soil mass. This water is termed as **free water**. The *bound* and *free* waters have different physical properties, For example, the free water has its boiling and freezing points, its viscosity and surface tension of what is ordinary termed as water, whereas, the boiling point of bound water is higher. its freezing point is lower, it is much more viscous and its surface tension is much greater than of ordinary water. Tergzahi (1) has pointed out that adsorbed water films smaller than 2/100,000 inch have the characteristics of semi-solid substances.

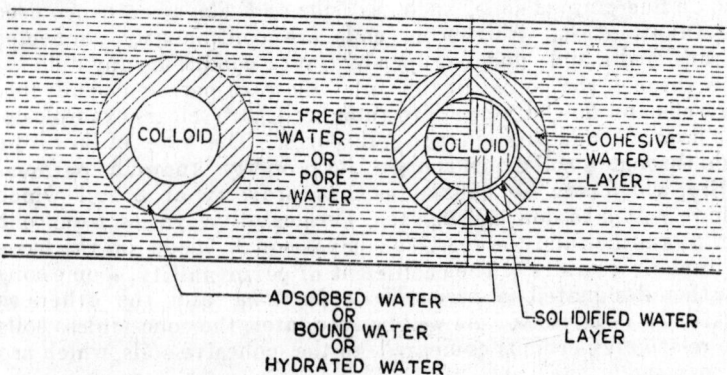

Classification of Soil Moisture based upon Adhesiveness of Water.

Fig. 5.1.

The free water remains outside the surfaces of adsorbed moisture films and it is this water which determines ordinarily the physical properties of soils. This water is usually referred to as **pore water**.

In deriving various soil parameters, it is normally assumed that the adsorbed moisture films do not affect these parameters. This assumption is only true so long as the conditions affecting these films remain constant. For example, a variation in temperature changes the thickness of these films and unless the temperature is kept constant, the assumption that the parameter is uninfluenced by these films cannot be made. This is one reason why in almost all the experimental work, the temperature at which the test is carried out is checked and an attempt made to keep it constant throughout the work.

The above classification of the soil moisture is based upon the adhesiveness of water to the soil grains.

Depending upon how the soil as a mass absorbs water, soil moisture can be classified as **hygroscopic moisture** and **capillary moisture**. **Hygroscopic moisture** is that water which the soil mass absorbs from the atmosphere. In the case of a dry soil in contact with atmosphere saturated with water vapours, the hygroscopic moisture absorption is of the tune of 2.3 percent by weight in case of soils of the type of sand and more than 13 percent by weight in case of fat clays at about 16°C. So the colloidal content of the soil has a property of absorbing more moisture from the atmosphere. **Capillary moisture** is that water which a soil sucks from the level of free water in a soil mass. As it rains, the rain water seeps into ground and due to the permeability of the soil goes deeper and deeper till it meets an impervious startum and starts accumulating over it and forms what is called *ground-water reservoir*. The surface of water within the ground, where the water has the atmospheric pressure is called the **phreatic surface** or more commonly the **water table**. The zone of soil below the water table is called the **zone of saturation** and the water contained in this zone is called the **ground water**. The zone of soil between the water table and the ground surface is called the **zone of aeration**[1]. The rain water which enters the soil mass to form the ground water is also sometimes called *gravitational water* since it enters the soil under the influence of gravity.

It has been observed that the soils above the water table are fully saturated upto a certain height and partially saturated a little further. This saturation is the result of *capillary rise* of water in the soil. Capillary rise of water in the soil is attributed to the effect of adhesive films which lift the water above the water table. The amount of lift is called **capillary rise**. The layer of soil in the zone of aeration moistened by capillary rise of water is called the **capillary fringe**.

The zones of *saturation, aeration* and *capillary fringe* are shown in Fig. 5.2.

Some minute drops of water remain suspended in the zone of aeration above the capillary fringe too, due to possibly the capillary forces in that region and it is this moisture which nourishes the plants. This moisture is called **contact moisture.**

---

1. Sometimes in literature, zone of aeration is defined as the soil between the ground surface and the capillary fringe.

The rise of water due to capillary action induces certain stresses in the soil mass known as **Capillary forces** and hence the need to study the capillary phenomenon in detail.   In order to understand the principle underlying the rise of water in soil pores, the case of a glass tube inserted in water is presented, in the succeeding article.

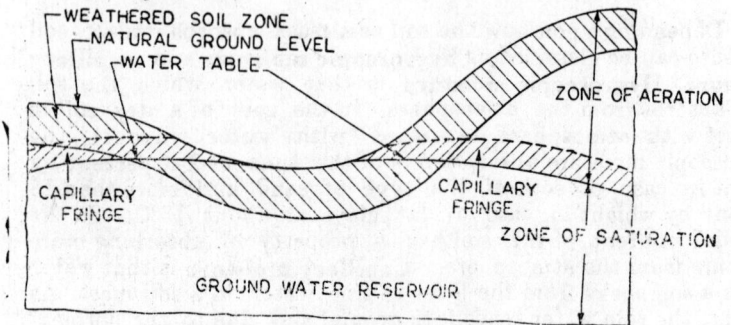

Zones of saturation, aeration and capillary fringe.
Fig. 5.2.

5.3.   Capillary rise in a glass tube of constant section

Consider a glass tube inserted in a trough of water as shown in Fig.   5.3.   As soon as the tube is inserted in the trough, water

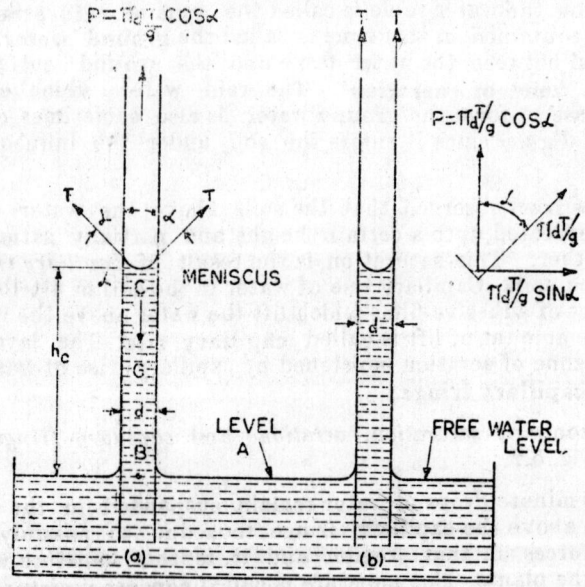

G = FORCE DUE TO GRAVITY = WEIGHT OF WATER COLUMN

Fig. 5 3.

rises in the tube above the level 'A' of the water in the trough and forms a meniscus as shown. The formation of the meniscus is explained as under :

The glass and water posses some inherent attraction for each other and as soon as the glass tube is inserted in water, a a water film of the thickness of a molecule, which originally forms a part of the topmost film in the trough rises along the sides of the tube due to this attraction. The rise above the surface of water in the trough depends upon the amount of attraction between the two. As this film rises inside the tube, it pulls up the adjacent water of which it itself is a part alongwith it and forms a meniscus in the tube.

The maximum height to which the water can rise in the tube depends upon the surface tension of water film covering the water and the subsequent lifting of the meniscus. In case the glass tube and water are not clean, this surface tension acting along the perimeter of the meniscus, makes an angle α, known as the *angle of contact*, with the vertical edge of the glass tube as shown in Fig. 5·3 (a). The meniscus in this case is not fully developed and is only an arc of a circle. In case the glass tube and the water are clean, then the meniscus so formed is a complete semicircle and the surface tension along the perimeter of the meniscus acts vertically upwards as shown in Fig. 5·3 (b).

Now if T in dynes per centimeter of the circumference of the meniscus is the surface tension ; then the total surface tension acting at the periphery of the meniscus in the direction shown in Fig. 5·3 (a) is given as ;

Total surface tension $=\pi d\ (T)$ dynes ...(5·1)

where d in cms. is the diameter of the glass tube. In terms of grams this total surface tension

$$=\pi d\left(\frac{T}{g}\right)\text{gms.} \qquad ...(5.2)$$

where $g = 1$ gram

$=980.7$ dynes.

As shown in Fig. 5.3 (c), the force represented by equation (5·2) can be split up into the horizontal component of $\pi d.\ T/g$ six α and the vertical component of $\pi d.\ T/g$ cos α. The vertical component $\pi d\ T/g$ cos α represents the lifting force P of the meniscus and must, therefore, be equal to the weight of the column of water standing in the tube above the level 'A'.

If h_c in cms. is the height of rise of water in the small tube, then.

weight of the column of water above the level.

$$'A' = G = h_c \left(\frac{\pi d^2}{4} \right) \gamma_w \qquad \qquad ...(5\cdot3)$$

where γ_w is the unit weight of water in c.g.s units. Equating (5·3) with the vertical component of total surface tension,

$$h_c \left(\frac{\pi d^2}{4} \right) \gamma_w = \pi d \, \frac{T}{g} \cos \alpha$$

$$\therefore \qquad h_c = \frac{4T \cos \alpha}{d.g. \; \gamma_w} \qquad \qquad ...(5\cdot4)$$

If the glass tube and water are chemically clean $\alpha = 0$ and the equation (5·4) reduces to

$$h_c = \frac{4T}{d.g. \; \gamma_w} \qquad \qquad ...(5.5)$$

At $20°C$ (about $67°F$), the value of $T = 72\cdot8$ dynes per cm. and taking $g = 980\cdot7$ dynes and $\gamma_w = 1$ gm./c.c. (approx.), h_c will work out as :

$$h_c = \frac{0\cdot3}{d} \text{ cms.} \qquad \qquad ...(5\cdot6)$$

The above equation shows that height of capillary rise is indirectly proportional to diameter of the tube at the meniscus level. The height of capillary rise of water does not depend upon the shape and size of the tube below the meniscus level, nor does it depend upon the inclination of the tube. The height only depends upon the tube diameter at the meniscus level. However, if the diameter of the tube at the meniscus level is different, then the height of capillary rise will be different. The point is clear from example 5·1

It may be noted that the lifting force at the surface carries the weight of the column of water by a tensile stress in the water column, as water has no shearing strength. This tension or negative hydrostatic pressure has its maximum value as $h_c \gamma_w$ just under the upper surface of the column. That the negative hydrostatic pressure exists in the column of water can also be seen by applying Bernoulli's theorem to the points B at the free water surface and C just at the meniscus. Take the datum at the level of B, then applying Bernoulli's theorem,

$$h_c + \frac{P_c}{\gamma_w} = 0,$$

so that $\qquad \dfrac{P_c}{\gamma_w} = -h_c \qquad \qquad ...(5\cdot7)$

or $\qquad P_c = -\gamma_w{}^{hc}. \qquad \qquad ...(5\cdot8)$

Equations (5·7) and (5·8) show that the pressure head at the level of the meniscus is equal to the capillary rise in water with a negative sign and the pressure at this level is negative.

At any other height say h_1 above the free water surface, the tension in water will correspond to $h_1 \gamma_w$.

Example 5·1· (*a*) The internal dimensions of the various sections and their respective heights of the three tubes immersed in a tub of water are shown in Fig. 5.4 (*a*). How high will the water rise in each tube, assuming the tubes and water are chemically clean ?

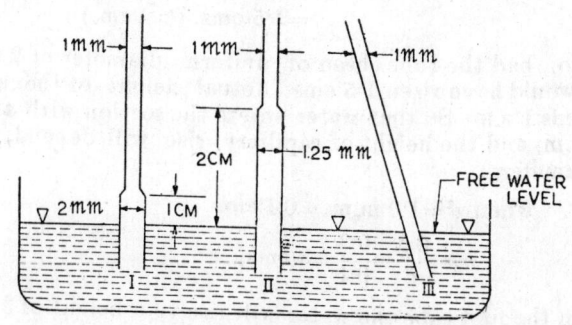

Fig. 5·4 (*a*)

(*b*) Fig. 5·4 (*b*) shows a tube of variable diameter at different sections. How high will the water rise in this tube, assuming again that the tube and water are chemically clean ?

(*c*) If the tube shown in Fig. 5·4 (*b*) is dipped in water and then withdrawn and allowed to stand as shown in this figure, how high will the water rise in the tube ?

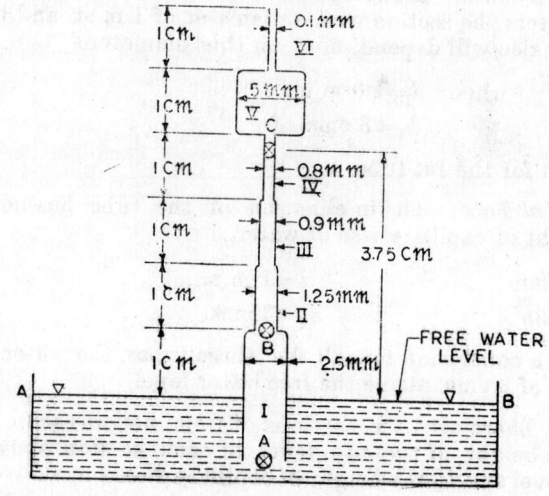

Fig. 5·4 (*b*)

(d) What are the water pressures in the tube shown in Fig. 5·4 (b) at the points A, B and C ?

Solution. (a) 1st tube,

When $d = 2$ m.m. $= 0.2$ cm.,

$$h_c = \frac{0.3}{d}$$

$$= 1.5 \text{ cms. } (>1 \text{ cm.})$$

So, had the tube been of uniform diameter of 2 m.m., the water would have risen 1·5 cms. Actual height of the section of 2 m.m. is 1 c.m. So that water enters the section with a diameter of 1 m.m. and the height of capillary rise will depend, now, on this diameter.

$\therefore$ when $d = 1.0$ m.m. $= 0.1$ cm.

$$h_c = \frac{0.3}{d} = \frac{0.3}{0.1} = 3 \text{ cms.}$$

In the first tube the water will rise to a height of 3 cms.

IInd Tube. When

$$d = 1.25 \text{ m,m.} = 0.125 \text{ cm.}$$

$$h_c = \frac{0.3}{d} = \frac{0.3}{0.125}$$

$$= 2.4 \text{ cms. } (>2 \text{ cm.})$$

By a similar argument as for the 1st tube, therefore, the water enters the section with a diameter of 1 m.m. and hence the capillary rise will depend, now, on this diameter.

$\therefore$ when $d = 1.0$ m.m,
$$h_c = 3 \text{ cms.}$$

as shown for the 1st tube.

IIIrd Tube. The inclination of the tube has no effect on the height of capillary rise of water.

When $d = 1$ m.m.,
again $h_c = 3$ cms.

As a conclusion for all the three tubes, the water stands at a height of 3 cms. above the free water level.

(b) There are six sections of 1 cm. height each. Let h denote the height of the top level of each section above the free water level and h_c the height of capillary rise.

Using $h_c = \dfrac{0\cdot3}{d}$, where d is the diameter of the tube in cms., prepare the following table starting at the lowest section assuming as if each section was individually immersed in water.

d (cms.)	h_c (cms.)	h (actual)	Remarks
0·25	1·20	1 cms.	$h_c > h$; so water enters section II
0·125	2·40	2 cms.	$h_c > h$; so water enters section III
0·090	3·33	3 cms.	$h_c > h$; so water enters section IV
0·080	3·75	4 cms.	$h_c < h$; so water does not enter section V

So, the water in the tube stands at a levels of $3\cdot75$ cms. above the free water level.

(c) If the tube is dipped in water, it will get completely filled while it is horizontal and when lifted vertically, it will remain completely filled as h_c for diameter of $0\cdot1$ m.m $= \dfrac{0\cdot30}{0\cdot01} = 30$ cms. which is more than the height of the tube itself.

(d) Let the water pressure at the three points A, B and C be denoted by p_A, p_B and p_C respectively.

The point A is 1 cm. below the free water level and hence the pressure here is positive.

$\therefore$ $p_A + \gamma_w h = 1$ gm/cm^2

The points B and C are inside the capillary tube where pressure is negative.

$\therefore$ $p_B = -\gamma_w h = -1$ gm/cm^2.

and $p_C = -\gamma_w h = -(1)(3.75) = -3.75$ gms./cm^2

Ans.

5·4 Capillarity and shrinkage in soils

The rise of water in the soils above the ground water table is analogous to the rise of water in a capillary tube above the free water surface. A soil mass may be assumed to carry bundles of capillary tubes of zig-zag shape in the form of interconnected voids. As soon as these so-called bundles of tubes come in contact with water, the water starts rising in soil pores. Size of an individual pore in a soil mass is a function of the grain size and the void ratio of the soil and it increases with the increase in grain size and void-ratio. So the rise of water in a soil will depend upon the size of grains of the soil and the void ratio. Application of equations (5·4), (5·5) and (5·6) to soils is not directly possible ; however, these equations indicate that the capillary rise is inversely proportional to the tube diameter at the meniscus level and as such, an inference can be draw from it, that the smaller is the pore size in soils, the greater will be the rise of water. That is why in the fine-grained soils the phenomenon of capillary rise of water is more pronounced than in coarse-grained soils.

As is seen from equation (5·8), the pore water in the soil due to capillary rise is in tension. The negative hydrostatic pressure in the soil induces a positive compressive stress on the soil phase *i.e.* soil particles (or in fact the soil skeleton) and this stress is equal to $\gamma_w h_c$. This pressure is designated as **Capillary Pressure**, The skeleton is under compression. This is exactly the reason why the clayey soil changes volume upto the shrinkage limit and not after it.

The shrinkage of a clayey soil can be explained in another way also. If a saturated soil sample above its shrinkage limit,

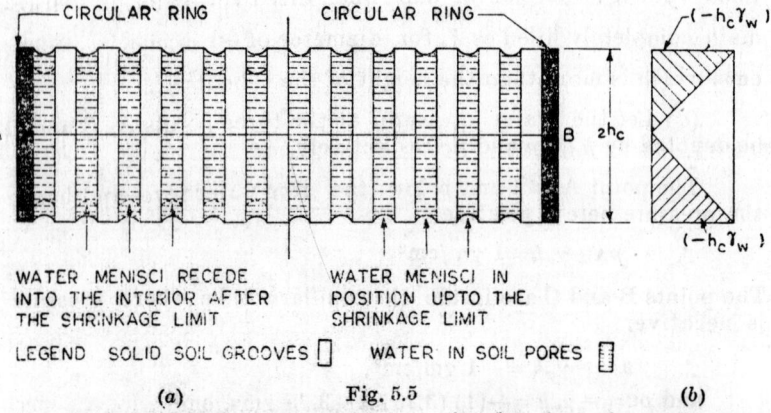

WATER MENISCI RECEDE INTO THE INTERIOR AFTER THE SHRINKAGE LIMIT

WATER MENISCI IN POSITION UPTO THE SHRINKAGE LIMIT

LEGEND SOLID SOIL GROOVES ☐   WATER IN SOIL PORES ☐

(a)   Fig. 5.5   (b)

say, at the liquid limit is cut into a circular ring, the surfaces of soil are exposed to the atmosphere, The soil may be imagined to consist of solid soil grains and capillary pores, the capillary pores extending from one end of the ring to the other end and filled with water. The hypothetical situation is illustrated in Fig. 5.5 (a) where the solid soil grains are shown separately and the capillary water in the soil standing in small capillary tubes[1] separately.

Surface tension exists at meniscus exposed to the atmosphere. The water at the level of meniscus is in tension *i.e.* the pore water pressure in the soil is negative at both the ends of the ring. Exactly at the middle section AB, this hydrostatic presure, neglecting the weight of the soil, is zero. So, a pressure difference exists between the interior of the soil mass and the outer surfaces as shown in Fig. 5 4 (*b*). The water starts flowing from the middle towards the two ends and at the ends it gets evaporated due to its contact with the amtosphere. Due to flow of water from the interior towards the ends and its evaporation, the soil mass decreases in volume. The meniscii at the ends do not tear off and remain in position till the shrinkage limit is reached, when the meniscii start receding inside the soil pores and air starts filling them. At this point the volume of the soil becomes constant.

Due to the negative pressure in pore-water, the intergranular stress in the soil increases, by an amount equal to this negative pressure. So if h_c is the head corresponding to the capillary rise, usually called the capillary head, then the negative hydrostatic pressure is $h_c \gamma_w$ as given in equation (5·8) and the increase in intergranular pressure will $h_c \gamma_w$. The capillary head for a soil sample can experimentally be determined in the laboratory as given in article 5.14 in this chapter.

5.5 Types of head and head lost

Total head of water at any point in, a fluid medium is the sum of three heads—the **elevation**, the **pressure** and the **velocity head**. The velocity of water flowing through a soil is so small that the velocity head given by $\frac{V^2}{2g}$ where V is the velocity and g the acceleration due to gravity is insignificant. The only two heads of any consequence in most of the problems in soil mechanics are the elevation head and the pressure head. The elevation head at any point in a saturated soil mass is the height of the point above a datum and the pressure head at the point is the quotient of the water pressure at that point and the unit weight of water (This pressure head was also referred to in calculating the height of capillary rise of water in a tube in article 5.3). So, the total head constitutes of these two components only.

In order to calculate the discharge through a layer of soil, determination of head loss across the laywer becomes necessary, so as to find out the gradient across the layer, for use in Darcy's

---

1. The exact situation in soils is not alike. The capillary rise of water in an actual soil mass is in the zig-zag passages formed by the voids enclosed by the soil grains. However, as a diagramatic illustration Fig. 5.5 (a) is a right.

Law given by equation (5·14) in one of the next sections. The difference of the total heads at the top and bottom of the layer gives the head lost, which can be used in finding out the gradient.

In order to find out the elevation head, datum is chosen at a convenient point in the medium comprising of a free water reservoir from which the supply is drawn into tne soil, the saturated soil layer and the free water reservoir to which the water is discharged. Then the elevation and pressure heads for the top and bottom of soil layer are determined separately and their respective sum totals give the total heads at the top and bottom of the layer. Difference between these total heads gives the head lost across the layer. However, another and direct method to find out the head lost is to find out the difference between the *head water level*[1] and the *tail water level*[1]. The level of the topmost point to which the water stands in the receiving reservoir, is called the *tail water level*, and the difference in the two values is the head lost as already stated above.

Example 5.2. Figure 5.6. shows a soil profile taken across a reservoir. The clayey silt is relatively impervious. Indicate the values of pressure head, elevation head and the total head at the points *A* and *B* shown on the profile and the total loss of head between these points assuming that the ground water table lies at relatively large distance below the clayey silt layer,. to have any significant effect.

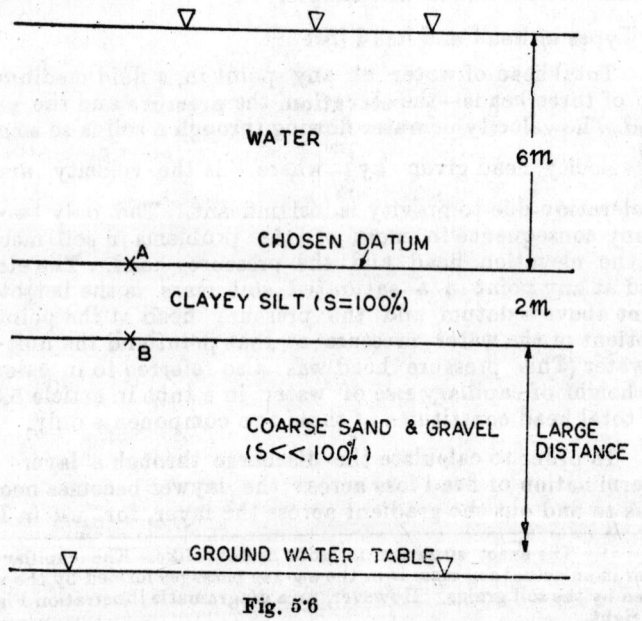

Fig. 5·6

1· Refer to Fig. 5.8 for illustration

Solution. Choose the top level of the clayey silt layer as the datum level. Denote the elevation head as h_1, the pressure as $h_2 \left[= \dfrac{\text{pressure } p}{\gamma_w} \right]$ and the total head as h at any point. Then the following table gives the various heads :

Point	h_1	h_2	$h = h_1 + h_2$	$\triangle h$	
A	0·0m	6.0m	6·0m		**Ans.**
B	—2.0m	0·0m	—2·0m	8.0m	

5.6. State of stresses in a soil mass

Widely accepted failure conditions in a soil mass emerge from the state of stresses at a point in a solid body. These stresses and their representation on a Mohr's circle are simple and elementary concepts and can be found in any book on *Elementary Strength of Material*[1]

However, what needs to be stressed in the soil mechanics field is the clear distinction between the *total, effective* and *neutral stresses*. Stress, *by definition*, is the force per unit area. If the soil is dry, there will be stress transmission at the points of contact between the various soil grains. Evaluation of grain to grain stresses is not possible. As such, stress in the dry soil at any cross section is considered as the applied load divided by the area of the cross section including both solid particles and void spaces. However, when the soil is saturated and the water table lies above the cross-section being considered the applied stress from loads is not only transmitted at the points of contact between the various soil grains, but in addition, there is some stress in the water filling the soil voids. In this latter case, it is essential to distinguish between what are called *total, effective* and *neutral stresses.*

Total stress at any point in the mass, in this case, is the applied load divided by the area of the cross-section including both solid particles and void spaces passing through this point. This total stress can further be sub-divided into effective or intergranular stress and neutral stress or pore-water pressure. **Effective** or **intergranular** stress may be defined as the total load on a cross section of a soil mass which is transmitted from grain of the soil divided by the area of cross section, including both soild particles and void spaces. **Neutral stress** is defined as the unit stress carried by the pore water in the voids in a cross section. This is sometimes called *pore water pressure.* In a static case of equilibrium the neutral stress or pore water pressure would have the same intensity in all directions. If the cross section being considered is h units below the ground water table, this neutral stress is nothing but the hydrostatic pressure $\gamma_w h$, where γ_w is the unit weight of water.

---

1. Students may also refer to chapter 10 for brief details on this subject.

The relation between the total, effective and neutral stresses is given by what is called the **principle of effective stress**. As far as the response of the soil to load is concerned, it is generally accepted that the behaviour under stress is governed by the equation.

$$\sigma = \bar{\sigma} + u \qquad \qquad ...(5\cdot9)$$

where σ is the the total stress, $\bar{\sigma}$ is the effective stress and u is the neutral stress or pore water pressure. The relationship is empirical. However, at the present state of knowledge of the subject, this is the only equation that is generally used to evaluate stresses in soils.

It may be mentioned here that the effective or integranular stress will, in general, have normal and shearing components. While evaluating the failure in soils due to shear, it is intergranular or effective stress which is of importance, rather than the total stress. Also it is important to note that the neutral stresses cannot have any shearing components.

Now assume the water table as if it is at the ground level and consider a cross section at a depth of h, below the ground level. Further, assume that the soil has a specific gravity of G, a void ratio of e and degree of saturation S as unity. Then, at this cross-section under consideration :

Total stress $\qquad = \dfrac{G+e^{1}}{1+e} \gamma_w \, h = \sigma$

and neutral stress $\qquad = \gamma_w \, h = u$, as

indicated in the preceding paras.

By equation (5·9), the effective stress or intergranular pressure $\bar{\sigma}$ is given by

$$\bar{\sigma} = \sigma - u = \frac{G+e}{1+e} \gamma_w \, h - \gamma_w \, h$$

or $\qquad\qquad\qquad \bar{\sigma} = \dfrac{G-1}{1+e} \gamma_w \, h \qquad ...(5\cdot10)$

The element $\dfrac{G-1}{1+e} \gamma_w$ in the above equation is nothing

but the submerged unit weight of the soil, as indicated in chapter 2. The above considerations indicate that *the effective or neutral stress at a point in the ground below the water table, depends upon the submerged unit weight of the soil.*

---

1. Degree of saturation S in unity.

Example 5·3. A sand deposit 12 metres thick has a specific gravity of solids as 2·70 and a void ratio e of 0·75. It overlies a bed of soft clay. The ground water table normally stands 3 metres below the ground level and the average degree of saturation above the water table is 50 per cent.

In connection with the construction of a building, the ground water table will be lowered by another 6 metres. Plot the diagrams showing the variation in the total, intergranular and pore pressure (*i.e.* $\bar{\sigma}$, σ and u) for the following condition :

(a) The original conditions.

(b) After the ground water table has been lowered, assuming that the sand from 3 to 9 metres is fully saturated.

Solution. The mass unit weight of a soil is given by

$$\gamma = \frac{G+Se}{1+e} \gamma_w$$

In the c.g.s. units,

$$\gamma_w = 1 \text{ gm./c.c.}$$

$\therefore$

$$\gamma = \frac{G+Se}{1+e} \text{ gms./c.c.}$$

or

$$\gamma = \frac{G+Se}{1+e} \times \frac{(100)^3}{1000} \text{ kgs./c. metre.}$$

For

$$S = 50\%.$$

$$\gamma = \frac{2·70+0·5\times0·75}{1+0·75} \times 10^3 \text{ kgs./c. metre}$$

or

$$\gamma = \frac{3·075}{1·75} \times 10^3 \text{ kgs./c. metre}$$

$$= 1760 \text{ kgs./c. metre,}$$

For

$$S = 100\%,$$

$$\gamma = \frac{2·70+1·0\times0·75}{1+0·75} \times 10^3 \text{ kgs./c. metre}$$

or

$$\gamma = \frac{3·45}{1·75} \; 10^3 \text{ kgs,/c. metre}$$

$$= 1970 \text{ kgs./c. metre.}$$

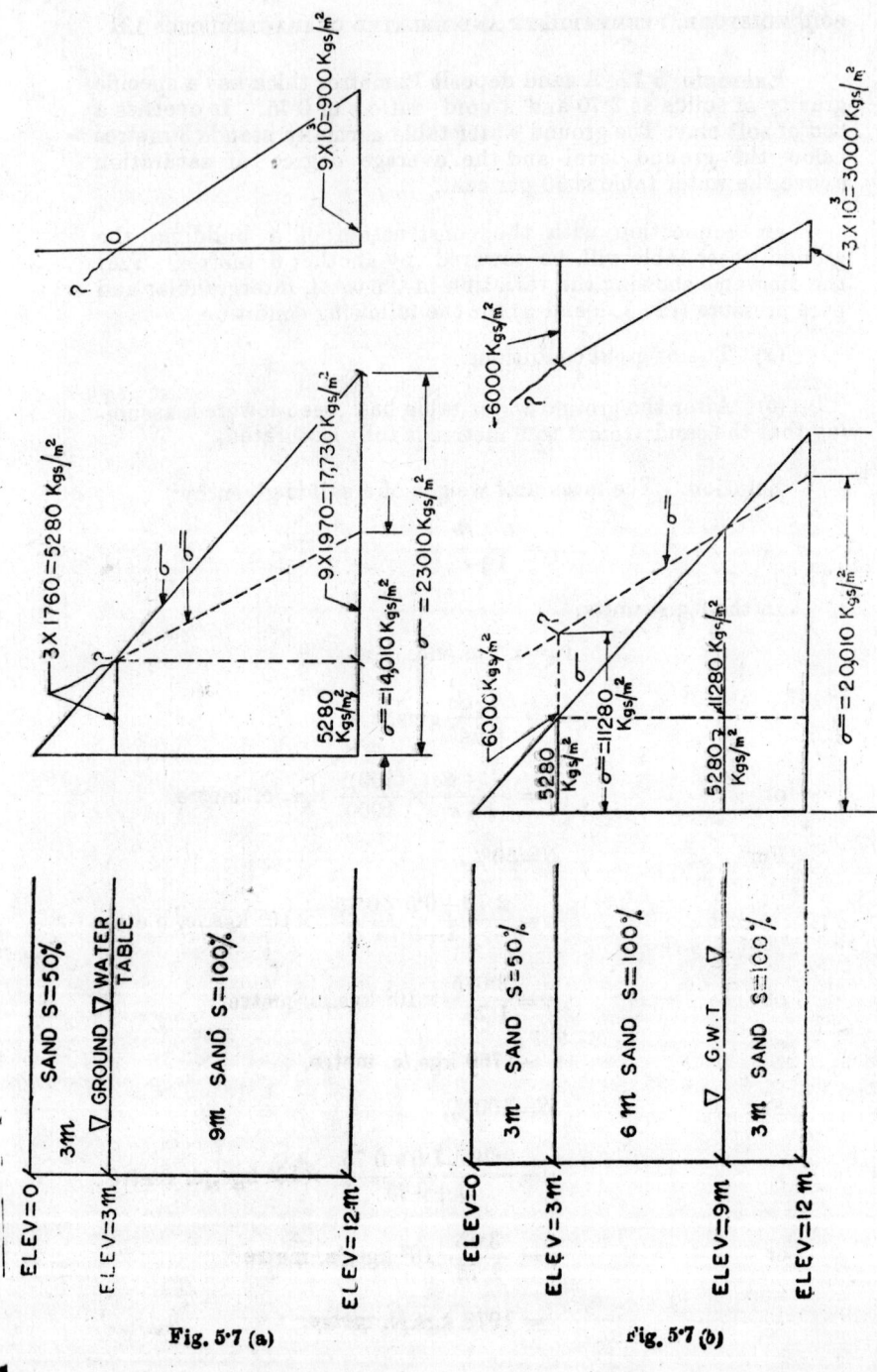

Fig. 5·7 (a)

Fig. 5·7 (b)

(In Fig. 5·7 (a), slope of pressure line above and below elev. 3 m. for the total pressure σ are different, although in the figure, the difference cannot be significantly presented. Same is true for the total pressure line (σ—line) above and bolow elev. 9 m. in Fig. 5·7 (b).

The pressure diagrams for the two situations are drawn as shown in Fig. 5·7. The stress at any point is obtained by multiplying γ corresponding to the appropriate degree of saturation with the depth at which the pressure or stress is sought.

Comparison of the water pressure figures for the two situations indicates that a lowering of the water table induces negative porepressure in the layer between Elev. 3 m and Elev. 9 m and consequently the inter granular stress increases as seen by the comparison of σ̄ diagrams for the two situations.

It may be indicated here that the pore pressures above Elev. 3 m in (a) and ·b) cannot be predicted and hence the inter granular pressures above this level are also doubtful. In case any negative pressure exists above this level, the intergranular pressure will naturally be increased by an equal amount.

Example 5.4 The water table at a building site is at a depth of 6.0 meters below the ground surface. The soil strata has an average Liquidity Index of 0.75 and $L.L.=60\%$ and $P.L.=38\%$. Assume saturated soil conditions and a specific gravity of solids as 2.74.

Calculate the effective stresses at a depth of nine meters below the ground level. Draw a profile of total, effective and neutral stresses to this depth. Note neutral stress is the same as water pressure.

Solution : By definition

Plasticity Index=Liquid Limit—Plastic Limit

In symbolic form

$I_w=L_w-P_w$, all quantities in percentages

$I_w=60-38=22\%$

Again by definition,

$$\text{Liquidity Index} = \frac{W_n-P_w}{I_w} = I_L$$

where $\qquad W_n=$natural moisture content in the field.

$\qquad\qquad P_w=$Plastic Limit as above

and $\qquad I_w=$Plasticity Index

But it is given that $I_L = 0.75$,

$$P_w = 38\% \text{ and } I_w = 22\% \text{ (as above)}$$

$$\therefore \quad 0.75 = \frac{W_n - 38}{22}$$

$$W_n = 22 \times 0.75 + 38 = 16.5 + 38 = 54.5\%$$

or natural moisture content of the soil is $54.5\% w$ (say).

The relationship between the degree of saturation, void ratio, specific gravity and natural moisture content is given by

$$S.e = G.w \qquad \text{where } S, e, G \text{ and}$$

w have their usual significance.

Since the soil is saturated $S = 1.0$

Also $G = 2.74$

Using the value of w as found in the preceding paragraphs

$$e = \frac{G.w}{S} = 2 \cdot 74 \times .545 = 1.495$$

Now γ_{sat} = unit saturated weight

$$= \frac{G + e}{1 + e} \gamma_w$$

where γ_w is the unit weight of water and the other symbols have their usual significance.

Take $\gamma_w = 1$ gm/c.c.

$$\gamma_{sat} = \frac{2.74 + 1.495}{1 + 1.495} \times 1$$

$$= \frac{4.235}{2.495} = 1.695 \text{ gms/c.c.}$$

Also below the water table $\gamma_{sub} = \gamma_{sat} - \gamma_w$

$$\gamma_{sub} = 1.695 - 1.0 = 0.695 \text{ gms/c.c.}$$

Now the water table lies at a depth of six meter below the ground surface.

$\therefore$ up to six meters $u = 0$ in the

equation $\bar{\sigma} = \sigma - u$ where

$\bar{\sigma}$ = Effective stress at any point

σ = Total stress at any point

and u = water pressure

∴   up to six meters $\bar{\sigma} = \sigma$ *i.e.* the effective stress
equals the total stress.

∴ $\bar{\sigma}_6 = \sigma_6 = 1.695 \times 600 = 1017$ gms/cm²

where $\bar{\sigma}_6$ and σ_6 represent respectively the effective and
total stresses at six meters below the ground surface.

Below six meter, water is present.

Water pressure at six meters = Nil

Water pressure at nine meters = $1.0 \times (300)$

$$= 300 \text{ gms/cm}^2$$

$U_9 = 300$ gms/cm²

Total stress at nine meters $= 1.695 \times 900$

or σ_9 $= 1525.5 - 300$

$$= 1225.5 \text{ gms/cm}^2$$

Ans.

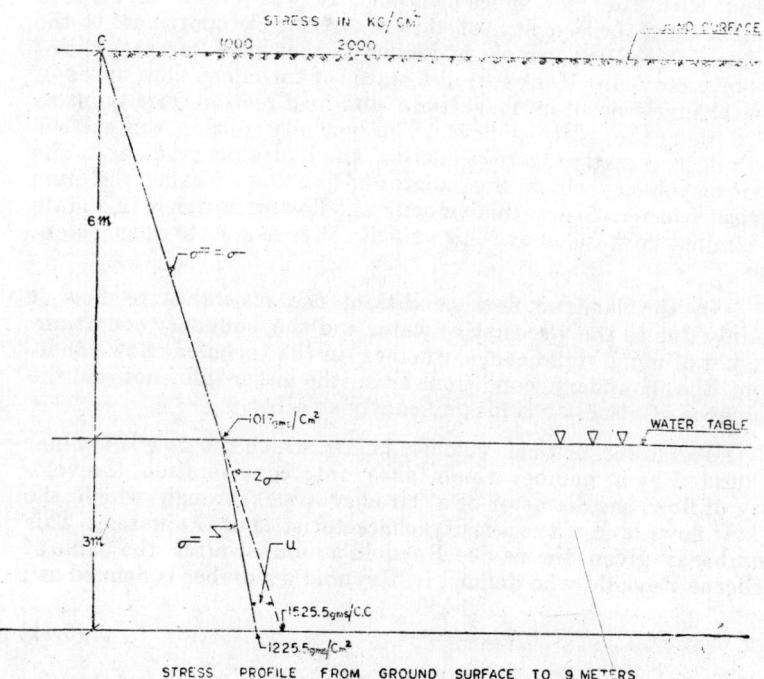

STRESS PROFILE FROM GROUND SURFACE TO 9 METERS

Fig. 5.8.

Fig. 5·8 shows the stress profile from the ground surface to
a depth of nine meters.

Check : To the effective stress $\bar{\sigma}_6$ at six meters, add the stress induced due to the submerged unit weight between elevations of six meters and nine meters.

$$\bar{\sigma}_9 = \sigma_6 + \text{effective unit weight between six and nine}$$
$$\text{meters} \times 300$$
$$= 1017 + \cdot 695 \times 300$$
$$= 1017 + 208.5$$
$$= 1225.5 \text{ gms/c.c.}$$

Ans.

5 7 Flow of water in soils

In a mass of water flowing through a pipe two types of flows called the *laminar flow* and *turbulent flow* exist. In the state of **laminar flow** a particle of water starts at a point, follows a well-defined path and does not cross the paths of other particles ; *whereas* in the state of **turbulent flow**, particles of water follow ill-defined paths and cross the paths of other flowing particles. Osborne Reynold had investigated the conditions that are associated with the two types of flows. It was found that for a laminar flow, the velocity of flow is directly proportional to the **hydraulic gradient**, which is defined as the head lost per unit of distance moved. However, in a state of turbulent flow, speed of flow changes both in magnitude and in direction from point to point along the path of flow. The velocity during this state is no longer directly proportional to the hydraulic gradient. The limit of velocity which demarcates he two flows is called the *lower critical velocity*. Below this velocity the flowing water is in a state of laminar flow and above this velocity it is in a state of turbulent flow.

In the laminar flow condition, the resistance to flow is mainly due to the viscosity of water and the boundary conditions are not of much significance, whereas in the turbulent flow condition, the boundary conditions have the major influence and the influence of viscosity is insignificantly small.

The lower critical velocity below which the flow is laminar is limited by a number which takes into consideration, the velocity of flow, the diameter of a circular pipe through which the water flows and the viscosity characteristics of the water. This number is given the name Reynold's number after the name of Osborne Reynold who defined it. Reynold's number is defined as :

$$N = \frac{V.D.}{v} \qquad \qquad ...(5.11)$$

where V is the velocity of flow, D is the diameter of the pipe through which the water flows and v is the *Kinematic Viscosity* of the liquid, in this case, water. **Kinematic viscosity** is the viscosity of water per unit mass density and hence is equal to, $v = \dfrac{\mu g}{\gamma_w}$ where μ is the viscosity of water, g is the acceleration due to gravity and γ_w is the unit weight of water. So the Reynold's number may also be written as,

$$N = \frac{V.D\gamma_w}{\mu.g} \qquad \qquad ...(5.12)$$

For laminar flow, the Reylond's number must not exceed 2000.

Conditions of flow of water in an individual soil pore cannot be studied for obvious reasons. Only average conditions existing in a soil mass can be studied and this, of course, is the main concern of a civil engineer who has to deal with the *seepage in soil* problems in the field. Pores of most soils are usually very small and the flow through them is usually laminar ; except in case of some coarse-grained soils, in which the flow may be turbulent. For uniform size materials, the laminar flow may be assumed upto an equivalent particle diameter of 0·5 m.m.

For most of the soils, velocity of flow V and the hydraulic gradient have a direct proportion

i.e., $\qquad \qquad V \infty i \qquad \qquad ...(5.13)$

5·8. Darcy's Law

H. Darcy in 1856 demonstrated experimentally that the rate of flow in isotropic homogeneous[1] soils is proportional to the hydraulic gradient. This law of flow has been named after him as Darcy's Law and can be stated as,

$$q = A.k.i \qquad \qquad ...(5.14)$$

where q is the rate of flow through a total cross-sectional area A of the soil mass, k is a constant and i in the hydraulic gradient. Since q is the rate of flow and A is the cross-sectional area, the quotient must represent the velocity of flow at that cross-section.

$\therefore \qquad \qquad V = \dfrac{q}{A} \qquad \qquad ...(5.15)$

Substituting in (5·14), $V = k.i.$ $\qquad \qquad ...(5·16)$

Comparison of equations (5·13) and (5·16) shows that k in equation (5·16) is a constant of proportionality. This constant

---

1. A material is said to be isotropic if its properties at a point within its mass are the same in all directions and it is said to be homogeneous, if its properties at different points within its mass are the same.

of proportionality in equation (5·16) is called **Darcy's coefficient of permeability** of the soil mass. The property to allow the flow of water through the soil mass varies from one soil to another and this Darcy's coefficient of permeability provides a means of comparison. This coefficient is almost universally applied and as such the word "Darcy's" is usually omitted and it is simply called the *coefficient of permeability* of soil or even as *permeability* of soil.

The hydraulic gradient *i* represents the head lost in a unit distance travelled . So if *h* is the head lost in a distance *l*, h/l gives the hydraulic gradient and it is a diemensionles quantity, as both *h* and *l* have he units of length: So from equation (5·16) it can be seen that has the dimensions of velocity. If *g* in (5.14) is represented n cubic centimetres per second and *A* in square centimetres, then *k* is represented in centimetres per second.[1]

As stated above, *A* is the total cross-sectional area of the soil and this total area includes the area of solids and area of voids. The water, in fact, flows only through the void area of the cross-section. As such, the velocity *V* represented by equation (5·15), as the quotient of *q* and *A*, must, therefore, be the *superficial velocity*. The actual velocity of flow is different from this superficial value.

If A_s represents the area of solids in the cross-sectional area *A*, then the area of voids denoted as *A*, would be given by,

$$A_v = A - A_s$$

The void ratio *e* and porosity η of the soil mass were defined in article 2·3 of this book as volumetric quantities. However, from the above discussion. it is possible to write down the void and porosity in terms of cross-sectional areas too. So,

$$\text{void ratio } e = \frac{A_v}{A_s} \qquad \qquad ...(5·17)$$

and $$\text{porosity } \eta = \frac{A_v}{A} \qquad \qquad ...(5·18)$$

Now the discharge *q* per unit time is given by (5·15) as

$$q = A.V$$

where *V* is the superficial velocity. If V_s represents the actual velocity of flow through the void area A_v, even then

$$q = A_v. V_s$$

---

1. In the C.G.S. system, *k* is usually mentioned as 10^{-4} cms. per second or microns per second. In the F.P.S. system it is usually measured as ft./day.

Equating the above two equations,
$$A.V = A_v.V_s$$
But A_v, from equation (5·18) is $\eta.A$.

$$\therefore \qquad A.V = \eta.A.V_s$$

$$\therefore \qquad V = \eta.V_s \qquad\qquad ...(5.19)$$

which gives the relationship between the superficial velocity V and the actual velocity V_s. This actual velocity of flow is called the **seepage velocity.**[1] It can be seen from equation (5·19) that the superficial velocity of flow given by the Darcy's Law is only a fraction of the actual seepage velocity, since η the porosity of the soil is always a fraction.

5·9. Determination of Coefficient of Permeability

As described in the last article, the coefficient of permea-bility varies from one soil to another. In general, it varies with the increasing size of voids, which in turn increases with the increasing particle size. Shape of the soil grains has also a marked effect on the value of k. Co-efficient of permeability k is also a function of unit weight of water and its viscosity as shown in next article 5·10 which in turn depends upon the temperature. However, as can be seen from Table 3·1, temperature viscosity of water, will not make too much difference in the value of k. Coeffi-cient of permeability for all practical purposes, is, therefore, taken as the property of the soil.

Any apparatus used for determination of the coefficient of permeability is called a **permeameter.** The coefficients of permeability of relatively permeable soils such as gravels, coarse sands, sand silt mixtures and coarse silts are found with the help of *constant and variable head permeameters*, whereas those of rela-tively impermeable soils such as fine silts and silty-clays are found with the help of consolidometer permeability apparatus and those of highly impervious soils such as homogeneous clays are better determined indirectly from the consolidation tests for such soils. The indirect determination of the value of k in case of clean sands and gravels is made, sometimes, from empirical relationships such as the one given by Allen Hazen, commonly called Hazen's for-mulae given in equation (5·26) in the next article, and in case of some coarse and fine sands it is determined indirectly from the horizontal capillarity test described in article 5·14. The values of k got from the capillarity tests are usually rough estimates. The constant and variable head permeameter methods are described in this article. The consolidometer permeability appa-ratus is considered outside the scope of this book. The indirect method of determining the coefficient of permeability of clays, from consolidation tests is given in Chapter. 7.

---

1. This is also sometimes called percolation velocity.

(a) **Constant Head Permeameter.** Many designs of this apparatus are available in the market. The principle on which

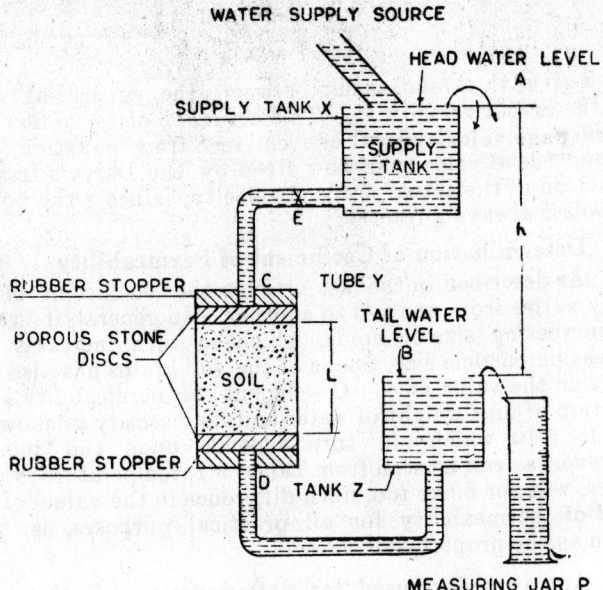

Fig. 5·9.¶ Constant Head Permeameter.

they are designed is the same and can be explained with the half of Fig. 5·9 In the arrangement shown in the figure, the water from a supply tank X flows into a tube Y, which contains the soil sample for which the co-efficient of permeability is desired and after passing through this tube the water goes to another tank and overflows into a measuring jar P. The soil in the tube is contained between two porous stone discs. It must be ensured that the permeability of these discs is much more then the expected permeability of the soil, so that permeability of these discs is much more than the expected permeability of the soil, and there is no restriction to the free flow of water into and out of the soil sample. In fact, there should be no restriction to the flow of water in any part of the apparatus and the apparatus should be able to supply much more water than can flow through the soil.

A water supply source keeps the head water level A of supply tank constant. Since the water flows out at the tail water level B of tank Z, if the positions of tank X and Z are kept fixed, the difference h in the head water level A and tail water level B would be constant. It is because of this constant head h that the apparatus is called a *constant head permeameter*. The permeability co-efficient depends upon the void space enclosed by the soil grains and hence would depend upon how the material is put into this tube. The soil in a dry state can either be put in the tube in a loose[1] state by gently pouring the soil or in a dense state, by compacting after placement. It should be put in this tube in the state in which the permeability is desired. After placing the soil in the tube, with the bottom porous disc in position, water is gently passed through it in the upward direction so as saturate the soil sample thoroughly and to remove all air bubbles. The air bubbles must be removed very gently so as not to disturb the original void space that the particles enclose. Presence of any air bubbles reduces the quantity of flow and hence affects the coefficient of permeability. Connections C and D are made as shown, as soon as the de-aired soil sample is ready for testing. Before the water is allowed to flow from the supply tank X into the tube Y and to the tank Z, the levels A and B of the tanks are checked in order to see that the exact difference h exists between the head and the tail water levels. The stop cock E, when opened, allows the water to flow into the soil and to the measuring jar placed at the tail end.

If Q is the quantity of water that flows through the soil sample of length L and cross-section area A in time t, then from Darcy's Law,

$$Q = A.k.i.t$$

$$\therefore \qquad k = \frac{Q}{A.i.t} \qquad (...5.20)$$

The time t needed to collect the quantity Q is found out with a stop watch and i the hydraulic gradient is given by h/L where L is length in the test sample. So all the quantities on the right hand side of equation (5.20) are known and hence the value of k can be determined.

(b) **Variable Head Permeameter.** Principal parts of a variable head type permeameter are type shown in Fig. 5.10. It essentially consists of a tube Y of the type shown in the constant head permeameter, although the size of this tube is usually smaller than the one used in the previous apparatus. This tube is connected to a burette X on one side and a tank Z on the other side. The burette X supplies the water to the soil and the water after passing through the soil goes to the tank Z. The head of water in the burette X above the constant water level B of tank Z varies from a value h_1 to h_2, the difference usually is 60 cms. It is because of this variable head that the apparatus is known as the *variable head permeameter*.

---

1. For loose densities, the material may be deposited with the help of a funnel to which a flexible tube is attached so that the material falls the least distance. For larger densities the material is deposited and then compacted or when it is a fine material, it is deposited under water.

This apparatus more suitable for test on soils of low permeability such as silts and silty clays. However, it can be used over a wide range of soils because the dimensions of the apparatus can be adjusted so as to carry out the observations of head and time more accurately over a wide range of the co-efficients of permeability. If a soil is more permeable, it will need more water and a bigger diameter burette can be used. If it is less permeable, a smaller diameter burette can be employed, as it will need less quantity of water.

The soil sample is prepared in the same way and held between two porous stone discs as shown. The saturation of the sample is done very carefully in the same manner as for the constant head permeameter soil sample. Sometimes it is useful to provide a stop cock F above the rubber stopper of the tube so as to allow any entrapped air to pass when the tube is in position. After the de-airation is complete, the stop cock F is

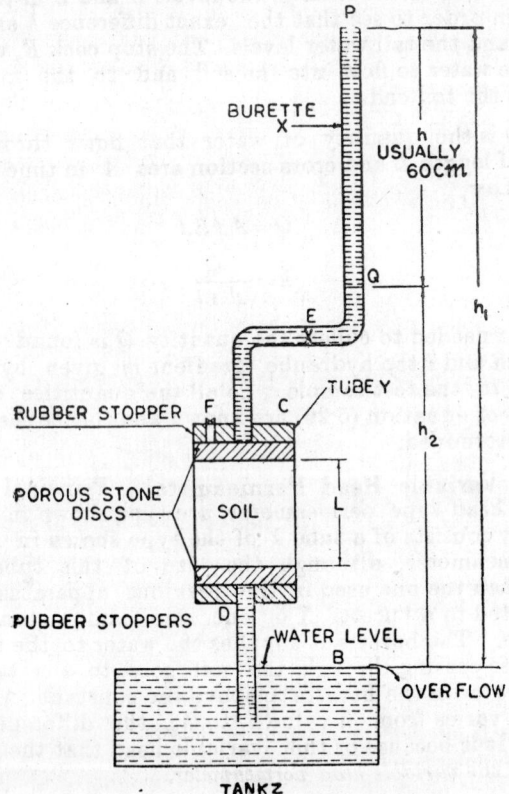

Fig. 5·10.   Variable Head Permeameter.

closed and the stop cock E is opened which allows the water in the burette to fall from a head h_1 to h_2 at which stage the stop cock E is closed. The time t required by the level of water to fall from P to Q i.e. a distance equal to (h_1-h_2) is noted on a stop watch.

If "a" is the inside area of the burette X and $\left(-\dfrac{dh}{dt}\right)$ is the rate of the fall of the head, then the rate of flow in the burette is $a\left(-\dfrac{dh}{dt}\right)$

The negative sign with the rate of fall of head only indicates the fact that the head is decreasing with the increase in time. If the difference (h_1-h_2) is denoted by h, then from Darcy's Law, the rate of flow in the soil is : $q=Ak.\dfrac{h}{L}$ where A is the area of the tube Y containing the soil, L the lentgh of this soil sample, k is the co-efficieut of permeability and $\dfrac{h}{L}$ the hydraulic gradient.

The above two quantities i.e. rate of flow in the burette and that in the soil must be equal.

$$\therefore \qquad -a\frac{dh}{dt}=A.k.\ \frac{h}{L}$$

or
$$-a\frac{dh}{h}=\frac{A.k.}{L}.\ dt$$

Integrate the above equation between the limits $h=h_1$ and $h=h_2$ corresponding to times $t_1=0$ and $t_2=t$ Then,

$$-a\int_{h_1}^{h_2}\frac{dh}{h}=\frac{A.k.}{L}\int_{t_1}^{t_2}dt=\frac{A.k.}{L}\int_0^t dt$$

or
$$-a\left[\log_e h\right]_{h_1}^{h_2}=\frac{A.k.}{L}\left[\ t\ \right]_0^t$$

or
$$a.\ \log_e\frac{h_1}{h_2}=\frac{A.k.}{L}.\ t$$

$$\therefore \qquad k=\frac{a.L}{A.t}.\ \log_e\frac{h_1}{h_2}$$

or
$$k=2{\cdot}303\frac{a.\ L}{A.t}\log_{10}\frac{h_1}{h_2}\qquad\qquad ...(5{\cdot}21)$$

All quantities on the right hand side equation (5·21) are known and hence it gives the coefficient of permeability for the soil.

As a general comment on the values of the coefficients of permeability obtained in the laboratory by these or any other procedures, it may be said the values so obtained are at best rough estimates of the coefficients of permeability of natural soil deposits.

Example 5.5 (a) Soil sample in a constant head permeameter is 4 cms in diameter and 12 cms long. Under a head of 20 cms the discharge was found to be 60 c.c. in 15 minutes. Calculate the coefficient of permeability of the soil.

(b) What is the type of the soil used in the above test ?

(P.U. April, 1965)

Solution. (a) By Darcy's Law.

$$Q = A.k.i.t$$

where the various terms have their usual signifiance.
Hence in this problem,

$$Q = 60 \text{ c.c.}$$

$$A = \pi(2)^2 = \pi = 12 \cdot 58 \text{ cm}^2$$

$$k = \text{Required quantity}$$

$$i = \frac{h}{L} = \frac{20}{12}$$

and

$$t = 15 \text{ minutes} = 15 \times 60 = 900 \text{ seconds,}$$

∴

$$k = \frac{Q}{A.i.t} = \frac{60 \times 12}{12 \cdot 58 \times 20 \times 900} \text{ cms/sec.}$$

or

$$k = 3 \cdot 16 \times 10^{-3} \text{ cms/sec.}$$

(b) On the A-Casagrande's Scale, this value of k falls between 10^{-2} cms/ sec and 10^{-3} cms/sec.
Most probably the soil used is clean sand with a little admixture of fine gravel.

Example 5·6. A falling head permeameter is used to determine the coefficient of permeability of soil sample 5 cms in diameter and 10 cms long. The water level in the burette above the the tail water level in the reservoir below falls from a height of 45 cms to 9 cms in exactly 5 minutes. The inside diameter of the burette is 0·5 cms. Find out the coefficient of permeability for this soil, What do you infer about the soil type.

Solution. Equation (5·20) gives

$$k = \frac{2 \cdot 303 aL}{A.t} \log_{10} \frac{h_1}{h_2}$$

where the various terms have their usual significance.
Here in this problem,

$$a = \frac{\pi}{4}(0.5)^2 = \frac{0.25\pi}{4} \text{sq. cms.}$$

$$A = \frac{\pi}{4}(5)^2 = \frac{25\pi}{4} \text{sq. cms.}$$

$$t = 5 \times 60 = 300 \text{ seconds.}$$
$$L = 10 \text{ cms.}$$

$$h_1 = 45 \text{ cms. and}$$
$$h_2 = 9 \text{ cms.}$$

$$\therefore \quad k = 2.303 \times 0.25 \frac{\pi}{4} \times \frac{4}{25\pi} \times \frac{10}{300} \log_{10} \frac{45}{9}$$

$$\therefore \quad k = 2.303 \times \tfrac{1}{3} \log_{10} 5 \times 10^{-3} \text{ cms/sec.}$$

or $$k = 2.303 \times \tfrac{1}{3} \times 0.6989 \times 10^{-3}.$$

$$= 4.363 \times 10^{-4} \text{ cms/sec.}$$

The soil is clean sand and gravel mixture. **Ans.**

Example 5·7 In a variable Head permeameter, the Darcy's coefficient of permeability was 1×10^{-4} cms/sec. If the head varied from 35 cms to 25 cms in a stand pipe 1 cm sq while draining through a sample of 10 cms depth, and of cross-sectional area 6 cm², what is time required in minutes.

(Mysore, 1965)

Solution. Equation (5·21) gives

$$k = \frac{2.303 \, aL}{A.t} \log_{10} \frac{h_1}{h_2}$$

or $$t = \frac{2.303 aL}{A.k} \log_{10} \frac{h_1}{h_2}$$

where the various terms have their usual significance.

Here in this problem,

$$k = 1 \times 10^{-4} \text{ cms/sec.}$$

$$h_1 = 35 \text{ cms.}$$

$$h_2 = 25 \text{ cms}$$

$$L = 10 \text{ cms}$$

$$A = 6 \text{ cm}^2$$

$$a = 1 \text{ cms}^2$$

$$t = \frac{2 \cdot 303 \times 1 \times 10}{6 \times 1 \times 10^{-4}} \log_{10} \frac{35}{25}$$

$$= \frac{2 \cdot 303 \times 10 \times 10^4}{6} \log_{10} 1 \cdot 4$$

$$= \frac{2 \cdot 303 \times 10^5 \times 0.146}{6}$$

$$= \frac{0 \cdot 336 \times 10^5}{6}$$

$$= 5660 \text{ Secs} = 94.5 \text{ minutes}$$

Ans.

5.10. Factors Affecting Coefficient of Permeability

A short reference was made to the factors that govern the coefficient of permeability of a soil, in the beginning of article 5·9. A discussion of these factors is presented in this article.

It was stated in article 5·7 that the flow of water in soils is laminar in nature. Poiseuille's equation[1] for the laminar flow through a round capillary tube may be written as

$$q = \frac{\pi . R^4 \gamma_w . i}{8 . \mu} \qquad \ldots (5.22)$$

where q is the rate of flow, R is the radius of the tube, γ_w is the unit weight of water, i the hydraulic gradient under which the flow takes place and μ the viscosity of the water. If the area of the tube πR^2 is represented by "a" and the hydraulic mean radius of the tube, defined as the ratio of area to the wetted permimeter, by m, i.e.

$$\pi R^2 / 2\pi R = R/2,$$

then the equation may be reduced to

$$q = \tfrac{1}{2} m^2 . a . i . \frac{\gamma_w}{\mu} \qquad \ldots (5 \cdot 23)$$

If the cross-section of the tube through which the water flows is not circular, by anology, it can be inferred that the equation for the rate of flow be similar to equation (5·23), the only variation being the numerical coefficient on the right handside. So, for th cross section of any other type, the rate of flow in a tube will b given by,

$$q = C_s m^2 . a . i \frac{\gamma_w}{\mu} \qquad \ldots (5 \cdot 2\,\text{t}$$

---

1. For derivation of this equation the reader may refer to any goo book on hydraulics.

where C_s is the coefficient depending upon the type or shape of the cross-section and hence may be called the *shape factor*.

Flow through soils is much more complicated than it is through regular tubes of any cross-sectional shape. In the first place the path of flow in soils is zigzag depending upon the configuration of particles in a soil mass, compared to much smoother path in tubes or pipes and secondly the concept of mean hydraulic radius for soils, presented in a subsequent paragraph, is based upon some hypothetical grain diameter. However, equation (5·24) can be used to indicate the factors on which the coefficient of permeability depends.

In the above equation "a" represents the area of the opening of the pipe. In the case of soils, the water flows through the pores and if A is the total cross-sectional area of a soil section as used in the Darcy's equation (5·14), the area of voids A_v is given by equation (5·18), as ηA. This $A_v = \eta A$ is the same as the area "a" of the pipe. η can be replaced by $\dfrac{e}{1+e}$ so that A_v the area of voids $= \dfrac{e}{1+e} A = $"$a$". Also the hydraulic mean radius "m" for pipes used in the equation is the ratio of area to the wetted perimeter *i.e.* $\dfrac{a}{p}$ where "a" denotes the cross-sectional area of the tube and p its perimeter. If the numerator and denominator in the ratio $\dfrac{a}{p}$ are multiplied by the length "l" of the pipe, the hydraulic mean radius may be defined as the ratio of the volume of flow channel to the surface area of the flow channel. This definition can be used to derive an expression for hydraulic mean radius for soils.

The volume of the flow channel in soils may be taken as the pore volume or the volume of voids which is equal to $e.V_s$ where e is the void ratio and V_s is the volume of solids ; and the surface area A_s for the soil grains may be worked out on the basis of a hypothetical spherical grains of diameter "d" having the same volume to surface-area ratio as the soil grains collectively have.

So, m for soils $= \dfrac{eV_s}{A_s} = e\ \dfrac{\pi d^3/6}{\pi d^2} = \tfrac{1}{6} d.e.$

Further the coefficient C_s in equation (5·24) may be replaced by a more general coefficient C_1 for soils to take into consideration the factor C_s and any other shape, grain size and configuration effects of the soil particles.

Substituting the values of 'a' and 'm' as found for soils and replacing C_s by C_1, equation (5·23) becomes

$$q = C_1 \left(\frac{1}{6} d.e \right)^2 . \left(\frac{e}{1+e} A \right) . i . \frac{\gamma_w}{\mu}$$

or

$$q = \left(C.d^2 . \frac{e^3}{1+e} . \frac{\gamma_w}{\mu} \right) i . A \qquad \dots (5·25)$$

where C absorbs the fraction 1/36.

If equation (5·25) is compared with the Darcy's equation (5·14), k can be written as

$$k = C.d^2 . \frac{e^3}{1+e} . \frac{\gamma_w}{\mu} \qquad \dots (5·26)$$

Equation (5·26) gives a clear indication of the factors on which the value of the coefficient of permeability for a soil depends. Conculsions from equation (5·26) can be enumerated as under :

(1) The permeability of a soil is a function of the shape of the soil pores which further depends upon the shape and size of the soil particles and their arrangement in the mass. In other words, the coefficient of permeability depends upon the structure of the soil mass represented by the factor C.

(2) d in the equation, though a hypothetical grain diameter is a measure of the grain size of the soil. It can be concluded, therefore, that the value of k depends upon the second power of some measure of the grain size of the soil. It is interesting to note that Allen Hazen (8) found that the permeability of filter sands could be represented by

$$k = 100 D_{10}^2 \qquad \dots (5·27)$$

where both k and D_{10} are in the c.g.s. units, k being the permeability in cms/sec and D_{10} in cms. being the 10 per cent size on the particle size distribution curve. D_{10} was designated as the effective size of the soil sample in article 3·11. Equation (5·27) gives a fairly good estimate of the permeability of uniform filter sands whose effective sizes lie between 0.1 and 3 m.m. and for which the coefficient of uniformity as given by equation (3·19) does not exceed 5. It may, however, be noted that the above equation only applies to filter sands and may not be good for other soils.

(3) Permeability of a soil depends upon the void ratio as evidenced by the factor $\frac{e^3}{1+e}$. The factor C depends upon the

shape of void space in the soil as stated in conclusion (1) above and consequently may be considered to depend to some extent on the void ratio. So $C.\dfrac{e^3}{1+e}$ represents in the above equation the effect of void ratio.

There is some evidence available at present, which indicates that a plot of e on arithmetical scale and k on log scale turns out to be a straight line. Since e is further a function of the density it can be stated that knowing the coefficient of permeability at two densities, coefficient of permeability at a third density can be found, using a straight line plot.

(4) $\dfrac{\gamma_w}{\mu}$ in the equation (5·26) represents the properties of

pore-water in the voids, γ_w being the unit weight and μ the viscosity of water. Both these quantities change with a change in the temperature, as can be seen from Tables 3·1 and 3·5 in this text. However, it may be noted that the unit weight of water does not change with temperature as much as does the viscosity. That is why sometimes the permeability of the soil determined experimentally at a temperature of $T°C$ is corrected by using equation (5·28) below to a temperature of $20°C$ considered a standard temperature for representing its value, assuming that $\gamma_w = 1$ gm/c.c. for all practical purposes for all temperatures.

$$\frac{k_{20}}{k_T} = \frac{\mu_T}{\mu_{20}} \qquad\qquad ...(5.28)$$

where k_{20} and k_T are the coefficients of permeability at $20°C$ ($68°F$) and at $T°C$ (temperature of the test) respectively and μ_T and μ_{20} are the respective viscosities of water at $T°C$ and $20°C$. As is seen from Table 3·5, the values of the viscosity of water decrease with the increase in temperature. Therefore, the coefficient of permeability increases with increasing temperature. However, as stated earlier also, the temperature correction is not considered essential for using values of k in practical civil engineering works.

(5) Air etrapped in the voids can obstruct the free flow of water in the soil mass. As such, the presence of air can be another factor that affects the value of the coefficient of permeability.

(6) It is a matter of common occurrence that most of the soils and foundation engineering problems involve water as the fluid that fills the voids. If however, some other fluid such as an oil were to fill the voids of the soil such as in the case of oil fields, it is interesting to note that the coefficient of permeability would vary depending upon the type of fluid. Some evidence for research work is presented is Fig. 5.11, which shows that the

reationship between coefficient of permeability and $\frac{e^3}{i+e}$ is depend upon the type of fluid for which coefficient of permeability is being measured.

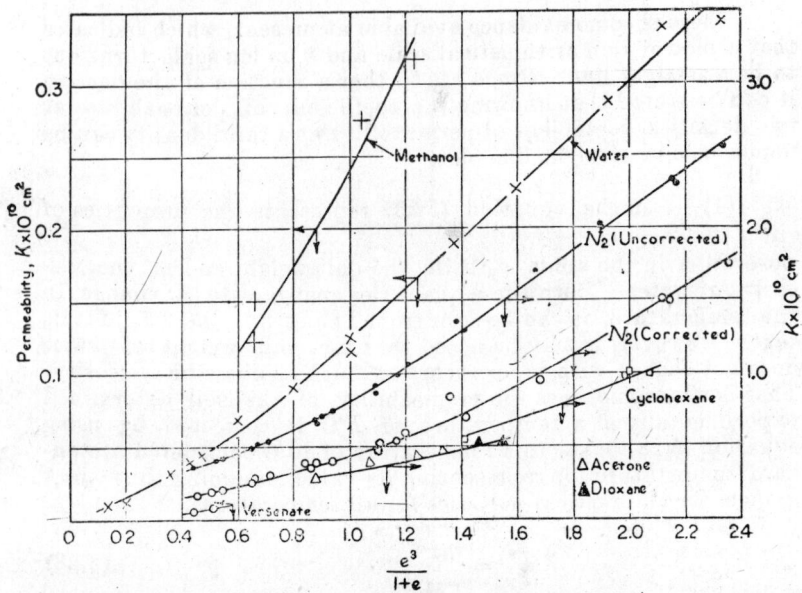

Permeability of kaolivite to various fluids as a function of $e^3/(1+e)$ [After Michaels & Lins]

Fig 5·11

5·11. Permeability of Layered Deposits

Natural deposits of soils usually exhibit stratification. Each soil layer, as such, has a different coefficient of permeability. The coefficient of permeability for the deposit as a whole is different both in the vertical as well as in the horizontal direction, however homgeneous each deposit may be. If the thickness of each layer and its coefficient of permeability is known, the average values of the coefficients of permeability in the vertical and horizontal directions can be found out.

As an illustration, a two-layered deposit may be taken to find out the values of cofficients of permeability in the vertical and horizontal directions. The analyses presented below can be extended to any number of layers in the deposit.

(a) *Coefficient of Permeability in the vertical direction.* Let l_1 and l_2 be the individual thicknesses of a two-layered deposit of total thickness l and k_1 and k_2 be their respective coefficients of permeability as shown in Fig. 5·12 Let k_v be the average coefficient of permeability in the vertical direction for this deposit and $\triangle h$ be the head lost in the total deposit. Then,

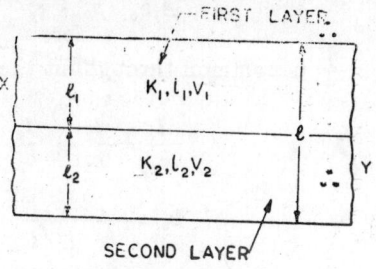

Permeability in vertical and Horizontal directions in a Two Layered System
Fig 5·12

Average hydraulic gradient,

$$i = \frac{\triangle h}{l}$$

Discharge through the deposit

$$q = A.k_v.i.$$

$$= A.k_v. \frac{\triangle h}{l} \qquad \qquad ...(5.29)$$

where A is the area through which the discharge is being considered.

Now, if $\triangle h_1$ and $\triangle h_2$ are the individual head losses in the two layers so that $\triangle h = \triangle h_1 + \triangle h_2$, then the respective hydraulic gradients for the two layers are $i_1 = \frac{\triangle h_1}{l_1}$ and $i_2 = \frac{\triangle h_2}{l_2}$. For continuity of flow, the discharge q is the same through the two layers whether considered individually or collectively.

So q is also equal to $A.k_1.i_1.$ and $A\ k_2.i_2.$

$$q = A.k_1.i_1 = A.k_1. \frac{\triangle h_1}{l_1}$$

∴ $$\triangle h_1 = \frac{q.l_1}{A.k_1}$$

and again $$q = A.k_2.i_2 = A.k_2. \frac{\triangle h_2}{l_2}.$$

∴ $$\triangle h_2 = \frac{q.l_2}{A.k_2}$$

From eqn. (5·29) $\triangle h = \frac{q\,l}{A.k_v}$

But $\triangle h = \triangle h_1 + \triangle h_2$

$\therefore$ $\dfrac{q.l}{A.k_v} = \dfrac{q.l_1}{A.k_1} + \dfrac{q.l_2}{A.k_2}$

$\dfrac{q}{A}$ is common throughout the above equation

$\therefore$ $\dfrac{l}{k_v} = \dfrac{l_1}{k_1} + \dfrac{l_2}{k_2}$

or $k_v = \dfrac{l}{\dfrac{l_1}{k_1} + \dfrac{l_2}{k_2}}$ ...(5·30)

For n number of layers, the analysis can be extended likewise and k_v will be given by,

$$k_v = \dfrac{l}{\dfrac{l_1}{k_1} + \dfrac{l_2}{k_2} + \dfrac{l_3}{k_3} \cdots\cdots + \dfrac{l_n}{k_n}} \qquad ...(5·31)$$

where l is the total thickness of the deposit, $l_1 \, l_2 \ldots l_n$ are the individual thickness and $k_1 \, k_2 \ldots\ldots k_n$ are their respective permeability coefficients.

The effect of two layers of soils is demonstrated by example 5·6.

(b) *Coefficient of Permeability in the horizontal direction.* Refer to Fig 15·10 again. Let the velocities of flow in the individual layers be $V_1 \, V_2$ and in the total deposit be V, in the horizontal direction. If k_h denotes the average coefficient of permeability in the horizontal direction, then

$V = k_h . i$

where i is the hydrualic gradient under which the flow takes place. Since the flow is in the horizontal direction, *say* from the section X to the Section Y, this hydraulic gradient is the same, whether the individual layers are considered or the deposit as a whole is considered.

$\therefore$ $V_1 = k_1 . i$ and
$V_2 = k_2 \, i$

Now, if a unit depth perpendicular to the plane of the paper is taken, then the end areas of individual layers at the sections X or Y, are l_1 and l_2 and the total area of the deposit at the section X or Y is l.

Discharge through the total deposit, $q = V . l = k_h \, i . l$ ...(5·32) and the individual discharges through the two layers are given by, discharge through the 1st layer, $q_1 = V_1 . l_1 = k_1 . i . l_1$ and discharge through the 2nd layer, $q_2 = V_2 . l_2 = k_2 . i . l_2$

But $(q_1 + q_2)$ must equal the total discharge q in equation ...(5·32)

$$\therefore \quad k_1 . i . l_1 + k_2 . i . l_2 = k_h . i . l.$$

Gradient i is common throughout.

$$\therefore \quad k_h = \frac{k_1 l_1 + k_2 . l_2}{l} \qquad \qquad ...(5.33)$$

For n number of layers, the analysis can be extended likewise and k_h will be given by,

$$k_h = \frac{k_1 l_1 + k_2 l_2 + ... + k_n l_n}{l}$$

where l is the total thickness of the deposit, $l_1, l_2 ... l_n$ are the individual thicknesses and $k_1, k_2 ... k_n$ are the respective permeability coefficients.

It may be noted that both the above analysis assume that the layers are of uniform thickness.

In case the layers are not deposited horizontally and are inclined at an angle to the horizontal, k_h may be considered parallel to the bedding planes and k_v normal to them.

Example 5·8. The figure 5·11 shows water flowing two soils of given permeabilities k_1 and k_2. Determine the head h in the

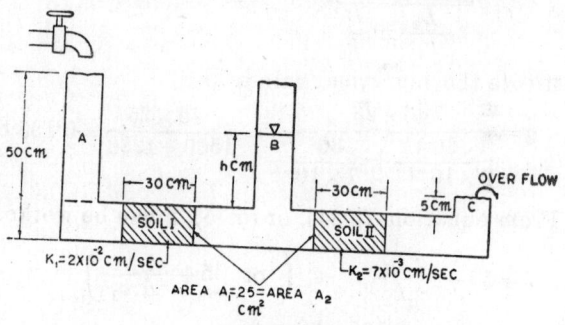

Fig. 5·13

region between the two materials, as indicated on the sketch. The other, quantities i.e. areas, lengths and heads are also shown on the sketch.

Solution. Head lost between the tubes A and B, $\triangle h_1 =$ $(50 - h)$ cms and head lost between the tubes B and C, $\triangle h_2 = (h - 5)$ cms.

Total head loss between tubes **A** and **C**, $\triangle h = \triangle h_1 + \triangle h_2$ = 45 cms.

If L_1 and L_2 denote the lengths of the soils samples (each= 30 cms), then the hydraulic gradients due to which the water flows are given by,

$$i_1 = \frac{\triangle h_1}{L_1} = \frac{(50-h)}{L_1} \text{ for the 1st soil and}$$

$$i_2 = \frac{\triangle h_2}{L_2} = \frac{(h-5)}{L_2} \text{ for the 2nd soil}$$

Darcy's Law says that $q = A\ k\ i$

For continuity of flow, the discharge q is the same through both soils.

$$\therefore \quad i_1 = \frac{50-h}{L_1} = \frac{q}{A.k_1} \qquad (A=A_1=A_2)$$

$$\text{or} \quad (50-h) = \frac{q.L_1}{A.k_1} = \triangle h_1 \qquad \dots(5.34)$$

$$\text{Again } i_2 = \frac{h-5}{L_2} = \frac{q}{A.k_2} \qquad (A=A_1=A_2)$$

$$\text{or} \quad (h-5) = \frac{q.L_2}{A.k_2} = \triangle h_2 \qquad \dots(5\cdot35)$$

But total head loss between A and C, $\triangle h = \triangle h_1 + \triangle h_2$.

$$\therefore \quad \triangle h = \frac{q.L_1}{A.k_1} + \frac{q\,L_2}{A.k_2} = \frac{q}{A}\left(\frac{L_1}{k_1} + \frac{L_2}{k_2}\right)$$

$$\therefore \quad q = \frac{\triangle h.A}{\dfrac{L_1}{k_1} + \dfrac{L_2}{k_2}}$$

substitute the numerical values

$$q = \frac{45 \times 25}{\dfrac{30}{2\times10^{-2}} + \dfrac{30}{7\times10^{-3}}} = \frac{45\times25}{1500+4286} = 0\cdot195 \text{ c.c/sec.}$$

$\therefore$ From equations (5·34), or (5·35), h can be worked out

$$h = 50 - \frac{q.L_1}{A.k_1} \qquad \left(\text{or} \quad 5 + \frac{q.L_2}{A.k_2}\right)$$

$$\text{or} \quad h = 50 - \frac{0\cdot195\times30}{25\times2\times10^{-2}} = 50-11\cdot8 = 38\cdot2 \text{ cms.}$$

Ans.

Note. The reader may check the answer by calculating

$$h = 5 + \frac{q.L_2}{A.k_2}$$

5·12 Determination of Field Permeability

The average permeability of the soil in the field can be determined by a pumping test. Pumping of water is carried out of either an existing well or a well dug into the soil specially for this purpose. In addition, minimum of two observation wells are needed to observe the level of water, as the pumping is under operation. These observation wells are located sufficiently away from the main well. This ensures accurate measurements, since the formula employed in calculations makes some assumptions and to satisfy these assumptions the above step is essential.

Two cases can be met with in the field :—

1. In the *first* case the ground water table lies within test stratum, as shown in Fig 5.14. This case is usually referred to as the case of *unconfined aquifer.*

2. In the *second* case the test stratum is overlain by an impervious layer of soil such as fat clay and the water table lies in the stratum above the impervious layer, or it lies inside the impervious layer or it is just above the stratum. This case is illustrated in Fig 5·15 and is usually referred to as the case of *confined aquifer.*

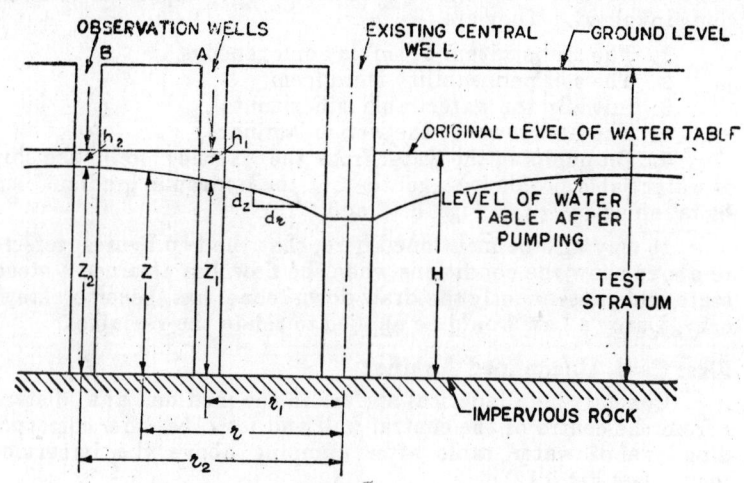

Flow towards wells—Case I, unconfined aquifer
Fig. 5·14

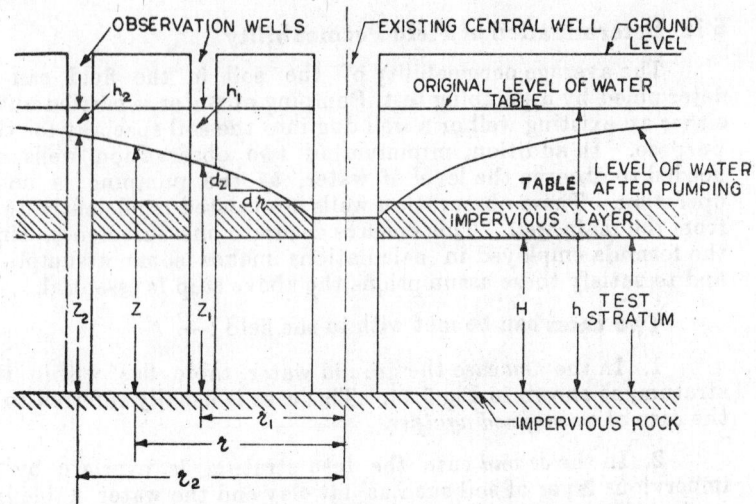

Flow towards wells—Case II, Confined aquifer
Fig. 5·13

Before derivations employed in calculations are made for the two cases, it is essential to mention the important assumptions involved. They are :—

1. The test stratum of soil is homogeneous.
2. The soil permeability is uniform.
3. Initially the water table is horizontal.
4. The flow towards the well is laminar.
5. On pumping the water from the existing well, the slope of water table profile is so gentle that the hydraulic gradient[1] may be taken as dz/dr (see Figs. 5·12 and 5·13).

It may also be mentioned here, that the two figures referred to above show the conditions when the flow has attained a steady state and consequently the draw down cone has become stationary. Darcy's Law would be applied to relate the variables.

First Case, Unconfined Aquifer

Consider a cylinderical surface in the medium at a distance r from the centre of the central well and let z be the corresponding level of water table after pumping above the impervious rock. (see Fig 5·12)

From Darcy's Law, discharge per unit time across this surface is given by,

$$q = A.k.\frac{dz}{dr},$$

---

2. This assumption is called the Dupit's assumption after the name of the person who derived these expressions.

where A is the area of the cylindrical surface $= 2\pi.r.H$, k is the required coefficient of permeability and $\dfrac{dz}{dr}$ is the hydraulic gradient.

$$\therefore \quad q = 2\pi r.z.k.\dfrac{dz}{dr}$$

or

$$k.z.dz. = \dfrac{q}{2\pi}.\dfrac{dr}{r}.$$

Integrate the above equation and let r vary from r_1 to r_2 when z from z_1 to z_1, as shown.

$$\therefore \quad k\left[\dfrac{z^2}{2}\right]_{z_1.}^{z_2} = \dfrac{q}{2\pi}\left[\log_e r\right]_{r_1}^{r_2}$$

or

$$k.\dfrac{z_2{}^2 - z_1{}^2}{2} = \dfrac{q}{2\pi}\log_e \dfrac{r_2}{r_1}$$

or

$$k = \dfrac{q}{\pi(z^2{}_2 - z_1{}^2)}\ \log_e\ \dfrac{r^2}{r_1} \qquad \ldots(5\cdot36)$$

If H denotes the thickness of the test stratum below original level of water table, then from the figure,

$$z_1 = H - h_1$$
and
$$z_2 = H - h_2,$$

where h_1 and h_2 are the respective falls in the water table level in the two observation wells, after pumping. These values are observed in the field, when steady state conditions have reached. All the quantities on the right hand side are known. Hence k can be evaluated.

Second case—Confined Aquifer.

Consider, again, a clylinderical surface in the medium at a distance r from the centre of the central well. (see Fig 5·13)

From Darcy's Law, discharge per unit time across this surface in the test stratum as shown, is given by

$$q = A.\ K.\ \dfrac{dz}{dr},$$

where A is the area of the cylinderical surface $= 2\pi r.H$, k is the required coefficient of permeability and $\dfrac{dz}{dr}$ is the hydraulic gradient.

$$\therefore \qquad q = 2\pi r.H.k.\frac{dz.}{dr}$$

or $$kdz = \frac{q}{2\pi H} \cdot \frac{dr}{r}$$

Intergate the above equation and let r vary from r_1 to r_2 when z varies from z_1 to z_2

$$\therefore \qquad k\left[\,z\,\right]_{z_1}^{z_2} = \frac{q}{2\pi H}\left[\,\log_e r\,\right]_{r_1}^{r_2}$$

or $$k(z_2 - z_1) = \frac{q}{2\pi H}\log_e\frac{r_2}{r_1}$$

or $$k = \frac{q}{2\pi.H(z_2 - z_1)}\cdot\log_e\frac{r_2}{r_1} \quad ...(5.37)$$

If h denotes the original level of the water table above the impervious rock, then, from the figure

$$z_1 = h - h_1$$

and $$z_2 = h - h_2$$

where h_1 and h_2 are the respective falls in the water table level in the observation wells after pumping, as in the first case.

$$\therefore \qquad z_2 - z_1 = h_1 - h_2$$

This value may be substituted in equation (5.27)

$$\therefore \qquad k = \frac{q}{2\pi.H.(h_1 - h_2)}\cdot\log_e\frac{r_2}{r_1} \quad ...(5.38)$$

Values of h_1 and h_2 are again measured in the field when the steady state conditions have reached.

Many engineers prefer to have more than two observation wells so as to arrive at better average results. As an illustration of the use of more then two observation wells refer to Fig. 5·14, in which six observation wells are shown. Each pair of wells A and A' ; B and B', C and C' are symmtrically placed about the centre of the test well. The level of the draw down curve above the impervious rock, in each well and the corresponding fall from the original level are marked. It would be found in the field that $z_1 \neq z'_1$ and $h_1 \neq h'_1$ for the symmetric pair A and A' and so on and

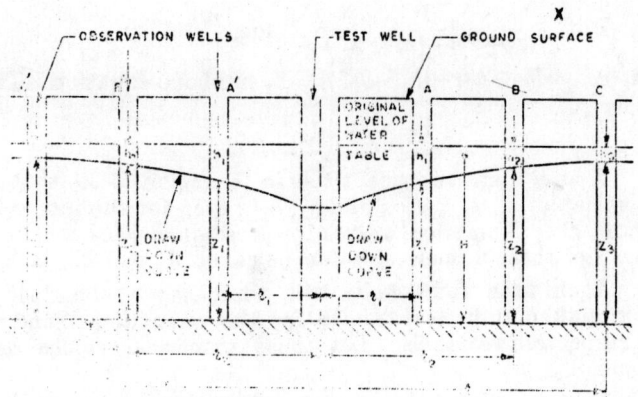

Use of more than two wells to find out the avarage Coefficient
of Permeability k in the field
Fig. 5·6.

and so forth for other symmetric pairs of wells. The data so obtained is processed in the following manner :

Equation (5.36) may be re-written as.

$$k = \frac{q}{\pi(z_2 - z_1)(z_2 + z_1)} \log_e \frac{r_2}{r_1} \text{ for the two wells.}$$

In this equation.

$$z_1 = H - h_1$$
and
$$z_2 = H - h_2$$
$$\therefore (z_2 - z_1) = (h_1 - h_2)$$

Substitute this in the above equation

$$\therefore K = \frac{q}{\pi(z_1 + z_2)(h_1 - h_2)} \cdot \log_e \frac{r_2}{r_1}$$

Now take two pairs of symmetrically placed wells A and A',
B and B' for consideration.

Since $z_1 \neq z_1'$, although radially the two wells are symmetric,
we substitute the average values $\dfrac{z_1 + z_1'}{2}$ for z_1.

Similarly, the average values $\dfrac{z_2 + z_2'}{2}$, $\dfrac{h_1 + h_1'}{2}$ and $\dfrac{h_2 + h_2'}{2}$ are
substituted in the above equation. This gives.

$$k = \frac{q}{\pi \left[\dfrac{z_1 + z_1'}{2} + \dfrac{z_2 + z_2'}{2} \right] \left[\dfrac{h_1 + h_1'}{2} - \dfrac{h_2 + h_2'}{2} \right]} \log_e \frac{r_2}{r_1}$$

or
$$k = \frac{2q}{\pi \left[(z_1 + z_2) + (z_1' + z_2') \right] \dfrac{[(h_1 - h_2) + h_1' - h_2']}{2}} \log_e \frac{r_2}{r_1}$$

Put $\dfrac{1}{(z_1+z_2)+(z_1'+z_2')} \cdot \log_e \dfrac{r_2}{r_1} = X$

and $\frac{1}{2}[h_1-h_2)+(h_1'-h_2')] = Y$, in above equation. This gives

$$k = \frac{2q}{\pi} \cdot \frac{X}{Y} \qquad \qquad ...(5\cdot39)$$

X and Y are two quantities in equation (5·39) that contain the variables z_1, z_1' ; etc., and h_1, h_1' ; etc., for the four wells A, B and A', B'. Numerical values for these quantities can be worked out. Let these numerical value be called X_1 and Y_1.

Again take four wells this time, say A, C and A', C' and work out the numerical results for the quantities X and Y with the respective variables. Let these numerical values be called X_2 and Y_2.

In this manner, taking two symmetric pairs of wells at a time, one can work out the following combinations and the corresponding values of X and Y :

Combination	Value of X and Y
A. B and A', B'	X_1, Y_1
A, C and A', C'	X_2, Y_2
B, C and B', C'	X_3, Y_3.

These values of X and Y are plotted as shown in Fig. 5·15 and a straight line is drawn so as to best fit these points. Then

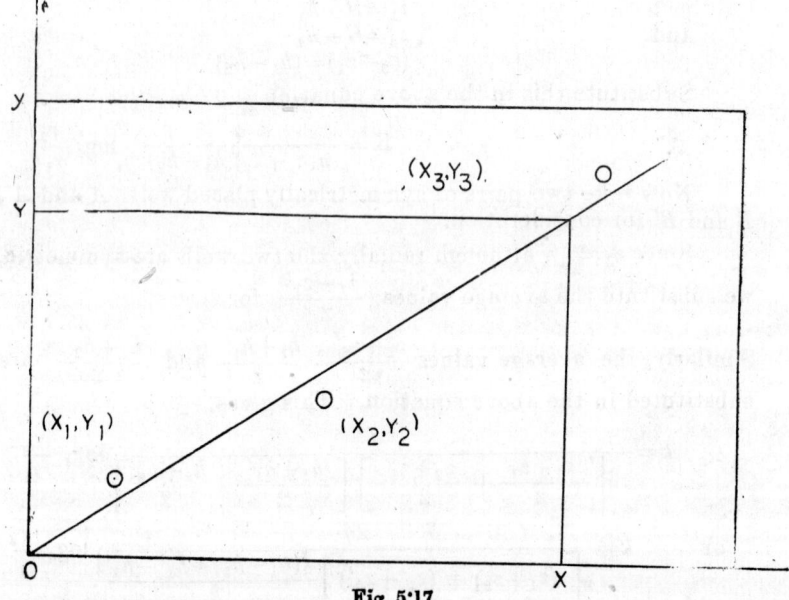

Fig. 5·17.

the value of X *versus* Y say for a point on this line is taken and substituted in equation (5·39) to get the required coefficient of permeability k.

This procedure is accurate compared to taking just two observation wells. However, it involves more work for the field worker and more calculations for the designer.

Example 5·9 A pumping test was made in a pervious soil extending to a depth of 18 m, where a bed of clay was encountered. The normal ground water level was 1 m below the ground level. Observation wells were located at distances of 4 m and 8 m from the pumping well. At a discharge of 9 cu. m. per minute from the pumping well, a steady state was attained in 24 hours. The drawdown at 4 m was 2 m and at 8 m was 0·5 m. Compute the coefficient of permeability in cms/sec. **(P.U. Nov. 1964)**

Solution. Refer to Fig. 5·12.

With the nomenclature used in the previous article and the data given in the example.

$$r_1 = 4 \text{ m and } h_1 = 2 \text{ m.}$$

$$r_2 = 8 \text{ m and } h_2 = 0·5.$$

$$H = 18 \text{ m.}$$

∴

$$z_1 = H - h_1 = 18 - 2 = 16 \text{ m} = 1600 \text{ cms}$$

and

$$z_2 = H - h_2 = 18 - 0·5 = 17·5 \text{ m} = 1750 \text{ cms}$$

Discharge $q = 9$ cu. m./min. $= \dfrac{9 \times (100)^3}{60} = 15 \times 10^4$ cm³/sec.

∴ From equation (5·36)

$$k = \frac{q}{\pi(z_2{}^2 - z^2{}_1)} \log_e \frac{[r_2}{r_1}$$

$$= \frac{15 \times 10^4}{\pi[(1750)^2 - (1600)^2]} \times \log_e \frac{8}{4}$$

or

$$k = \frac{15}{\pi[17·5)^2 - (16)^2]} \times \log_e 2$$

$$= \frac{15 \times 0·693}{\pi(306·25 - 256)}$$

$$= \frac{15 \times 0·693}{\pi \times 50·25} = 6·59 \times 10^{-2} \text{ cms/sec.} \quad \textbf{Ans.}$$

5·13. Coefficient of Permeability for Various Soils

As a guide relative values of the coefficients of permeability for various soils are given in Table 5·1 the degree of permeability mentioned in this table has been adopted after Terzaghi and Peck (5). These values may only be taken for relative comparison. Actual permeability for any soil must be found out experimentally for use in the design of hydraulic structures.

TABLE 5.1

Relative Coefficient of Permeability Value of for Various Soils Types

S. No.	Soil Type	Coeff. of permeability k in cm/sec.	Degree of Permeability
1.	Gravels. coarse and medium.	More than 10^{-1}	High
2.	Fine gravel to fine sand.	10^{-1} to 10^{-3}	Medium
3.	Very fine sand, silt-sand admixtures, loose silt, rock flour and loess.	10^{-3} to 10^{-5}	Low
4.	Dense silt loess, clay-silt admixtures, non-homogenous clays.	10^{-5} to 10^{-7}	Very low
5.	Homogeneous clays.	Less than 10^{-7}	Almost impervious

A. Casagrande and Fadum (4) have given the values of k to be used in earth dams. These values are presented in Table 5.2 along with the types of soils with which they are associated, the quality of drainage of these soils, method of determination of the k values and the reliability of results from these methods, along with the relative experience needed to carry out testing work.

(See Table 5.2 opposite page 163).

5·14. Determination of Capillary Head for Soils

A simple experiment, called the *Horizontal Capillarity Test*, is carried out to find out the capillary head for a soil. The apparatus shown in Fig. 5.17, consists of a tub of water, a glass tube *AB* fitted with a porous stone disc (or even 200 mesh screen

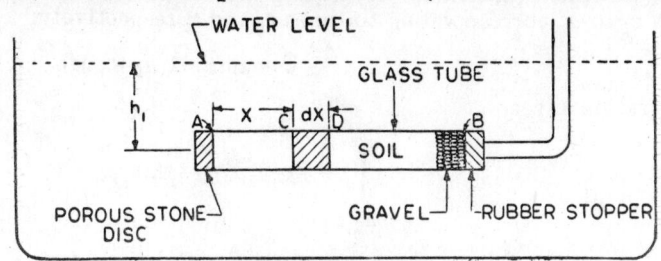

Horizontal Capillary Test
Fig. 5·17.

can be used) at one end *A* and a rubber stopper carrying a bent glass tube, at the other end *B*. The soil for which the capillary head is desired is packed into this glass tube in a dry state at the desired density and some gravel is placed at the end *B*, before placing the stopper and the bent glass tube into position. Gravel helps in keeping the bent glass tube open by preventing it from getting choked, due to the flow of the soil into it.

The tube *AB* filled with the dry material is lowered into the water bath in a horizontal position at a fixed depth with the open end of the bent tube outside the water as shown. As soon as this tube is lowered into water, the water starts flowing from the end *A* towards the end *B* of the tube, under a combined head $h=(h_1+h_c)$ where h_1 is the pressure head of water at *A* as shown and h_c is the capillary head for the soil.

If x denotes any distance say up to the point *C* that the water has travelled in the tube from the end *A* in time t and if dx is a small distance that the water travels from the point *C* to to the point *D*, in time dt, then $\frac{dx}{dt}$ represents the actual velocity of flow through the soil pores and it may be equated to V_s. This velocity of flow is the same at all points in the soil mass.

The superficial velocity of flow V as given by equation (5.19), therefore, is ηV_s where η is the known porosity of the soil sample.

$$\therefore \quad V=\eta V_s=\eta\frac{dx}{dt}$$

Also by Darcy's Law $V=k.i$ where k is the co-efficient of permeability of the soil and i is the hydraulic gradient under which the flow occurs, which in this case at the point *C* is nothing but $\frac{h.}{x}$

$$\therefore \quad \eta \cdot \frac{dx}{dt} = V = k \cdot \frac{h}{x} = k \cdot \frac{h_1 + h_c}{x}$$

Integrate this equation by separating the variables, between the limits x_1 to x_2 corresponding to times t_1 and t_2 respectively.

$$\therefore \quad \eta \cdot x \cdot dx = k(h_1 + h_c)dt$$

Intergration gives,

$$\int_{x_1}^{x_2} \eta \cdot x dx = \int_{t_1}^{t_2} k(h_1 + h_c)dt$$

or

$$\eta \left[\frac{x^2}{2}\right]_{x_1}^{x_2} = k(h_1 + h_c)\left[\ t\ \right]_{t_1}^{t_2}$$

or

$$\eta \frac{x_2{}^2 - x_1{}^2}{2} = k(h_1 + h_c)(t_2 - t_1.)$$

or

$$\frac{x_2{}^2 - x_1{}^2}{t_2 - t_1} = \frac{2k}{\eta}(h_1 + h_c.) \quad \ldots(5.40)$$

x_1 at the start is zero corresponding to zero time t_1 ; x_2 and t_2 correspond, therefore, to any distance x measured from the end A after time interval t. The distance x and time t in the experiment are noted at small *regular* intervals of time and curve between x^2 and t is plotted to obtain the slope of the straight line, which will give the value of the left hand side of equation (5·40).

On the right hand side of equation (5·40), value of h_1 and the porosity η are known. However, k and h_c are two unknowns and to solve for their values, two equations are needed. In order to have a second equation involving k and h_c, after a part of the soil has been saturated, the pressure head is changed by either dipping the horizontal tube AB deeper or by bringing it up towards the surface. keeping it horizontal again in either case. The values of x and t are again noted for the rest of the length of the soil sample, from which another x^2 *versus* t plot can be made and the value of the slope so determined is plugged in the left hand side of equation (5·40) with the appropriate value of h_1 on the right hand side.

The two equations so obtained from these readings can be simultaneously solved to get the values of h_c and k.

This method, indirectly, provides the value of the co-efficient of permeability k for the soil. However, it may be noted that this co-efficient cannot be compared in value to the one obtained from the other tests mentioned in section 5·9 for the obvious reasons, *namely*, the soil in this case is packed dry and is not saturated before the actual test is started and as a consequence the air present in the voids is likely to obstruct the flow as the water column advances in the tube and due to this reason the co-efficient of permeablity obtained may be lower than the actual value ; and secondly, the density when the soil is packed dry is quite different from the one obtained if the soil is placed under water for the permeability test and due to this reason the value of the co-efficient of the permeability obtained in the test may refer to a relatively large void ratio of the sample. However, as a rough estimate the value so obtained is alright.

CHAPTER-BIBLIOGRAPHY

1. American Society of Testing Materials, Procedure for Testing Soil, Philadelphia, 1958.

2. ASTM, "Annual Book of ASTM Standards," ASTM, Philadelphia, Pa., 1972.

3. Casagrande, A., and Fadum R.F.. Note on Soil Testing for Engineering purposes, Cambridge Massachusettes.

4. Hazen Allen, Discussion of Dams on Sand Foundations by Koening, A.C. Trans A.S.C.E. Volume 73, 1911.

5. Hugh, B.K., "Basic Soils Engineering," Second Edition Ronald Press Company, New York, 1969.

6. Jumikis, A.R., Soil Mechanics, D. Van Nostrand Company, Inc. Princeton, New Jersey, 1962.

7. Lambe, T.W., and Whitman, R.V., "Soil Mechanics,". Wiley & Sons Inc., N.Y., 1969.

8. Peck. R.B., Hanson W:E. and Thornburn T.H., "Foundation Engineering" New York, John Wiley & Sons, Inc. 1957.

9. Taylor. D.W., "Fundamentals of Soil Mechanics." John Wiley & Sons Inc., New York, N.Y. 1948.

10. Terzaghi, C., New Facts About Surface Friction, Physical Review, Volume 16, July-Dec. 1920.

11. Terzaghi, Charles, Principles of Soil Mechanics, 1-Phenomena of Cohesion of Clays, Engineering News Record, Vol. 95, No, 19, Nov. 5th, 1925.

12. Terzaghi, K., and Peck, R.B., Soil Mechanics in Engineering Practice, New York, John Wiley & Sons, 1968.

QUESTIONS

1, (a) A sample of dry sand is taken and an effort is made to roll
it in hand into a ball but the effort fails. Then a little water
is added to this sand and in this case it is possible to roll
it into a ball. What is the reason that it is possible to
make a ball from the mass of sand, when moist ?

   (b) The moist sand ball is thrown into a water bowl. The sand
particles fall apart ; why do the sand particles fall apart
when the ball is thrown in water ?

2. (a) What is permeability ? Discuss the variables on which the
permeability of a given soil depends.

   (b) A constant head permeability test has been run on a sand
sample 25 cms. in length and 30 sq, cm. in area. Under a
head of 40 cms, the discharge was found to be 200 c.c. in 1
minute and 56 seconds. The specific gravity of the solids
was 2·66 and the dry weight of the sample was 1320 gms.
Find out the permeability of the soil and the seepage
velocity during the test.   (P.U. 1959)

3. (a) Define the following terms :

   (i) Neutral pressure.   (ii) Intergranular pressure.

   (iii) Combined pressure.   (iv) Capillary pressure.

   Give the relationship between (i), (ii) and (iii) and (i) and
   (iv).

   (b) The water table is lowered from a depth of 10 feet to a
depth of 20 feet in a deposit of silt. All the silt is saturated
even after the water table is lowered. If the water content
is 26 per cent, estimate the increase in effective pressure at
a depth of 34 feet on account of lowering of the water
table ?   (P.U. 1959 Supp)

4. (a) How will you measure the permeability of a soil by variable head permeameter ? Mention the precautions to be
taken in the test.

   (b) In a falling head permeameter test, the hydraulic head at
$t=0$ is 40 cms. and drops 1 cm. in 3·5. minutes. It is
desired to run the test until this head is 20 cms. How
much longer must the test continue ?   (P.U. 1959 Supp)

5. Define the co-efficient of permeability. How is it determined in
a laboratory ? Under what field situations will you recommend the use
of permeable and less permeable soils ?

Water is percolating through the earth fill of a rectangular cofferdam
founded on an impervious soil. The cofferdam is 100 feet long and 40
feet thick. The depth of water on one side is 15 feet over the impervious soil. Compute the total amount of water that will flow through
the entire Co-efferdam in c. ft. per day. if the earth fill is a silty sand
with a coefficient of permeablity of 0·11 cm. per minute.   (P.U. 1960)

6. (a) Discuss briefly the effect of various factors on which the
coefficient of permeability depends.

   (b) A sample of soil 8 cms. in diameter and 4 cms. thick is
tested in a falling head permeamater. The elevation of

the water in the standpipe above the tail water level was observed to drop from 42 cms. to 37 cms ; in 5 minutes and 42 seconds. The inside diameter of the standpipe is 0·2 cms. Compute the co-efficient of permeability, using the average hydraulic gradient and Darcy's Law.   *(P.U. 1961)*

7. Define "Coefficient of permeability" of soil. How is it determined in the laboratory with the help of the constant head permeameter ?
(P.U. Apr, 1965)

8. *(a)* Describe a falling head permeameter and how it is used to determince the coefficient of permeablity of a fine grained soil. Derive the expression.

(b) Estimate the capillary rise of water in a clean glass tube of 0·075 m.m. inside diameter, if the value of surface tension is 75 dynes per cm.   *(P.U. Nov. 1965)*

9. How does shrinkage take place in cohesive soils ?

10. State Darcy's Law and indicate the significance of each of the terms in the equation. Is the superficial velocity of the soil more than actual velocity or is it the other way round ?

11. Derive an equation for calculating experimentally the capillary head of a soil sample. Comment on the value of the coefficient of permeability that can be obtained from this expression.

12. How is the coefficient of permeability of a soil evaluated in the field ? Derive the formula used ?   *(P.U. Nov., 1966)*

13. *(a)* List the factors that influence the Darcy's Cofficient of permeability ?

(b) What is shape factor ? Calculate the shape factor for a flow taking place through the medium of sand, given that equivalent diamter 2m.m, unit weight $8=1.8$ gms/cm specific gravity of solid 2·6 and water is flowing through the medium at a rate of 1 gallon per minute under a gradient of 0.8.   *(Mysore 1965)*

Seepage Through Soils

6.1. Introduction

Seepage through soils may be defined as the flow of a fluid through the soil pores under a pressure gradient. Flow of water through soils is a fundamental problem in the Soil Mechanics field. For *example*, seepage problems occur in all earthern dams and foundations placed under the ground water table. Seepage forces act on a soil mass and influence the behaviour of the soil. These forces are of considerable importance in the stability of slopes and retaining walls subject to seepage. As such, a study of the fundamental seepage characteristics forms the subject-matter of this chapter.

6.2. The Continuity Equation

The flow of water in a soil mass is three-dimensional. However, most of the flow problems can be solved by a two-dimen-

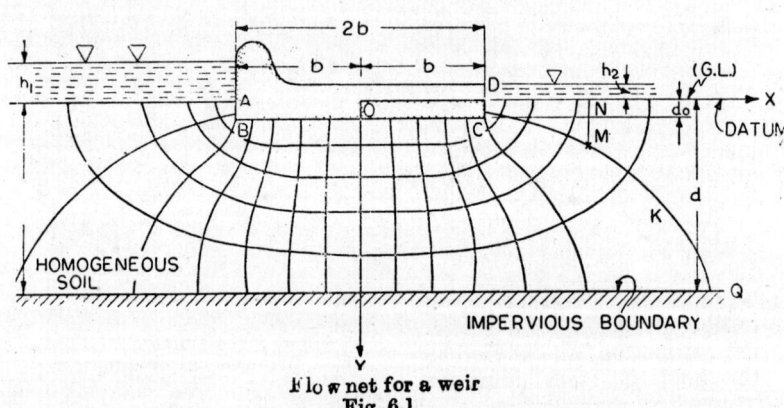

Flow net for a weir
Fig. 6.1.

sional analysis because the third dimension can be assumed to be a constant quantity. For *example*, for estimation of the total amount of flow or discharge underneath a spillway, it is sufficient to draw only a two-dimensional flow net[1] as shown in Fig. 6·1 and to consider the depth perpendicular to the plane of the paper as unity and to calculate the discharge on this basis. As such, an equation which ensures the continuity of flow in two dimensions is derived in this section. A three-dimensional continuity equation can either be independently derived in a similar manner or can be mathematically inferred from the two-dimensional equation.

The conditions of isotropy, homogeneity, and laminar flow are assumed in the derivation of this equation. As already mentioned in Article 5·7, in laminar flow a particle of water starts at a particular point and follows a particular path in the direction of flow, without crossing the paths of other particles and then ends up at a definite point in the medium. Darcy's Law is applicable to this type of flow. Firstly, we will derive the velocity components for flow in two directions x and y and then will proceed on to the derivation of the two-dimensional continuity equation.

Velocity Components. Ref. to Fig. 6·2. Let V_r be the velocity of flow of a particle along a radial direction OR inclined at an angle θ to the X-axis. Let this velocity correspond to a change in head $\triangle h$ in length $\triangle r$ of a soil sample, or in other words, let the flow along OR correspond to a pressure gradient of $\dfrac{\triangle h.}{\triangle r}$.

The velocity of flow

$$V_r = -k.\frac{\triangle h.}{\triangle r} \quad ...(6.1)$$

Minus sign in equation (6.1) only indicates that as r increases, the hydraulic head h decreases and that is how the flow is taking place. So the head h is variable in the radial direction OR, called the r-direction and therefore the gradient from point to

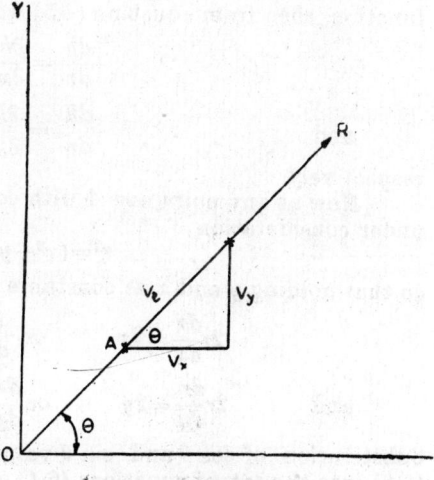

Velocity components for a two-diamensional field of fl w.
Fig. 6·2.

---

1. For *definition* of a flow net see Art. 7.3.

point is a variable quantity. As such, the velocity V_r in the r-direction may, in the limit be written as,

$$V_r = -k\frac{dh}{dr} \qquad \text{...(6.2)}$$

or
$$V_r = \frac{d}{dr}(-kh) \qquad \text{...(6.3)}$$

k is considered constant in this analysis. Resolve this velocity V_r along the x and y directions.

Then,
$$\left. \begin{aligned} V_x &= -k.\frac{dh}{dr}\cos\theta \\ V_v &= -k\frac{dh}{dr}\sin\theta \end{aligned} \right\} \qquad \text{...(6.4)}$$
and

Now, from a knowledge of the total and partial differentials, we know that,

$$\frac{dh}{dr} = \frac{\partial h}{\partial x}.\frac{dx}{dr} + \frac{\partial x}{\partial y}.\frac{dy}{dr} \qquad \text{...(6.5)}$$

The head h at any point in the medium being a variable in the r-direction, must be a function of distance r from the origin O and r is further a function of distances x and y from the origin O, so that h is in fact a function of x and y, i. e.,

$$h = f(r) = F(x, y) \qquad \text{...(6.6)}$$

Now, if y and x are held constants respectively in the above function, then from equation (6.5) we may write

$$\left. \begin{aligned} \frac{dh}{dr} &= \frac{\partial h}{\partial x}.\frac{dx}{dr} \\ \frac{dh}{dr} &= \frac{\partial h}{\partial y}.\frac{dy}{dr} \end{aligned} \right\} \qquad \text{...(6.7)}$$
and

respectively.

Now at any point say A with co-ordinates (x, y) in the figure under consideration,

$$r^2 = (x^2 + y^2)$$

so that holding y and x as constants respectively,

$$\left. \begin{aligned} 2r\frac{dr}{dx} &= 2x \qquad \text{or} \quad \frac{dr}{dx} = \frac{x}{r} = \cos\theta \\ 2r\frac{dr}{dy} &= 2y \qquad \text{or} \quad \frac{dr}{dy} = \frac{y}{r} = \sin x \end{aligned} \right\} \qquad \text{...(6.8)}$$
and

Substitution of $\cos\theta$ and $\sin\theta$ values from the set of equations (6.8) into the set of equations (6.4) gives

$$\left. \begin{aligned} V_x &= -k.\frac{dh}{dr}.\frac{dr}{dx} \\ V_v &= -k.\frac{dh}{dr}.\frac{dr}{dy} \end{aligned} \right\} \qquad \text{...(6.9)}$$
and

But from the set of equations (6.7),

$$\frac{dh}{dr} \cdot \frac{dr}{dx} = \frac{\partial h}{\partial x}$$

and

$$\frac{dh}{dr} \cdot \frac{dr}{dy} = \frac{\partial h}{\partial y},$$

and substitution of these values in the set of equations (6·9) gives,

$$V_x = -k\frac{\partial h}{\partial x} = \frac{\partial}{\partial x}(-kh) \qquad \ldots(6·10)$$

and

$$V_y = -k\frac{\partial h}{\partial y} = \frac{\partial}{\partial y}(-kh) \qquad \ldots(6·11)$$

From equations (6·3), (6·10) and (6·11), we may conclude that the velocity of flow in any direction say a may be written as the differential of $(-kh)$ in that direction or,

$$Va = \frac{d}{da}(-kh) \qquad \ldots(6·12)$$

Continuity Equation. Consider a small element of area $\triangle x \times \triangle y$ as a portion of a big isotropic and homogenous soil mass. The element is shown in Fig. 6·3. The third dimension of this element perpendicular to the plane of the paper may be taken as unity. The coefficient of permeability k is assumed a constant quantity.

Let the water enter this element through the faces AB and BC and get out of it at the faces AD, DC. Let the entrance velocity at the faces AB and BC be V_x and V_y respectively and the exit velocity at the faces DC and AD be V_x' and V_y' respectively, so that quantites of water that enter and come out of the element in unit time can be written as :—

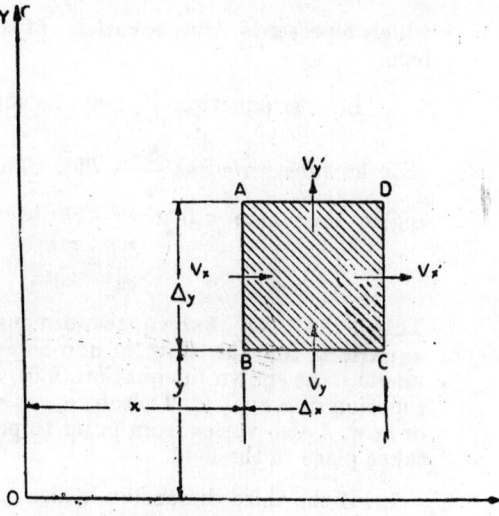

Fig. 6.3.

q = quantity of water entering the element

$$= V_x. \, (\triangle y.1) + V_y \, (\triangle x. \, 1) \qquad \qquad ...(6{\cdot}13)$$

and q' = quantity of water coming out of the element

$$= V_x'.(\triangle y.1) + V_y'(\triangle x.1) \qquad \qquad ...(6{\cdot}14)$$

Assume that no volume change is involved. Then the above two quantities q and q' must be equal.

$$\therefore \quad V_x.(\triangle y.1) + V_y. \, (\triangle x. \, 1) = V_x'.(\triangle y.1) + V_y'. \, (\triangle x. \, 1)$$

Rearrange the terms.

$$(V_x - V_x')\triangle y + (V_y - V_y'). \, \triangle x = 0 \qquad \qquad ...(6{\cdot}15)$$

where $(V_x - V_x')$ and $(V_y - V_y')$ represent the small changes in velocities V_x and V_y across the small distances $\triangle x$ and $\triangle y$. Let these small changes be represented as $\triangle V_x$ and $\triangle V_y$ respectively. The equation (6·15) becomes,

$$\triangle V_x. \, \triangle y + \triangle V_y. \, \triangle x = 0$$

Divide both sidess by $\triangle x. \, \triangle y$, the area of element

$$\frac{\triangle V_x}{\triangle x} = \frac{\triangle V_y}{\triangle y} = 0.$$

In the limit,

$$\frac{\partial V_x}{\partial x} + \frac{\partial V_y}{\partial y} = 0 \qquad \qquad ...(6{\cdot}16)$$

which represents the equation of continuity in two dimensional form.

In this equation V_x may be substituted as $\dfrac{\partial}{\partial y}(-kh)$ and V_y may be substituted as $\dfrac{\partial}{\partial y}(-kh)$. Then the equation after dividing by the common factor $(-kh)$ becomes,

$$\frac{\partial^2 h}{\partial x^2} = \frac{\partial^2 h}{\partial y^2} = 0 \qquad \qquad ...(6{\cdot}17)$$

This is the well known two-dimensional, **Laplace's differential equation**, for the flow of non-compressible fluid in a porous media. As shown in equation (6·6), h is a function of r or it is a function of x and y, Therefore, as r or x and y vary in the field or flow, h also varies from point to point and that is how the flow takes place in the field.

If the third dimension is also considered, the different equation for flow may be written as,

$$\frac{\partial^2 h}{\partial x^2} + \frac{\partial^2 h}{\partial y^2} + \frac{\partial^2 h}{\partial z^2} = 0 \qquad \qquad ...(6{\cdot}18)$$

where h is a function of x, y, z in these dimension co-ordinate system.

For the known boundary conditions, equations (6·17) and (6·18) can be solved and the solution will gave the hydraulic head h at any point in terms of the co-ordinates. Laplace's differential equation (6 17) or (6·18) does not contain the quantity k, which is the coefficient of permeability for the soil. So, the equation governing flow in porous media is *independent* of the coefficient of permeablity.

Sometimes it is convenient to work in polar co-ordinate system so that equation (6·17) may be written as,

$$\frac{\partial^2 h}{\partial r^2} + \frac{1}{r}\frac{\partial h}{\partial r} + \frac{1}{r^2}\frac{\partial^2 h}{\partial \theta^2} = 0 \qquad \qquad ...(6·19)$$

where (r, θ) are the polar co-ordinates of any point whose hydraulic head is h.

Definition. The quantity $(-kh)$ is defined as the **potential** of the field of flow or simply as **flow potential**. Then $\frac{\partial}{\partial a}(-kh)$ given in equation (6·12) may be called gradient of the potential in the a—direction.

6·3. Position of Moving Particle

At time $t=0$, a particle of water has some position in the field[1] of flow. As the time passes, since the particle moves, it assumes, different positions at different times and as such traces a path of movement. This path that the particle describes in the field of flow is called a **flow line**. In order to understand, therefore, the nature of flow in a porous medium, the position of the moving particle at any time and the path it describes must be known. One way to find out the position of the particle at any time t is to know its velocity component in the x and y directions and equate them to $\frac{dx}{dt}$ and $\frac{dy*}{dt}$ respectively and solve the two differential equations so formed, to get x and y co-ordinates for the moving particle in terms of the time t. Then, to find out the path of flow for the particle, the time may be eliminated from the equation for x and y. To illustrate this method. an example is taken below.

---

1. Field of flow, in general, sense may be defined as the extent of the medium in which the flow. takes place.

*If x and y are the co-ordinates of a moving point in a field at time t, then $\frac{dx}{dt}$ and $\frac{dy}{dt}$ represent the velocity components of this point at time t, in the x and y directions.

Let us assume that $h = h_1 - ax - by$ is the hydraulic head at any point whose co-ordinates are (x, y). The first quation that arises is this : "Does this function h satisfy the Laplace's equation ?"

Second partial derivatives of h with respect to x and y give,

$$\frac{\partial^2 h}{\partial x^2} = 0 \text{ and } \frac{\partial^2 h}{\partial y^2} = 0, \ a, b \text{ being constants.}$$

Summation shows that Laplace's equation is satisfied.

Now, the velocity components V_x and V_y from this function are

$$V_x = \frac{\partial}{\partial x}(-kh) = +k.a.$$

$$V_y = \frac{\partial}{\partial y}(-kh) = +k.b.$$

Also $V_x = \dfrac{dx}{dt} \text{ and } V_y = \dfrac{dy}{dt}.$

$\therefore$ $\dfrac{dx}{dt} = k.a. \text{ and } \dfrac{dy}{dt} = k.b.$

Integration of the above to differential equations gives,

$$\left. \begin{array}{c} x = kat + x_0 \\ y = kbt + y_0 \end{array} \right] \qquad \qquad ...(6.20)$$

and

where x_0 and y_0 are arbitrary constants of integration.

The set of equations (6.20) gives the position of the moving particle in the field of flow, with respect of time t.

The path of flow traced by the moving particle is determined by eliminating the time t from the above two equations.

So, $$\frac{x - x_0}{ka} = \frac{y - y_0}{kb}$$

or $$y - y_0 = \frac{b}{a}(x - x_0)$$

or $$y = y_0 + \frac{b}{a}(x - x_0) \qquad \qquad ...(6.21)$$

which is the equation for the path of flow. This is the equation of a straight line, with slope equal to $\tan^{-1} \dfrac{b}{a}$.

By giving different numerical values to the constants x_0 and y^0, chosen arbitrarily, a family of lines can be obtained represented by the same general equation (6·21).

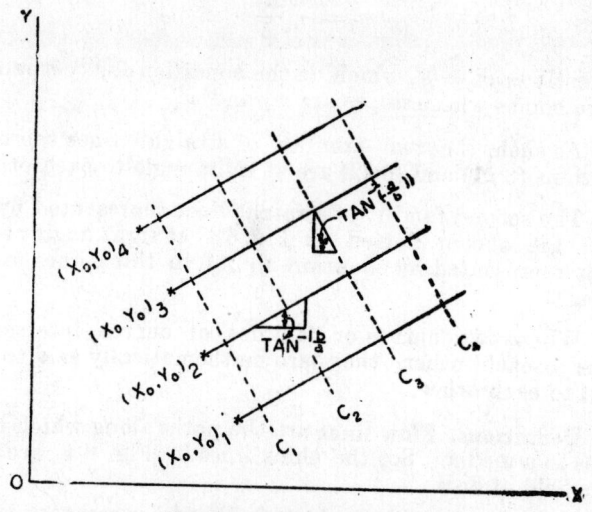

Simplest example of a flow net.
Fig. 6.4.

Fig. 6·4 represented a family of paralled and thick straight lines reprsesent the paths of flow of the water particles flowing through a medium.

The hydraulic head h_1 which equals $h_1 - ax - by$, at any point in the medium has some value, say C.

$$\therefore \qquad h = h_1 - ax - by = C \qquad \qquad ...(6·22)$$

Equation (6·22) is also the equation of a straight line. Futher on, by choosing different numerical values for C, equation (6·22) can be made to represent a second family of straight lines. Every point on a particular straigh: line out of this family has a constant head since h has a particular constant numerical value of C, for this particular line.

Now, we proceed to find out, if there is any relationship between the family of lines represented by (6·21) and the family of lines represented by (6·22).

$\dfrac{dy}{dx}$ from equation (6·21) gives the slope of this family of

straight lines and it is equal to $\dfrac{b}{a}$. Let it be designated by m_1.

$$\therefore \qquad m_1 = \frac{b}{a}$$

Similary, if the slope for the second family of lines represented by (6·22) is designated as m_2, then

$$m_2 = -\frac{a}{b}$$

Evidently $m_1 m_2 = -$, which is the condition for perpendicularity of two conics whose slopes are m_1 and m_2.

As such, the two families of straight lines represented by equations (6·21) and (6·22) are at right angle to each other.

The second family of straight lines represented by equation (6·22) are shown dotted in Fig. 6·4 at right angles to the first family represented by equation (6·21) in thick lines in the same figure.

When two curves or families of curves intersect at right angles to each other, they are mathematically said to be **orthogonal** to each other.

Definitions. Flow lines are the paths along which the water moves in a media. So, the thick lines in Fig. 6·4 are flow lines in the field of flow.

Locus of a point for which the head h is constant is called an **equipotential line** or simply an **equipotential**. So, the dotted lines in Fig. 6·4 are equipotentials in the field of flow and, as stated above, the equipotentials and flow lines areortho gonal.

Whatever may be the equations for flow lines and equipotential, in an isotropic and homogeneous medium, they are bound to be orthogonal. The net work formed by flow lines and equipotentials, as for *example*, in Fig 6·4 is called a **flow net**. Another example is Fig. 6·1

The above example was taken to illustrate the position of a moving particle with time and the definitions of flow lines, equipotentials and a flow net. Incidentally it may be mentioned here, that by a proper choice of the constants a and b in $h = h_1 - ax - by$, equations (6·21) and (6·22) can be made to represent the flow conditons between two parallel flat boundaries, for *example*, between two parallel flat plates.

Equation (6·21) can be rearranged and written as,

$$y - \frac{b}{a}x = (y_0 - \frac{b}{a}x_0)$$

The quantity on the left hand side of the above equation is a variable in x and y and on the right hand side the quantity is a constant depending upon particular numerical values chosen for x_0 and y_0. The quantity on the left hand side may be denoted by a functson ψ, so that

$$\psi = y - \frac{b}{a}x = (y_0 - \frac{b}{a}x_0) \qquad \qquad ...(6\cdot23)$$

If second partial derivatives of ψ with respect to x and y are taken and summed up, it results into the equation

$$\frac{\partial^2 \psi}{\partial x^2} + \frac{\partial^2 \psi}{\partial y^2} = 0,$$

which again is a Laplace's differential equation. So, the function ψ which represents the flow lines also satisfies the Laplace's differential equation.

From equations (6·22) and (6·23) we may conclude that the two functions $h(x, y)$ and $\psi(x, y)$ each equal to a constant representing the equipotentials and flow lines, satisfy the Laplace's differential equation.

6·4. Boundary Conditions in a Physical Problem

Laplace's differential equation (6·17) can be written in a more convenient form as,

$$\nabla^2 h = 0 \qquad \qquad ...(6\cdot24)$$

where $\qquad \qquad \nabla^2 = \frac{\partial}{\partial x^2} + \frac{\partial}{\partial y^2}.$

This equation has a number of solutions, as, besides governing the flow in a porous media, it is applicable in so many other fields. The solution of this equation, wherever applied, will need the knowledge of the boundary conditions. How these boundary conditions may be written can be visualised from. Fig. 6·1.

Let OX coinciding with the ground level (G.L.) be taken as the x-axis and OY as shown be taken as the y-axis. The point O lies at the centre of the length of the spillway so that $OD = OA = b$, in the figure. Assume OX as the datum for finding out the total head h at any point. Then following boundary conditions can easily be visualised.

(1) Since the flow takes place between the bottom of the spillway section and the impervious boundary at the bottom, $ABCD$ must represent *one limiting boundary* and PQ the *second limiting boundary*.

(2) The head all along the upstream line AE is the same and is equal to the water pressure head h_1.*

---

*Datum here is assumed as the line AE so that the elevation head here is zero. See also section 5·5 on how to find out the head at any point.

So, *AE* forms *one limiting equipotential line* for the system.

(3) Again the head all along the downstream line *DX* is the same and equal to the water pressure head h_2

So, *DX* forms the *second limiting equipotential line* for the system.

These limiting conditions of the boundaries can be expressed mathematically as under :

When | x | $<b$ and $0 <y<d_0, \dfrac{\partial h}{\partial x}=0=\dfrac{\partial h}{\partial y}$

[No flow across *ABCD*]

(2) When $y=d, \dfrac{\partial h}{\partial y}=0$ [No flow across *PQ*]

(3) When $x<-b, h=h_1$ [Constant head h_1 along *AE*]

(4) When $x>b, h=h_2$ [Constant head h_2 along *DX*]

These boundary conditions can be used for a mathematical solution of the Laplace's differential equation (6·24) for the actual physical conditions of the problem. Mathematical solutions of the physical problems are intricate and are considered outside the scope of this book. However, the above illustration gives an insight into the mathematical methods of attack.

6.5. Properties and Uses of Flow Nets

The definition of a flow net has already been given in section 6·3. A flow net is a convenient device to find out the quantity of seepage through soils.

Refer to Fig. 6·1 again. A particle of water can enter the surface *AE* at any point. During its flow through the soil mass, it traces a path of flow called a flow line and at the exit surface *DX*, it will get out at a definite, point, corresponding to the flow line that it traces out. As the particle can enter *at any point*, there can be innumerable points along the surface *AE*, where it can enter and as such correspondingly there can be inumerable flow lines. By a similar argument, it can be said that there can be innumerable equipotential lines. However, in actual practice, it is convenient to draw a reasonable number of flow lines and correspondingly a reasonable number of equipotential lines and calculate the discharge from the flow net so obtained.

Now consider a portion of the flow net of Fig. 6·1, drawn to a larger scale in Fig. 6·5,, showing three flow lines f_1, f_2 and f_3 and three equipotential lines $e_1 e_2$ e_3. The channel of flow between any two flow lines may be called a **flow channel**. Let l_1 and l_2 be the average lengths of the flow channels A and B between two sets of equipotential e_1, e_2 and $e_2 e_3$ respectively ; and a_1 and a_2 be the average widths of the channels between the same two sets of equipotential lines, as shown in the figure. Let the head drop (or many times called the potential drop) between the equipotentials e_1 and e_2 be $\triangle h_1$ and between e_2 and e_3 be $\triangle h_2$. Then, hydraulic gradient between e_1 and $e_2 = \dfrac{\triangle h_1}{l_1}$ and hydraulic gradient between e_2 and $e_3 = \dfrac{\triangle h_2}{l_2}$

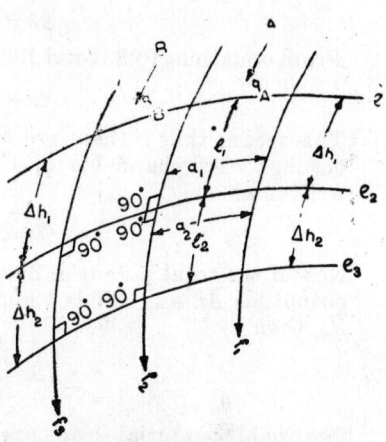

Flow net to a bigger
scale.
Fig. 6·5.

Taking the depth perpendicular to the plane of the paper as unity, the discharge q_A through the channel A, may be written as,

$$q_A = a_1.k.\frac{\triangle h_1}{l_1} = a_2.k.\frac{\triangle h_2}{l_2} \qquad \qquad ...(6·25)$$

where k is the permeability for the soil, assumed constant.

Now, if the ratios $\dfrac{a_1}{l_1}$ and $\dfrac{a_2}{l_2}$ are made unity, then from equation (6·25) $\triangle h_1$ will equal $\triangle h_2$.

So, in case the spaces enclosed by the equipotentials and the flow lines are *elementary squares*. the potential drop between the successive equipotential lines must be equal. If this potential drop is denoted by $\triangle h$ and the ratio $\dfrac{a_1}{l_1}$ and $\dfrac{a_2}{l_2}$ taken as unity, then the discharge through the channel under consideration, from equation (6·25) will be given by

$$q_A = k.\triangle h \qquad \qquad ...(6·26)$$

Again, if the channel B is considered, the potential drop between the succesive equipotentials is the same *i.e.* $\triangle h$ and if the spaces occupied between the equipotentials and the flow lines are again elementary *squares*, the equation of discharge through channel B, may be written as

$$q_B = k. \triangle h \qquad \qquad ...(6\cdot 27)$$

From equations (6·26) and (6.27) it is seen that

$$q_A = q_B$$

This means that if there are N_f number of channels and discharge through each channel is q, then the total discharge Q may be written as

$$Q = N_f q = N_f \ (k. \triangle h)$$

Now if the total potential drop between the two extreme equipotentials AE and DX is h and the number of potential drops is N_d, then

$$\triangle h = \frac{h}{N_d}$$

because all potential drops are equal.

Substitute this value of $\triangle h$ in the above equation gives.

$$Q = k.h. \frac{N_f}{N_d} \qquad \qquad ...(6\cdot 28)$$

Equation (6·28) gives the discharge *per unit time per unit length* (or depth perpendicular to the plane of the paper in Fig. 6·1) of the spill way, calculated from the flow net.

After recognition of the boundary conditions *square flow nets* can be drawn for any seepage field and discharge calculated as shown above.

In summary, the properties of a flow net for isotropic and homogeneous soils may be stated as uuder :—

(1)   The flow lines and equipotential lines are orthogonal *i.e.* they intersect at 90°.

(2)   The spaces occupied between the successive equipotentials and flow lines are *elementary squares, i.e.,* average dimensions lengthwise and breadthwise of particular enclosed space is the same.

(3)   Because of the second property above, potential drop between any two successive equipotential lines is the same.

(4)   Because of the second and third properties above, the flow through one channel is equal to the flow through the second channel.

It may be mentioned here that the head at any point on an equipotential is the total head *i.e.* the sum of the elevation and pressure heads, *the velocity head being negligible.*

If the datum in Fig, 6·1. is taken as coinciding with the ground line AE on the left side and the ground line DX on the other side, then the head for the limiting equipotential AE is h_1 and the head for the limiting equipotential DX is h_2 There are 14 potential drops and the total potential drop is (h_1-h_2). So, if the total head for the equipotential K is desired, then it will be equal to.

$$h_1-\frac{13}{14}(h_1-h_2) \text{ or}\left(\frac{h_1}{14}+h_2\right)$$

For a particular point M on this equipotential, the elevation head is $(-MN)$ so that the pressure head would be equal to

$$\frac{h_1}{14}+h_2-(-MN) \text{ i.e.,} \left(\frac{h_1}{14}+h_2+MN\right)$$

Being an equipotential, *total head* at any point on this curve

will be the *same i.e.* $\left(\dfrac{h_1}{14}+h_2\right)$

6.6. Patterns of Flow Nets

The flow net shown in Fig. 6·4 is the simplest because it just constitutes of two families of straight lines. However, for most of the cases in practice, the boundary conditions for flow are not so simple. So, the flow nets usually consitute of two families of curves and these curves may be parabolas, hyperbolas or ellipses. Analytical solutions of many practical flow problems have revealed the above patterns. What type of conic section is applicable will mostly depend upon the boundary conditions.

As already indicated, analytical solutions to the flow problems are outside the scope of this book. However, it is essential that the student has an idea of the types of flow net pattern, so that drawing a flow net becomes easy for him.

6.7. Critical Hydraulic Gradient, Quick Sand Condition and Seepage Forces

Consider a small sand sample of length L and area A subjected to upward flow conditions shown in Fig. 6·6. The head lost in seepage throught lenght L of the sample is h so that the hydraulic gradient is $\dfrac{h}{L}$. The upward force at the bottom surface of the sample is $(h+L).\gamma_w.A$, were γ_w is the unit weight of water. The downward weight of the saturated sand mass is $\dfrac{G+e}{1+e}.\gamma_w.L.A$, where G and e are respectively the specific gravity of solids and the void ratio of the sand sample. If the frictional losses are neglected it can be imagined that under a large head loss h, the upward force $(h+L).\gamma_w.A$, may be so large as to create unstable conditions and ultimate washing off of the sand sample, as the downward force equal to the weight of the saturated soil sample is constant with respect to the variable head loss h,

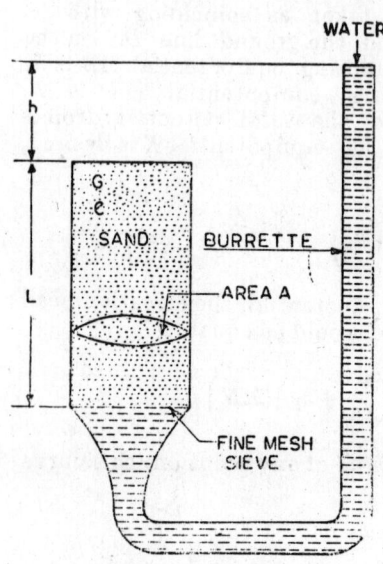

WATER

G
e
SAND   BURRETTE

———AREA A

———FINE MESH
   SIEVE

Demonstration of Quick Sand
Conditions in Sands.
Fig. 6·6.

When this unstable condition is imminent, the upward and down-ward forces must be equal.

$$\therefore \qquad (h+L)\cdot\gamma_w A = \frac{G+e}{1+e}.\gamma_w.L.A \qquad \qquad \text{...(6·29)}$$

Divide both sides L and re-arrange the terms

$$\therefore \qquad \frac{h}{L} = \frac{G-1}{1+e} \qquad \qquad \text{...(6·30)}$$

which gives the magnitude of hydraulic gradient for a *boiling phenomena* to occur. This gradient represented by equation (6·30) is called the **critical hydraulic gradient** or simply the **critical gradient.** The boiling condition of the soil corresponding to this critical gradient is given the special name, **quick sand condition,** since it makes the material a quicksand. So quicksand is only a condition of the sand and not a particular type of sand.

Quicksand conditions are observed to occur when the gradient is in the neighbourhood of unity.

Darcy's Law says that velocity V is given by $(-k.i)$ where k is the coefficient of permeability of the material and i is the hydraulic gradient under which the flow takes place. If i is given a constant value of unity in order to expect quicksand conditions, then the velocity V is simply proportional to the coefficient of permeability k. If the coefficient of permeability k for a soil is low, the water may flow with a low velocity in order that quicksand condition occurs and if k has a high value, then water *must* flow with a high velocity in order that quicksand condition may occur. This is the reason why quicksand condition is mainly observed in fine sands which have low value of k and as such do not require flow with a high velocity for the occurrence of this condition, whereas in coarse sands and gravels a very high velocity of flow would be needed to create a quicksand condition, which velocity normally is not possible to attain.

The equation (6·30) can be derived from another type of analysis, too. The head h is lost in a length L of the sample. This dissipation of head may be attributed to the viscous friction in the pore-water. Whenever any energy is lost in friction, it is accompanied by a force in the direction of motion. In the case of soils, this force is exerted on the soil grains in the direction of flow. Consider now only the individual soil particles and their frame work. The effective weight of the submerged sand mass is $\frac{G-1}{1+e}\gamma_w.L.A$, and the upward force lost in the pore water is $h.\gamma_w.A$. Quicksand condition occurs when these two forces are equal.

$$\therefore \qquad h.\gamma_w.A = \frac{G-1}{1+e}.\gamma_w.L.A \qquad ...(6·31)$$

$$\text{or} \qquad \frac{h}{L} = \frac{G-1}{1+e} \qquad ...(6.32)$$

The upward force $h.\gamma_w A$ on the left hand side of equation (6.31) is, in this case, uniformly distributed in the soil mass of volume LA. So the force per unit volume may be written as $h.\gamma_w.A/L.A$ or $\frac{h}{L}.\gamma_w$. Now whether the quicksand condition has reached or not, in an isotropic and homogeneous soil mass through which the water flows, there always exists a force in the direction of flow and its magnitude per unit volume at any point in the soil mass is equal to $i.\gamma_w$ where i is the hydraulic gradient at that point and γ_w is the unit weight of water. When the quicksand condition is imminent, i assumes the value i_c (in this case h/L), which denotes the *critical hydraulic gradient.*

Where the hydraulic gradient varies from point to point as it happens in the case of any hydraulic structures, beneath whose foundations seepage occurs, the seepage force may be written in a slightly different manner. In such a case, let dh be the head lost in a soil mass of depth dz (see Fig 6·7 showing the case of a sheet pile). If a unit area of flow is considered, then the volume of this mass is $(dz.1)$. The hydraulic gradient $i = \dfrac{dh}{dz}$ so that the seepage force per unit volume may be written as :

$$\text{Seepage force per unit volume} = i.\gamma_w = \frac{dh}{dz}.\ \gamma_w$$

$\therefore$ Total seepage force in the soil mass of volume $(dz.1)$

$$= \frac{dh}{dz}.\ \gamma_w.(dz.1) = \gamma_w dh \qquad \qquad ...(6·33)$$

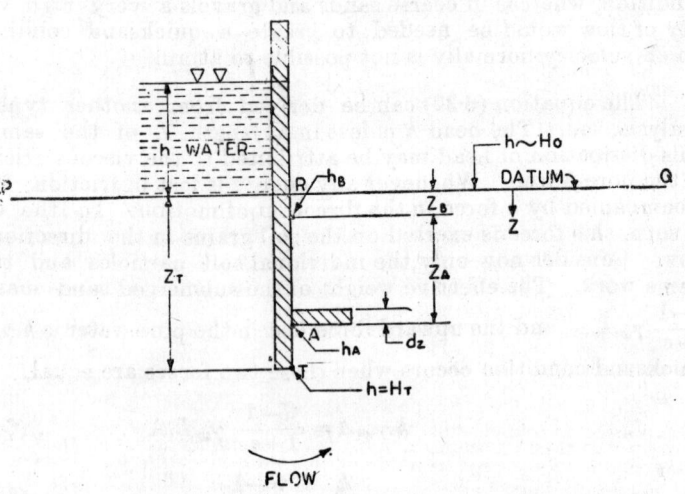

Seepage Force.
Fig. 6·7.

If now the seepage force is needed between any two points say A and B corresponding to elevations $*z'_A = (z_A + z_B)$ and z_B (with respect to a fixed line say RQ) on the downstream side of the sheet pile within the soil mass, where the heads are h_A and h_B

$*z'_A$ is not shown in the figure.

respectively, then all that we need to do is to integrate equation
(6·33) between the limits h_A and h_B.

Seepage force between A and $B = \int\limits_{h_A}^{h_B} \gamma_w.dh$ ...(6·34)

or Seepage force between A and $B = \gamma_w(h_A - h_B)$, neglecting
sign   ...(6·35)

Equation (6·35) shows that if the toe heads between any two
points could be found, the seepage force between these points can
be evaluated.

As a particular example, if RQ is taken as the datum for
evaluating heads also and if the head h at the equipotential RQ is
denoted as H_0 and the head h at the toe T corresponds to H_T,
then the seepage force exerted upwards at the toe may be written
as

$$F_T = \gamma_w(H_T - H_O)^*   ...(6.36)$$

where H_T is naturally more than H_O.

6·8 Fffect of Seeepage Force on the State of Stresses in a Soil Mass

In order to illustrate the effect of the seepage force on intergranular and total stresses, we shall consider the case of a sheet
pile embedded in a soil mass with a pressure head h on one side
and zero on the other side as shown in Fig. 6'7.

Consider a small thickness dz and a unit area at the level of
point A on the down strem side of the sheet pile as shown.
When see page is not taking place, the bouyant weight of this
saturated mass represents the effective stress or the intergranular
stress.

Effective force without seepage

$$= \frac{G-1}{1+e} \cdot \gamma_w(1.dz)   ...(6·37)$$

Since the area being considered is unity, the above force represents the effective stress without seepage, at the level of A.

When seepage occurs, an upward force equal to $\left(\dfrac{dh}{dz}\gamma_w\right)(dz.1)$
acts at the same level, so that the net intergranular stress is
given by :

$$\sigma = \frac{G-1}{1+e} \cdot \gamma_w.dz - \gamma_w dh   ...(6.38)$$

*If the equipotential RQ is taken as the datum and if there is no pre-
ssure head due to standing water on the downstream side as shown, Ho will
be zero. But we keep the equation as such just to maintain its generality.

If the material between the points A and B is considered, then the total intergranular stress at the level of A is obtained by integratiug equation (6.38) between proper limits :

$$\therefore \qquad \sigma = \gamma_w \frac{G-1}{1+e} \int_{z_A+z_B}^{z_B} dz - \gamma_w \int_{h_A}^{h_B} dh$$

$$\therefore \qquad \bar{\sigma} = \gamma_w \frac{G-1}{1+e} \left[(z_A+z_B)-z_B \right] - \gamma_w(h_A - h_B).$$

neglecting—ve signs

Put $\qquad z_A+z_B = z_A'{}^*$

$$\therefore \qquad \bar{\sigma} = \gamma_w \frac{G-1}{1+\epsilon} \left[z'_A - z_B \right] - \gamma_w(h_A - h_B).$$

Now if $z_B = O$ i.e., if the material between the point A and the eqnipotential RQ is considered, then

$$\bar{\sigma} = \gamma_w \frac{G-1}{1+e} z'_A - \gamma_w(h_A - H_O) \qquad \dots(6\cdot39)$$

For the datum chosen in Fig. 6·7, $H_o = O$. Therefore, for the conditions shown in the Figure.

$$\bar{\sigma} = \gamma_w \frac{G-1}{1+e} z'_A - \gamma_w h_A \qquad \dots (6\cdot40)$$

Now the elevation head at the point A is $:-z'_A [=-(z_A+z_B)]$ for the datum shown. So the pressure head at this point will be $h_A - (-z'_A)$ i.e. $h_A + z'_A$ according to this figure.

Water pressure at the level of

$$A = \gamma_w(h_A - z'_A)$$

or $\qquad \sigma_w = \gamma_w h_A + \gamma_w z'_A$

or $\qquad \sigma_w = \gamma_w z'_A + \gamma_w h_A \qquad \dots(6\cdot41)$

In equation (6.40) $\gamma_w . z'_A$ represents the force due to static head and $\gamma_w' h_A'$ the seepage force.

The total stress at the level of A, therefore is

$$\sigma = \bar{\sigma} + \gamma_w$$

Substitute the values of $\bar{\sigma}$ any σ_w from equation (6·39) and (6·40) into equations (6·41), then the value of the total stress at the level A is given by,

$$\sigma = \gamma_w \frac{G-1}{1+e} z'_A - \gamma_w h_A + . \gamma_w z'_A + \gamma_w h_A$$

---

* z_A is the distance of the point A from the datum RQ. This distance is not shown in the figure.

or
$$\sigma = \gamma_w \left(1 + \frac{G-1}{1+e} \right) z'_A$$

$$= \gamma_w \frac{G+e}{1+e} z'_A \qquad \qquad ...(6.42)$$

Equation (6·42) shows that the total stress at any level corresponds to the submerged unit weight of the soil. Equations (6·40) and (6·41) and (6·42) indicate three things.

(1) When the seepage occurs in the upward direction as in this case, it *decreases* the intergranular stress at any level by an amount equal to the seepage force at that level.

(2) When the seepage occurs in the upward direction as :n this case, it *increases* the water pressure at any level by an amount equal to the seepage force at that level.

(3) For the upward flow, the total stress at any level remains *unchanged* and corresponds to saturated unit weight of the soil.

If an analysis is carried out for the soil mass on the upstream side of the sheet pile, where the flow is in the downward direction, we will arrive at the following conclusions.

(1) When the seepage occurs in the downward direction, it *increases* the intergranular stress at any level by an amount .equal to the seepage force at that level.

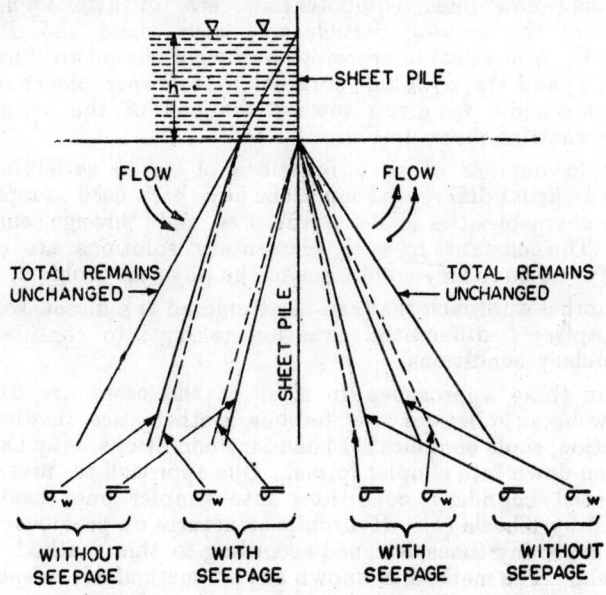

Effect of Seepage on the State of Stresses in a Soil Mass

Fig. 6·8.

(2) When the seepage occurs in the downward direction, it *decreases* the water pressure at any level by an amount equal to the seepage force at that level.

(3) For the downward flow, the total stress at any level remains *unchanged* and corresponds to the saturated unit weight of the soil.

Figure 6·8 shows a qualitative diagram indicating the effect of seepage forces on the integranular, neutral and total stresses on the two sides of a sheet pile. The effect of seepage in increasing or decreasing the intergranular and water pressures is shown by dotted lines.

6.9. Obtaining of Flow Nets

There are a number of ways by which the flow problems can be solved and flow nets obtained. They may be enumerated as :—

1. Mathematical solutions of Laplace's Differential Equation.

2. Graphical drawing of flow nets.

3. Numerical solution by relaxation technique.

4. Electrical analogy technique.

1. **Mathematical Solutions.** There are a number of mathematical funtions of $h(x, y)$ that satisfy the Laplace's differential equation. One such function was used as an example to define the terms : flow lines, equipotentials. etc. in article 6·3. The position of the moving particle was also found and denoted as $\psi(x, y)$. A physical interpretation of what these two functions *i.e.*, $\psi(x, y)$ and $h(x, y)$ could represent by a proper choice of the constants a and b was given toward the end of the article. A flow net was also shown in Fig. 6·4.

So, a number of such functions of $h(x, y)$ satisfying the Laplace's partial differential equations have been used to represent the flow characteristics under structures and through embankments. The constants in such elementary solutions are chosen to satisfy the boundary conditions of the physical problem.

Another approach that can be attempted is a direct solution of the Laplace's differential equation taking into consideration the boundary conditions.

But these approaches in most of the cases are difficult to follow because besides the tedious mathematics involved in the solution, some complicated boundary conditions may have to be broken down into simpler forms. One approach to break the complicated boundary conditions into simpler ones has been suggested by Khosla (1). Hydraulic structures on previous foundations are many times designed according to this method given by Khosla. The method is known as the method of *independent variables*.

Mathematical solutions to the flow problems are outside the scope of this book. There is a limitation of these mathematical solutions and that is that the soil has to be an isotropic and homogeneous medium.

2. Graphical drawing of flow nets. Graphical drawing of of flow nets is a convenient technique to solve the flow problems. The method is simple but needs lot of practice on the part of an engineer. After recognising the limiting boundary conditions *i.e.*, the limiting equipotentials and limiting flow lines, all that is needed is to draw a square flow net underneath the structure, keeping in mind the following points :—

(*a*) Any combination of two flow lines and two equipotentials must enclose an elementary squre *i.e.*, the average length and breadth of each square must be equal.

(*b*) Each flow line must cut an equipotential at right angles.

(*c*) All changes in direction of the flow and equipotential lines must be gentle and gradual.

The method is essentially a *trial and error* procedure. However, it is the quickest and the most practical method for determination of the seepage discharge. For a beginner, it is obligatory to study some available flow nets and to have a general idea of the nature of flow nets. For this purpose a few flow nets are presented in Fig. 6.9. Reference in this connection may also be made to Fig. 6·1.

Taylor suggests a method which he calls the *procedure by explicit trials*, for drawing a flow or equipotential line close to

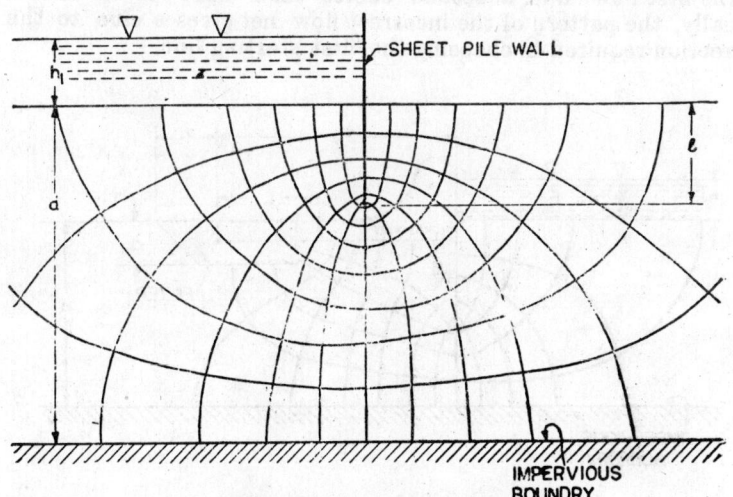

IMPERVIOUS
BOUNDRY

Flow net under a sheet pile.
Fig. 6·9 (*a*).

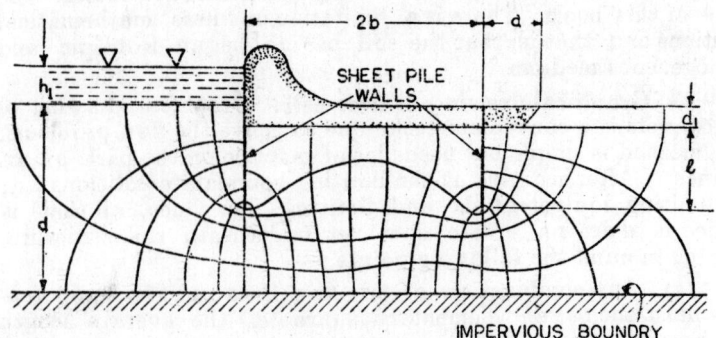

Flow net under a masonary weir with two sheet piles,
Fig. 6˙9 (b).

the known limiting flow or equipotential line. If a flow line, for *example*, is chosen, then the space between this flow line and the limiting flow line above it is divided into whole number of elementary squares and then the pattern of these squares is extended further towards other bundaries for the flow net, keeping, of course, in view the above-mentioned three points (a), (b) and (c), If the choice of the first flow lines is incorrect, the last flow line will cut the second boundary for flow, otherwise if the choice were correct, it would conform to the second boundary flow line. In case there is an indication of incorrect choice of the first flow line, a second choice shall have to be made. Usually, the pattern of the incorrect flow net gives a clue to the correction required in choosing the first starting line.

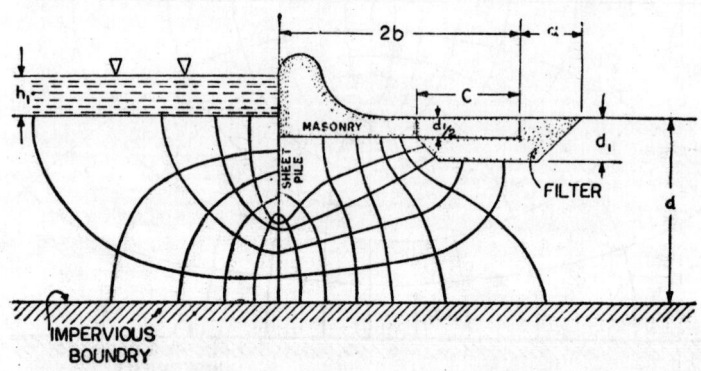

Flow net under a masonry weir with a sheet pile and a filter.
Fig. 6˙9 (c).

An advantage of a flow net is that it can take into consideration the anisotropy and non-homogeneity of the soil too.

3. Numerical Solution. The numerical solution of a differential equation can be found for conditions where an analytical approach fails or is difficult. This method of solution is called *relaxation method.* Essentially, it involves the conversion of a differential equation. If the co-efficient of permeability k is considered a constant, then the Laplace's differential equation for flow is given by equation (6·17). This equation can be converted into a finite difference equation as follows :

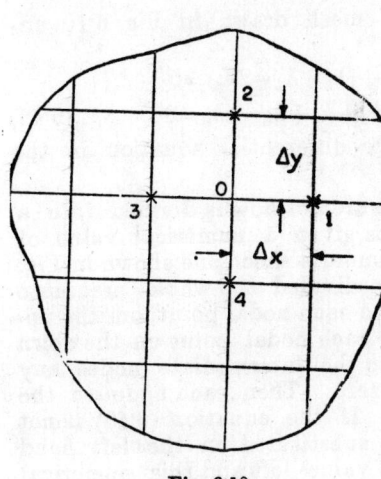

Fig. 6·10.

Refer to Fig. 6·10, which represents the portion of a flow zone in two dimensional co-ordinate system. Divide this zone into small portions of size $(\triangle x \times \triangle y)$. Consider the nodes marked, 0, 1, 2, 3 and 4 and let the head h at these nodes have values h_0, h_1, h_2, h_3 and h_4 respectively.

Then we may write,

$$h_1 = h_0 + \triangle x \cdot \left(\frac{\partial h}{\partial x}\right)_0 + \frac{(\triangle x)^2}{2!}\left(\frac{\partial^2 h}{\partial x^2}\right)_0 + \ldots \ldots \quad \ldots(6·43)$$

and

$$h_3 = h_0 - \triangle x \cdot \left(\frac{\partial h}{\partial x}\right)_0 + \frac{(\triangle x)^2}{2!}\left(\frac{\partial^2 h}{\partial x^2}\right)_0 + \ldots \ldots \quad \ldots(6·44)$$

where equations (6·43) and (6·44) are just Taylor's expansions. Neglecting terms higher than the second order differential, add equations (6·43) and (6·44).

$$\therefore \quad h_1 + h_3 = 2h_0 + (\triangle x)^2 \left(\frac{\partial^2 h}{\partial x^2}\right)$$

or

$$\frac{\partial^2 h}{\partial x^2} = \frac{h_1 + h_3 - 2h_0}{(\triangle x)^2} \quad \ldots(6·45)$$

Similarly, if we write Taylor's expansions for heads h_2 and h_4 and add similarly, we get,

$$\frac{\partial^2 h}{\partial y^2} = \frac{h_2 + h_4 - 2h_0}{(\triangle x)^2} \quad \ldots(6·46)$$

Adding equations (6·45) and (6·46), we get

$$\frac{\partial^2 h}{\partial x^2} + \frac{\partial^2 h}{\partial y^2} = \frac{h_1 + h_3 - 2h_0}{(\triangle x)^2} + \frac{h_2 + h_4 - 2h_0}{(\triangle y)^2} \qquad ...(6·47)$$

Left hand side of equation (6·47) is the expression on the left hand side of Laplace's differential equation (6·17). So, if this is substituted equal to zero, then the right hand side will also be zero. Also substitute $\triangle x = \triangle y$ in equation (6·47).

So, for the flow through this mesh drawn in Fig 6·10 enclosed by the nodes,

$$(h_1 + h_3 - 2h_0) + (h_2 + h_4 - 2h_0) = 0$$

or $$h_1 + h_2 + h_3 + h_4 - 4h_0 = 0 \qquad ...(6·48)$$

which may be called the Laplace's differential equation in the finite difference form.

In the practical problem, the area of flow is divided into a number of squares and each node is given a numerical value of head h. For *example*, if the flow under a sheet pile shown in Fig. 6·9 (a) is to be evaluated, we may divided the whole area into small squares and give the values to each nodal point on the upstream equipotential as 100 and to each nodal point on the down stream equipotential as zero and to the intermediate nodes any reasonable value between 100 and zero. Then each node in the grid must satisfy equation (6·48). If the equation (6·48) is not satisfied, then when h_0, h_1, etc. are substituted in the left hand side, there may be some numerical value[1] left and this numerical value is not zero. Then suitable changes may be made in the nodal values looking at this residue and a distribution carried out again on the whole grid. A repetition of this procedure will yield some value at each node such that equation (6·48) is satisfied simultaneously.

Then a flow net can be drawn by drawing equipotentials through these nodes and flow lines at right angles to these equipotentials. The process is laborious and time-consuming but can yield good results where analytical solution are not available or are tedious.

4. Electrical Analogy Technique. Laplace's differential equation (6·17) represents also the distribution of electrical potential (as also heat flow) in a homogeneous medium. This fact is made use of in developing an electrical model representing the structure with the help of which the distribution of electrical potential can be measured, using a potentiometer. The pervious medium used is a weak solution of acid in water, called an *electrolyte*. A null point detector is placed in series with the jockey on the potentiometer and a metal probe in the electrolyte. The jockey on the potentiometer and the metal probe in the solution are moved till there is no current, indicating that

---

1.   This numerical value is called *residue* for the node.

the two points are at the same potential. In this way, the potential at various points in the electrolyte can be found out and the flow net drawn for the structural model, by first drawing the equipotentials from the potential measurements and then the flow lines at right angles to these equipotentials.

6·10 Flow Net for Anisotropic Soils

Soils in nature, usually show stratifications in the horizontal direction and due to these stratifications, the horizontal permeability of a soil stratum as a whole is usually more than its vertical permeability. So, the stratum cannot be called isotropic. On the other hand, it is anisotropic (or sometimes this condition is called transverse isotropy).

If k_x and k_y denote the coefficients of permeability in the horizontal and vertical directions $(k_x > k_y.)$. then it can be shown that the Laplace's differential equation for flow is as under[1] :

$$\frac{k_x}{k_y} \frac{\partial^2 h}{\partial x^2} + \frac{\partial^2 h}{\partial y^2} = 0 \qquad ...(6·49)$$

and it is satisfied by a function $h(x,y)$.

Now if $\dfrac{k_x}{k_y}$ is substituted as "a^2", the equation becomes,

$$a^2 \frac{\partial^2 h}{\partial x^2} + \frac{\partial^2 h}{\partial y^2}$$

In the above equation if a substitution $x = aX$ is made, the equation may be written as[1],

$$\frac{\partial^2 h}{\partial X^2} + \frac{\partial^2 h}{\partial y^2} = 0 \qquad ...(6·50)$$

which is satisfied by the function $h = H(X,y)$

Equation (6·50) is the Laplace's regular differential equation as shown in equation (6·17).

In order to draw the flow net for the anisotropic conditions, the geometry of the structure and area of flow underneath it is distorted by substituting $x = aX$ where $a = \sqrt{\dfrac{k_x}{k_y}}$ and on this distorted geometrical figure a square flow net is drawn, in the usual manner. Then the discharge is calculated from the following equation.

$$Q = k_e h. \frac{N_f}{N_d} \qquad ...(6·51)$$

where Q is the discharge per unit time,

k_e is the equivalent coefficient of permeability,
N_f is the number of flow lines and
N_d is the number of equipotential drops.

---

1. Detailed proof is not considered necessary here.

It can be shown mathematically be taking the flow in the horizontal and vertical directions into consideration, that

$$k_e = \sqrt{k_x . k_y}$$

In case the stratifications are inclined at an angle and are not horizontal, then k_x and k_y may be looked upon as permeability coefficients in the directions parallel to and perpendicular to the stratification planes, respectively, and the distorted geometrical figure drawn accordingly.

It may be noted here, that the above observations are applicable theoretically to anisotropic and homogeneous media.

6·11. Seepage through an Earthen Dam

Flow through an earthen dam or an embankment is bounded by the upstream face of the dam or the embankment, the uppermost seepage line in the field of flow, called the **Phreatic surface** and the conditions of discharge at the downstream end. The main problem in drawing a flow net is the location of the topmost seepage line. Once this line has been located, the field of flow can be divided into elementary squares and the flow net drawn. The seepage discharge may then be calculated.

In this article, a method is presented for the location of the topmost seepage line for two simple cases.

1. When the discharge at the downstream end takes place into a horizontal filter provided inside the downstream toe.

(2) When the downstream slope of the dam forms in itself a medium for discharge of water and a horizontal filter is outside the downstream toe.

The practical cases may not be as simple as the above two cases, but they provide a good start for the beginner.

Case 1. *When the discharge at the downstream end takes place into a horizontal filter provided inside the downstream toe.*

Through a mathematical treatment, it has been established that the topmost seepage line in an earthen dam is parabolic in shape for most of its part. Corrections may be

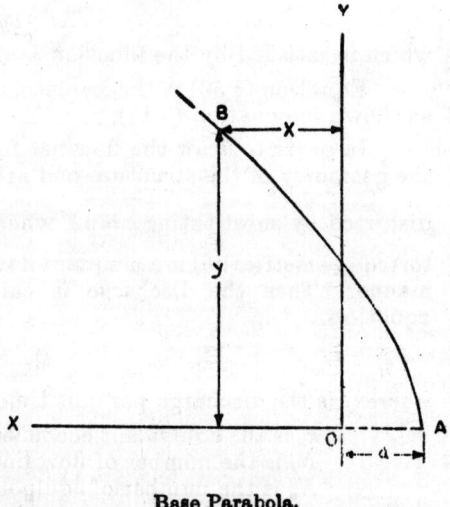

Base Parabola.
Fig. 6·11

needed at the upstream or downstream faces where this parabolic seepage line meets them.

Refer to Fig 6·11. point O represents the focus of a parabola and A represents its vertex. Then if B is a point whose co-ordinates with respect to the axes shown are (x,y), then the equation of this parabola may be written as

$$y^2 = 4a(x+a) \qquad ...(6·52)$$

where "a" is the distance between the vertex A and the focus O

Now in the earthen dam this parabola, which henceforth would be called the *base parabola*, represents the topmost seepage line for most of its part. In the case of the dam where a horizon-

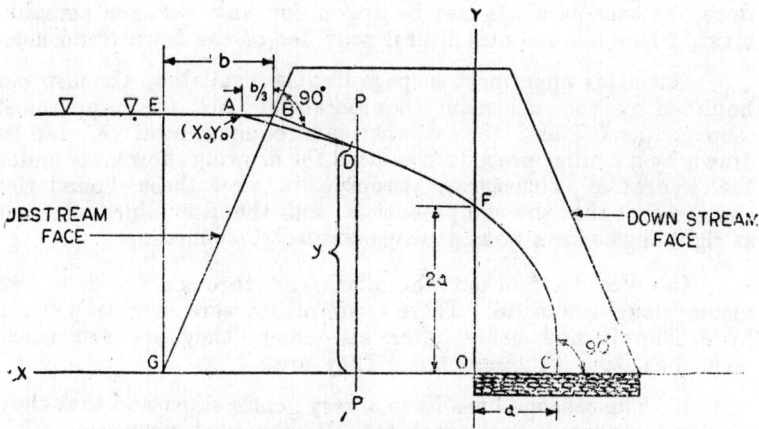

Dam with a filter inside the downsteam toe.
Fig. 6.12·

tal filter is provided at the down-stream end, this parabola ends at some point say C within the horizontal filter, as shown in Fig. 6·12 after starting from a point A in the free surface of water on the upstream side. The actual seepage line separates out from this parabola at some point D and meets the upstream face of the dam at a point B, as shown. Since the upstream face of the dam and the filter on the downstream side represent the two limiting equipotentials, the seepage line at the points B and C is normal to the equipotentials. Casagrande, found that the distance BA, i.e., the distance between the points where the seepage line meets the upstream face of the dam and where the parabola meets the free water surface is approximately one-third of the dam and the point where the free surface meets the upstream face of the dam.

i.e. $$BA = \frac{1}{3} BE = \frac{b}{3} \text{ in the figure.}$$

The other seepage lines below this top seepage line will also end in the horizontal filter, so that the last seepage line may be considered to end at the point O which is the start of the horizontal filter. In order to locate the topmost seepage line, take O as the origin. The co-ordinates of the point A can be known from the profile of the dam. Let its co-ordinates be $(x_0.y_0)$. Substitution of these known values of x_0 and y_0 in equation (6·52) would yield the value of "a" and, consequently, the distance OC in Fig. 6·12. Again when $x=0$, $y=2a$ t.e., the point F where this base parabola cuts the y-axis is known by this substitution of $x=0$ in equation (6·52). Similarly, a few more points can also be found, by substituting different values for x in equation (6·52) and finding out the corresponding values for y. By this procedure, the base parabola may be drawn for any earthen embankment, where a horizontal filter is provided on the downstream side.

Once this uppermost seepage line is available, the flow net bounded by the upstream equipotential BG, the uppermost seepage line BC and the downstream equipotential OC can be drawn by a similar procedure as used for drawing flow nets under the hydraulic structures, keeping in view these boundaries and the fact that the equipotentials and the flow lines intersect at right angles for a homogeneous and isotropic medium

In order to find out the discharge through the dam, two assumptions[1] are made. These assumptions were originally made by J. Dupuit and hence after his name, they are sometimes termed as *Dupuit's assumptions*. They are :

1. The seepage lines have a very gentle slope and that they may be considered horizontal for all practical purposes. This simply means that if a function $h(x,y)$ represents the family of equipotential lines, then gradient is given by $\dfrac{dh}{dx}$ and the y-component of the gradient is negligible.

2. The hydraulic gradient at the surface of a section may be taken as the gradient for the entire depth at that section and discharge calculated accordingly.

Now if OX as shown in the Fig. 6·12 is taken as the datum, then since the top seepage line is under atmospheric pressure the only type of head that can exist at any point is the elevation head at that point i.e., the y-co-ordinate of the seepage line in this case. So the gradient for discharge, by Dupuit's assumption will be given by $\dfrac{dy}{dx}$ and it will be applicable throughout the depth.

---

1. Discussion on the validity or non-validity of these assumptions is outside the scope of this book.

Consider now any section PP in this dam, so that the gradient $\dfrac{dy}{dx}$ at this section is obtained by differentiating equation (6.52), i.e.,

$$2y.\frac{dy}{dx}=4a$$

or
$$\frac{dy}{dx}=\frac{2a}{y} \qquad\qquad ...(6.53)$$

If one unit length of the dam is considered, then area of seepage at this section PP

$$=1.y=A \text{ (say)}$$

This discharge $\qquad q=k.\dfrac{dy}{dx}.A$

or
$$q=k.\frac{2a}{y}.y=2.k.a \qquad\qquad ...(6.54)$$

Case II. *When the down-stream slope of the dam forms in itself a medium for discharge and a horizontal filter is outside the downstream toe.*

Where the above situation[1] exists, a correction to the base parabola is required on the down stream side also, to obtain top seepage line. The situation is illustrated in Fig. 6.13.

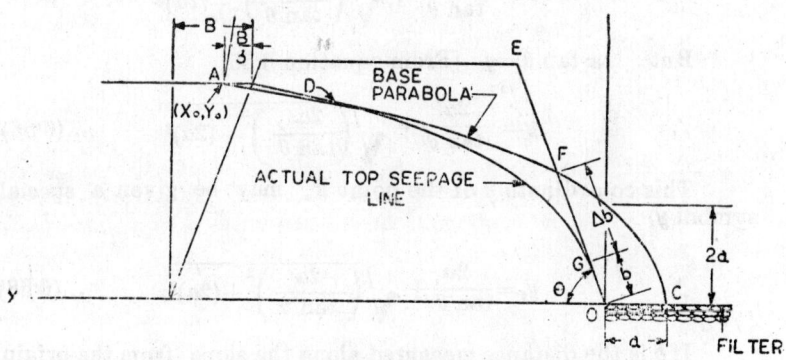

Dam with a filter outside the down-stream toe.
Fig. 6.13.

1. There is another situation where the correction to the base parabola on the down-stream end is needed and this is the case when the dam has a rock filled toe on the down-stream side. However, this is not discussed here.

Let θ be the angle of inclination of the down-stream slope of the dam. Take the intersection of the down-stream slope and horizontal as the origin. The actual base parabola is ADC but the seepage line is below this parabola towards the end of the downstream slope of the dam, it merges with it.

The equation of the base parabola is

$$y^2 = 4a(x+a) \qquad \qquad \text{...(6.55)}$$

and the equation of the line OE can be written as

$$y = x \tan \theta \qquad \qquad (\text{...(6.56)}$$

Substitution of the value of y from equation (6·56) in the equation (6·55) will give the point of intersection F of the line OE and the parabola ADC.

$$x^2 \tan^2 \theta = 4a(x+a)$$
$$\text{or} \quad x^2 \tan^2 \theta - 4ax - 4a^2 = 0.$$

Solve this quadratic in x.

$$x = \frac{4a + \sqrt{16a^2 + 16a^2 \tan^2 \theta}}{2 \tan^2 \theta}$$

$$\text{or} \quad x = \frac{2x}{\tan^2 \theta} + \frac{\sqrt{\left(\dfrac{2}{\tan \theta}\right)^2 + (2a)^2}}{\tan \theta}$$

$$\text{or} \quad x \tan \theta = \frac{2a}{\tan \theta} + \sqrt{\left(\frac{2a}{\tan \theta}\right)^2 + (2a)^2}$$

But $x \tan \theta = y$ [From equation 6·56]

$$\therefore \quad y = \frac{2a}{\tan \theta} + \sqrt{\left(\frac{2a}{\tan \theta}\right)^2 + (2a)^2} \qquad \text{...(6·58)}$$

This co-ordinate y of the point F, may be given a special symbol y_i

$$\therefore \quad y_i = \frac{2a}{\tan \theta} + \sqrt{\left(\frac{2a}{\tan \theta}\right)^2 + (2a)^2} \qquad \text{...(6·58)}$$

If b is the distance measured along the slope from the origin O upto the point G where the actual seepage line meets this slope and $\triangle b$ is the distance from G to the point F, then y_i can also be written as

$$y_i = (b + \triangle b) \sin \theta \qquad \text{...(6·59)}$$

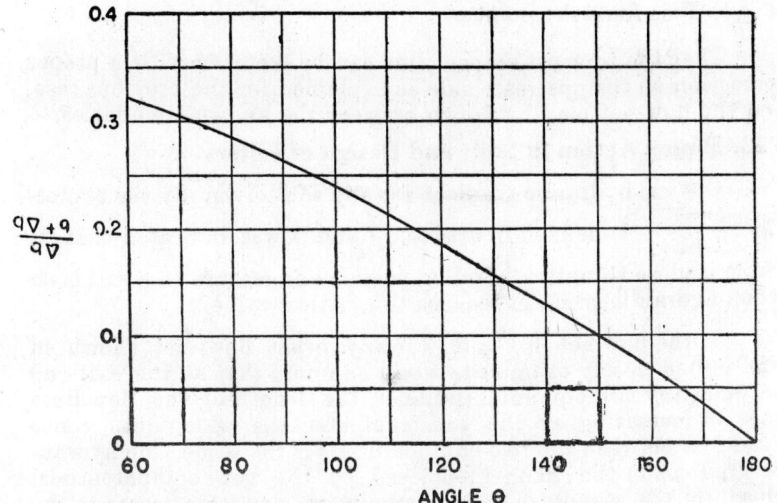

Correction to base parabola for a dam with a filter outside
the down-stream face.

Fig. 6·14.

From some known flow nets for dams with different values
of θ the values of $\dfrac{\Delta b}{(b+\Delta b)}$ versus θ are known. A plot of θ versus

$\dfrac{\Delta b}{(b+\Delta b)}$ is shown in Fig. 6·14. With this background, the following steps may now be taken to obtain the corrected seepage line.

1. From the profile of the dam, locate the point A and find
out its co-ordinates (x_0, y_0).

2. Substitute (x_0, y_0) in equations (6·55) and get the value
of "a".

3. With this value of "a" locate the point C in the filter.

4. Find out y_i, from equation (6·58). This fixes the
position of the point F.

5. Substitute the value of y_i obtained above in equation
(6·59) along with the value of θ for the down stream slope, to
obtain $(b+\Delta b)$.

6. Against the known value of θ, read the value of
$\dfrac{\Delta b}{(b+\Delta b)}$ from Fig. 6·14 and by substituting the value of $(b+\Delta b)$

obtained above, in this parameter from the graph, get the value of $\triangle b$. This fixes the point G.

Then the topmost seepage line can be sketched with a proper correction on the upstream side as explained for the previous case, and the flow net may be drawn between the known boundaries.

6·12. Piping Action in Soils and Design of Filters

Critical hydraulic gradient for any soil given by the expression $\dfrac{G-1}{1+e}$ was defined in article 6·7 and it was indicated that its value is close to unity. Boiling phenomena occurs in a soil mass when hydraulic gradient reaches this critical value.

If the flow net in Fig. 6·1 (or any other flow net shown in Fig. 6·9) is closely examined, it will be found that at the exit end between any two equipotential lines, the length of the flow lines goes on increasing as the two equipotentials widen out. Since head loss between the two equipotentials is the same, the hydraulic gradient in the channel (enclosed by the two equipotentials) close to the structure is the maximum and it decreases in the other channels away from the structure but enclosed between the same two equipotentials. If this hydraulic gradient in the channel close to the downstream end of the structure exceeds the critical value of unity, quick sand conditions would occur and, as the water flows outwards, it will erode the soil near this end. With the passage of time, erosion would travel backwards underneath the structure and ultimately what might happen is that a hollow channel or a number of channels may be formed underneath the structure which may lead to its failure. This phenomena of erosion and consequent formation of flow channels is called piping. Such a situation is guarded against at the downstream end by providing an inverted filter. All that the inverted filter does is that it reduces the exit gradient of flow and hence the seepage force. Consequently, erosion does not take place.

An inverted filter is nothing but a number of layers of processed material. The permeability of each layer is more than that of the layer below it. As the water emerges from the soil into these layers of increasing permeability, the hydraulic gradient falls sharply in each successive layer from the bottom towards top and so does the seepage force.

It is appropriate to mention here that the soils that are susceptible to piping action are neither the very uniformly graded nor the very coarse types. Their gradation curves usually lie between 0·04 mm. and 0·4 mm. equivalent diameter sizes.

There are certain requirements that must be met with in selecting the filter material. These requirements are asunder :—

Piping ratio, defined as

$$\frac{D_{15}}{d_{85}} = \frac{15\% \text{ diameter size of filter material}}{85\% \text{ diameter size of base material}} < N$$

and permeability ratio, defined as

$$\frac{D_{15}}{d_{15}} = \frac{15\% \text{ diameter size of filter material}}{15\% \text{ diameter size of base material}} > N$$

where N is a number whose value is 4 according to Terzaghi (8) and it is 5 according to the specifications of the U.S. Waterways Experiment Station at Vicksburg (3).

The piping ratio requirement is essential to prevent the flow of silt or fine sand into the filter material and the permeability in ration requiremnt is meant to provide sufficient permeability in the filter meterial.

These two limits define a range for the D_{15} size of the filter material. It is normally recommended that the particle size distribution curve for the base material and for the filter material are parallel.

Sometimes the filters are needed for the underdrains where pipes are provided to carry the water longitudinally as for example in the case of sub-surface drainage provisions for highways (4). In such cases, a limit is also placed on the coarser particle of the filter material around the drains. This limit on the coarser particle size is based upon the size of holes in the drainage pipes in case of perforated pipes and on the gap at the joints in the case of open-jointed pipes. D_{85} of the filter material in the such cases must be larger than twice the size of the diameter of the holes or the size of the gap, as the case may be.

CHAPTER—BIBLIOGRAPHY

1. Casagrande, A., seepage through Dams, Published in Contribution to Soil Mechanics, 1925 to 1940, Boston Society of Civil Engineers, Boston.

2. Gregg, L.E., Typical Flow Nets for the Solutions of Problems in Ground Water Flow and seepage, proceedings of the 1st International Conference on Soil Mechanics and Foundation Engineering, 1936.

3. Jumkis, A.R., Soil Mechanics, New Delhi, Van Nostrand/ East-West Press, 1965.

4. Khosla and Co-workers, Design weirs on Pervious Foundation, Central Board of Irrigation, India Publication No. 12.

5. Lambe, T.W., and Whitman, R.V., "Soil Mechanics," J. Wiley and Sons Inc., N.Y., 1969.

6. Road Research Laboratory, England, Soil Mechanics for Road Engineers, 1955, Her Majesty's Stationary Office.

7. Terzaghi, K, Theortical Soil Mechancis, New York, John Wiley and Sons, 1943.

8. Tomlinson, M.J., "Foundation Design and Construction," Wiley Interscience, John Wiley and Sons, Inc., New York, N.Y., 1969.

9. U.S. Waterways Experiment Station, Wickburg, Mississippi, Investigation of Filter Requirements for Underdrains, Technical Memo No. 183-1.

10. Wu. T.H., Soil Mechanics, Boston, Allyn and Boston, Inc., 1966.

QUESTIONS

1. What is a flow net ? State its properties and applications. Describe the different methods used to construct the flownets.
                                                               (P.U. 1959)

2. Draw the flownet for seepage under a vertical pile wall penetrating 25 ft. into a uniform stratum of sand of 50 ft. thick, overlying an impervious layer.

If water level on one side of the wall is 35 ft. above the sand and on the other side of the wall it is 5 ft. above the sand, compute the quantity of seepage for foot length of wall. Co-efficient of permeability of sand is 0·03 cms. per sec.   (P.U. 1960 Supp.)

3. Determine the total head, elevation head and pressure head of water at the entering end, the exist and point A in the sample of soil shown in the sketch below.

Find the factor of safety against quick condition if the soil has a void ratio of 0·85 and specific gravity of solids 2·71.   (P.U. 1962)

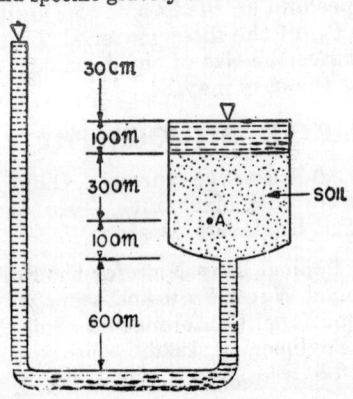

30 cm

100m

300m

100m

600m

SOIL

·A

Fig. 6·15

4. (a) What is a flow net ? What are its salient characteristics ?

(b) A weir shown below is built on a permeable stratum 22 ft. thick and underlain by rock. Plot graphically a flow net for the permeable foundation of the weir and estimate the seepage loss per foot length of weir if the co-efficient of permeability is 2×10^{-4} ft. per sec. Also calculate the hydrostatic uplift pressure at 3, 10·5 and 18 feet measured from the upstream edge of the floor. Use a scale of 1 in.= 3 ft. for plotting.

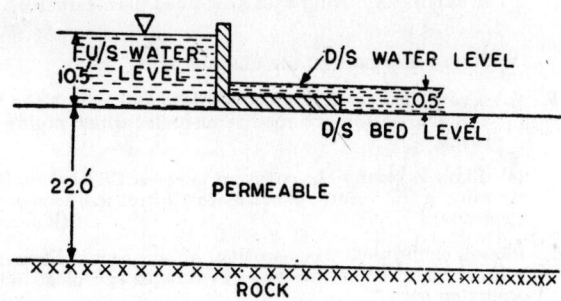

Fig. 6'16

5. Why is the quantity of seepage between two successive flow lines equal ? *(P.U. April, 1964)*

6. *(a)* What is a flow net and what are its uses in Civil Engineering practice ? By means of a simple illustration, prove that for a square flow net, there must be the same quantity of flow that passes through each figure and the same head drop in crossing each figure. How will you determine the quantity of seepage ?

(b) A concrete dam, 150 m. long has lines of sheet piling at both heel and toe which extend half-way down to an impervious stratum. The head on the dam is 10 m. From a flow net made up of square figures, on transformed scale, it is found that there are five seepage paths and sixteen equipotentials drops. The average value of co-efficient of permeability may be taken as 5×10^{-4} cms./sec. horizontally and 1×10^{-4} cm./sec. vertically. What will be the quantity of seepage ɔu. metres per day. *(P.U. Nov. 1964)*

7. Explain the mechanics of ;piping in hydraulic structures. What methods can be used to increase the factor of safety against piping ? *(P.U· April 1965)*

8. *(a)* What is a flow net ? Where do we ues it in Engineering Practice ? Illustrate with sketches.

(b) A coarse grained soil has a void ratio of 0'8 and a specific gravity of solids as 2'68. Calculate the critical gradient at which piping will occur ?

9. What is quicksand ? What hydraulic gradient is required to create quicksand conditions, in a sample of sand whose void ratio is 0'75 and specific gravity of soilds in 2'65 ?

10. Why many times a filter is considered necessary on the downstream end of an earthen dam ?

11. What is a flow net ? What information can be derived from a flow net drawn through an earthen dam resting on sound rock ? Illustrate with sketches. *(P.U. April 1966)*

12. *(a)* Derive an expression for "critical gradient".

(b) A natural deposit of sand has void ratio of 0'67 and a specific gravity of solids of 2 ʊ7. D.termine critical gradient for this soil.

(c) State Darcy's Law governing Flow of Water through soil. Explain how co-efficient of permeability is determined in laboratory using available head permeameter.

(Punjab 1967)

Note : For part (c) see chapter 5.

13. (a) Draw the sketch of an earth dam and sketch the flow net pattern. State the conditions under which piping will take place.

(b) Explain briefly the constant permeability set up for determining 'K' value. What factors affect the accuracy of "K" value. *(Warangal 1967)*

14. Sketch the typical cross-section of an Earth Dam about 50 meters high. What are the advantages and disadvantages of a sloping core ? *(Baroda 1967)*

15. Calculate the seepage loss and hydrostatic excess pressure at a point x after drawing the flow net for the case shown in the figure. *(Mysore 1968)*

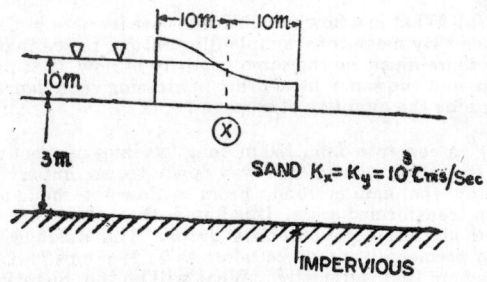

7

Consolidation Characteristics of soils

7·1. Introduction

When a structure such as a building is placed on a soil stratum, the layers of soil underlying the foundations have to withstand all the load that the building brings on the surface in touch with the bottom of the foundations. Two situations have to be guarded against

(1) The settlement of the foundations is not excessive.

(2) The soil underneath the foundations has sufficient strength to withstand the stresses induced by the loads.

The **settlement** of a foundation may be *defined* as its vertical deformation below the original level of placement. The soil properties responsible for this vertical deformation would be discussed in this chapter. The soil strength and its bearing capacity will be dealt with in a subsequent chapter,[1]

7·2. Compressibility of Soil

As indicated in chapter 2, a mass of soil consists of the solid particles and voids enclosed by these particles. The voids may be completely filled with air (when the soil is in a dry state) or they may be partially filled with air and partially with water (when the soil is in a ttate of partial saturation), or they may be completely filled with water (when the soil is in a state of complete saturation) When a foundation is placed on a stratum, settlement can take place due to :—

(1) Lateral displacement of the soil particles from their original position, or

(2) Decreases in volume of soil mass, or

(3) A combination of (1) and (2).

---

1. See Chapter 12.

In a natural deposit, there is not much scope for lateral displacement of the soil, since whatever soil layers come underneath the foundation, they are confined on all sides, and due to this confinement, the material cannot squeeze out from below the foundation.

The only possibility, therefore, of settlement lies in the volume decrease of the soil. Volume decrease of the soil mass can be due to three possible reasons :—

(1) Compression of the solid grains,

(2) Compression of the pore water and pore air,

(3) Expulsion of the water from the voids and consequent decrease in the void ratio.

Under stresses usually induced by loads in soil masses the solid grains and the pore water may be considered practically incompressible. Therefore, the only other reason for volume decrease is the compression of pore air and the expulsion of pore water from the voids. In case the soil is completely saturated, it is reasonable to assume that the volume decrease occurs due to the expulsion of water from the voids and consequently settlement takes place. In case of partially saturated soils, however, an application of a small amount of stress may compress the air or gas in the voids considerable and cause appreciable volume decrease and consequent settlement, without expelling any water from the voids.

Sedimentary soil deposits are normally 100 percent saturated. Moreover, the settlement problems are normally associated with submerged saturated clays and hence in all settlement problems, hundred percent saturation is assumed.

If an all round pressure is applied to a saturated granular soil sample without allowing any material to escape and precision measure ments on volume change made, it will be found that the volume decreases although the decrease is very small and inappreciable for all practical purposes as stated in the preceding paragraphs. If, now, under similar conditions the all round pressure is released voluntary, expansion of the sample will take place, though the expanion will not be as much as the previous volume decrease. A similar behaviour was noticed by Terzaghi, (2) on yellow residual pottery clay, when 2 cm and 4 cm cubes of this clay were tested to determine the stress-strain characteristics. The stress-strain diagrams so obtained were quite comparable to the ones that are obtained on concrete or natural stone. This behaviour of the soils indicates that the strain so introduced is both elastic and non-elastic in nature. It is impossible to separate out this compresion into elastic and in eleastic components but it is sufficient to know that the elastic tendency in soils is very very small and any elastic strain under normal loads is recovered on removal of the loads.

Another important strain-rebound occurs in fine grained soils, like clays and silt-clay admixtures. Between the flaky soil particles, water is held up by some forces, the quantity of water depending on the load on the soil skeleton. When the load is increased on such a soil sample, some of this water is expelled out and as soon as the load is brought back to its original value, these forces suck the water in and swelling takes place. Expulsion and sucking in the water may take considerable time.

As a conclusion, it can be said that the compression of saturated fine-grained soils, mainly occurs due to the expulsion of water from the soil mass and some of the water gets sucked in on removal of the pressure and consequently strain-rebound occurs, Process which involves escape of water from pores of the soil mass and its consequent gradual and simultaneous compression is called **consolidation.**

In the laboratory, a sample of soil can be consolidated in two ways, for determination of its compressibility characteristics :

(1) The sample is placed in metal ring and no lateral movement of the sample is allowed during loading. Here the cross-sectional area remains constant and the compression is uni-axial only. This type of compression is called one-dimensional consolidation.

(2) The sample is placed in a triaxial cell and pressuse is applied both axially as well as radially. The type of compression is called trianial compression or consolidation.

In this chapter, only one dimensional consolidation will be dealt with. The testing is usually done in the undisturbed state of the sample, unless compressibility characteristics of remoulded soils are needed.

7·3. Measurement of Compressibility Characteristics

Refer to Figure 7·1. A saturated soil sample is contained within a gun-metal ring. Two deaired and saturated porous

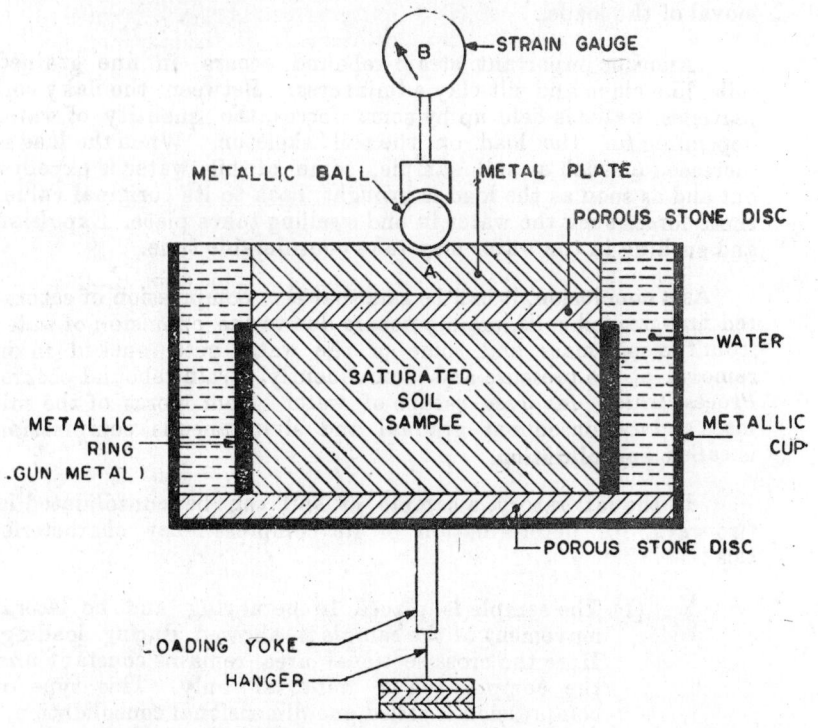

Diagrammatic representation of the Consolidation Device

Fig.7

stone discs are placed, one touching the lower surface of the soil and the other touching its upper surface. The assembly is placed in a metallic cup, to which water is added to the level of upper porous stone in order to avoid any surface tension in the soil and to allow free flow of water only under the external loads. The sample can be loaded through a metal plate A which carries a socket in the centre for a metallic ball on which a loading yoke can be hung as shown and the desired loads placed on the hanger below. Alternatively, a lever arrangement is used for loading the sample. A dial gauge B with a minimum reading of 0.0001 inch (=0·000254 cms) measures the strain under any load.

An apparatus, like the one shown in Fig 7·1 is called a *conshdometer*. Many patent designs of the apparatus are available in the market, but the principal parts are similar.

When the load is placed upon the hanger, the sample of soil starts compressing first rapidly and then after some time slowly, as seen from the strain dial gauge readings. The reading of the dial gauge are noted at 15,30 secs, 1, 2, 4, 8, 15, 30 minitues, 1, 2, 4, 8, 16, and 24 hour (and even afterwards at regular intervals of time, if needed), after placeintervals of the load on the sample.

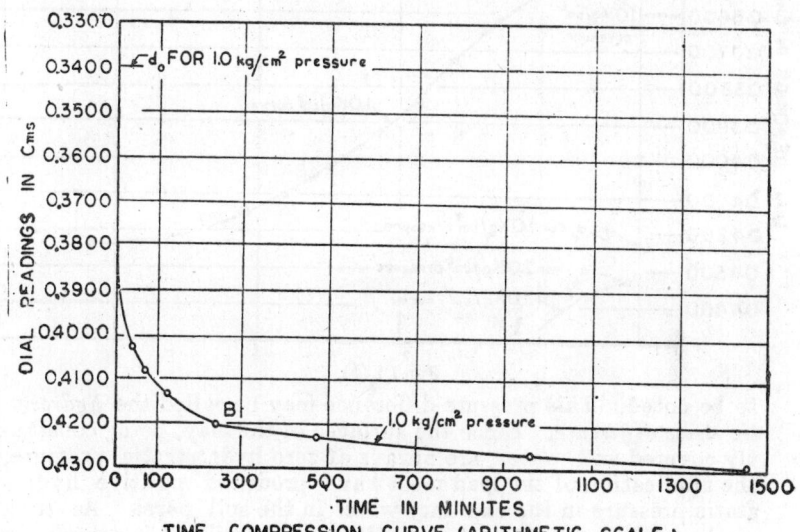

TIME COMPRESSION CURVE (ARITHMETIC SCALE)

Fig 7·2 (a)

The moisture-content and pressure in clayey soils are as definitely related as the stress and strain in soild bodies. The only difference is that compression of a clay that takes place on application of pressure, and consequent release of moisture from the soil mass takes time, as seen from the strain dial gauge readings, plotted in Fig 7·2 (a) and (b), against time since loading, whereas the application of pressure on a solid body is immediately followed by the corresponding strain.

Two points in the case of soils are noteworthy.

The flow water from a soil, when the pressure on it is increased can take place only if a hydraulic gradient exists. The existence of hydraulic gradient in the soil sample means the presence of pressure difference between the interior of the soil sample and its two surfaces where the water oozes out into the metallic cup through the porous stone discs. That a pressure difference exists between the centre and surface is the *first point*

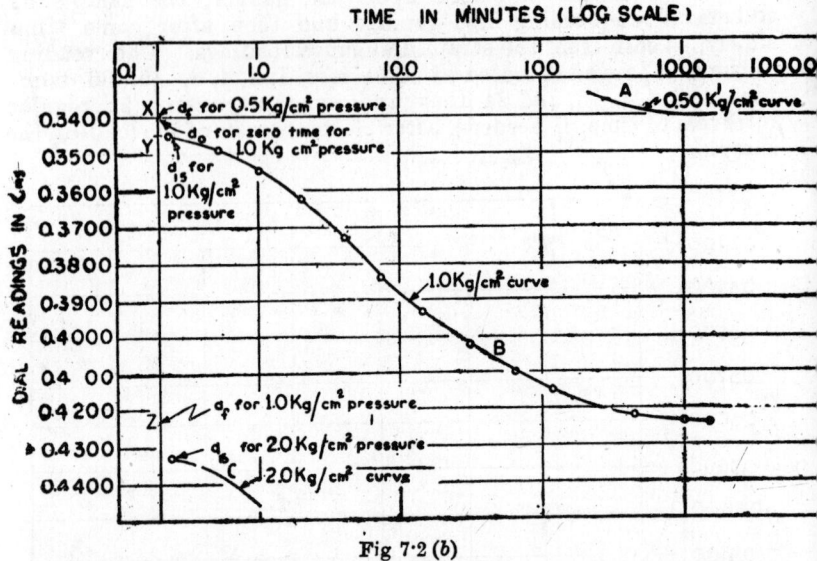

Fig 7·2 (b)

to be noted. This pressure difference may be called the *hydrosta-tic stress difference*. Since the surfaces of the clay, being constan-tly covered with water, are always at zero hydrostatic pressure, the application of the load must have produced positive hydro-static pressure in the capillary water in the soil pores. As soon as the load is placed on the sample, momentarily, it is completely supported by the pore water ; the pore-water pressure through-out the sample is equal, at this instance and is at its maximum value equal to the pressure exerted by the load on the sample, and the intergranular stress is zero throughout the soil sample. At this moment, hydrostatic stress difference between the surface and any point even an infinitesmally small distance inside the soil sample is maximum, so that the water, starts flowing out at the surface at a rapid rate. As the time passes, this hydrostatic stress difference is slowly consumed by the resistance which the water meets in the narrow voids of the clay on its way to the surface or alternatively it can be said that with passage of time the pressure is gradually shifted on to the soil grains and the pore pressure decreases. As the pore pressure decreases, the intergranular pressure increases by an equal amount, till such time that hydrostatic equilibrium is reached, *i.e.*, the pore pressure becomes zero and the whole of the pressure due to the load is taken up by the grains as intergranular stress. At this moment the sample has attained its full compres-sion under the load. That a time lag exists between the application of a load and attainment of full compression under it, is the *second point* to be noted.

The hydrostatic pressure referred to above, which exists under the aforesaid transient conditions, is designated at *hydrostatic excess pressure.*

The variation of the hydrostatic excess pressure with time will further be discussed under *mechanical analogy to the consolidation process* on soils, in Section 7·7.

The dial readings taken above are plotted against the corresponding time intervals. A plot of these readings on a natural scale when the load on a sample has been increased from 0·50 kgs. to 1·0 kg /cm., is shown in Fig 7·2 (*a*) and on a semi-logarithmic scale, with the dial readings on natural scale and time on the log scale are shown in Fig. 7·2 (*b*) (Curve B).

Fig. 7·2 (*a*) clearly indicates that the compression is very rapid in the beginning and then, as the time passes, the rate of compression becomes less. In most of the soils, at the end of 24 hours, the rate of compression becomes negligibly small. At this time the sample is assumed to have attained almost full compression under the load and the hydrostatic excess pressure inside the sample is considered almost zero at all points. The integranular stress at all points in the sample, after complete compression, equals the applied pressure.

A close examination of Fig. 7·2 (*a*) shows that if the dial readings corresponding to 15 and 30 seconds, 1, 2, 4, 8 and 15 minutes were to be plotted, there would have been to much congestion of these points in the early part of the time-compression curve, whereas these readings when plotted on a semi-logarithmic scale given in Fig 7·2 (*b*) are all spread out. Also the slow rate of compression towards the end of compression under the load is not as much prominent in an arithmetically plotted curve as in a semi-logarithmic curve, as seen from comparison of the latter part of the time-compression curves in the two plots.

At the end of the loading time for each load, the load on the sample is increased in such a manner that the *load increment ratio i.e.,* the ratio of the amount by which the load is increased to the original load is always unity. This procedure of loading gives a proper curve between the void ratio and the intergranular pressure[1] p. The first load is usually about $\frac{1}{4}$ kg./cm². and the last load depends upon the degree of compressibility to which the testing is required to be done, invariably depending upon the

---

1. The intergranular pressure is also denoted as σ.

field problem. The above law, of course, governs each load increment. Placement of complete set of loads on the sample till the desired pressure is reached. is usually referred to as a *cycle of loading*. After the desired loading is over, and in case the rebound of the e versus $\bar{p}$ curve is needed, the sample is unloaded by removing loads from the hanger one by one in the reverse sequence in which they were placed on it and for each load in this cycle of unloading, dial readings corresponding to similar time intervals are taken for 24 hours after removal of a load. During this period the water enters the soil, swelling takes place and the dial readings decrease. This process will yield the rebound leg of the e versus $\bar{p}$ curve, shown in Fig. 7·3 (b).

After the loading and unloading cycles are over, the sample is taken out of the ring and its water content determined. In order to know the void ratio of the sample at the end of the test, specific gravity G of its solids is needed. This is determined separately by taking a sample from the same deposit just close to the sample used for the consolidation test. Knowing the value of w and G, the void ratio e at the end of the test is given by,

$$e = Gw \qquad \qquad ...(7\cdot1)$$

since the soil is completely saturated and degree of saturation $S=1$. The inner diameter of the gun-metal ring gives the area of the sample. The height of the sample at any stage of testing can be measured accurately by taking two dial gauge readings, one obtained on the top of the metallic disc and the other on the top of the ring. The actual sample thickness at that stage of the test in then obtained as under.

Actual sample thickness $= h =$ Difference of the above two dial gauge readings $+$ the height of the ring $-$ the thickness of metal plate and the porous stone disc. Usually, the sample thickness h is found out by dial gauge measurements at the end of the test, as the corresponding void ratio e is also known at this stage. The dial readings, then, furnish data to determine the sample thickness at any other stage of the test.

If V is the volume of the sample at any stage, when the void ratio is e and its thickness is h, then,

$$V = A.h = V_s (1+e) \qquad \qquad ...(7\cdot2)$$

where A is the area of the sample and V_s the volume of soilds in the soil mass.

Now if, $\triangle h$ is the change in the height of the sample and $\triangle e$ the corresponding change in its void ratio, then

$$A.\triangle h = V_s.\triangle e \qquad \qquad ...(7\cdot3)$$

Divide equation (7·3) by equation (7·2),

$$\frac{\triangle h}{h} = \frac{\triangle e}{1+e} \qquad \qquad ...(7·4)$$

If e and h at any stage of the experiment are known and $\triangle h$ is the difference in height given by the dial gauge readings corresponding to a subsequent change in void ratio $\triangle e$, then $\triangle e$ can be worked out from equation (7·4) and the new void ratio found out.

7·4. Pressure Void-Ratio Diagram

Each load in the cycle of loading yields a curve of the type shown in Fig. 7·2 (a) or (b). This curve shows that before putting a load from a series of loads for which the consolidation cycle is to be run, the sample is at an initial void ratio say e_i corresponding to some dial reading which may be called d_e, and after compression under each load, as the time passes, the sample attains a final void ratio say e_f corresponding to some other dial reading which may be called d_f. The initial void ratio e_i. corresponding to a load is the final void ratio for the previous load. So, for

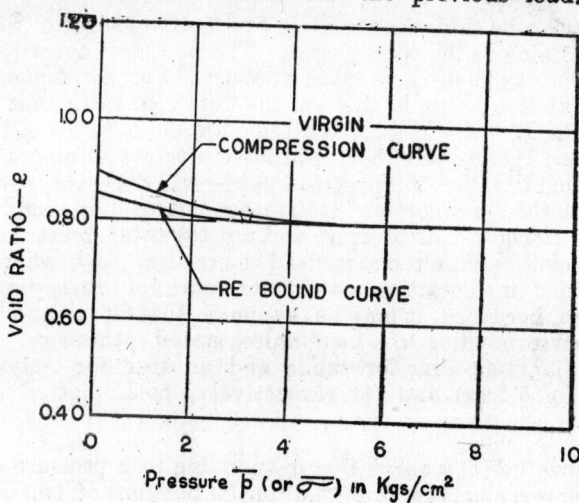

Typical pressure void-ratio diagram for sands.

Fig. 7·3 (a).

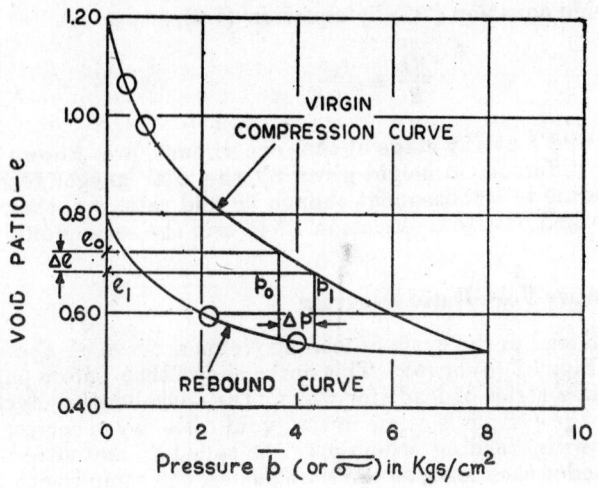

Typical pressure void-ratio diagram for clay.

Fig. 7·3 (b).

each load, it can be said, that there is one final void ratio which the sample attains as the time passes. The pressure corresponding to this load is the intergranular pressure. So, corresponding to one integranular pressure, there is one final void ratio that the sample attains. If these pressures attained under a series of loading are plotted against the final void ratio attained, a diagram is obtained which is called the **pressure void-ratio diagram**, usually designated as the *e-p diagram*. It may be noted, however, that the p is the inter-granular pressure and not the total pressure and mere nomenclature should not mislead the reader. Also whenever the void ratio e in connection with the pressure void-ratio diagram is referred to, hereafter, it may be assumed that it is the final void ratio corresponding to a load, unless stated otherwise. Two typical *e-p* diagrams, one for sands and another for clays are shown in Fig. 7·3 (a) and (b) respectively, both plotted on a natural scale.

In figure 7·2 (b) a curve B corresponding to a pressure of 1·0 kg/cm² is shown and the final and initial portions of two curves designated as A and C corresponding to 0·5 kg/cm² and 2.0 kg/cm² pressures respectively are also shown. If a horizantal line is drawn from the last point on the curve A, it is seen that it does not meet curve B at all, and the curve B falls below this line. Theoretically, the final dial readings d_f for 0.50 kg/cm² load should be the initial dial reading for the next pressure of 1.0 kg/cm² and so on. But the final dial reading d_f for 0.50 kg/cm² pressure and the apparent initial dial reading d_{15} for 1.0 kg/cm² pressure have some difference. This difference shown as XY on the vertical axis occur either due to the immediate escape of any dissolved air in the sample when the pressure of 1.0 kg/cm is placed on the sample or due to the rapid compression that takes place within the first 15 seconds when the reading cannot be taken or due to both. As such, the total compression under a load of 1.0 kg/cm² is not the ordinate YZ, it is the ordinate XZ. The difference XY between the final dial reading d_f for 0.5 kg/cm² pressure and the apparent initial dial reading d_{15} for 1.0 kg/cm² pressure is called the initial compression. As such, at zero time corresponding to 1.0 kg/cm pressure, the true initial dial reading d_o may be taken as 0.3400 cms. corresponding to the final dial reading d_f for 0.5 kg/cm² pressure.

The calculations for the final values of e corresponding to the various pressures is based upon the total compression XZ including the initial compression XY.

Fig. 7.3 (a) is a typical pressure void ratio diagram for sands and Fig. 7.3 (b) is a typical pressure void ratio diagram for clays. A comparison of the two diagrams shows that, for the same range of pressures, the compression in sands is much less than it is in clays. Furtheron, it has been observed that the small compression in sands occurs immediately on placement of loads and does not prolong for a long time as in the case of clays. In fact the settlement problems are never associated with the foundations below which the strata are of granular material. They are always associated with the foundations below which clay stratifications exist.

The two diagrams also indicate that the compression is not recovered completely, when the loads are removed. The curve obtained during the loading cycle may be called **virgin compression curve** or simply *virgin curve* and the one obtained during the unloading cycle, the **rebound curve**. The virgin compression and rebound curves obviously are different, as seen in the two diagrams.

If after complete removal of all the loads, the sample is reloaded with the same series of loads as in the initial loading cycle, another curve called the re-compression curve is obtained. The virgin, rebound and re-compression curves for a clay sample are shown Fig. 7.4. The void ratios and corresponding pressures in this diagram are both plotted on a natural or arithmetical scale. It is seen from these curves that the **rebound** started at a pressure of 8 kg/cm² and at zero pressure, the sample attained a void ratio of 0.80, much less than the original void ratio of 1.15.

The re-compression curve obtained on rebounding of the sample at this void ratio of 0·80 is different, though difference in void

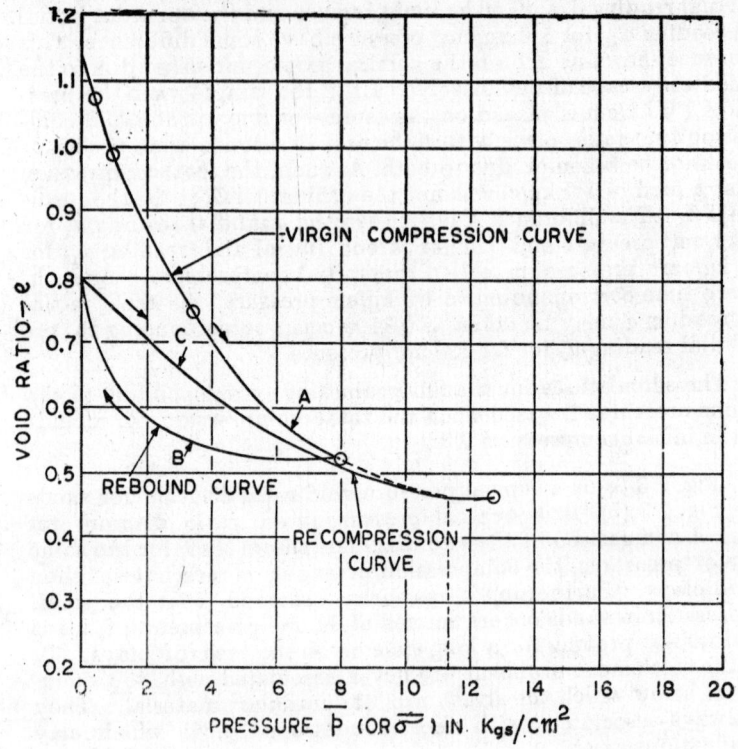

Virgin Compression, Rebound and Recompression Curves.
(Natural Scale)

Fig. 7·4.

ratios at any pressure obtained on the re-compression and rebound curves is not much. However, the difference in void ratios against a pressure, obtained on the re-compression curve and the virgin curve is quite large at lower pressures but as the gap between these two curves narrows down with increasing pressure the difference in void ratios becomes less and less gradually till at about 8·0 kg/cm² pressure the two curves start joining and there is no difference in void ratio obtained on the two curves.

The three curves shown in Fig 7·4 can also be plotted on a semilogarithmic scale, with the void ratios on the vertical arithmetical scale and the pressures on the horizontal log scale. A

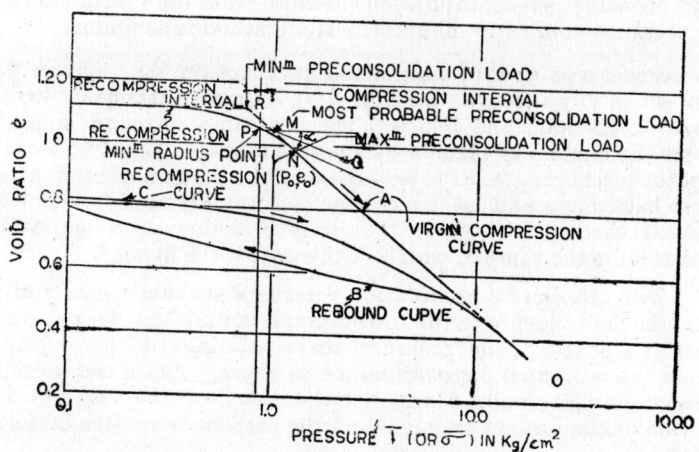

Virgin Compression, Rebound and Recompression Curves,

(Log Scale)

Fig. 7·5.

semi-log plot gives a better understanding of the nature of the soil. A plot of the same three curves on a semi-log scale is shown in Fig. 7·5. The virgin curve on the semi-log plot is convex at low pressure and as the pressure increases, it becomes a straight line. If the re-compression curve is examined, it is found that below a pressure of 8 kg/cm² the curve is convex and above this pressure it tends to become a straight line. It may be stated here that the sample was compressed to 8 kg/cm² before allowing it to rebound and recompress. The identity between the shape of the recompression and the virgin curves, suggests that possibly the original soil sample had been compressed to some pressure in the field and then due to some field conditions, the pressure had been released and the laboratory virgin curve is only another version of a curve for a pre-compressed material. The pressure to which the sample has attained initial compression is called the **pre-consolidation pressure** or the **pre-consolidation stress,** or the **pre-consolidation load.** Normally, when the pre-consolidation pressure for a soil is mentioned, it implies that the pre-consolidation pressure of original soil on the virgin curve, although for the laboratory recompression curve also, this term could be used and for this curve C shown in Fig. 7·5, the pre-consolidation load is 8 kg/cm².

A Casagrande[1] has devised a method for determining the most probable pre-consolidation pressure from the virgin curve of the pressure void ratio diagram. The method is as under.

Select a point P of maximum curvature on the convex portion of the virgin curve (See Fig. 7·5). Draw a tangent to the curve at this point and also a horizontal line through a point. Bisect the angle α so formed by these two lines. Let PQ be the bisector of this angle. Project the straight portion of the virgin curve backwards and let it meet the bisector PQ in M. Then the vertical through M gives the most probable preconsolidation pressure for the sample, which in this case is 0·9 kg/cm^2.

Two more pre-consolidation pressures are sometimes indicated. If the tangents to the re-compression and the virgin curve meet at a point R, the pressure corresponding to this point is called the *minimum preconsolidation pressure*. Again the pressure corresponding to point N which is the start of the straight line portion of the virgin curve is called the *maximum pre-consolidation pressure*.

A knowledge of the most probable pre-consolidation pressure is helpful in settlement predictions. Since in the convex portion of the virgin curve, the change in void ratio is insignificantly small, any foundations inducing a pressure less than the pre-consolidation pressure will result in very little settlement and in almost all such cases there is no need for settlement analysis.

If (p_0, e_0) are the co-ordinates of the point of start N of the straight line portion NO of the virgin curve, then the equation of this straight line can be written as,

$$e_0 - e = C_c(\log_{10} p - \log_{10} p_0) \qquad ...(7·5)$$

where (p, e) are the co-ordinates of any other point on the line and C_c is a constant depending upon the slope of the line.

Re-arrangement of equation (7·5) gives,

$$e = e_0 - C_c(\log_{10} p - \log_{10} p_0)$$

$$= e_0 - C_0 \log_{10} \frac{p}{p_0} \qquad ...(7·6)$$

C_c is negative here.

The virgin curve as seen in Fig. 7·5 has a negative slope.

Sometimes it is not possible to mark exactly the point N where the virgin curve becomes a straight line. So the straight

---

1. A Casagrande's procedure is one out of a number of procedures, that are used to determine the pre-consolidation pressure.

line portion of the virgin curve is produced backwards and the values of p_0 and e_0 are picked up for the point where this straight line cuts the ordinate of 1·0 kg/cm² pressure.

In order to determine the value of C_c, the pressure and void ratio over one cycle of the log scale may be chosen and substituted in equation (7·6). *For example*, for p_0 equal to 1·0 kg/cm², $e_0 = 1·01$ and p equal to 10·0 kg/cm² e is equal to 0·48, in the figure under reference.

Substituting in equation (7·6) gives,

$$0·48 = 1·01 - C_c \ \log_{10} \frac{10}{1}$$

or $\qquad\qquad C_c = 0·53.$

It can be seen that the rebound or expansion curve B is also a straight line over a large range of pressure, and its equation can be written as,

$$e_0 - e = C_e (\log_{10} p - \log_{10} p_0) \qquad\qquad ...(7.7)$$

where ($p_0 \ e_0$) may be taken as co-ordinates of point of intersection of the 1 kg/cm² pressure ordinate with the curve B and (p, e) are the co-ordinates of any other point on the curve and C_e is a constant depending upon the slope of the line.

Rearrangement of equation (7·7) gives

$$e = e_0 - C_e (\log_{10} p - \log_{10} p_0)$$

$$= e_0 - C_e \ \log_{10} \frac{p}{p_0} \qquad\qquad ...(7·8)$$

The value of C_e may be determined again by taking the pressures and void ratios over one cycle of the log scale.

For the curve B shown in Fig 7·5,

when $\qquad\qquad p_0 = 1$ kg/cm², $e_0 = 0·635$ and

when $\qquad\qquad p = 0·1$ kg/cm², $e = 0·790$

Substitution in equation (7·8) gives,

$$0·790 = 0·635 - C_e \ \log_{10} \frac{0·1}{1} = 0·635 = C_e$$

or $\qquad\qquad C = 0·790 - 0·635 = 0·155.$

Pressure void ratio diagrams can be drawn from this undisturned and remoluded states of the same soil and the two diagrams invariable differ as shown in Fig. 7·6 and explained in the next section.

7·5 Definitions

Before any further details of consolidation of soils are given it seems essential at this point to give a few definitions and present some explanations.

(a) **Normally Consolidated Soil.** A soil is said to be *normally consolidated*, if the present load on it has not been exceeded any time during its stress history in the past. In the laboratory consolidation curves, shown in Fig. 7·5. the strnight line portion of the virgin curve may be called as corresponding to a normally consolidated soil. *Normally loaded soil,* is another name for this normally consolidated soil. In this case of a normally consolidated soil, the change in void ratio corresponding to a change in pressure is substantial and one must expect a good deal of settlement in case the pressure from a foundation on a soil is more than the preconsolidation pressure of the soil.

This is evident from curve A of Fig. 7·5. Experience shows that the natural water content of a normally consolidated clay is equal to its liquid limit.

(h) **Overconsolidated Soil.** A soil is said to be *over consolidated* if the present load on it has been exceeded at any time during its stress history in the past. In the laboratory consolidation curves, shown in Fig. 7·5, the branch of the virgin compression curve before the pre-consolidation pressure *i.e.,* the branch corresponding to the precompression interval may be considered as corresponding to an over consolidated soil. *Precompressed soil* is another name for the over consolidated soil. All it means is that the soil has previously been compressed to a pressure higher than the pressure it carries at the present moment. In the figure under reference the soil has been compressed to a pressure of 0·9 kg/cm² in the field and the branch of the virgin curve in the pre-compression interval only represent the recompression of the soil to 0·9 kg/cm².

In the case of over-consolidated soils, the change in void ratio corresponding to a change in pressure (less than the preconsolidation pressure), is very small and if a foundation which exerts a pressure lesser than the preconsolidation pressure is laid on it, the settlement will be negligible. That is why many times settlement analysis is not necessary if the foundation pressure is less than the preconsilidation pressure. If the foundation pressure is more than the preconsolidation pressure, we enter the normally consolidated state of the soil and the settlement can be substantial, as explained earlier.

As a conclusion, it can be said that a normally consolidated soil is much more compressible than an over consolidated soil.

It may be noted, here, that phrases "normally consolidated" and "over-consolidated" are not used to describe any special type of a soil but they just indicate the state in which the soil exists under existing pressure.

In the field there are many reasons for the soil to get over-consolidated. They can be enumerated as :—

(1) In the geological history of a soil deposit, it might have been under the weight of an ice sheet or a glacier, which after the soil got consolidated under its pressure, melted away, leaving the soil deposit in a precompressed state.

(2) Thick layers of over-burden over a soil might have eroded, leaving it precompressed.

(3) A soil deposit might have had heavy construction over it in the past, which might have been pulled down for some reason. Under the weight of the heavy structures, the soil might have got preconsolidated.

(4) Evaporation of water from the clay layers leads to pre-consolidation.

(5) A fine grained soil might have been subjected to capillary pressure in the past and this pressure might have been removed for some reasons.

(c) **Time Lag During Compression.** It has been mentioned in Section 7·4 that a time lag exists between the application of a load and attainment of full compression under it, as seen in Fig. 7·2 (a) or (b). This time lag is a special characteristic of clays under compression and it is due to this time lag that the settlements in building foundations do not occur immediately either during or after construction. It takes, in many cases, years for the complete settlement to take place. This can be be seen from the time settlement curve shown is Fig. 9·6 in Chapter 9.

There are mainly two reasons for this time lag. The *first* is that it always takes time for the pore-water to escape from the voids of the soil and time for flow is dependent upon the co-efficient of permeability of the soil and the path of flow. The more permeable a soil is, the less is the time required by the pore-water to escape and *vice versa*. This portion for the total time lag is called *hydro dynamic lag*. The second reason for the time lag is due to the plastic action[1] in the adsorbed water layers between the soil grains in contact with each other. Due to this

1. Terzaghi's Theory of one-dimensional consolidation which is explained in article 7·7 does not take into account this plastic action of the adsorbed water layers.

action, more time is needed for a soil layer to consolidate under a pressure. Under this plastic action the intergranular pressure corresponding to a load is not a constant quantity as assumed in drawing the pressure void ratio diagrams previously. However, the pressure void ratio for simplicity and for want of any definite knowledge about the exact effect of plastic action, is considered a constant quantity. The time lag due to this plastic action is defined as *plastic lag*.

The time required by a sand sample to consolidate is mainly of frictional nature, as it takes only a minute or so for the water to escape from the pores and the balances of the time may be attributed to the frictional resistance to adjustment of the grains. The *frictional lag* in sands may be considered the simplest form of plastic lag in soils.

(d) **Compression and Swelling Indices.** A virgin compression curve is represented in Fig 7·5 on a semi-log scale and it so happens that beyond the preconsolidation pressure, it is almost a straight line. Equation 7·6 represents the void ratios and pressures on this straight line, related by a constant C_c dependent upon the slope of the line. This constant is called the **compression index** of the soil. It may be noted that C_c is a dimensionless quantity because from equation (7·6),

$$C_c = \frac{e_o - e}{\log_{10} \dfrac{p}{p_o}} . \qquad \qquad ...(7·9)$$

In equation (7·9), both the numerator and denomenator on the right hand side are dimensionless quantities and hence C_c is diamensionless.

The value of C_c is a relative measure of the compression in a soil. The smaller is the value, the less will be the compression in the soil and *vice versa*. This is obvious, because a flatter slope of the virgin curve and hence relatively lesser change in void ratio for a given change in pressure and *vice versa*.

The pressure void-ratio diagrams can be drawn both for the undisturbed state of a soil as well as for its remoulded state. The values of the compression indices for the two curves are invariably different.

Refer to figure 7·6. Supposing a clay sample for consolidation test is drawn from a depth of z below the surface level. The sample in position in the ground has the void ratio e_f corresponding to a pressure p_f from the over-burden on it. This point (p_f, e_f) is plotted on a semilog scale in this figure as point a. Draw a line $e_f a$ as shown. An undisturbed sample is brought from this depth and tested in the laboratory. During the sampling procedure, the over-burden pressure on the clay from which the sample is taken is reduced considerably, while the, water content remains the same. The laboratory pressure void ratio curve for this undisturbed soil sample corresponds, threrefore, to curve *B* in this figure. Now, if this soil sample is remoulded at the same water content, and a consolidation test carried out on it, another pressure void ratio curve designated as *C* is obtained. These two curves fall sufficiently apart.

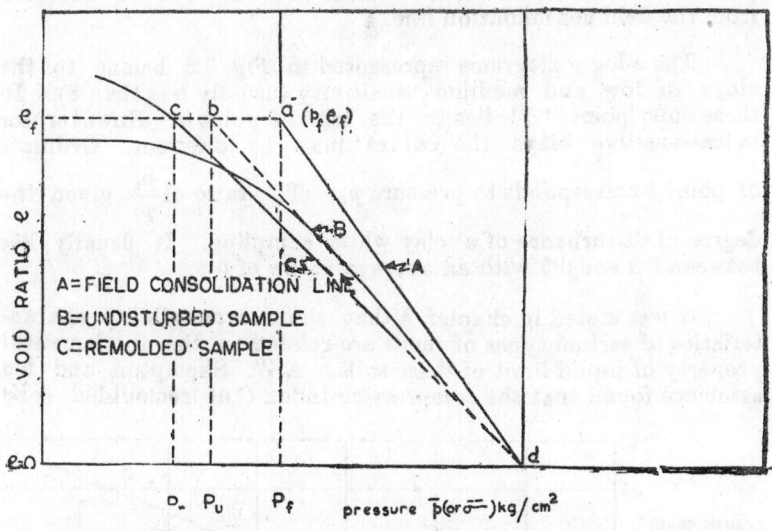

Pressure void ratio diagrams for a sample in the field, for an undisturbed sample and for a remoulded sample of clay.
Fig. 7·6.

The curve *C* cuts the line $e_f a$ at a point *c*. Below this point *c*, the curve *C* is almost a straight line.

If the straight line portion of the cure *B* is produced upwards, it cuts the line $e_f a$ in a point *b* and if it is produced downwards, it cuts the horizontal axis where void ratio is zero, at .a point *d*. It is found that, although the curve *C* corresponding to the remoulded state of the soil sample falls away from the curve *B* corresponding to the undisturbed state, the projection of the straight line portion of curve *C*, backwards also meets the horizontal axis either exactly at the point *d* or very near it.

It is, therefore, reasonable to assume that the *e*-log *p* curve corresponding to field consolidation of the in-situ soil also is a straight line and this straight line must also pass through the point *d*. The point *a* which represents the pressure and void ratio for the in-situ sample must also be on the field consolidation curve.

So, the curve A, e.i. $e_f ad$ represents the field consolidation. The line "ad" is called *field consolidation line.*

It is seen from the above discussion that the slopes of the e-log p curves for a clay tested in the undisturbed and remoulded states are different and hence the values of compression indices must differ, Also, it may be noted that both the vales of the compression indices are different from the one that can be obtained from the field consolidation line.

The e-log p diagrams represented in Fig. 7·6 belong to the clays of low and medium sensitivity (usually less than 8). In these soils point "b" lies to the left of point a. However, for extra-sensitive clays the curves may be different. Ordinate of point b corresponds to pressure p_u. This ratio of $\dfrac{p_u}{p_f}$ given the degree of disturbance of a clay while sampling. It usually lies between 0·3 and 0·7 with an average value of 0·5.

It was stated in chapter 4 that the compressibility characteristics of certain types of soils are related to the fundamental property of liquid limit of these soils. A.W. Skempton and his associate found that the compression index C of remoulded soils

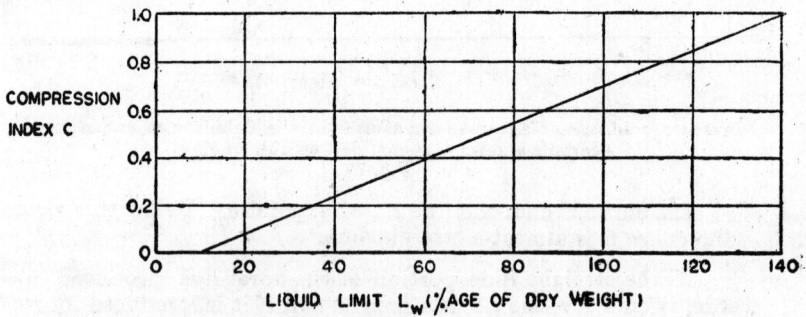

Relationship between Compression Index C and liquid limit L_w of remoulded clays (After A.W. Skempton etc.)

Fig. 7·7.

of all types including ordinary as well as clays bears a straight line relationship with their liquid limits, of type shown in Fig. 7·7 The equation of this straight line can be written as

$$C = 0·007 \ (L_w - 10\%) \qquad \qquad ...(7·10)$$

where C is the compression index of a soil and L_w is its liquid
limit, as a percentage of dry weight. It is also known from expe-
perience that the compression index C_e corresponding to the field
consolidation line A shown in Fig. 7·6 for the clays of low and
medium sensitivity is related to the compression index of such
soils in the remoulded state, by,

$$C_e = 1\cdot30C \qquad \qquad ...(7\cdot11)$$

substitution in equation (7·10) gives

$$C_e = 1\cdot30 [0\cdot007 (L_w - 10\%)] = 0\cdot009 (L_w - 10\%) \qquad ...(7\cdot12)$$

Equation (7·12) can give a fairly good estimate of the settlement
of a structure founded on clays of low and medium sensitivity.

Equation (7·8) represents the rebound leg of the pressure-
void ratio diagram on log scale as shown in Fig. 7·5 provided it
is assumed as a straight line. The constant C_s in this equation also
depends upon the the slope of this rebounded curve (assumed as
a straight line). This constant is called the **swelling index** for

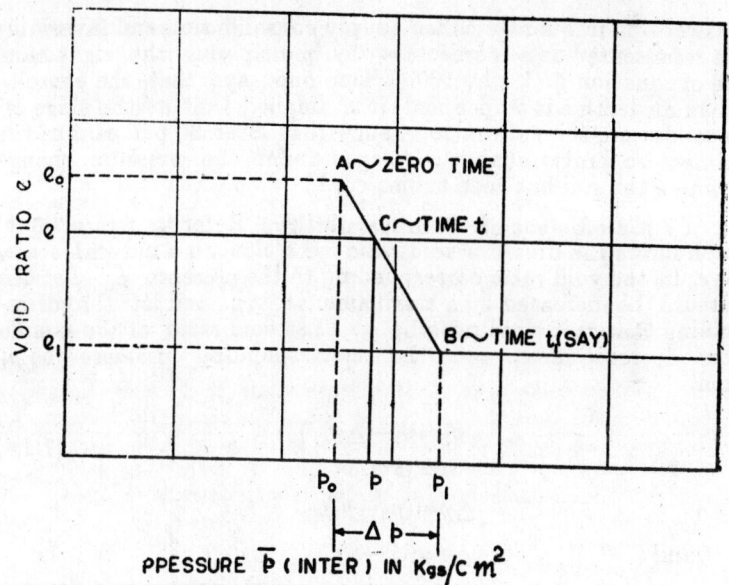

Pressure void-ratio diagram for a normally consolidated soil, assumed
as a straight line during the load increment $\triangle p$ (Terzaghi's Theory)

Fig. 7·8.

the soil. Like the compression index, it is also a diamensionless quantity and its value represents a relative increase of the elastic tendency in the soil.

(e) **Consolidation Ratio.** Refer to Fig. 7·8, which represents the pressure void-ratio diagram for a normally consolidated soil sample[1] for a load increment $\triangle p$. A is the point that corresponds to the initial pressure p_0, corresponding to the initial void ratio e_0 and B is a point that corresponds to the final pressure p_1 and the corresponding void ratio e_1, where $p_1 - p_0 = \triangle p$. Suppose it is required to find out at any time after this load increment, the degree of consolidation at any point distance z from the surface of the sample. Let the point C in Fig. 7·8 represent this point at a distance z from the surface let e and p be the void ratio and pressure corresponding to this point at that time. The degree of consolidation at that stage is, then expressed by the consolidation ratio, which for the point C in Fig. 7·8 is given by

$$U^*_z = \frac{e_0 - e}{e_0 - e_1} \qquad ...(7·13)$$

This ratio is frequently called simply consolidation and invariably it is represented as a percentage by multiplying the right hand side of equation (7·13) by 100. When one says that the consolidation at depth z is 50 per cent, it is implied that at the stage of consolidation, the void ratio change has been 50 per cent of the ultimate void ratio change envisaged under the pressure change to which the soil has been subjected.

(f) **Co-efficient of Compressibility.** Refer to figure 7·3 (b) which shows the pressure void ratio for a clay on a natural scale. Let e_0 be the void ratio corresponding to the pressure p_0. Let the pressure be increased by a small amount $\triangle p$ and let the corresponding change in void ratio be $\triangle e$. Let void ratio of the sample after decrease be e_1 and the corresponding pressure be p_1 Then.

$$\begin{array}{ll} \text{and} & \left. \begin{array}{l} p_1 = p_0 + \triangle p \\ e_1 = e_0 - \triangle e \end{array} \right] \qquad ...(7·14) \end{array}$$

$$\triangle p = (p_1 - p_0)$$

$$\text{and} \qquad \triangle e = (e_0 - e_1)$$

---

1. The definition is equally applicable to over consolidated soils too.

*. Alternatively this ratio also represents the extent to which the hydrostatic excess pressure has dissipated from its original magnitude say u_1 at the point A to u_2 at the point B. See equation (7·35).

Divide $\triangle e$ by $\triangle p$.

$$\therefore \qquad \frac{\triangle e}{\triangle p} = \frac{e_0 - e_1}{p_1 - p_0} \qquad \qquad ...(7.15)$$

The left hand side of equation (7.15) in the limit gives the slope of the $e-p$ curve drawn on a natural scale. This slope is negative at any point as can be seen from the figure[1] under reference The void ratio decreases as the pressure increases. Hence a negative sign must be used for the decrease in void ratio per unit increase in pressure. This slope is usually denoted by the letter a_v so that,

$$a_v = -\frac{\triangle e}{\triangle p} \qquad \qquad ...(7.16)$$

a_v in equation (7.16) represent the **co-effieint of compressibility** of the soil. Its units usually are cm²/gm., *i.e.* reciprocal of the pressure units of gms/cm². This is because the void ratio is a dimensionless quantity. As can be seen from the figure, the value of slope at any point will decrease as the point advances on the curve with the increase in the pressure. Hence for a given pressure change $\triangle p$, the coefficient of compressibility a_v decreases with the increase in pressure. While expressing the coefficient of compressibility numerically, the negative sign is usually omitted, because negative is implied by the word *compressibility*.

(g) Coefficient of Volume Compressibility. Refer to Fig. 7.9. It represents a layer of a saturated clay of thickness H. The volume of solids in the layer corresponds to a thickness of h_s. At

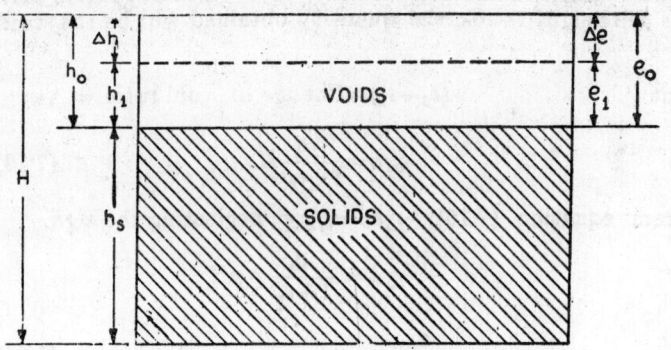

Settlement of saturated clay layer of thickness H
Fig. 7.9.

1. This is also evident from the right hand side of equation (7.15) $(p_1 - p_0)$ is positive close and $(e_0 - e_1)$ is negative.

any pressure p_0, the volume of voids corresponds to a thickness of h_0 and on increase of pressure from p_0 to p_1, by an amount $\triangle p$, the volume of voids reduces and the voids at pressure p_1 correspond to thickness of h_1. The change in thickness due to the increase in pressure $\triangle p$ is $\triangle h$. Let the void ratio corresponding to voids thickness of h_0 be e_0 and the void ratio corresponding to voids thickness of h_1 be e_1.

Now $$\triangle h = h_0 - h_1 \qquad \qquad ...(7{\cdot}17)$$

and $$H = h_s + h_0 \qquad \qquad ...(7{\cdot}18)$$

Divide equation (7·17) by equation (7·18)

$$\frac{\triangle h}{H} = \frac{h_0 - h_1}{h_s + h_0}$$

or $$\triangle h = \left(\frac{h_0 - h_1}{h_s + h_0} \right) H.$$

$$\triangle h = \frac{\dfrac{h_0}{h_s} - \dfrac{h_1}{h_s}}{1 + \dfrac{h_0}{h_s}} H$$

or $$\triangle h = \frac{e_0 - e_1}{1 + e_0} . H.$$

because the area A for each height, whether of solids or voids is the same and, therefore, whether the volume ratios are taken or the height ratios, the quantity obtained will be the void ratios.

But $$(e_0 - e_1) = \text{Change in void ratio} = \triangle e$$

$\therefore$ $$\triangle h = \frac{\triangle e}{1 + e_0} . H. \qquad \qquad ...(7{\cdot}19)$$

From equation (7·16), $\triangle e = a_v . \triangle p$, neglecting the sign.

$\therefore$ $$\triangle h = \frac{a_v}{1 + e_0} . \triangle p H$$

Put $$\frac{a_v}{1 + e_0} = m_v \qquad \qquad ...(7{\cdot}20)$$

$\therefore$ $$\triangle h = m_v . \triangle p . H. \qquad \qquad ...(7{\cdot}21)$$

The coefficient m_v in equation (7·21) is called the **co-efficient of volume compressibility** or it is also designated as **modulus of volume change.** Equation (7·20) shows that the units of m_v are identical with the units of a_v. Recall that the units of a_v as shown in the last sub para (f) of this section are cm²/gm, *i.e.* reciprocal of the units for pressure. Like a_v, m_v is also negative and is not a constant quantity, but decreases with the increase in pressure.

Incidentally equation (7·19) showing change $\triangle h$ in the thickness H of a layer of soil, due to the change in void ratio, is indentical with equation (7·4) derived for a laboratory sample. Equations (7·19) and (7·21) give the settlement under the change in pressure.

From equation (7·21), it can also be said that m_v is the change in thickness per unit thickness per unit applied pressure.

7·6. Reconstruction of Field Pressure Void Ratio Diagram

It is seen from figure 7·6 that even if an undistorbed sample of soil is tested in the laboratory the pressure void diagram so obtained does not represent the true field conditions because the true field consolidation line A as shown in this figure falls sufficiently above the laboratory virgin curve B, obtained on the undisturbed soil sample. This variation is due to the fact that in spite of all the care that is taken to take out an undisturbed sample, from a soil deposit and to transfer this sample from a sampling tube to the consolidation ring, there is always some amount of structural distrurbance of the sample. Terzaghi and Peck (3) have suggested the use of the field consolidation line A drawn in Fig. 7·6 and the point b (which represents the maximum pre-consolidation load for the soil) as the basis for compilation of settlements of structures founded on normally loaded clays.

Another method of reconstructing the true pressure void ratio diagram has been suggested by John H. Schmertmann (5). The method is fairly simple, and an interested student may look up the relevant reference. It is important, however, to realise that the field e-log p curve is quite different from the laboratory e-log p curve and for accurate settlement analysis, it is essential to reconstruct the field curve from the laboratory curve.

7.7. Mechanical Analogy to the Consolidation Process

It was concluded in Section 7.2 that the compression of fine-grained soils occurs mainly due the expulsion of water from the soil grains. It was also stated that a pressure difference exists between the centre of a soil sample in a consolidometer and its upper and lower surfaces and due to this pressure difference the flow takes plaçe. Also, as soon as the load is placed on ths sample, the whole of the pressure momentarily is supported by the water and integranular stress is zero at that moment and as the time passes, the pressure is gradually transferred onto the soil grains as intergranlular pressure and the additional pressure on the water becomes zero. These facts can be verified by the following mechanical analogy.

Fig. 7·10 shows a cylinder fitted with a number of pistons (marked P) connected by springs in between. Each of the four compartments made by these pistons is connected to the atmosphere with the help of L-shaped standpipes. The compartments are full of water and it is assumed that there is no leakage from the cylinder and water can move from one compartment to ano-

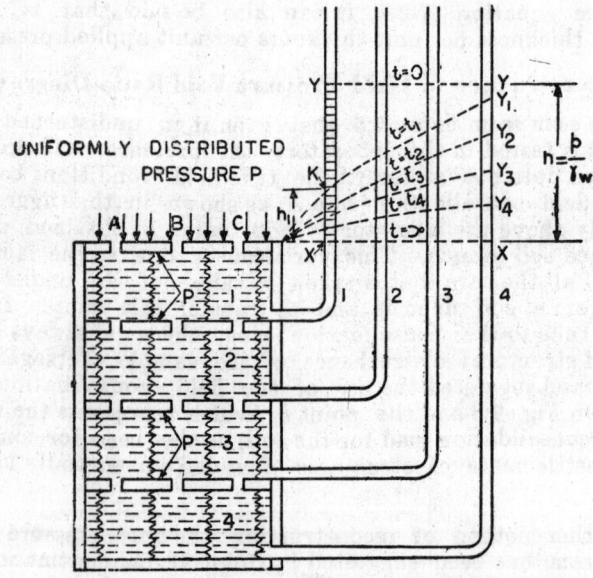

Mechanical analogy to the consolidation process
Fig. 7.10

ther only through the perforation in the pistons. It is also assumed that if a pressure is applied to the topmost piston, it is transmitted undiminished to the medium of water, springs and pistons below. The uppermost piston is fitted with valves at the perforations and v ves can open and close these openings

To start with, the cylinder is full of water and there is no pressure on the top piston. The weights of the pistons are taken up by the springs and there is no pressure on the water except the atmospheric pressure. Under this atmosphere, when the valves A, B, and C are open, the water stands in the four stand-pipes at the level of X-X. Now the valves A, B and C are closed and the water in all the standpipes still holds its level at the section X-X. The uppermost piston is now loaded with a uniform pressure of p units per unit area and the valves A, B and C are simultaneously opened at the same moment at which the pressure is placed on the piston. It is noticed that the moment the pressure is applied to the top piston, the water in all the standpipes rises to a level

Y-Y. The difference between the X-X and Y-Y levels is $h = \dfrac{p}{\gamma_w}$.

where γ_w is the unit weight of water. At this instant, the height of the springs is unchanged. The time of loading may be called zero time, i. e., $t = 0$. An equal rise of water in all the standpipes shows that the hydrostatic excess pressure at all points in the medium at $t = 0$. is the same. There is no pressure on the springs at this moment. As the time passes, the water level in the stand pipes starts falling, the pistons start moving downwards gradually and the water starts coming out of the cylinder at the openings A B, and C. At any stage when $t = t_1$, the water pressure in compartment l is the least as seen from the level of water in the corresponding stand-pipe no. 1, the water pressure in the second stand-pipe stands a little higher, so the pressure in the second compartment is slightly more than it is in the first and so on for the respective compartments. If a curve is drawn through the levels of the water in the various stand pipes, it turns out to be a curve ending at Y_1. At any other instant, say $t = t_2$, another pressure curve can be obtained, as shown by the curve ending at Y_2. For various times, t_1, t_2, t_3 and t_4, variations of excess hydrostatic pressure in the various compartments are shown by the pressure curves ending at points Y_1, Y_2, Y_3 and Y_4, till ultimately the excess hydrostatic pressure in each compartment becomes zero. As the excess hydrostatic pressure decreases, in each compartment pressure equal to the amount of this decrease in hydrostatic excess pressure is taken up by the springs. For example, at $t = t_1$, the excess hydrostatic pressure in compartment l is $XK(= h_1)$ so that pressure equal to the column of water $KY(= h - h_1)$ is taken up by the spring.

If the excess hydrostatic pressure at any stage of consolidation is denoted by u and the pressure on the springs by the symbol $\bar{p}$, then for compartment 1,

when $t = 0$, $u = h\gamma_w = p$ and $\bar{p} = 0$, so that
$$\bar{p} + u = h\gamma_w = p$$

when $t = t_1$, $u = h_1\gamma_w$ and $\bar{p} = (h - h_1)\gamma_w$, so that   ...(7.22)
$$\bar{p} + u = (h - h_1)\gamma_w + h_1\gamma_w = h\gamma_w = p$$

and so on till such times that $t = \infty$

when $t = \infty$. $u = 0$ and $\bar{p} = h\gamma_w = p$ and gives
$$\bar{p} + u = h\gamma_w = p$$

The curve KY_1, therefore, gives distribution of the hydrostatic excess pressure and the pressure on the springs for all compartments at $t = t_1$. Similar is the interpretation of other curves.

Equation (7.22) holds throughout the entire time interval from $t=0$ to $t=\infty$

The behaviour of saturated soils under pressure is analogous to this mechanical model if the plastic lag or the plastic action of the particles is ignored. The soil grains may be equated to springs in the model. When pressure is applied to this soil sample in a consolidometer, initial hydrostatic excess pressure at all points within the mass is equal to the applied pressure and as the time passes drainage occurs and the hydrostatic excess pressure is slowly transferred to the soil grains as intergranular pressure.

Equation (7.22) holds in the case of soils subjected to the consolidation process. At any stage during consolidation, u is the excess hydrostatic pressure at a point $\bar{p}$ is the corresponding intergranular stress at this point, so that the sum of the two always is the total pressure p,

∴   For soils,

$$p = \bar{p} + u \qquad \qquad ...(7.23)$$

where the symbols have the above meanings.

If the soil sample in a consolidometer draining both sides is considered, curves can be drawn to show the distribution of the hydrostatic excess pressure for a soil sample corresponding to time intervals $t=0$, $t=t_1$,......, $t=\infty$, as shown in Fig. 7.11. These time curves showing the distribution between hydrostatic excess pressure and the intergranular stress are called **isochrones**.

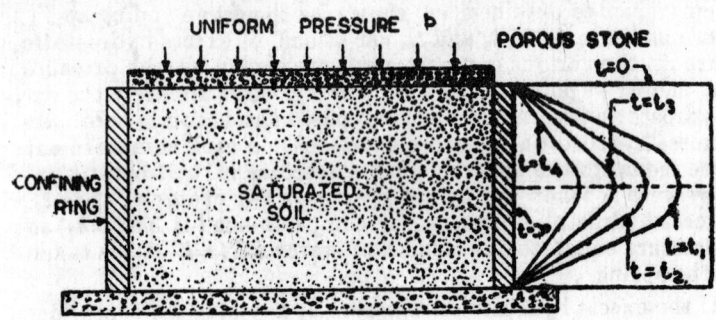

Isochrones showing the distribution of excess hydrostatic pressure and intergranular pressure for a double draining soil sample.
Fig. 7.11.

7.8. Terzaghi's One Dimensional Theory of Consolidation

Compression of the few top layers of a soil strata under a foundation is usually three-dimensional. However, the deeper layers are confined by more soil layers on all sides. Overburden on these confining layers prevents their upheaving. So, for these deeper layers, consolidation is truly one-dimensional.

Terzaghi has given a theoretical analysis for the one dimensional consolidation process. This analysis provides a relationship between the hydrostatic excess pressure, the void ratio and time after loading for a consolidating layer of a soil. From the solution of Terzaghi's equation, curves can be plotted between what is called the degree of consolidation and a parameter called the time factor. A comparison of these theoretical results with the experimental data obtained from a consolidation test on a soil sample taken from a clay layer provides a means of drawing a time settlement curve for the soil layer of any thickness in the field.

The theory advanced by Terzaghi (7) is based upon certain simplifying assumptions, which are stated below :—

1. The soil is homogeneous material.

2. The soil is completely saturated.

3. The solid soil particles and water can be assumed as incompressible.

4. Compression is one dimensional.

5. Flow of water in the soil is only in one direction and Darcy's Law governs this flow.

6. Flow of water in the soil masses of infinitesimally small size behave exactly like larger representative masses.

7. When the pressure on a layer is increased by an amount Δp, the pressure distribution between the hydrostatic excess pressure and the intergranular pressure is governed by equation (7.23). Also during an increment, pressure and void ratio are represented on a straight line, ac shown in Fig. 7.12.

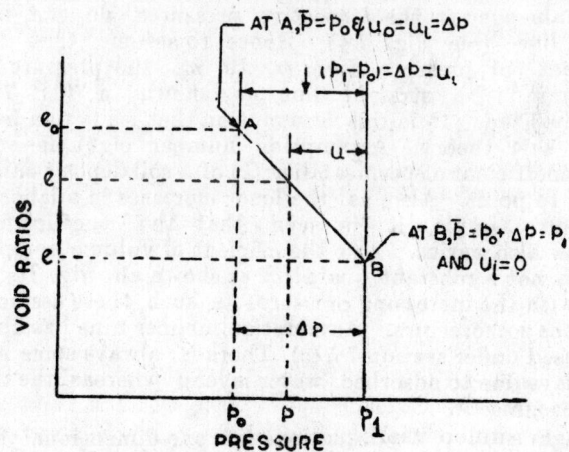

Idealized pressure void ratio diagram during an increment of pressure from p_1 to p_0 (Natural Scale)
Fig. 7·12.

8. Some soil properties such as permeability and volume compressibility are constant.

9. There is no plastic action within the soil mass.

The assumptions of homogeneity, complete saturation and negligible compressibility are not far from the actul conditions in clays. Most of the clay deposits are homogeneous, and due to the high water table position normally, met with in the field, they are completey saturated. The solid soil particles and water have negligible compressibilitely characteristics. As stated earlier, the deeper soil layers consolidate uni-directionally due to lateral confinement by other layers. So the fourth assumption for such layers is quite valid. For the laboratory consolidation test, the application of Darcy's law and uni-directional flow of water (vertical in this case) are quite alright. However, applicability of these assumptions in the filed should be checked and will depend on the field conditions. The sixth assumption has been given to cover the theoretical derivation, wherein infinitesimally small distances are used to derive the equation. For the laboratory or the field soil data, this assumption has no real significance.

The validity of the seventh assumption cannot be questioned as far as the application of eqnation (7·23) goes, because there are only two phases i.e., soil grains and water in the saturated clay and the total external pressure must equal the sum of the pressures on the individual phases, at any time during an increment of load. But it has been shown that the final void ratios attained under different pressures, when plotted on natural scale against the respective pressures, do not lie on a straight line (see Fig. 7·4). Hence to assume that during an increment of pressure from p_o to p_1, the pressure void ratio diagram is a straight line as shown in Fig. 7·12 is little too idealized. It is this assumption that leads to a limited validity of this theory. Assumption number eight is erroneous as the coefficient of permeability (k) of a soil deposit can vary from point to point. Also, as the load increases in a laboratory consolidation sample, it is seen that the coefficient of permeability also varies. Also the cfficient of volume compressibility (m_v) is not a constant quanstity as shown already. Its value decreases with the increasing pressure. As such, these assumption can introduce some errors. Assumption number nine has already been discussed under section 7·5 (c). There is always some plastic action in clays due to adsorbed water layers, whereas the theory does not recognise it,

Terzaghi's differential equations for one-dimensional flow is derived below, taking into consideration the assumptions given above. The boundary conditions under which the solution can be found are also given. The solution of the differential equation is outside the scope of this book.

Refer to Fig. 7·13, which represents a layer of clay of thickness H, loaded uniformly with a load transmitting a uniform stress equal to $\triangle p$ units, to the top of the layer. The layer is assumed pervious at the top and impervious at the bottom. It is subjected to one dimensional consolidation only, in this case in the upward vertical direction. It is assumed that, at the moment the load is placed on the soil mass the initial hydrostatic excess pressure u_i at any point is equal to the stress at that point. This initial hydrostatic excess pressure is the same at all points within the mass of a laboratory sample at the instant of loading and equals $\triangle p$. Consider a small element of unit area and of thickness dz. Immediately on loading water will start flowing across this element. Assume that the water flows in the upward direction across this element, as shown by an arrow. Let ∂h be the head lost across the two faces of this element corresponding to a change in hydrostatic excess pressure ∂u

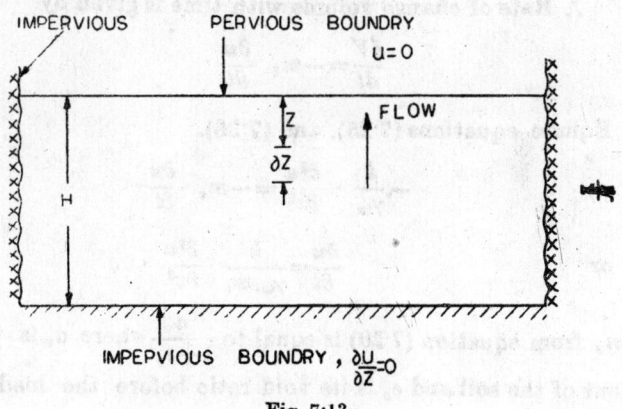

Fig. 7·13.

Then the velocity of flow is given by,

$$v = -k\,\frac{\partial h}{\partial z} = -\frac{k}{\gamma_w} \cdot \frac{\partial u}{\partial z} \qquad \ldots (7\cdot24)$$

where v is the velocity of flow in the upward direction, k is the coefficient of permeability of the soil and γ_w is the unit weight of water.

If ∂V represents the change in volume per unit volume due to the applied stress, in time ∂t, then $\dfrac{dV}{dt}$ gives the rate of change of volume, per unit volume of the soil. The continuity equation, therefore, for this one dimensional case is given by,

$$-\frac{k}{\gamma_w} \cdot \frac{\partial^2 u}{\partial z^2} = \frac{dV}{dt} \qquad \text{...(7·25)}$$

The change in hydrostatic excess pressure ∂u, ultimately gets transferred as increease in the intergranular stress ∂p, on the soil particles within the element ∂z.

So, the volume change dV under this increase in intergranular stress is given by,

$$dV = m_v.\partial p = -m_v.\partial u,$$

where m_v is the coefficient of volume compressibility of the soil as defined earlier.

∴ Rate of change volume with time is given by

$$\frac{dV}{dt} = -m_v \frac{\partial u}{dt} \qquad \text{...(7.26)}$$

Equate equations (7·25). and (7·26).

$$-\frac{k}{\gamma_w} \cdot \frac{\partial^2 u}{\partial z^2} = -m_v \cdot \frac{\partial u}{\partial t} \cdot$$

or

$$\frac{\partial u}{\partial t} = \frac{k}{\gamma_w.m_v} \cdot \frac{\partial^2 u}{\partial z^2} \cdot$$

But m_v from equation (7·20) is equal to $\dfrac{a_v}{1+e_o}$ where a_v is the coefficient of the soil and e_o is its void ratio before the load increment $\triangle p$.

Substitute $\qquad m_v = \dfrac{a_v}{1+e_o}$ in the above equation

$$\therefore \qquad \frac{\partial u}{\partial t} = \frac{k(1+e_o)}{a_v. \gamma_w} \cdot \frac{\partial^2 u}{\partial z^2} \qquad \text{...(7·27)}$$

Put $\qquad \dfrac{k(1+e_o)}{a_v.\gamma_w} = c_v \qquad \text{...(7·28)}$

$$\therefore \qquad \frac{\partial u}{\partial t} = c_v \cdot \frac{\partial^2 u}{\partial z^2} \qquad \text{...(7·29)}$$

c_v in equation (7·28) or (7·29) is called the **coefficient of consolidation**. In the differential equation (7·29), u represents the hydrostatic excess pressure at distance z measured downwards from the surface of soil mass corresponding to any time t after the consolidation process.

Equation (7·29) is called *Terzaghi's differental equation for one dimensional consolidation*.

Value of c_v for a soil as is seen from eqaution (7·28), depends upon its coefficient of permeability k, its coefficient of compressibility a_v, its initial void ratio e_0 aud unit weight of water γ_w.

Since drainge is allowed only at one face of the sample (top surface here), the longest path of drainage is H. The boundary conditions for the solution of eduation (7.29) can be written as :

1. There is complete drainage at the top surface of the sample, so that

when $t=0$

and $z=0,$

                        $u=0$ (Free drainage).

2. There is no drainage at the bottom surface. so that

when $t=0$

and $z=H,$

                        $du/dz=0$ (No drainage).

3. The initial hydrostatic excess pressure u_i is equal to the applied stress $\triangle p$, at the instant of loading, so that

At $t=0,$ $u=u_i=\triangle p.$

For the special case of the thin laboratory sample,

At $t=0$

and $0<z<H,$ $u=u_i=\triangle p,$

because, as stated earlier, in the thin laboratory sample it can be assumed that at zero time the initial hydrostatic excess pressure u_i is constant at all points in the soil mass and equals $\triangle p$, as $\triangle p$ the applied stress at the surface gets transmitted undiminished to the bottom surface. However, in thick clay layers in the field, the applied stress may not be transmitted undiminished to the bottom layers, so that the initial hydrostatic excess pressure may not be constant throughout the depth.

It may be noted in equation (7·29) that z is the distance from the surface to the soil particle under consideration in a soil mass which constitutes of material grains and voids filled with water. As the consolidation progresses, the voids decrease and hence the thickness H and the distance z, both decrease. For convenience, therefore, sometimes H is replaced by $H_o (1+e_o)$ and z by $z=h (1+e_o)$: where H_o is the total thickness of soil grains and h is the distance from the surface to the soil grain under consideration as if only solid were present.

If this substitution is made in equation (7·27), the equation of consolidation works out as :

$$\frac{\partial u}{\partial t} = \frac{k}{a_v \gamma_w (1+e_o)} \cdot \frac{\partial^2 u}{\partial h^2}$$

or
$$\frac{\partial u}{\partial t} = c_v' \frac{\partial^2 u}{\partial h^2} \qquad \qquad ...(7·30)$$

where
$$c_v' = \frac{k}{a_v . \gamma_w . (1+e_o)} \qquad ...(7·31)$$

Equation (7·29) or (7·30) can be solved for the constant initial hydrostatic excess pressure conditions as well as for the variable hydrostatic excess pressure conditions, provided c_v or c'_v is considered a constant quantity. In actual effect, as void ratio decreases under load, both a_v and k diminish. So c_v, or c_v' actually is not a constant quantity over the entire range of loads that are normally used in a consolidation test. But for one load increment, it can be considered a constant quantity and hence the theoretical solution based upon considering c_v or c'_v as a constant can be applied.

For the boundary conditions where u_i is considered constant and equal to $\triangle p$, the solution of equation (7·29) can be written as,

$$u = f(z,t)$$

$$= \frac{4}{\pi} (\triangle p) \sum_{n=0}^{n=\infty} \frac{1}{(2n+1)} \varepsilon^{\frac{-(2n+1)^2}{4H^2} \pi^2 c_v t} \sin(2n+1) \frac{\pi z}{H} ...(7·32)$$

Some other boundary conditions will lead to a different solution. In this solution make the following substitution.

$$T = \frac{c_v t}{H^2} \qquad \qquad ...(7·33)$$

and divided by $\triangle p$ on both sides, equation will turn out as

$$\frac{u}{\triangle p} = f\left(\frac{z}{H}, T\right)$$

$$= \frac{4}{\pi} \sum_{n=0}^{n=\infty} \frac{1}{(2n+1} \epsilon^{\frac{-(2n+1)^2}{4} \pi^2 T} \sin(2n+1)\frac{\pi z}{2H} \quad ...(7·34)$$

T, in the above equation, is called the time factor, c_v in the equation has been considered a constant quantity. H, the length of the drainage path decreases as the void ratio changes. But for one load increment an average value of H before and after loading is taken and it is assumed a constant quantity. So in equation (7·33), since c_v and H have been assumed as constant quantities, the time factor T is proportional to the time t, after the start of the consolidation process.

In equation (7·28) the factor $(1+e_0)$ is a diamensionless quantity, k has the units of velocity, i.e., cm/sec., (in the c.g.s. system for example) γ_w in gms/cm³ and a_v has the units of cm²/gm, (reciprocal of the units for pressure) so that c_v has the units of cm²/sec. Therefore from equation (7·33) since t can be represented in seconds and H is cms, the time factor T works out to be a dimensionless quantity.

Now the degree of consolidation at any stage during the consolidation process at the point z below the top surface is given by,

$$U_z=\frac{e_0-e}{e_0-e_1}, \text{ see equation (7·13)}$$

If the idealized pressure void ratio diagram shown in Fig. 7·12 is taken into account, then the above equation can be written as,

$$U_z=\frac{p-p_0}{p_1-p_0}=\frac{u_0-u}{u_0-u_1}=\frac{u_0-u}{u_0-0}$$

$$=\frac{u_i-u}{u_i}=1-\frac{u}{u_i} \qquad ...(7·35)$$

With reference to Fig. 7·12, in equation (7·35), u_0 designates the hydrostatic excess pressure corresponding to pressure p_0 at a point A at zero time and u_1, the hydrostatic pressure corresponding to pressure p_1 at point B at the end of the consolidation process due to load increment $\triangle p$, and u, the hydrostatic excess pressure at an intermediate point with void ratio e and corresponding pressure p at antermediate time t after the state of consolidation. Evidentally $u_0=u_i=\triangle p$ and $u_1=0$, as shown at points A and B in the figure.

Since $u_t=\triangle p$, substitution of $u/\triangle p=u/u_i$ from equation (7·34) in equation (7·35) gives the degree of consolidation at any point z can be written as.

$$U_z=1-\frac{4}{\pi}\sum_{n=0}^{n=\infty}\frac{1}{(2n+1}\epsilon^{\frac{-(2n+1)^2}{4}\pi^2T}\sin(2n+1)\frac{\pi z}{2H}$$

$$...(7·36)$$

Normally what an engineer is interested in is the average degree of consolidation of the entire layer of clay and not the degree of consolidation U_z at a point. Whenever the consolidation of a layer is stated, it is always stated in terms of the average consolidation of the layer and never as the consolidation at a point. Hence, the necessity to determine the average consolidation of the entire layer.

The average initial hydrostatic excess pressure may be expressed as,

$$\frac{1}{H}\int_0^H u_i \, dz$$

and the average hydrostatic excess pressure at any intermediate time t may be expressed as,

$$\frac{1}{H}\int_0^H u \, dz,$$

so that the average degree of aonsolidation U, i e, average value of U_z over the depth can be written as

$$U = 1 - \frac{\dfrac{1}{H}\int_0^H u \, dz}{\dfrac{1}{H}\int_0^H u_i \, dz}$$

or

$$U = 1 - \frac{1}{H}\int_0^H \frac{u}{u_i} \, dz$$

$$= 1 - \frac{1}{H}\int_0^H \frac{u}{\triangle p} \, dz \qquad \ldots(7{\cdot}37)$$

Substitution of $u/\triangle p$ from equation (7·34) in the above equation gives,

$$U = 1 - \int_0^H \frac{4}{\pi} \sum_{n=0}^{n=\infty} \frac{1}{(2n+1)} \epsilon^{\frac{-(2n+1)^2}{4}\pi^2 T} \sin (2n+1)\frac{\pi z}{2H} \, dz$$

or

$$U = 1 - \frac{8}{\pi^2} \sum_{n=0}^{n=\infty} \frac{1}{(2n+1)} \epsilon^{\frac{-(2n+1)^2}{4}\pi^v T} \qquad \ldots(7{\cdot}38)$$

Equation (7·38) is evidently applicable to the constant initial hydrostatic excess pressure condition, since equation, (7·32) relates to this boundary condition.

Equations (7·32), (7·34) and (7·36) are quite complicated and their significance is easily studied by plotting the variables. As an illustration U_z is plotted against z/H for different numerical values of T in Fig. 7·14. This is the plot of equation (7·36) and shows how the degree of consolidation at various depths varies at a constant time factor T. Alternatively one may interpret this diagram as showing the effect of the variation of time on the degree of consolidation U_z at a point. For example, if a horizontal line is drawn for a z/H ratio of 0·20 (shown by the arrow) then this horizontal line cuts the various curves corresponding to $T=0·05$, $T=0·10$, $T=0·15$, etc. and so on at some points. Vertical ordinates from such points give the degree of consolidation U_z

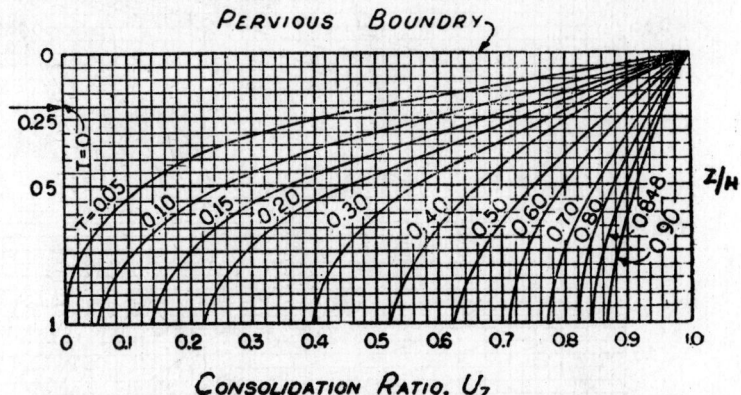

Degree of consolidation U_z verses depth ratio z/H
(Single drainage case)
Fig. 7·14.

with variation in time at a depth equal to 0·20 times the longest drainage path (in this case H) from the surface of drainage.

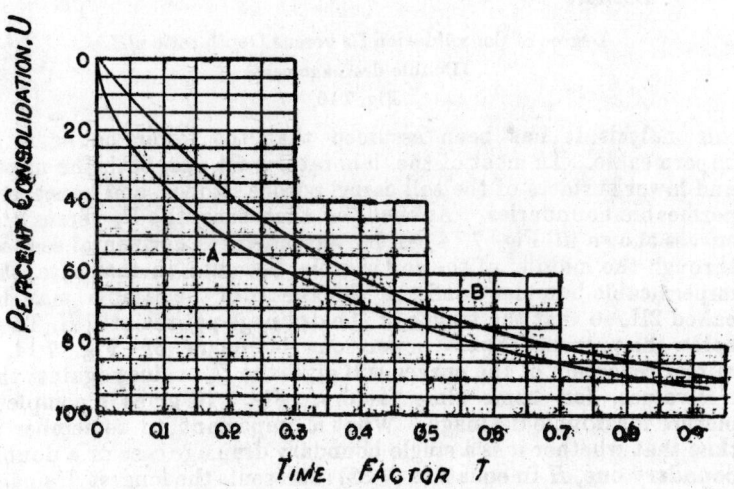

Percent Consolidation U versus time factor T.
Fig. 7·15.

Equation (7·38) is plotted as curve A in Fig. 7·15. This figure shows the variation of the average degree of consolidation U for a layer, versus time factor T. This is the plot which is most commonly used in solving practical settlement problems. In all

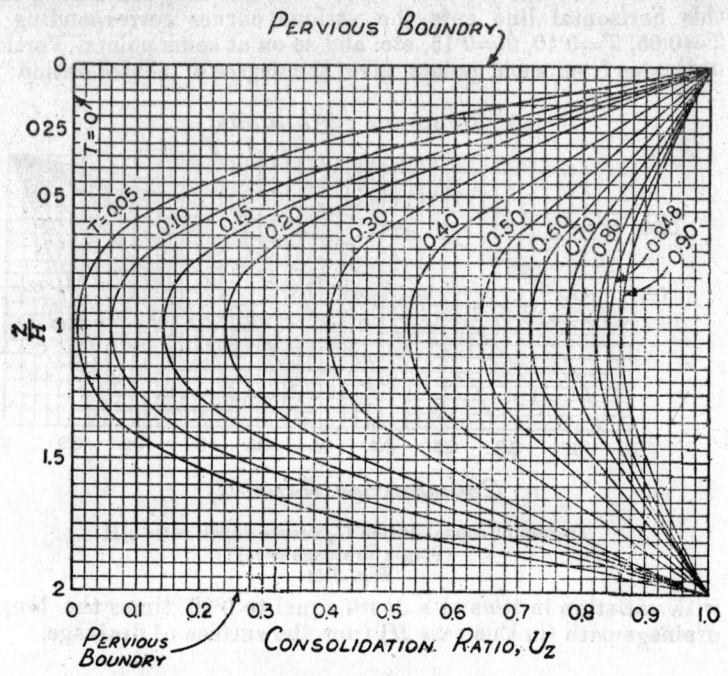

Degree of Consolidation U_z versus Depth ratio z/H.

(Double drainage case)

Fig. 7·16.

this analysis, it has been assumed that the lower boundary is impermeable. In most of the laboratory set-ups, both the upper and lower surfaces of the soil carry porous stones and hence are permeable boundaries. As such, in that case, the U_z versus z/H curves shown in Fig. 7.14 can be extended. A horizontal section through the middle of the soil sample becomes, in that case, the impermeable boundary and the thickness of the sample may be called $2H$, so that the length of the drainage path is still H. Then below the present lower impervious boundary of Fig. 7·14, a mirror reflection of the curves will give the U_z values against the z/H values, for same time factors. Fig. 7·16 gives a complete picture for double drainage. What is important to remember is this, that whether it is a single boundary drainage case or a double boundary one, H in equation (7·33) represents the longest drainage path.

7·9. Initial Hydrostatic Excess Pressure

The mechanical analogy discussed in article 7·7 shows that on loading, the initial hydrostatic excess pressure u_i is the same at all points in the water enclosed in the cylinder. This condition, *i.e.*, constant initial hydrostatic excess pressure at all points can be assumed to exist in the laboratory consolidation sample, because the sample is thin and load can be assumed to be transferred undiminished throughout the depth of the soil sample. However, in natural soil deposits, it cannot be assumed that the pressure distribution due to a load placed upon the surface is transferred undiminished to the deeper layers. As will be dicussed in detail in the next chapter, the vertical stress due to an external load on the soil diminishes with depth and due to reduction in the stress in the deeper layers, it cannot be assumed that at zero time when load is placed on the surface of a natural soil deposit, the initial hydrostatic excess pressure is the same throughout the depth. Equation (7·38) may not, as such. be directly applicable to practical field problems, in case stress distribution is to be taken into account. The initial hydrostatic excess pressure u_1 will naturally follow the same pattern of distribution along the depth as the stress distribution along it.

The actual distribution of vertical stress at the centre of a loaded area along the depth of the soil is usually curvilinear

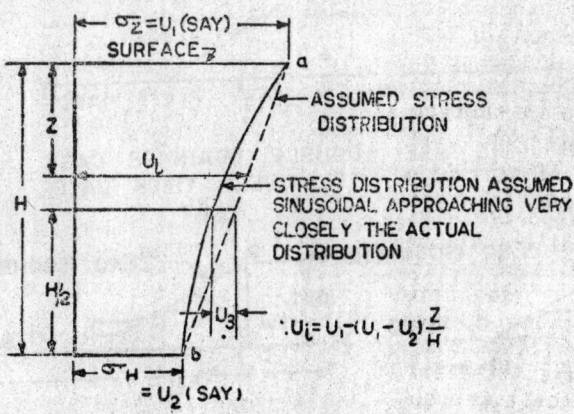

Fig. 7·17.

starting with some value say σ_z at the surface and ending with another value σ_H at a depth H, as shown in Fig. 7·17. In an actual design problem, the actual curvilinear stress distribution may be replaced by a straight line ab shown dotted in the figure, and the settlements due to loads calculated accordingly. The initial hydrostatic excess pressure, will, therefore, also be identical with the assumed stress distribution at any point z units below the surface and is given by ;

$$u_i = u_1 - (u_1 - u_2)\frac{z}{H}$$

where the symbols have the meanings shown in the diagram. The average consolidation U and the time factor T when the initial hydrostatic excess pressure is constant are given, theoretically by equation (7·38) and the plot of U versus T corresponding to this condition is shown as curve A in Fig. 7·16. It is found that if the initial hydrostatic excess pressure u_i is not constant but varies in the manner shown by the assumed stress distribution, i.e. dotted line "ab" in Fig. 7.17, the plot of average consolidation U and time factor T remains unchanged, i.e. the assumed condition illustrated by Fig. 7·17. is also represented by the curve A of Fig 7·16.

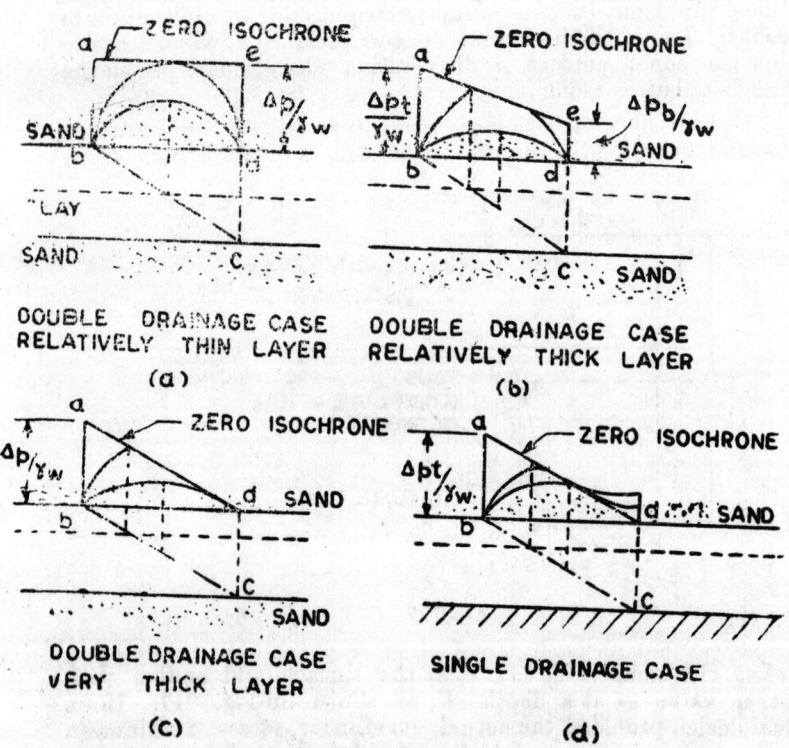

Variation of slopes of zero isochrones.

Fig. 7.18.

If the actual curvilinear stress distribution (assumed sinusoidal) as shown by curve ab is Fig, 7·17 is considered and if the corresponding initial hydrostatic excess pressure at 'any depth z below the surface is taken as.

$$u_i = u_1 - (u_1 - u_2)\,\frac{z}{H} - u_3 \sin \frac{\pi z}{H} \qquad \qquad ...(7\cdot39)$$

where u_3 as shown in Fig. 7·17. is the difference between the abscissas of the curve and the straight line at the horizontal section at the middle of the sample, then it is seen that the plot of average consolidation U versus T for this condition is the curve B in Fig. 7·15. The two curves A and B are fairly identical and this clearly indicates that even if the stress distribution due to an external load is considered linear rather than sinusoidal, the error involved is not excessive.

Another way of looking at the variation in the development of excess hydrostatic pressure and its effect on U and T lies in observing the slope of the zero isochrone. Fig 7·18(a) to 7.18.(d) give the variations in the slopes of the zero isochrones for four different conditions. $\triangle p_t$ and $\triangle p_b$ represents the pressures at the top and botom of the consolidating clay layer. Subsequent distribution of excess hydrostatic pressure is also shown by other isochrones. The relationship between the time factor "T" and the degree of consolidation "U" under vertical drainage conditions in the cases (a), (b) and (c) is given by curve C_1 in Fig 7·19 irrespective of the slope of the zero isochrone. Curve C_2 in the figure shows "T" versus "U" in a single drainage case shown as "d" in Fig. 7·18. incidentally it may be observed that "T" is drawn on the log scale in Fig 7·19.

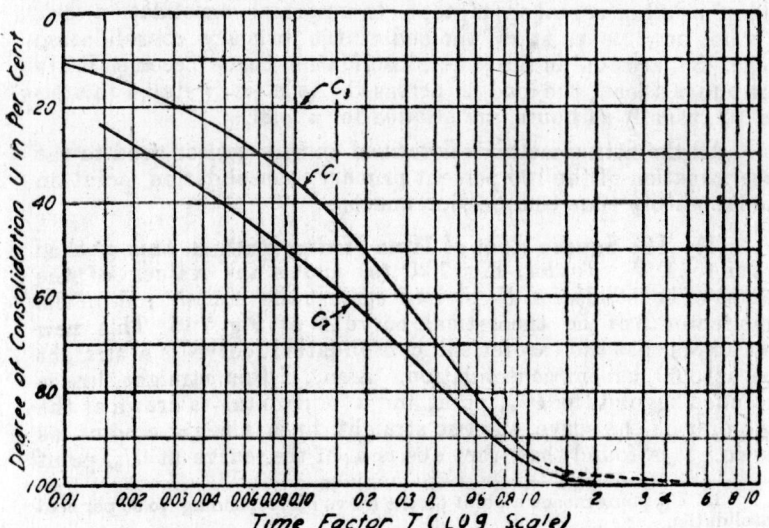

"T" versus "U" for conditions shown in Fig 7·19.

7.10. Evaluating c_v

There are a number of methods that are available for evaluating the co-efficient of consolidation from laboratory consolidation data. These methods are called the *fitting methods* because they involve a comparison of the theory with practice and the laboratory curve in a sense is fitted into the theoretical curve to evaluate c_v. Three such methods of evaluating c_v are presented in this Section.

(1) The square root of time fitting method, by D.W. Taylor.

(2) T_{60} method, by H. Gray.

(3) The logarithm of the time fitting method, by A. Casagrande.

Before any one of these methods is explained, it is essential to know that the amount of consolidation that occurs due to the dissipation of pore-pressure (hydrostatic excess pressure) in a soil is called the **primary consolidation**. Any further consolidation of the layer may be attributed to the plastic flow in cohesive soils or in other words to internal structural re-arrangement in such soils. The plastic flow in soils occurs at a very slow rate and it can continue to occur after the dissipation of pore-pressures This behaviour of the soil is usually referred to as the **secondary consolidation effect**. Due to this secondary consolidation, many times the shape of the laboratory time compression curve is very much different from the theoretical curve A of Fig. 7·15. because the theory does not recognise the plastic flow effects. Due to secondary consolidation effect in a soil sample. the laboratory time compression curve has a slope at large elapsed time, instead of ending as a horizontal asymptote. In a normal consolidation test, it is not possible to know the end of the primary consolidation stage. So, as soon as the time consolidation curve becomes fairly flat under a load, the load is increased. As already stated in article 7·3, usually 24 hours, are allowed for a load.

All the fitting methods described in this section lead to the determination of the 100 percent primary consolidation point on the laboratory time compression curve.

(1) The Square Root of Time Fitting Method. This method is due to D.W. Taylor. Fig 7·20 (a) shows the values of the degree of consolidation U plotted against the square root of the time factor T of the theoretical curve A of Fig. 7·15. This new plot shows that the theoretical consolidation curve is a straight line upto 60 percent consolidation. Also, if this straight line is extended beyond the U^1_{60} point and a horizontal is drawn at the U_{90} point on the curve, the two straight lines meet at a point, as shown. It is found that the abscissa of the curve at U_{90} point

---

1. U_{60} point means a point on the curve corresponding to 60 per cent consolidation.

is 1·15. times the abscissa of the above point of intersection. This characteristic of the theoretical curve is made use of in

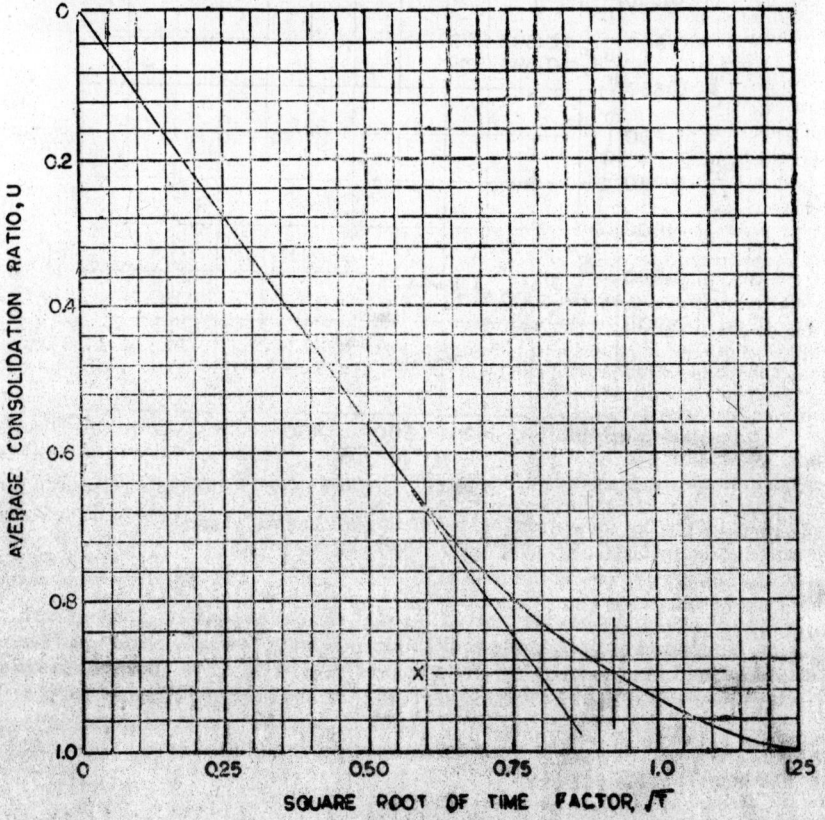

Fig. 7·20 (a).

finding out a point of 90 per cent consolidation on the empirical laboratory time compression curve.

The laboratory dial *versus* square root of time (in minutes) readings may be plotted as shown in Fig. 7.20 (b) This data relates to a pressure of 1 kg./cm² on a soil sample, when a load increment corresponding to 0 5. kg./cm² was applied to the sample (The previous pressure on the sample was 0·5 kg/cm²). Through the early part of this curve, a straight line can be drawn representing all these points. Then *another straight line* with the first line at zero time is drawn, and its abscissas are 1·15 times those of the first line. The common point on these lines at $t=0$ is usually below the start of the compression curve. This common point shown as B in the figure is called the corrected zero point and the compression that has occurred before this point is called the *initial compression*. The dial reading at point B is

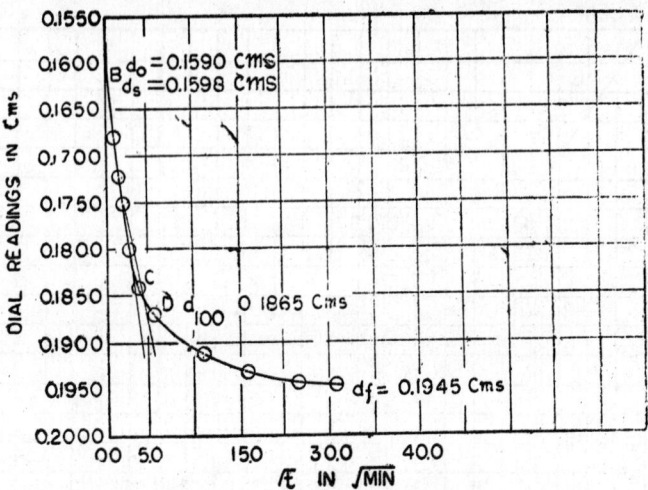

The square root of time fitting method *(a)* theoretical
curve *(b)* laboratory curve
(After Taylor).
7·20 *(b)*

0·1598 and may be designated as d_s. The dial reading at the start of the compression curve is 0·1590 and may be called the d_0. The point of intersection of the *second line* with the compression curve gives the dial reading for 90 per cent consolidation. This point is shown as point C and the dial reading d_{90} at this point is 0·1840. The time at this point may be designated as t_{90} and $\sqrt{t_{90}}$ is equal to 3·5 units.

From equation (7·33), therefore,

$$T_{90} = \frac{C_v \cdot t_{90}}{H^2} \qquad \qquad ...(7·40)$$

T_{90} for the theoretical curve is 0·848 and for the laboratory curve from above $t_{90} = (3·5)^2$ minutes. If the average value of the sample thickness *i.e.*, average of the thickness values before and after the load increment is 2·44 cms., then the value of H, for double drainage case, is 1·22 cms.

$$\therefore \qquad c_v = \frac{T_{90}.H^2}{t_{90}} = \frac{0.848 \times (1·22)^2}{(3·5)^2 \times 60}$$

or $\qquad c_v = 1·72 \times 10^{-3} \text{ cm}^2/\text{sec}.$

The final dial reading d_f is 0·1245. So is the total compression is $(d_f - d_o) = 0·0355$ cms. The dial reading corresponding to 100% of primary consolidation can be found as

$$d_{100} = d_s + \frac{10}{9}(d_{90} - d_s) = 0.1598 = \frac{10}{9}(0.1840 - 0.1598)$$

or $d_{100} = 0.1865$.

The primary compression point corresponding to this d_{100} is shown as point D. So, the primary compression ratio, i.e., the ratio of primary compression to total compression is given by,

$$r = \frac{d_{100} - d_s}{d_f - d_o} = 0.754 \text{ or } 75.4\%, \text{ in this case.}$$

The above discussion reveals that there are three portions to the compression curve, *initial compression* ranging from d_o to d_s *primary compression* from d_s to d_{100} and the *secondary compression* from d_{100} to d_f.

(2) T_{60} **Method.** The theoretical curve shown as curve A in Fig. 7.15 is parabolic upto $U = 60\%$. In fact, the degree of consolidation U upto 60% can be represented by the empirical equation,

$$T = \frac{\pi}{4}U^2. \qquad \qquad ...(7.41)$$

The fact that the early portion of the theoretical curve is parabolic is made use of in finding out the 60 per cent consolidation point on an empirical laboratory time compression curve as it is responsible to assume that the early part of the laboratory time compression curve is also parabolic, whether plotted on a natural scale or on a semi-logarithmic scale. Fig. 7.19 shows a semi-log plot of the time compression readings under a load of 1kg./cm². (same readings as are plotted in Fig. 7.18)

Let the equation of the early parabolic part of the laboratory empirical curve[2] be,

$$y = y_0 + a\sqrt{x} \qquad \qquad ...(7.42)$$

Consider two points A (x_1, y_1) and B (x_2, y_2) as lying on this parabolic portion. Then, substitution gives,

and
$$y_1 = y_0 + a\sqrt{x_1} \qquad \qquad ...(7.43)$$
$$y_2 = y_0 + a\sqrt{x_2} \qquad \qquad ...(7.44)$$

From the set of above two equations, both the constant y_0 and a can be found.

Substracting the two equations,

$$(y_2 - y_1) = a(\sqrt{x_2} - \sqrt{x_1})$$

or
$$a = \frac{y_2 - y_1}{\sqrt{x_2} - \sqrt{x_1}}$$

---

1. This value of d_0 for the pressure of 1.0 kg/cm.₂ is the last reading of the dial gauge under 0.5 kg/cm pressure. The various readings on the figure may not be a: clear since it is not possible to show such small variations.

2. The theoretical treatment upto the point where the numerical values are substituted in the equations hereafter, applies equally to the theoretical T versus U curve, i.e., curve A of Fig. 7.15.

Substituting this value of a in equation (7·43)

$$y_1 = y_0 + \frac{y_2 - y_1}{\sqrt{x_2} - \sqrt{x_1}} \cdot \sqrt{x_1}$$

or

$$(y_1 - y_0) = \frac{y_2 - y_1}{\sqrt{x_2} - \sqrt{x_1}} \sqrt{x_1} \qquad \ldots (7·45)$$

In this equation, if x_2 is substituted as $4x_1$, then

$$(y_1 - y_0) = \frac{y_2 - y_1}{\sqrt{4x_1} - \sqrt{x_1}} \sqrt{x_1} = (y_2 - y_1) \ldots (7·46)$$

So the choice of $x_2 = 4x_1$, makes the equation (7·45) much simpler and y_0 can be found, if needed, from equation (7·46).

Designate the difference $(y_1 - y_0) = (y_2 - y_1)$ as $\triangle$. Then, if the points A and B are so chosen that their abscissas are in the ratio of 4, then a point C can be located as the intersection of a horizontal line through A and a vertical line through B. Then BC represents $(y_2 - y_1) = \triangle$. Extend this line BC upwards and on this extended line cut and intercept $CD = \triangle = (y_0 - y_1)$. Ordinate point D is then y_0.

In a similar manner take a few more pairs of points[1] having their abscissa ratios as 4 and locate a series of points E, F, G. etc., as shown in Fig. 7·21. Since y_0 is a constant, all these

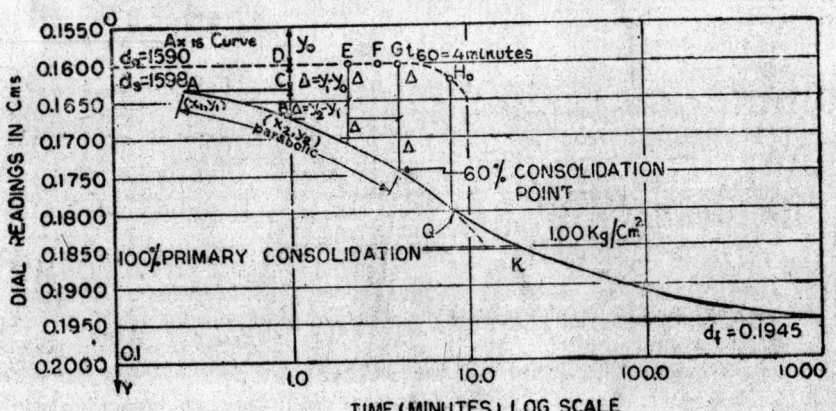

T_{60} Method of determining C_v.

Fig. 7·21

points i.e. D, E, F, etc. shall be on a straight line, *so long as* the pairs of points chosen on the time compression curve lie on its parabolic portion. As soon as the time compression curve departs from a parabola, further points located similarly as the points D, E, F etc. no more lie on a curve. One such point is shown as H_0. The curve obtained through the points D, E, F, G and H_0, etc., may be termed as the *axis curve*, since its straight line portion is the axis of the parabola. The point where the curvature starts on this axis curve is, therefore, the point that belongs to 60 per cent consolidation, because the compression curve ceases to be parabolic at this point and this point on the empirical curve, therefore, must correspond to 60% consolidation point on the theoretical curve.

On the theoretical consolidation curve A of Fig. 7·15, when $U=60$ per cent, $T=0·287$. So we may write,

$$T_{60} = \frac{c_v t_{60}}{H^2} \qquad \qquad ...(7·47)$$

Knowing t_{60} from the empirical curve of Fig. 7·19, C_v can be found out. For the data shown in this figure, $t_{60}=4$ minutes, and H as in the previous method may be taken as 1·22 cms., then

$$C_v = \frac{T_{60}. H^2}{t_{60}} = \frac{0·287 \times (1·22)^2}{4 \times 60}$$

or $C_v = 1·77 \times 10^{-3}$ cm^2./sec.,

which is in fairly close agreement with the value obtained by Taylor's method.

In this method, the point of zero compression of the time compression curve is assumed to lie on the horizontal portion of the axis curve. The ordinate value of the horizontal portion of the axis curve is 0·1598 cms. The last dial reading on the previous method of 0·5 kg/cm^2., as indicated in the previous method was 0·1590 cms. So there is an instantaneous compression of $(0·1598-0·1590)=0·0008$ cms., before the time consolidation commenced.

The dial reading against 60% consolidation point i.e., against 4 minutes time reading is 0·1752 cms. So the time compression upto this point is $(0·1752-0·1598)=0·0540$ cms. The total primary compression, therefore, is 0·0154/0·60=0·0257 cms. as shown by the point K, because the dial reading corresponding to 100 per cent primary consolidation becomes 0·1598+0·0257=0·1855 cms.

The primary compression ratio by this method is

$$r = \frac{d_{100}-d_s}{d_f-d_0} = 0·725 \text{ or } 72·5\%.$$

The utility of the axis curve lies in the fact that as soon as the compression curve ceases to be parabolic, there is a steep curvature in the axis curve and it is possible to detect this departure easily, from the axis curve than from the time compression curve itself.

If the axis curve is plotted on the theoretical T versus U curve, the axis curve is strictly a straight line upto the 60% consolidation and then it sharply becomes a curve. In the empirical curve, in practice, however, sometimes it is found that before falling from the horizontal the axis curve shows some rise. This rise indicates that the material is consolidating more rapidly than the theory suggests and it may be due to the secondary consolidation effects, which are not accountable in the theory. Another characteristic is also sometimes found that the axis curve instead of being a horizontal line initially and then a curve, shows actually a mild curvature downwards right from the beginning. This is an indication that something is retarding the consolidation process.

(3) **The Logarithm of Time Fitting Method.** If the theoretical consolidation curve A of Fig 7·15 is plotted on a semi-log

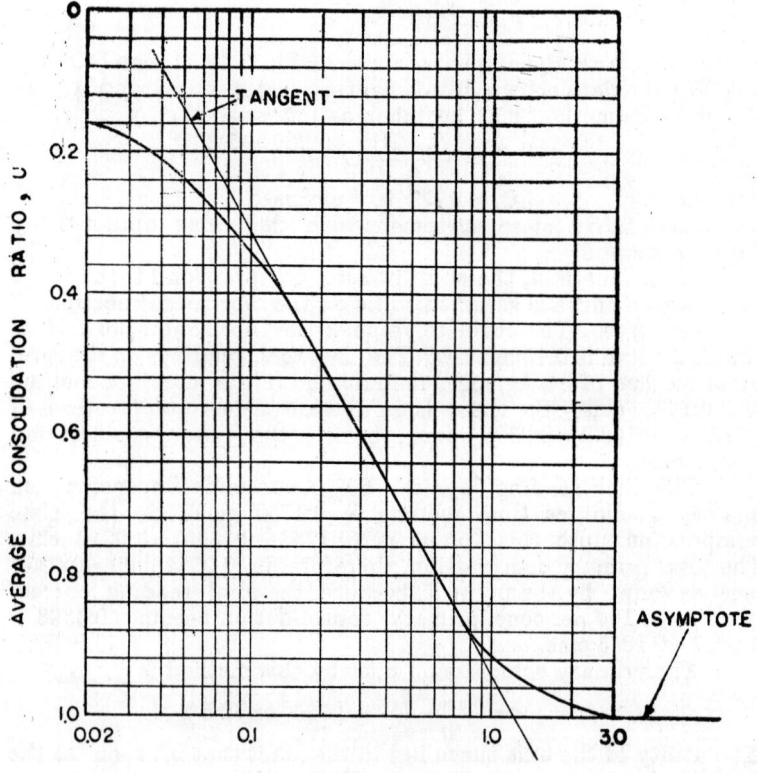

TIME FACTOR, T

Logarithm of Time Fitting Method, theoretical curve of U versus T (semi-log scale).

Fig. 7·22.

scale as shown in Fig. 7·22, it is found that the tangent and the asymptote meet at the ordinate of 100 per cent consolidation. This fact is employed to locate the 100 per cent primary consolidation point on the laboratory curve, by drawing the tangent and the asymptote to it and locating their intersection.

The corrected zero point, in this case, is found out by assuming the initial portion of the time compression curve as a parabola and using the method for drawing the axis curve. The horizontal portion of the axis curve produced backwards to meet the dial gauge readings axis gives the corrected zero point.

After the zero and 100 per cent consolidation points are located, the time corresponding to any intermediate consolidation precentage say 50 per cent can be located and c_v compiled from.

$$T_{50} = \frac{c_v\, t_{50}}{H^2} \qquad \ldots(7 \cdot 48)$$

From the theoretical curve A, 50 per cent consolidation occurs when $T = 0 \cdot 197$ i.e., T_{50} in the above equation is $0 \cdot 197$. Knowing t_{50} from the empirical laboratory curve H, the longest drainage path, c_v can be found out by using equation (7·48).

From the empirical laboratory curve shown in Fig. 7·23 for the same data as presented in Figs. 7·18 and 7·19, $t_{50} = 2 \cdot 9$ minutes and H as mentioned previously is $1 \cdot 22$ cms.

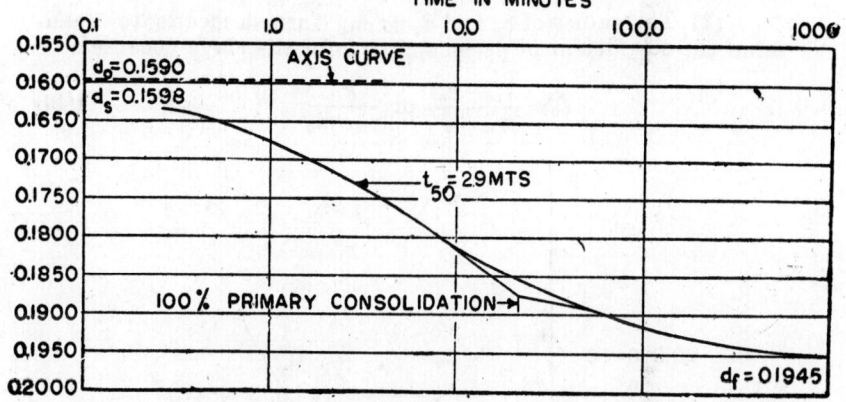

Logarithm of Time Fitting Method, laboratory curve
Fig. 7·23.

$$\therefore \qquad c_v = T_{50}\, \frac{H^2}{t_{50}}$$

$$= \frac{0 \cdot 197 \times (1 \cdot 22)^2}{2 \cdot 9 \times 60}$$

or $\qquad = c_v = 1 \cdot 69 \times 10^{-3}$ cm²./sec.

Primary compression ratio by this method is

$$r = \frac{d_{100} - d_s}{d_f - d_0} = 0 \cdot 766 \text{ or } 76 \cdot 6\%$$

As a concluding comment, it can be said that, in general, there is always a fair amount of agreement in the values of c_v and r obtained from various methods. It is not possible to advocate the use of one method in preference to the other. If there is any small difference in the numerical values of c_v and r obtained from these methods, it is possibly due to human judgment in plotting and obtaining these values.

7·11. Utility of Consolidation Data

There are many ways in which the consolidation data is used :

(1) Co-efficient of consolidation c_v is found out for different pressures applied to a representative soil sample from a deposit and is made use of in the settlement analysis, as indicated in Chapter 9.

(2) From the e-log p curve, the compression index can be determined and the settlement under load may be expressed in terms of compression index.

(3) The values of c_v and a_v or m_v furnish means to determine the co-efficient of permeability k for the clays because,

$$c_v = \frac{k(1+e_0)}{\gamma_w \, a_v} \text{ or } \frac{k}{\gamma_w \cdot m_v} \qquad \ldots(7\cdot49)$$

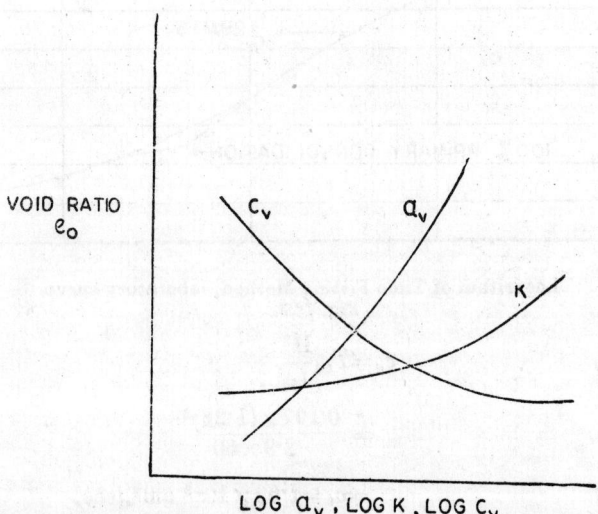

Qualitative plot of log a_v, log k and log c_v versus void ratio e_0.

Fig. 7·24.

so that knowing c_v and a_v or m_v along with the value of e_0, the co-efficient of permeability for the corresponding pressure can be found.

Some observations with regard to the values of a_v, k and c_v may be made here. With the decrease in void ratio, due to increase in pressure,

(a) a_v decreases quite rapidly.

(b) k decreases but not as rapidly as a_v.

(c) as a consequence of (a) and (b) above c_v increases.

In a typical test, if a semi log plot is drawn with e on the natural scale and a_v, k and c_v on the log scale, the plot will look like the one shown in Fig. 7·24.

Example 7·1. In a consolidation test on an undisturbed soil sample, the following readings were obtained :—

Pressure in kg./cm².	Void ratio
0·25	0·953
0·50	0·942
1·00	0·938
2·00	0·913
4·00	0·878
8·00	0·835
16·00	0·715
8·00	0·740
4·00	0·752
2·00	0·768
1·00	0·785
0·50	0·802
0·25	0·820

(a) Plot these results on a semi-log plot. Determine the equation for the virgin curve from this plot.

(b) If the sample belongs to the top of the clay stratum shown in Fig. 7·25 indicate whether the clay is normally consolidated or it is over-consolidated.

Solution. On a three-cycle semi-log paper, the e-log p curve is shown in Fig. 7·26.

Let the equation of the virgin curve be,

$$e = e_0 - C_c \log_{10} p/p_0$$

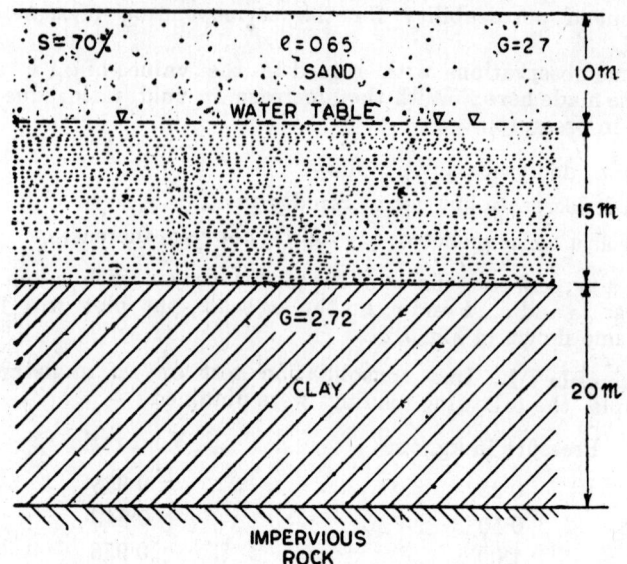

Fig. 7·25.

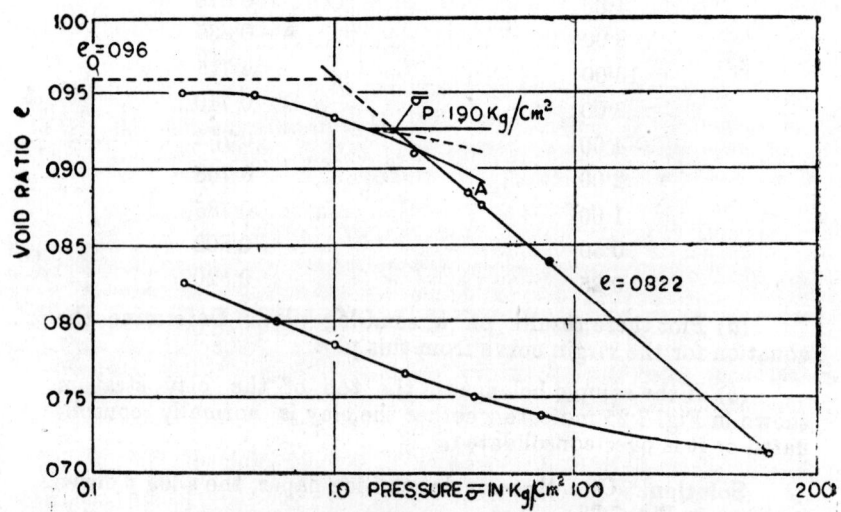

Pressure void ratio diagram (semi-log scale).
Fig. 7·26.

Let $p = 10$ kg./cm².

and $p_0 = 1$ kg./cm². so that log $p/p_0 = 1$

and $C_c = e_0 - e,$

where e_0 and e correspond to the 1 kg./cm². and 10 kg./cm². pressure.

From the Fig. 7·24.

$$e_0 = 0·96 \text{ and } e = 0·822$$

$$\therefore \quad C_c = 0·138$$

$$\therefore \quad e = e_0 - 0·138 \log_{10} \frac{p}{p_0} \qquad \ldots(2)$$

is the equation of virgin curve.

(b) The pre-consolidation load from the Fig 7·24 is

$$\overline{\sigma}_p = 1·90 \text{ kg/cm}^2$$

If the sample is taken from the top of the clay layer in Fig. 7·23, the present over-burden can be calculated from the given soil properties.

So $\overline{\sigma}_{present}$ = unit weight of 10 metres of sand + submerged unit weight of 15 metres of sand.

$$\overline{\sigma}_{present} = 1\left(\frac{G+Se}{1+e} \times 1000 + \frac{G-e}{1+e} \times 1500\right) \times \frac{1}{1000} \text{ kg/cm}^2$$

or $\quad \overline{\sigma}_{present} = \dfrac{2·70+0·7 \times 0·65}{1+0·65} + \dfrac{2·70-0·65}{1+0·65} \times 1·5$

or $\quad \overline{\sigma}_{present} = 3·02 \text{ kg/cm}^2$

Since $\overline{\sigma}_{present}$ is more than the pre-consolidation pressure of 1·90 kg/cm², the soil is normally consolidated. The point corresponding to $\overline{\sigma}_{present}$ is shown as point A in the figure.

Example 7·2. The time to reach 60 percent consolidation is 32·5 seconds for a sample 1 cm thick, tested in the laboratory under conditions of double drainage. How long will the corresponding layer in nature require to reach the same degree of consolidation, if it is 10 metres thick and drained on one side only ? **(P.U. 1962)**

Solution. For the laboratory sample, longest path of drainage

$$H = 0·5 \text{ cms.}$$

and t_{60} corresponding to $U_{60} = 32·5$ seconds.

$$\therefore \quad T_{60} = \frac{c_v t_{60}}{H^2}$$

or $\quad c_v = T_{60} \dfrac{(0·5)^2}{32·5} \qquad \ldots(1)$

From the field stratum,

longest path of drainage H=10 metres=1000 cms.

$$\therefore \quad T_{60}=c_v\ \frac{t_{60}}{H^2}=T_{60}\frac{(0\cdot5)^2}{32\cdot5}\cdot\frac{t_{60}}{(1000)^2}$$

$$\therefore \quad t_{60}=\frac{32\cdot5\times(1000)^2}{0\cdot25}\ \text{seconds}$$

$$=4\cdot125\ \text{years} \qquad \textbf{Ans.}$$

Example 7.3. A 7-metre thick fine silt stratum has submerged course sand above it and overlies a 0·15 meter thick fine and layer under which there is clay stratum of 3-metre thickness. Previous sand-stone lies underneath the stratum of clay. The following laboratory data was obtained for these soils in series of tests :

For fine silt, $k=2\times10^{-7}$ cms/sec.

when $\sigma_1=3\cdot50$ kg/cm^2, $e_1=0\cdot80$

when $\sigma_2=5\cdot50$ kg/cm^2, $e_2=0\cdot70$

For clay. $k=1\times10^{-8}$ cms/sec.

when $\sigma_1=3\cdot20$ kg/cm^2, $e_1=1\cdot20$

when $\sigma_2=4\cdot20$ kg/cm^2, $e_2=1\cdot00$

The fine sand layer is connected to free water and acts as a drainage channel. When the clay reaches a degree of consolidation of 20%, what would be the degree of consolidation of the slit layer. Curve A of Fig. 7·15 is supplied with the problem ?

Solution. *For the clay layer,*

Permeability $k=1\times10^{-8}$ cms/sec.

Pressure increment $\triangle\sigma=(\sigma_2-\sigma_1)=1\cdot00$ kg/cm^2
$$=1000\ \text{gms/cm}^2$$

Under this pressure increment, t change in void ratio

$$\triangle e=e_2-e_1=0\cdot20$$

$$\therefore \quad a_v=\left|\frac{\triangle e}{\triangle\sigma}\right|=\frac{0\cdot20}{1000}=2\times10^{-4}\ \text{cm}^2/\text{gm.}$$

Coefficient of consolidation corresponding to this pressure increment

$$c_v=\frac{k(1+e_0)}{a_v\cdot\gamma_w}=\frac{10^{-8}(1+1\cdot20)}{2\times10^{-4}\times1}$$

or $c_v=1\cdot1\times10^{-4}$ cm^2/sec.

Note that the clay layer has double drainage.

Now apply $T_{20}=\dfrac{c_v\cdot t_{20}}{H^2}$

$$\therefore \quad t_{20}=T_{20}\frac{H^2}{c_v}=\frac{0\cdot0313\times(1\cdot5\times100)^2}{1\cdot1\times10^{-4}}\ \text{secs}$$

or $=74.3$ days

For the silt layer.

Permeability $k=2 \times 10^{-7}$ cms/sec.

Pressure increment $\triangle \sigma = (\sigma_2 - \sigma_1) = 2.0$ kg/cm^3.

$= 2000$ gms/cm^2.

Under this pressure increment, change in void ratio

$$\triangle e = e_2 - e_1 = 0.10.$$

$\therefore$ $a_v = \left| \dfrac{\triangle e}{\triangle \rho} \right| = \dfrac{0.10}{2000} = 5 \times 10^{-5}$ cm^2/gm.

$\therefore$ Coefficient of consolidation corresponding to this pressure increment,

$$c_v = \frac{k(1 + e_0)}{a_v. \ \gamma_w} = \frac{2 \times 10^{-7}(1 + 0.8)}{5 \times 10^{-5} \times 1}$$

or $c_v = 7.2 \times 10^{-3}$ cm^2/sec.

When the clay reaches 20% consolidation,

$$t = 74.3 \text{ days.}$$

$\therefore$ $T = \dfrac{c_v \times 74.3}{H^2}$

$$= \frac{7.2 \times 10^{-3} \times (74.3 \times 24 \times 60 \times 60)}{(3.5 \times 100)^2}$$

or $T = 0.392.$

$\therefore$ From the theoretical time consolidation curve, degree of consolidation U is 69 percent.   **Ans.**

CHAPTER—BIBLIOGRAPHY

1. Bowls, E., "Foundation Analysis and Design", McGraw Hill Book Co., New York, N. Y., 1968.

2. Casagrande, A., Determination of Preconsolidation Load and its Practical Significance. 1st International Conference on Soil Mechanics and Foundation Engineering Vol. III 1936.

3. Leonard, G. A., "Foundation Engineering", McGraw Hill Book Co., Inc., New York, N. Y., 1962.

4. Schmertmenn, John H., The Undisturbed Consolidation Behaviour of Clay, A.S.C.E., Transactions, Vol. 129, 1955 (Paper No. 2775).

5. Scott, R. F., "Principles of Soil Mechanics", Addison—Wesley Publishing Co., Inc., Mass., 1963.

6. Taylor, D. W., "Fundamentals of Soil Mechanics", New York, John Wiley & Sons, Inc., 1948.

7. Terzaghi, C., "Principles of Soil Mechanics", I-Phenomena of Cohesim of Clay, Engineering News Record, Volume 95, No. 19, November 5th, 1925.

8. Terzaghi, C., "Principles of Soil Mechanics" IV—Settlement and Consolidation of Clats, Engineering News Records, Vol. 95, No. 22, Nov. 26, 1925.

9. Terzaghi K. and Peck R. B., Soil Mechanics in Engineering Practice, Asia Publication House, 1948.

QUESTIONS

1. (a) Describe briefly with a sketch the consolidometer test as done in the laboratory. Show how the results of this test are used to predict the settlement of a structure founded on a clay layer.

(b) A compressible soil layer is 30 ft. thick and its critical void ratio is 1·038. Test indicates that its final void ratio after the construction of a building will be 0·981. What will be the probable settlement of the building over a long period of time ?

2. Write short notes on :—

(a) Precompression in clays.   (P. U. 1959 Supp.)
(b) Hydrostatic excess pressure.   (P. U. 1961)
(c) Hpdrodynamic time lag.   (P. U. 1961)
(d) Frictional time lag.   (P. U. 1961)
(e) Preconsolidation stress.   (P. U. 1961)
(f) Degree of consolidation.   (P. U. 1961)
(g) Coefficient of consolidation.   (P. U. 1962 and 1963)
(h) Compression Index.   (P. U. 1962 and 1963)

3. (a) Derive the basic differential equation for the consolidation of soils clearly indicating the assumptions involved.

(b) Representative samples were obtained from a layer of clay 20 ft. thick, located between the two layers of sand. By means of consolidation tests, it was found that the average value of the coefficient of consolidation of these samples was $4·29 \times 10^{-4}$, sq. cm. per sec. By constructing a building above the layer, the average vertical pressure in the lower layer was everywhere increased and the building began to settle. Within how many days did half the settlement occur ?   (P. U. 1960)

4. (a) Describe briefly with a neat sketch, the consolidation test for determining the consolidation properties of s saturated clay. Show how the results of this test can be used to predict the settlement of a structure founded on a clayey soil.

(b) At a certain depth below the foundation of a building, there is a stratum of clay of 20 ft. with relatively incompressible permeable soil above and below. In a consolidation test on this clay, a sample initially 1·030 inch thick was compressed under a steady pressure. Half of the final settlement took place in the first 11 miniutes after the application of the pressure.

Estimate how long will it take for the settlement of the building to reach 50 per cent of its ultimate value.   (P. U. 1960 Supp.)

5. A sample of clay tested in a consolidometer (drained top and bottom) was 1·2 inches thick at the beginning of a load increment and 20 per cent consolldation was reached in 20 minutes. How long will it take a layer of the clay 30 ft. thick drained on the upper surface only to reach 60 per cent consolidation ?

(**Note.** Numerical U versus T tables are not needed).   (*P. U. 1961*)

6. State the assumptions made in deriving the differential equation for one dimensional consolidation.

Explain the symbols used in the above equation. What are the boundary conditions used to solve this equation ?   (*P. U. 1962*)

7. Sketch the e-log p curves for consolidation tests on an undisturbed sample on a partly disturbed sample and on a remoulded sample of normally consolidated clay.   (*P. U. 1963*)

8. *(a)* In the theory of consolidation, what is meant by

    (i)   Time factor.   (*P. U. 1963*)

    (ii) Coefficient of compressibility.
                             (*P. U. 1963 & 1963 Supplementary*)

    (iii) Coefficient of volume compressibility.
                             (*P.U. 1963 & 1963 Supplementary*)

(b) A building was constructed above a clay layer 20 feet thick located above rock stratum. The coefficient of consolidation of this clay was found to be 4.92×10^{-4} sq. cms. per sec. Within how many days did half the ultimate settlement occur ?   (*P.U. 1963*)

9. Discuss briefly the "Log curve fitting" method in consolidation test ?   (*P. U. 1964*)

10. *(a)* State the assumptions and derive Terzaghi's differential equation for one dimensional consolidation.

(b) Define :

    (i) Normally consolidated clay,

    (ii) Preconsolidated clay, and

    (iii) Under-consolidated clay.   (*P. U. 1964*)

11. Complete the statement :

"The compression index for an insentive normally loaded clay having a liquid equal to natural moisture content of 40% will be . . . .

12. *(a)* Describe the consolidation test. Explain with the help of assumed data the time fitting methods. How are the values of C_c and C_b obtained from the consolidation test ?

(b) State the Terzaghi's differential equation of dimensional consolidation. Explain the symbols used.   (*P. U. April, 1965*)

13. *(a)* Define "compression index" and "coefficient of consolidation". How would you proceed to determine both of them ?

(b) How is the preconsolidation pressure of an over-consolidated clay determined ? State the causes of preconsolidation of a soil in nature.   (*P. U. Nov. 1965*)

14. Explain the meaning of the terms :

    (i) Secondary compression.

    (ii) Coefficient of consolidation.

    (iii) Coefficient of compressibility.

    (iv) Compression index and expansion index. (*P. U. Nov. 1966*)

15. Why do the field lime, the consolidation curve for an undistressed sample and the consolidation curve for a remoulded sample differ and full apart ?

16. Is it reasonable to assume that the initial hydrostatic pressure in a thick layer of clay, constant. If not, why ?

17. The priliminary settlement analysis of a proposed building to be founded above a clay startum indi·ates a settlement of 3 inches in 5 years and an ultimate settlement of 12 inches at typical interior column. The estimated average increment of pressure in the clay below this column is 0.133 tons per sq. ft. Data that has recently been obtained indicate the following variations from the assumptions used in the preliminary analysis :

(a) A 3' 00" lowering of water table will take place during construction and this change is likely to be permanent.

(b) The loading period will be a full 2 years, where as no loading was considered in the previous estimate.

(c) The compeessible clay stratum is 20 p·c. thicker than the assumed in original analysis.

Obtain estimates of the ultimate settlement, the settlement at the middle and at the end of the loading period and settlement 2 years after the completion of the building.   (*Mysore 1966*)

18. A fill which is to serve as structural foundation is placed hydrolically. Tests run on material form a certain layer in the fill show that the preconsolidation pressure is 1/2 of that value which one would figure for the effective overburden pressure. When the structural load is added, the vertical pressure will be $1\frac{1}{2}$ times the effective overburden pressure.

What is the ratio of settlement (caused by this particular layer) without structural load to settlement with structural load on embankment ?   (*Mysore 1967*)

19. (a) Distinguish between "coefficient of compressibility" and "Coefficient of Consolidation".

(b) Sketch and describe the "Consolidometer" used for conducting consolidation tests in the laboratory.   (*Warrangal 1968*)

20. (a) An additional floor is to be added to an existing 30 year old house. The earlier column loads were 1 ton/sq. foot. After the construction of new floor, the load intensity on the soil under the column increases to 2·5 ton/sq. ft. Determine the total settlement of the column after construction of additional floor. The compressible layer below the columns is 4 feet thick. Load intensity at the centre of the layer is 0.9 times the load at the bottom of the footing. Void ratio of the soil strata at 0.5 ton is 2.0. Liquid limit of the soil is 80%. The soil is normally consolidated.

(b) Draw $e - log_e P$ curve for a normally consolidated clay and also an over consolidated clay. How do you estimate the compression under from this curve.

Stress Distribution in Soils

8.1. Introduction

When a load is placed on the surface of the ground or below the ground level, some stress is transmittted to the deeper layers of the soil underneath. Theoretical solutions for distribution of stresses induced by loads applied to elastic materials are available. A number of persons have worked ou the problems of stress distribution in elastic bodies, notable among them being Kelvin, Ceruti, Mindlin, Boussinesq, Newmark and Westergaard. Solutions evolved by them, based upon the theory of elasticity, are of great practical value to a Civil Engineer, as they help in understanding the stress-strain relationships in many statically indeterminate structures of steel. Of late, these theoretical solutions have more and more been used in estimating stresses in soil masses. The fundamental requirement[1] for applicaiion of any of of these solutions is that the stresses and corresponding strains are proportional. So, for any material that satisfies this requirement, these solutions can be applied.

Soils, in fact, are not elastic in nature. However, for some problems in soils where it is resonable to assume that the stresses and corresponding strains are proportional. it may be worthwhile to determine the distribution of stresses by these theoretical relationships. How far this assumption of proportionality of stresses and strains can be made is questionable. In the consolidation theory, for example, it was assumed that for *a load increment*, the increase in intergranular pressure is proportional to the void ratio change and since void ratio change is approximately proportional to the vertical strain in one dimensional consolidation, this assumption of proportionality between stresses and strains for *an increment* may be may be made with resonable accuracy In estimating stresses induced by loads for calculating the settlements under foundations in clayey soils, therefore, it may be worthwhile to use the formulae based upon the elastic theory. However, it may be noted that if plastic lag and secondary consolidation effects are also taken into consideration, this assumption

---

1. Other requirements have not been mentioned here

may appear to be invalid, because such effects are not accounted for in the elastic theory. Furtheron, a decrease in load does not show a proportionality between stresses and corresponding strains.

Similarly, if the stress and strain curve for a sample subjected to axial compression is examined (See Fig. 10·10 a) it is seen that the stress ratio increase is approximately proportional to the axial strain for low values of strain, whereas, the proportionality does not hold at higher strains. Thus it may generally be concluded that a formula based upon elastic theory may be applied to evaluate distribution of stresses in a soil mass to a case where the stresses is only increase and the strains are low.

Another assumption in the derivation of the simple expression given in the subsequent section is that the elastic material is isotropic and homogeneous. These conditions are far from reality in soils. A natural soil deposit is anything but homogeneous and it is impossible to expect it to show isotropy. However, for pure clay deposits, we may make the assumptions that a deposit is homogeneous and isotropic, if too much variation in consistency is not found ; but in sand deposits these assumptions cannot be assumed isotropic because in natural deposits it has been found that the pressures on vertical planes are much smaller than those on horizontal planes and hence nonisotropic conditions always exist. Cummings (15) has given modified forms' of for_____ based upon the elastic theory, for application to sand deposits. It is sufficient to note here that when stresses are transmitted to a clay deposit, through a small sand layer on top of it, any effect due to the presence of sand layer is neglected and the stresses in clay layer are found based upon the elastic theory.

8·2. General Expression for Stresses

The state of stress at any point within an elastic, homogeneous and isotropic medium can be described by six components of stress—three of them being normal stresses and the other three being shear stresses. The normal stress in any direction is denoted by σ with a subscript of the respective direction tion and the shearing stress by τ with two subscripts, first denotes the direction of the normal to the plane on which τ acts and the second denotes the direction in which τ acts. A general expression for any component of stress or τ at any point includes a numerical constant K, a load factor P which is function of the applied load, inverse of the square of distance R of the point where or τ act from the origin and a factor F, which is a function of the co-ordinates $(x, y. z)$ of the point and the Poissons ratio μ.

Thus, $\qquad \sigma$ or $\tau = K . \dfrac{P}{R^2} F (x, y, z, \mu) \qquad ...(8·1)$

_____

1. These modified formulae and a discussion on them is outside the scope of this book.

For purposes of calculating settlements, we are only interested in the vertical stress σ_z due to any loading system. So, attention will mainly be given to its determination under vertical loads of different types, applied at the surface. Furtheron, expressions for vertical stress given by Boussinesq, Westergaard and Newmark will only be presented in this book as its scope is limited to the undergraduate work.

8·3. Expressions when Point-load is Applied to the Surface

Two cases will be mentioned here—one given by Boussinesq for ideally elastic, isotropic and homogeneous medium and second given by Westergaard's for elastic and non-isotropic medium.

(a) **Boussinesq Case.** Boussinesq gave a solution for evaluation of stresses inside an elastic homogeneous and isotropic medium extending infinitely in all directions from a level surface, when a point load is applied to the surface.

If A is a point inside the above medium with co-ordinates (x, y, z) from the origin which is taken as the point of application of a point load P (see Fig. 8·1), then the vertical stress σ_z at this point P due to this load is given by,

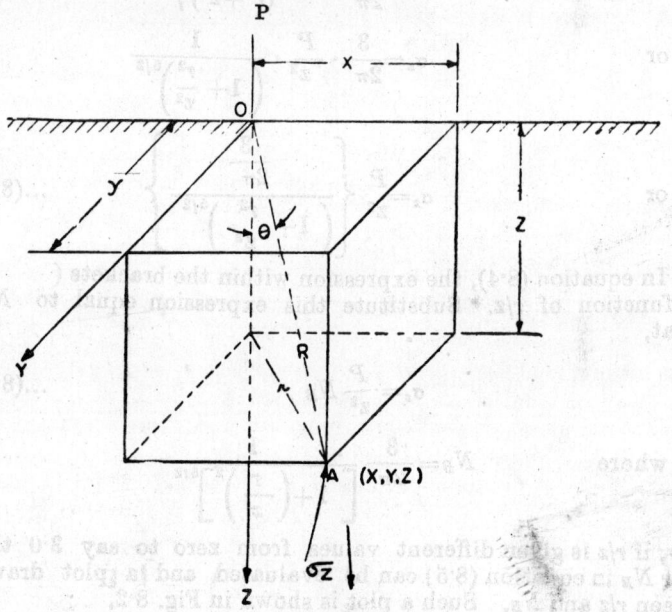

Fig. 8·1.

$$\sigma_z = \frac{3}{2\pi} \cdot \frac{P}{R^2}\left(\frac{z}{R}\right)^3 \qquad \ldots(8\cdot2)$$

where $R^2 = x^2 + y^2 + z^2$

If the angle made OA with the vertical through O is taken as θ, then $\dfrac{z}{R} = \cos\theta$ and equation (8·2) becomes,

$$\sigma_z = \frac{3}{2\pi} \cdot \frac{P}{R^2} \cos^3\theta \qquad \qquad ...(8\cdot3)$$

where $\theta = \tan^{-1}\dfrac{r}{z}$, $r^2 = x^2 + y^2$

as shown in the figure.

Equation (8·2) can be written in another form also.

From the figure,

$$R^2 = x^2 + y^2 + z^2$$
$$= r^2 + z^2$$
$$\therefore \qquad R = (r^2 + z^2)^{1/2}$$

Substitute in equation (8·2) to get σ_z.

$$\sigma_z = \frac{3}{2\pi} P z^3 \frac{1}{(r^2 + z^2)^{5/2}}$$

or

$$\sigma_z = \frac{3}{2\pi} \cdot \frac{P}{z^2} \cdot \frac{1}{\left(1 + \dfrac{r^2}{z^2}\right)^{5/2}}$$

or

$$\sigma_z = \frac{P}{z^2} \left\{ \frac{\dfrac{3}{2\pi}}{\left(1 + \dfrac{r^2}{z^2}\right)^{5/2}} \right\} \qquad ...(8\cdot4)$$

In equation (8·4), the expression within the brackets (   ) is a function of r/z. Substitute this expression equal to N_B. so that,

$$\sigma_z = \frac{P}{z^2} N_B \qquad \qquad ...(8\cdot5)$$

where $N_B = \dfrac{3}{2\pi} \cdot \dfrac{1}{\left[1 + \left(\dfrac{r}{z}\right)^2\right]^{5/2}}$

Now, if r/z is given different values from zero to say 3·0 the factor N_B in equation (8·5) can be evaluated and a plot drawn between r/z and N_B. Such a plot is shown in Fig. 8·2.

N_B may be called Boussinesq index. The plot of N_B versus r/z is a convenient device to evaluate σ_z. It eliminates cumbersome calculations.

(b) **Westergaard's Case.** Westergaard gave a solution for evalution of stresses inside an elastic material which is reinforced

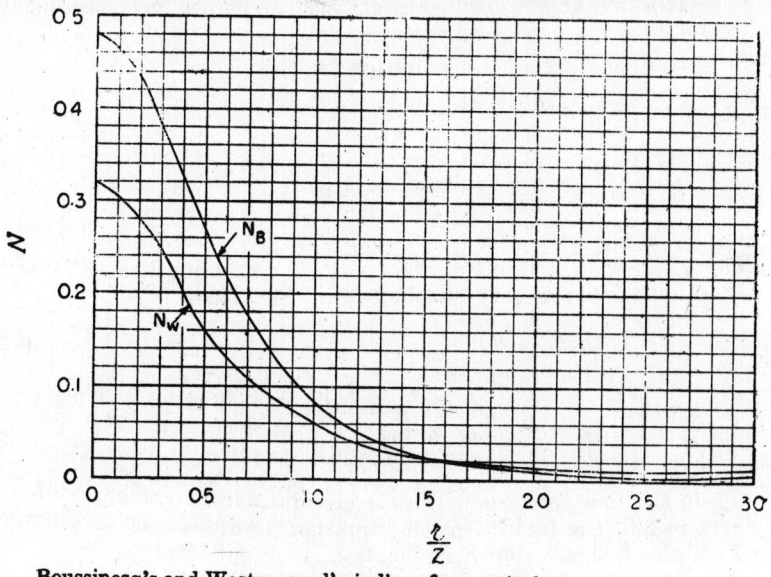

Boussinesq's and Westergaard's indices for vertical stress due to a point load on the surface of an elastic medium.

Fig. 8·2.

horizontally by closely spaced sheets of infinite rigidity but negligible thickness, which do not allow the material to undergo any lateral strain. The case can be applied to a clay stratum in which thin lenses of coarser material are interposed. This condition is often found in sedimentry soils. Due to the presence of these thin layers of coarse material the soil strata can no more be assumed as isotropic, as in the Boussinesq case given above. This non-isotropy in the soil leads to a great resistance to lateral strain.

Westergaard's solution for this type of a medium gives the following expression for vertical stress σ_z when a point load P is placed at the surface.

$$\sigma_z = \frac{P}{z^2} \; \frac{\dfrac{1}{2\pi}\sqrt{\dfrac{1-2\mu}{2-2\mu}}}{\left[\dfrac{1-2\mu}{2-2\mu}+\left(\dfrac{r}{z}\right)^2\right]^{3/2}} \qquad ...(8.6)$$

All terms except μ in the above equation can be visualized in figure 8·1 and carry the same significance as given in equation (8·4). μ is the Poisson's ratio for the material. For elastic materials the value of μ varies from zero to 0·5 the latter value being applicable to materials that show no volume change characteristics under load.

If μ is taken as zero[1], the stresses at points directly below the load have a minimum value.

Substituting $\mu = 0$ in equation (8·6),

σ_z is given by,

$$\sigma_z = \frac{P}{z^2}\left[\frac{\dfrac{1}{\pi}}{\left\{1+2\left(\dfrac{r}{z}\right)^2\right\}^{3/2}}\right] \qquad \ldots(8·7)$$

The expression within the brackets [   ] in equation (8·7) is a function of r/z and may be denoted by N_W. Then,

$$\sigma_z = \frac{P}{z^2}N_W \qquad \ldots(8·8)$$

where

$$N_W = \frac{1}{\pi}\frac{1}{\left[1+z\left(\dfrac{r}{z}\right)^2\right]^{3/2}}$$

Again as done previously if r/z is given different values, say from zero to 3·0, the factor N_W in equation (8·8) can be evaluated. This plot is again shown in Fig. 8·2.

This curve of N_W is flatter than that of N_B. A flat curve is logical for the case of large lateral restraint,

8·4. Expressions when Uniformity Distributed Load is Applied to the surface.

(a) **Newmark's Case.** Newmark (3) gave a solution for evaluation of st esses induced beneath the corner of a rectangle loaded uniformly by a load having an intensity q. The load is assumed as having been applied to an elastic, homogeneous and isotropic material extending infinitely in all directions from a level surface where the load rests.

If mz and nz denote the dimensions of the rectangular area on which the uniformly disturbed load acts, where z is the depth at which the vertical stress under the corner is needed, then the expression for the vertical stress σ_z is

$$\sigma_z = \frac{g}{4\pi}\left[\frac{2mn\sqrt{m^2+n^2+1}}{m^2+n^2+1+m^2n^2}\frac{m^2+n^2+2}{m^2+n^2+1}\right.$$

$$\left. +\sin^{-1}\frac{2mn\sqrt{m^2+n^2+1}}{m^2+n^2+1+m^2n^2}\right] \qquad \ldots(8·9)$$

---

1. $u=0$ gives the flatest curve for stress distribution under a point ad in a medium where large lateral restraint is present.

A couple of points with respect to equation (8·9) are worth noting.

(1) If m and n are interchanged, the equation remains unchanged, as it is symmetrical with respect to m and n. This implies that there is no rigidity as to which dimension of the loaded rectangle is called mz and which nz.

(2) The value of σ_z depends only on m and n and the intensity of loading q. Factors m and n are further functions of the depth z and the dimensions of the rectangle.

If the rectangle has the dimension "a" and "b" and the stress is required at a depth z below one of its corners then n may be called (a/z) and m may be called (b/z) or they may be intercharged, $i.e.$ n may be called (b/z) and m may be called (a/z), according to (1) above.

Now, if the stress is desired at a point which does not lie directly below the corner of the loaded rectangle, then in that case the loaded area may be split up into a number of rectangles. For example, if the stress is needed at a point directly below the point K, in a loaded rectangle $ABCD$, as shown in Fig. 8·3, then the rectangle $ABCD$ may be split up into four rectangles such that the point K forms the corner of each of the four rectangles $AEKH$, $BGKE$, $GCFK$ and $DFKH$. The stress at a point distance z below the point K is then the sum of the stresses induced by these four loaded rectangles.

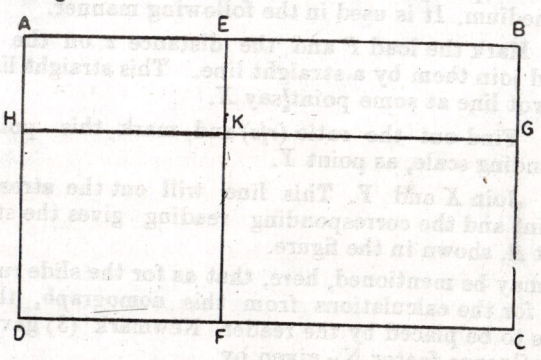

Fig. 8·3.

(b) **Westergaard's Case.** For the non-isotropic conditions as given earlier for the point load, Westergaard's analysis can be extended to cover the case of a uniformly loaded rectangle, if the load intensity is q and the dimensions for the loaded rectangle are mz and nz (as in the Boussinesq's case) then the vertical stress σ_z for the Westergaard's case is given by,

$$\sigma_z = \frac{q}{2\pi}\cot^{-1}\sqrt{\left(\frac{1-2\mu}{2-2\mu}\right)\left(\frac{1}{m^2}+\frac{1}{n^2}\right)+\left(\frac{1-2\mu}{2-2\mu}\right)^2\frac{1}{m^2 n^2}} \quad \ldots(8·10)$$

In case μ is put equal to zero, equation (8·10) becomes.

$$\sigma_z = \frac{q}{2\pi} \cot^{-1} \sqrt{\frac{1}{2m^2} + \frac{1}{2n^2} + \frac{1}{4m^2n^2}} \qquad \ldots(8·11)$$

The observations mentioned under (a) in respect of equation (8·9) are applicable to equation (8·10) or (8·11).

There are many other cases of loading that can be discussed for stress distribution but they are outside the scope of this book. In the next article, simple charts and monographs are presented which help in determining the stresses under loaded areas quickly.

8·5. Influence Charts and Nomographs

Fig. 8·2 shows an influence diagram for the values of N_B and N_W to be used in calculating stresses induced in strata below, when a point load P is placed at the surface in the Bouesinesq's and Westergaard's analysis respectively. The value of N_B is also availab.e in a table published by Gilboy[1] (2) from which the value of N_B corresponding to the value of (r/z) can be picked up. Both the influence charts and the table given by Gilboy reduce the time required for calculations.

Gray (7) has given a nomograph shown in Fig. 8·4 which helps in further reducing the time for calculations of σ_z as given by Boussinesq's equation for the point load and the semi-infinite elastic medium. It is used in the following manner.

(1) Mark the load P and the distance z on the respective scale and join them by a straight line. This straight line will cut the x-pivot line at some point say X.

(2) Find out the ratio (r/z) and mark this point on the corresponding scale, as point Y.

(3) Join X and Y. This line will cut the stress scale at some point and the corresponding reading gives the stress σ_z for the point A, shown in the figure.

It may be mentioned, here, that as for the slide rule calculations, so for the calculations from this nomograph, the decimal point has to be placed by the reader. Newmark (3) gave a table[1] for the influence factor N_N given by

$$N_N = \frac{1}{4\pi} \left[\frac{2mn\sqrt{m^2+n^2+1}}{m^2+n^2+1+m^2n^2} \cdot \frac{m^2+n^2+2}{m^2+n^2+1} \right.$$
$$\left. + \sin^{-1}\frac{2mn\sqrt{m^2+n_2+1}}{m^2+n^2+1+m^2n^2} \right] \qquad \ldots(8·12)$$

which reduces equation 8·9 to,

$$\sigma_z = q, N_N \qquad \ldots(8·13)$$

---

1. This table is published in this book in the Appendix, as Table 2.

or the known values of m and n, the value of N_N can be picked up from this table and the time for computations reduces. Alternatively, the influence chart given in Fig. 8·5 can be used for the purpose. Once the influence N_N is known from the chart, the stress σz at a depth z due to a uniformly loaded rectangle, over a semi-infinite elastic medium can be found.

Fig. 8·6 shows the influence chart for the equation (8·11), from which the value of an influence factor N_{W_1}, given by

$$N_{W_r} = \frac{1}{2\pi} \, \text{cot}^{-1} \sqrt{\frac{1}{2m^2} + \frac{1}{2n^2} + \frac{1}{4m^2n^2}} \qquad \ldots(8·14)$$

can be found. This influence factor reduces the Westergaard's equation (8·11) to.

$$\sigma z = q N_{W_1} \qquad \ldots(8·14a)$$

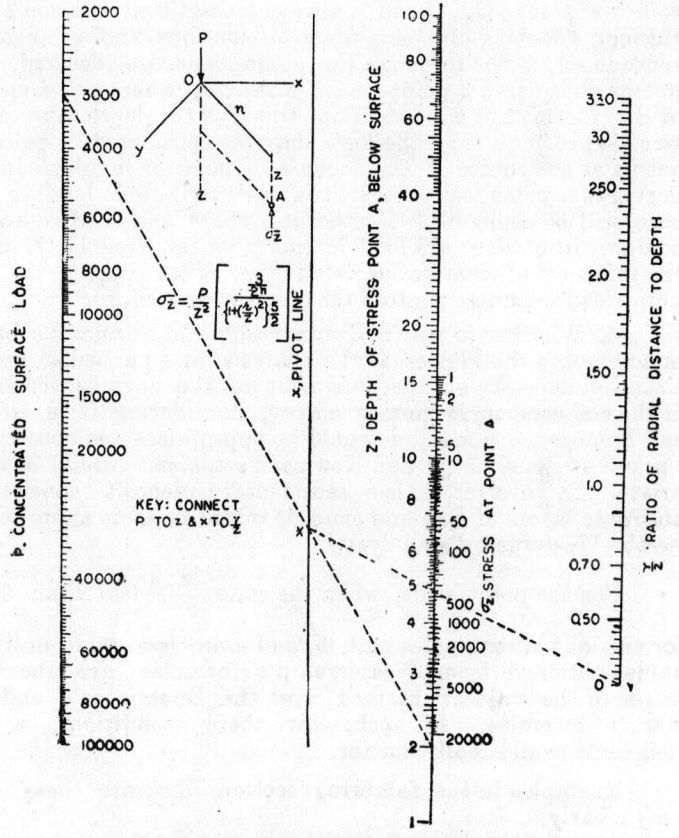

Nomograph for Boussinesq's point load formula
(After Gray)
Fig. 8·4.

1. This table is published in this book in the Appendix as Table 2.

Knowing the value of Nw_1, σz can be found from this chart and this reduces the time for computations.

8·6. Some Important Statements and Observtaions

Some statements made below are helpful in the use of the formulae given in the previous sections.

1. If a load is split up into a number of convenient units, the stresses (and displacements) induced in an elastic medium by these individual loads are simply additive and the sum gives the stress (and displacement[1]) due to the composite load. This law in known as the principle of superposition.

2. At *greater* depths, the stress induced in an elastic medium is the same, irrespective of the type of the load placed on the surface. This statement gives the **Saint Venant's** principle.

3. One would hardly find a load being transmitted to the soil as a point load. Load is always transmitted to a soil stratum through a footing of some finite dimensions and as such in the sense in which the formulae for point loads, are derived, it cannot be considered a point load. However, where the dimensions of the footing are less than one third of the depth at which the stresses are required, the load may be considered a point load acting at the centre of the footing. The error involved in considering it a point load rather than a distributed load, in such a case, will be negligible. Furtheron. where the loaded areas are complex in plan, or the load intensity varies irregularly, or when the influence of remote isolated footing is sought, the use of point load formulae is often the simplest procedure.

4. Whether to use the Boussinesq's and Newmark's formulae or to employ the Westergaard's analysis for a particular practical situation depends on the judgment of the engineer concerned. If the soil stratum is purely clayey, considering it an isotropic and homogeneous medium would be appropriate and consequently the use of Boussinesq's and Newmark's analysis would be appropriate. If, however, it is a sedimentary deposit consisting of alternate layers of clay and sand, it would be more appropriate to usethe Westergaard's analysis.

For the point loads, when the ratio $\dfrac{r}{z}$ is less than 0·8 and for the loaded rectangles with m and n are less than unity, the values obtained from Westergaard's formulae are about two-thirds of the values obtained from the Boussinesq's and Newmark's formulae. As such, for these conditions, a sound judgment would really matter.

Examples in the following section illustrate these points very clearly.

---

1. Formulae for displacements have not be given in this book.

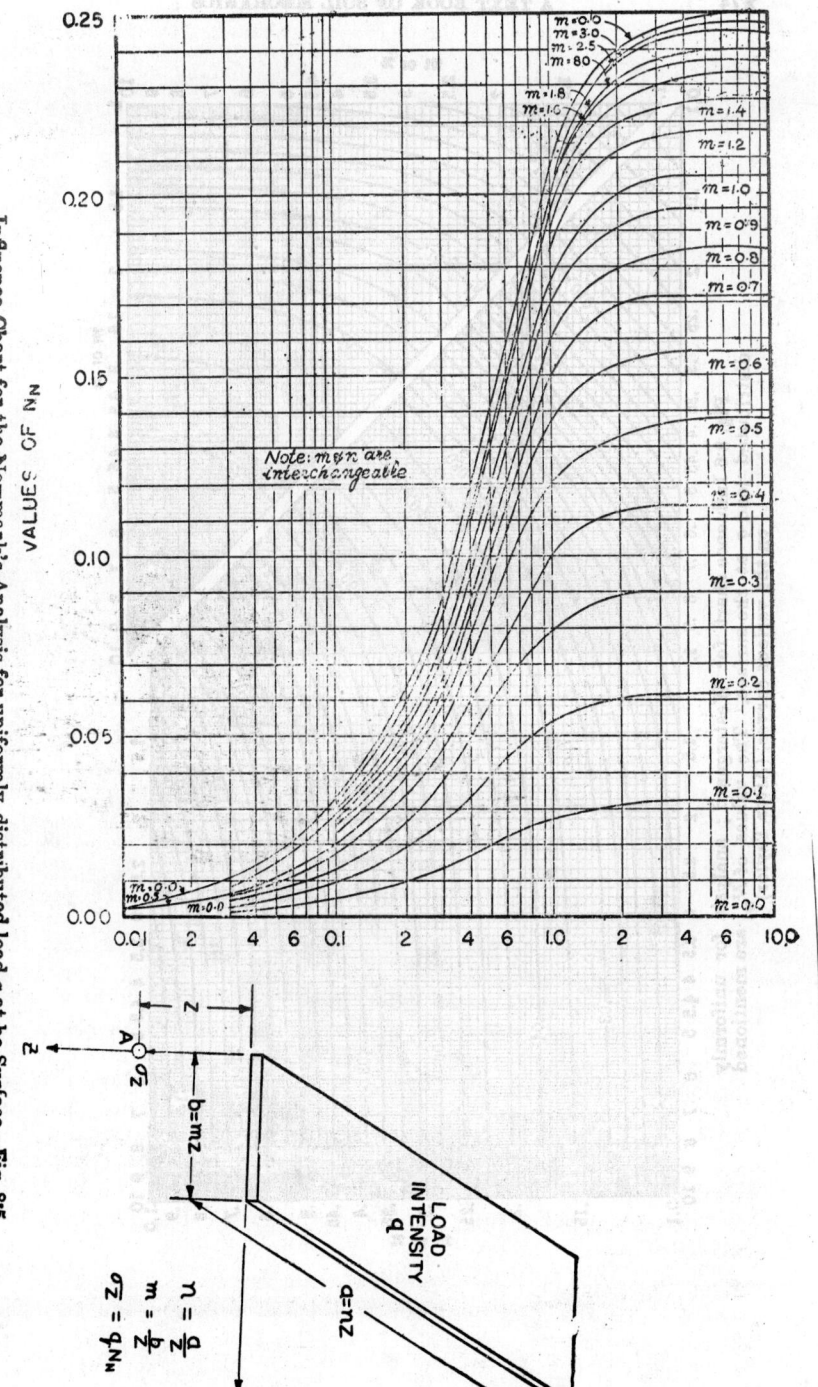

Influence Chart for the Newmark's Analysis for uniformly distributed load q at the Surface. Fig. 8·5.

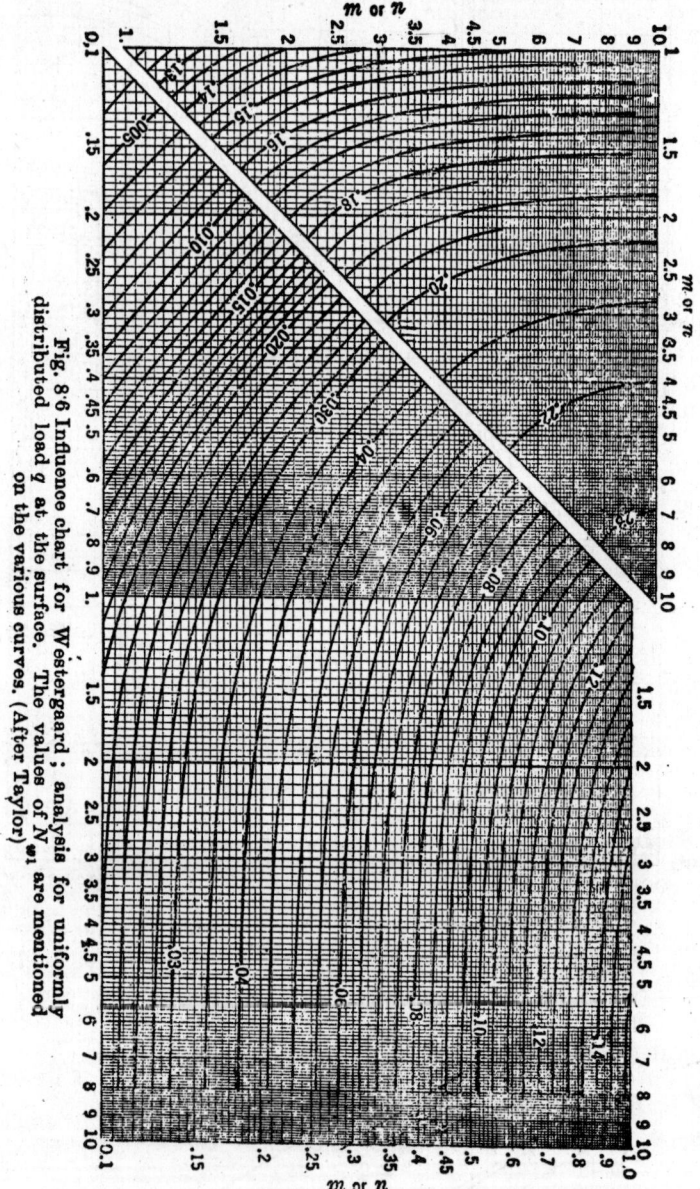

Fig. 8·6 Influence chart for Westergaard; analysis for uniformly distributed load q at the surface. The values of N_{wi} are mentioned on the various curves. (After Taylor)

8·7. Sixty Degree Distribution of Load

In contrast to the Boussnesq's and Westergaard's analysis, it is quite common in practice to use sixty degree dispersion of the load from the plane of loading as shown in Fig 8·7. Under this concept, it is assumed that the stresses at successive depth beneath the plane of application of the load are also distributed uniformly. The area A_z at any plane below the plane of loading is defined by the planes decending at an inclination of 60^0 with the horizontal from area A_1 at the plane of loading. Thus, the stress intensity at any plane below the plane of loading becomes $\dfrac{P}{A_2}$ where P is the total load at the plane of loading..

For purposes of simplicity, sometimes it is convenient to assume the dispersion at an angle of $63\ 1/2^0$. For example, if a square footing of size (b x b) square meters is assumed to disperse the P load at an angle of $63\ 1/2^0$, then the area A_1 at the plane of loading is b^2 square meters and area A_2 at a plane z meters below the plane of loading will be $(b+z)^2$ because the $63\ 1/2^0$ dispersion gives the slopes AC and BD as two vertical to one horizontal. Then the stress intensity at a depth of z meters for the footing could be assumed as $\dfrac{P}{(b+z)^2}.$ This simplified procedure is illustra

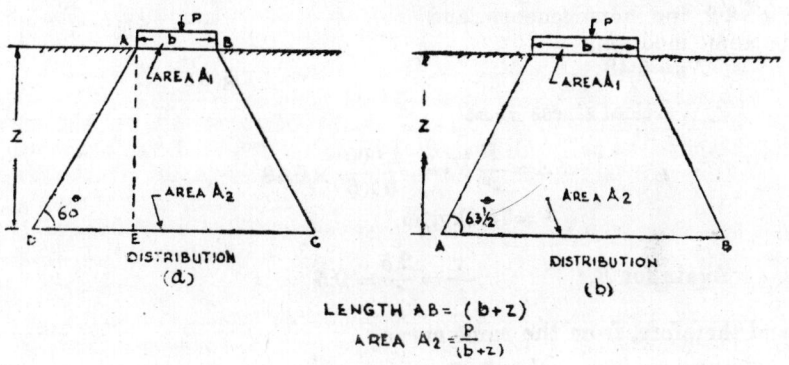

LENGTH AB= $(b+z)$

AREA $A_2 = \dfrac{P}{(b+z)}$

Fig 8.7.

ted in example. Reference may also be made to article 12-9 in Chapter 12.

8·8· Examples.

Example 8·1. A concentrated load of 4, 000 kg. acts on the surface of a homogeneous soil mass of large extent. Determine the stress intensity at a depth of 5 metres (*i*) directly under the load, (*ii*) at a horizontal distance of 2·5 metres.

Solution. The soil is homogeneous. Although, not mentioned, is implied that the soil mass is isotropic too.

Figure 8·7 shows the two points A and B at a distance of 5 metres directly below the load point and at a horizontal distance of 2.5 metres.

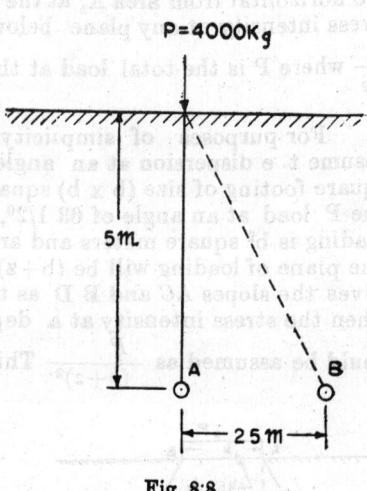

Fig. 8·8

Hence, $z=5m$ for both the points

$r=0·0m$ for the point A

and $r=2·5m$ for the point B,

For $A, \dfrac{r}{z}=0$

and, therefore, from Boussinesq's influence curve given in Fig. 8·2. for homogeneous and isotropic medium,

$N_B=0·48$

∴ Vertical stress σ_A at

$$A=\frac{P}{z^2} N_B= \frac{4,000}{5\times5} \times 0.48$$
$$=76.8kg./m.^2$$

Again for B, $\dfrac{r}{z}= \dfrac{2.5}{5}= 0.5$

and therefore, from the same curve,

$N_B=0·27$

∴ Vertical stress σ_B at

$$B=\frac{P}{z^2} N_B= \frac{4,000}{5\times5} \times 0.27$$
$$=43.2kg./m^2.$$

So, the required stresses are 76·8kg./m.² and 43.2 kg/m². **Ans.**

---

1. This may be verified mathematically from the value of N_B in the equation (8·5).

Comments. This example is an indication of the fact that the stress at a point directly below the load point and another point laterally away from the load point but in the same horizontal plane at the previous point, are *very much different*. Also, it indicates that the load applied at a point induces stresses not only directly below it, but at points lying horizontally away from the load point.

Example 8.2. A point load 3.0 tonnes acts on the surface of the ground. Assuming the soil mass to be uniform, compute the intensity of vertical stress due to the load at depths of 0, 1, 2, 5, 10, 50 and 100 metres.

(a) along the vertical line of action of a point load

i. e. $r = 0$

and (b) for $r = 5.0$ metres.   **(P. U. April, 1964)**

Solution. The soil is homogeneous and isotropic. This is the conclusion drawn from the assumption that the soil mass is uniform.

Boussinesq's influence curve given in Fig. 8.2 will be applicable, here, too. Let z represent the depths of the points and r represent their radial displacement with respect to the load point.

(a) Equation (8.5) gives the vertical stress at a depth z as

$$\sigma_z = \frac{P}{z^2} N_B.$$

P in this problem is 3.0 tonnes and z assumes various values from 0 to 100 metres. The radial distance in each case is zero.

When r and z are both zero, N_B will assume some numerical value[1] but σ_z at the surface from the above equation will be ∞, since $z = 0$.

TABLE 8·1

S. No.	Depth of z in metres	$\frac{P}{z^2}$	$\sigma_z = \frac{P}{z^2} \times 0.48$ tonnes/m²
1	1	3.0	1.44
2	2	0.75	0.36
3	5	0.12	0.057
4	10	0.03	0.0144
5	50	0.0012	0.00057
6	100	0.0003	0.000144

For all other values of z, (r/z) ratio is zero and the corresponding influence factor from Fig. 8.2 is 0.48. Stresses so calculated are given in Table 8.1.

(b) When $r=5$ metres, σ_z at the surface will be zero[1] since $z=0$.

For all other values of z, (r/z) values are calculated and the corresponding influence factor read from Fig. 8.2. Stresses so calculated are given in Table 8.2.

TABLE 8·2

S. No.	Depth z in metres	Ratio (r/z)	Influeuce Factor N_B	P/z^2	$\sigma_z=(P/z^2)N_B$ tonnes/m²
1	1	5.0	0.005(say)	3.0	0.015
2	2	2.5	0.005	0.75	0.00375
3	5	1.0	0.08	0.12	0.0096
4	10	0.5	0.27	0.03	0.0081
5	50	0.1	0.465	0.0012	0.000558
6	100	0.05	0.470	0.0003	0.000141

Conclusion. Table 8·1 and 8·2 indicate that as the depth below the load increases, the stress due to the load decreases. Also, the stress immediately under the load at any depth is more than the stress at the same depth but at a radial distance of 5 metres from the axis of the load.

Example 8.3. The figure 8.9(a) shows the plane of a footing 12×6 metres in dimensions loaded by a uniform load of one metric tonne per square metre. Find out the vertical stresses at 10 metres below the point A, by

(a) considering the rectangle as uniformly loaded as stated above.

(b) dividing the rectangle into 12 panels of 3×2 metres each and considering the load on each panel as a point load acting at the centre of the panel.

---

1. This may be verified from equation (8·2).

Compare the results from (a) and (b) and comment on them.
Assume the footing placed on the surface of isotropic, as homogeneous and an elastic material.

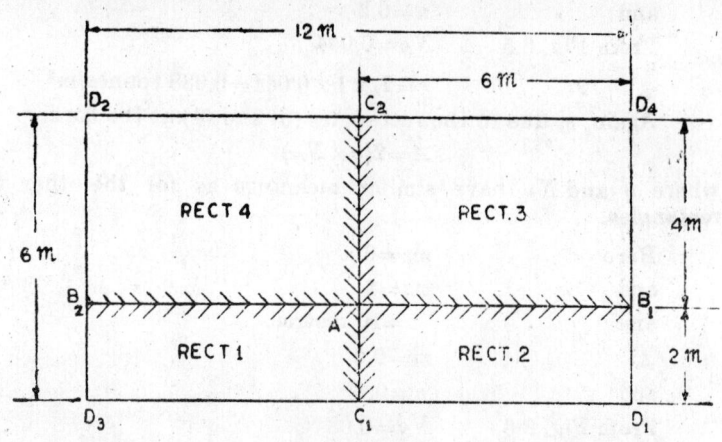

Fig. 8.9 (a)

Solution. (a) The vertical stresses at a poin inside an isotropic and homogeneous medium due to a uniformly distributed load on the surface are given by Newmark's analysis as given in A t, 8.4 (a) and equation (8.3).

The footing in Fig. 8.8(a) can be considered to be composed of four rectangles :

(1) Rectangle $AB_1D_1C_1$
(2) Rectangle $AB_2D_3C_1$
(3) Rectangle $AB_1D_4C_2$
(4) Rectangle $AB_2D_2C_2$

Point A lies at the corner of each of the above four rectangles. Further note that the dimensions of the first and second rectangles are the same and so is true of third and fourth rectangles ; and consequently these similar rectangles will have the same effect as far as stresses are concerned, since they are loaded similarly.

σ_z due to rectangles (1) and (2) at the corner

$$A = 2(q \times N_N)$$

where q is the intensity of loading and N_N is the Newmark's influence factor.

For these two rectangles,

$$mz = 6 \text{ and } nz = 2$$

Also $\qquad z = 10$ metres

$\qquad\qquad m = 0.6$

and $\qquad\qquad n = 0.2$

From Fig. 8.5 $\qquad N_N = 0.044$

$\therefore \qquad\qquad \sigma = 2_z \times 1 \times 0.044 = 0.088$ tonnes/m²

Again, σ_z due to the rectangles (3) and (4) at the corner

$$A = 2(q \times N_N)$$

where q and N_N have similar meanings as for the other two rectangles.

Here $\qquad\qquad mz = 6$

and $\qquad\qquad nz = 4$

and $\qquad\qquad z = 10$ metres.

$\therefore \qquad\qquad m = 0.6$

and $\qquad\qquad n = 0.4$

From Fig. 8·5, $\qquad N_N = 0.08$

$\therefore \qquad\qquad \sigma_z = 2 \times 1 \times 0.308 = 0.16$ tonnes/m²

Total vertical stress at 10 metres below the point

$$A = 0.088 + 0.16 = 0.248 \text{ tonnes/m}^2. \quad \textbf{Ans.}$$

(b) Total vertical stress at a point inside an isotropic and homo-geneous medium due to a concentrated load on the surface are given by Bossinesq's analysis as given in Art. 8.3 (a) and equation 8.4 or 8.5.

The loaded rectangle 12×6 metres is divided into 12 panels of 3×2 metres each as shown in Fig. 8·9 (b). with the load placed at the centre as shown by a dot in each panel. The load at each

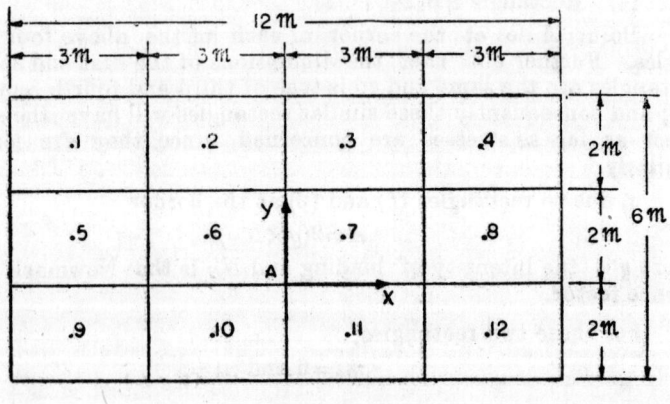

Fig. 8·9 (b)

of the load points as $(3 \times 2) \times 1 = 6$ tonnes and shall be treated as a point load directly resting upon the surface of the isotropic, homogeneous and elastic medium. The combinations of loads 1 and 4, 2 and 3 ; 5, 8, 9 and 12 ; 6, 7, 10 and 11 are symmetric with respect to point A, beneath which the vertical stress is desired. So, out of each combination we need only consider one load and find out the stress due to it and multiply it by two or four, as the case may be, to find out the stress due to the symmetric loads. This procedure will be followed in Table 8·3, x and y will denote the co-ordinates of a load on the surface plane (disregarding the sign), with respect to A as the origin so that $r = \sqrt{x^2 + y^2}$ and z would denote the depth below the surface. Boussinesq's influence factor N_B can be read from the curve given in Fig. 8·2 and the vertical stress due to a load p determined from,

$$\sigma_z = \frac{P}{z^2} N_B$$

Table 8·3 gives the details of the calculations according to the above procedure.

TABLE 8·3

S. No.	Load Nos.	Total No. of loads in a combination	x in metres	y in metres	$r = \sqrt{x^2+y^2}$ metres	z in metres	Ratio r/z	N_B from Fig. 8·2	$\frac{P}{z^2}$	σ_z for one load $(9) \times (10)$	σ_z for the combination of loads $(3) \times (11)$
(1)	(2)	(3)	(4)	(5)	(6)	(7)	(8)	(9)	(10)	(11)	(12)
1.	1 & 4	2	4·5	3	5·4	10	0·54	0·251	0·06	0·01506	0·03012
2.	2 & 3	2	1·5	3	3.355	10	0·3355	0·366	0·06	0·02196	0·04392
3.	5, 8, 9 & 12	4	4·5	1	4·61	10	0·461	0·295	0·06	0·0177	0·07080
4.	6, 7, 10 & 11	4	1·5	1	1·80	10	0·18	0·441	0·06	0·02646	0·10584

$$\Sigma \sigma_z = 0.25068$$
$$\approx 0.250$$

So, the total vertical stress treating the load on each panel as a concentrated load is

$$\sigma_z = 0.250 \text{ tonnes/m}^2. \quad \textbf{Ans.}$$

Comparison and comments. The two results obtained in (a) and (b) are identical. The dimensions of each panel of 3×2

metres are less than one-third of the depth $z=10$ metres, where stress distribution was needed. This proves the validity of the observation (3) under Art. 8·6. The percentage error in using point load formula instead of the uniformly loaded formula, in this case works out as :

$$\text{Percent error} = \frac{(0·250-0·248)}{0·248} \times 100$$

$$= \frac{2}{248} \times 100 < 1\%.$$

Example 8·4. In example 8·2, find out the stress induced at a depth of 20 metres, below one corner of the loaded rectangle using Newmark's analysis for an isotropic and homogeneous material and Westergaard's analysis for anisotropic material. Compare the results.

Solution. For a uniformly loaded rectangle the stress at a depth z below the surface is given by :

$$\sigma_z = q \times I$$

where q is the load intensity and I is the influence factor, whether it is Newmark's case (where $I=N_N$) or it is Westergaard's case (where $I=N\omega_1$)

In this problem,

$$z = 20 \text{ metres}$$

and $\qquad \left.\begin{array}{l} mz = 12 \text{ metres} \\ nz = 6 \text{ metres} \end{array}\right]$ for any corner

$$\therefore \qquad m = 0·60$$

and $\qquad n = 0·30$

For the Newmark's case of isotropic and homogeneous material,

$$\sigma_z = qI = qN_N$$

where N_N from Fig. 8·5 for the above values of m and n equals 0·062.

$$\therefore \qquad \sigma_z = 1 \times 0·062 = 0·062 \text{ tonnes/M}^2.$$

Again for the Westergaard's case of an istropic material,

$$\sigma_z = qI = qNw_1$$

where Nw_1 from Fig. 8·6 for the above values of m and n equals 0·041.

$$\sigma_z = 1 \times 0·041 = 0·041 \text{ tonnes/m}^2.$$

Comparison and comments. Comparison of the above two values of σ_z indicates that the vertical stress, obtained by Westergaard's analysis, is approximately $\frac{2}{3}$rd of the vertical stress obtained by the Newmark's analysis i.e., 0.041 tonnes/m² is approximately equal to $\frac{2}{3}$ of (0·062) tonnes/m².

This clearly points out as to how important it is to explore the sub-soil conditions at a site by properly taking suitable number of borings so as to make the correct judgment as to which of the two analysis would be more appropriate to the work-site.

This example also proves the validity of observation (4) under Art. 8·6.

Example 8·5. The contact pressure for a 2m × 2m footing is 2k g./cm². Using as 60° and 63½° dispersion, find out the depth at which the contant pressure will be 0·5 kg./cm².

Solution. Refer to Fig 8·7 (a)

Area A_1 at the loading plane $=(200)^2$ sq cms.

∴ Total load on the footing $=2 \times (200)^2$ kg $=p$ (say)

Let the distance $ED=x$ and the depth $BE=z$

then, $\dfrac{x}{z}=\cot 60°$

∴ $x=z \cot 60°$

This means that $CD=b+2x=b+2z \cot 60°$

$$=b+\frac{2}{\sqrt{3}}z=(b+1·15_z)^2$$

where b and z see in cms.   Here $b=200$ cms.

∴ Area A_2 at a depth $z=(b+1·15_z)$

Stress at a depth $z=\dfrac{p}{(b+1·5_z)^2}$

But this stress should equal 0·5 kg/cm²

∴ $0·5=\dfrac{p}{(b+1·15_z)^2}$ where

and $p=2 \times (200)^2$ kg

                 $b=200$ cms

∴ $0·5=\dfrac{2 \times (200)^2}{(200+1·15z)^2}$

Solving for z,z epuals 174 cms.

∴ The stress at a depth of 1·74 metres will be equal 0·5 kg/cm².   **Ans.**

Now refer to Fig. 8·7 (b) where the dispersion is at an angle of 63½°.

Area A_2 $=\left(b+\dfrac{z}{2}+\dfrac{z}{2} \right)$, because the slope of dispression is 2 vertical to 1 horizontal

∴ By similar procedure as in the preceding case, stress at

a depth of z would be : $\dfrac{2 \times (200)^2}{(b+z)^2}$ where $b = 200$ cms.

If this stress is equal to 0·5 kg/cm², then $0·5 = \dfrac{2 \times (200)^2}{(200+z)^2}$

Solving for z, z equals 200 cms.

∴ The stress at a depth of 2 meters will equal 0·5 kg/cm².

Ans.

Comments. The procedure of evaluation is very much simplified compared to any method based upon the elastic theory.

Example 8·6. It is anticipated that a square footing would be constructed at a distance of 1·5 m above a weak soil layer. Bearing capacity analysis indicates that the average stress on the weak layer must not exceed 0·5 kg/cm². What size of the footing should be used to support a 50 tonne load from a column, to meet the above requirement.

Solution. Refer to example 8·5 in this chapter.

If the size of the footing is assumed as "$b \times b$" and if $z = 1·5$ meters, using a 60° dispersion, the area at the level of the layer will be $(b+1·15z)^2$. Here, b and z can be substituted in cms. or in meters.

∴ Stress at the level of the weak layer is $\dfrac{50,000}{(b+1·15z)^2}$.

Substitute for z in terms of cms and equate this stress to 0·5 kg/cm².

∴ $$0·5 = \dfrac{50,000}{(b+1·15 \times 150)^2}$$

Solve for the value of b.

∴ $b = 142·5$ cms. (say) or 1·425 meters.   **Ans.**

If the dispersion is assumed at $63\tfrac{1}{2}°$, area at the level of the weak layer will be $(b+z)^2$.

By a similar reasoning,

$$0·5 = \dfrac{50,000}{(b+150)^2}$$

∴ $b = 166$ cms. or 1·66 meters   **Ans.**

CHAPTER BIBLIOGRAPHY

1. Burmister, D.M., Graphical Distribution of Vertical Pressure Beneath Foundations, Transactions A.S.C.E., Vol. 103, app 308-313.

2. Cumings, A.E., Distribution of Stresses Under a Foundation, Transaction A.S.C.E. 1936.

3. Gilbay, G., Inforence tables for solution of Boussinesq Equation in Earth and Foundations, Progress Report of Special Committee, Procedings A. S. C. E. Vol. 59, P. 781.

4. Gray, H., Stress Distribution in Elastic Solids, Proceedings, First International Conference on Soil Mechanics and Foundation Engineering 1936, Vol. 2, Cambridge, Mass., pp. 161—174.

5. Gray, H., Charts Facilitate Determination of Stresses Under Loaded Areas, Civil Engineering, June, 1948, pp. 49—51.

6. Gray, H., Stresses and Displacements from Loads over Rectangular Areas, Civil Engineering, May 1943, pp. 227—229.

7. Gray, H., Nomograph Aids Use of Boussinesq Equation, Civil Engineering, June 1947, pp. 46—47.

8. Lambe, T.W., and Whitman, R.V., "Soil Mechanics", J. Wiley & Sons, Inc., N.Y., 1969.

9. Leonard, G.A., "Foundation Engineering", McGraw Hill Book Co., Inc., New York, N.Y. 1962.

10. Love, A.E.H., Mathematical Theory of Elasticity, 4th Ed., Cambridge University Press.

11. Mindling, R.D., Force at a point in the interior of a semi-infinite solid, Physics Vol. 7, p-195, 1936.

12. Newmark, N.M., simplified Computations of Vertical Pressures in Elastic Foundations, Circular No. 24, Engg. Stations, University of Illinoil, 1935.

13. Osterberg, J.O., Influence Chart for Vertical Stresses, Embankment Loading Infinite Extent, Boussinesq Case, Proceedings, Third International Conference on Soil Mechanics and Foundation Engineering.

14. Scott, R.F., Principles of Soil Mechanics. Addison Wesley Publishing Co., Inc., Mass., 1966.

15. Terzaghi, K., Theoretical Soil Mechanics, New York, John Wiley & Sons, Inc. 1962.

16. Tschebotarioff, G.P., "Soil Mechanics, Foundations, and Earth Structures", McGraw Hill Book Co., Inc., N.Y., 1951.

17. Westergaard, H.M., A problem of elasticity suggested by a problem in soil mechanics, soft Material Reinforced by Numerous Strong Horizontal Sheets, in Contributions of the Mechanics of Solids, Stephen Timosheko 60th University Volume ; New York, MacMillan, 1968.

QUESTIONS

1. A footing three metre square supports a load of 100 tonnes. Calculate the vertical stress at a depth of 5 metres below the centre of the footing. Use both Boussinesq's and Westergaard's point-load analysis and compare the results.

2. Another footing four metre square supporting a load of 100 tonnes is placed at a distance of 5 metres centre to centre of the footing in question No. 1. Find out the effect of the placement of this footing on the vertical stress at a depth of 5 metres below the centre of the first footing using Boussinesq's analysis.

3. In question No. 1 above, draw a stress distribution profile due to the given load at the centre of the loaded area by taking different depths and finding out the corresponding stresses ?

4. A bridge pier 40 feet by 10 feet in plan rests on the ground surface transmitting an average of 4000 lbs. per square foot pressure. Compute the intensity of pressure due to this load at ·5 feet and 10 feet below the centre of the pier. (*P.U. 1960 Supp.*)

5. Write a short note on "Boussinesq equation with its limitations".

6. An elevated structure is supported on a tower with four legs. The legs rest on piers located at the corners of a square 7 m. on a side. If the value of the vertical stress increment due to this loading (considering four equal concentrated loads) is 0·25 kg/cm², at a point 8 metres beneath the centre of the structure, what will be the stress increment at 10 m. below one of the legs ?

Ratio	Influence value for concentrated load
0	0·48
0·50	0·27
0·60	0·22
0·62	0·21
0·70	0·18
0·80	0·14
0·90	0·11
0·99	0·085
1·00	0·084
1·20	0·051

(*P.U. Nov. 1964*)

7. (a) State the assumptions of Boussinesq's theory of stress distribution in a soil mass.

(b) How far are these conditions realized in practice ?

(c) A concentrated load of 100 kg. acts on the surface of homogeneous, isotropic and elastic soil mass of large extent. Determine the stress intensity at a depth of 5 metres directly below the load.

8. Discuss the assumptions made in computing stresses below the ground surface due to a point load acting on it. Discuss their validity in practice.   (*P. U. 1966*)

9. A point load of 3·0 tonnes acts on the surface of the ground. Assuming the soil mass to be uniform, compute intensity of vertical stress due to the load at depths of 0, 1, 2, 5, 10, 50, and 100 meters.

   (*a*) Along the vertical line of action of point load *i.e.* $r=0$, and

   (*b*) for $r=5·0$ meters.   (*P.U. Ap. 1966*)

10. (*a*) List assumptions made in Boussinesq theory for determining stresses in a soil mass due to a concentrated load.

    (*b*) Calculate the stresses in a soil mass below the centre of a uniformly loaded circular area of radius 5 ft. with a unit load of 2 kips/sq: ft. and thus obtain the exact depth at which the stress reduces to 10% applied unit load.
                                                         (*Mysore 1966*)

11. (*a*) State the assumptions made in computing stresses below the ground surface due to a point load acting on it. Discuss their validity in practice.

    (*b*) Concentrated load of 4000 kg acts on the surface of a homogeneous soil mass of large extent. Determine the stress intensity at a depth of 5 meters and (1) directly under the load (ii) at a horizontal distance of 2·5 meters.
                                          (*Jammu & Kashmir 1966*)

9

Settlement Analysis

9·1. Introduction

All structures have their foundations placed on natural soil. In case the soil below a foundation is compressible, the structure above is likely to settle due to the compression in the compressible stratum. In clayey soils, time is required for the water in the pores to squeeze out and as a consequence the settlement continues over a number of years. It is important, therefore, to know the total settlement expected due to the loads transmitted by the structure as well as the rate of settlement. In case it can be assumed that the water travels only in the vertical direction, then use can be made of Terzaghi's one-dimensional consolidation theory explained in the Chapter 7, in order to predict the time-rate of settlement and also the total settlement. The compressible stratum may be lying in a soil profile at a considerable distance from the **ground surface and it may,** therefore, be essential to find out the stresses transmitted

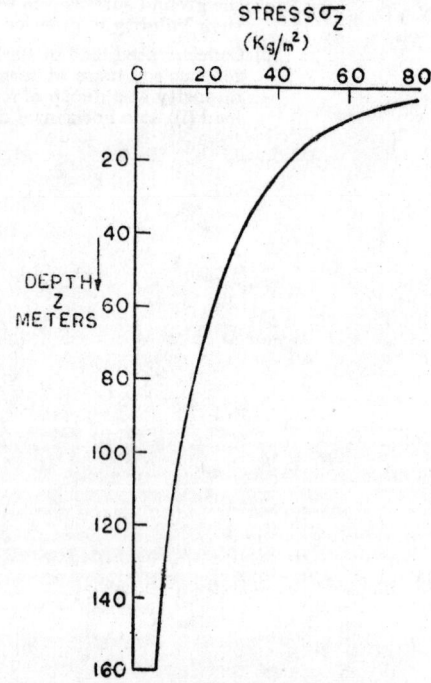

Fig. 9·1.

by the loads from the structure, to various points within the compressible stratum, For this purpose, use is made of the stress-distribution theories described in Chapter 8. The stress due to a surface load transmitted to various depths decreases as one goes down and down. The pattern of stress transmission as a function of depth directly below a surface load is indicated is Fig. 9·1. It may in general be said that any compressible stratum lying within three times the shorter dimension of a footing, would certainly contribute to the settlement of the footing.

It may be indicated here that a uniform settlement of the foundations of a building is not harmful but if there is a *differential settlement* of the various parts of the foundations on which a rigidly connected structure rests, it may lead to cracking of the structure and in case the differential settlement is excessive, the building may ultimately collapse.

A *procedure* that allows the calculations of settlements is called **settlement analysis** or consolidation analysis.

9·2. Requirements for Settlement Analysis

Before the settlements can be predicted or the rate of settlement could be determined, the following steps have to be taken.

1. Drawing a Soil Profile. It is essential that reliable data is obtained regarding the thickness of the stratum for each type of soil underneath the area on which the foundation is to rest. For this purpose, it is essential that bore-holes at suitable points be made and boring logs prepared from the data so obtained. From the boring holes drilled in a line, it is possible, then, to plot a soil-profile, which gives an idea of the thickness of the stratum for each type of soil. The number of bore-holes that must be drilled depends upon the size of the project. No hard and fast rules can be given here. However, for a building site say 75 metres × 40 metres in area, five bore-holes—one at centre and four at the four corners—are quite sufficient. The depth of each bore-hole again depends upon the type of material encountered and no rules can be laid down for it (See chapter 16). It is recommended, however, that the minimum depth to which one must go in a bore-hole should not be less than 4 to 5 times the shortest dimension of the foundation.

The thickness of a stratum in the soil-profile may not be constant for the length of the profile. For settlement calculations, if need be, some suitable and sensible simplifications may have to be made.

The existing level of the water table must be marked on the boring logs and the soil profiles. This is helpful in calculating the existing overburden pressure on the compressible stratum.

2. Other Pertinent Soil Data. In order to know the existing void ratio of the compressible soil, it is essential to know its moisture content and the specific gravity of solids. It is invariably found that such compressible strata are fully saturated in the field and it is always beneficial to make this check also. Then using the formula given in equation (2·13) the existing void ratio $e_0 = G w$ can be found out, where it is assumed that degree of saturation S is unity.

In addition to this, the consolidation characteristics of the representative sample from this stratum are required for use in the settlement calculations. For this purpose, undisturbed soil samples are taken, usually in 4″ diameter thin wall tubes with the least amount of disturbance in the drilling procedure. Then samples out from the centre of the above field samples are used in a consolidometer, in order to determine their consolidation behaviour. The e—log p diagram is drawn for each sample and values of the coefficients of consolidation for respective load increments are determined.

All the above data is then used to compute the settlements.

9·3 Settlement Under a Foundation

The total settlement under a foundation, in fact, constitutes of

(1) Initial settlement due to the elastic compression of the soil underneath, usually [called the **Distortion** or **Contact** settlement.

(2) Consolidation settlement, which is due to the compression of the compressible soil layers underneath and which is derived from the consolidation theory.

(3) Secondary settlement, which takes place due to the secondary consolidation of the compressible layers.

Relationships based on the elastic theory which give the initial settlement of the material are available. For example. for Boussineqs's case of uniformly distributed load on a rectangle, the distortion settlement is given by,

$$\rho = \frac{qb(1-\mu)}{E} I \qquad \qquad ...(9·1)$$

where
$\rho =$ Distortion settlement
$q =$ Intensity of loading
$b =$ Width of the loaded area
$\mu =$ Poisson's ratio
$E =$ Modulus of Elasticity for the soil
$I =$ Influence factor depending upon the location of the point where settlement is required and on the shape and size of the loaded area.

Influence factors for some conditions of loading are given in Table 9·1. just for illustration.

The distortion settlement takes place very rapidly on application of load and as such in case of saturated clays and other similar soils, chances of any volume change are remote. So a Poisson's ratio μ of 0·5 is ordinarily used, in equation (9.1).

The analysis of distortion settlement in case of sands is complicated because the modulus of elasticity E of such soils increases with confining pressure. Also E increases with the depth below the foundation and hence the distortion settlement cannot be found rigorously.

TABLE 9·1.

Influence Factors for Distortion Settlement for Uniform loads on a semi-infinite elastic homogeneous and isotropic medium

Shape of loaded area		Influence Factor		
		Centre of loaded area	Corner of loaded area	Average
Square		1·12	0·56	0·5
Rectangle	$L/b=2$	1·52	0·76	1·30
	$L/b=5$	2·10	1·05	1·83
	$L/b=10$	2·54	1·27	2·20
Circle[1]	(put $d=b$ in Equation 9·1)	10	0·64 (edge)	0·85

In fact, distortion settlements are quite negligible compared to the consolidation settlements and in practice, whenever it is required to find out the settlement for a building foundation, the distortion settlement is neglected.

Concepts of one dimensional consolidation theory advanced by Terzaghi are then employed to compute consolidation settlements at various points of the foundation. It is felt necessary to emphasize that the consolidation theory is based upon so many assumptions and any calculations based upon it is approximate. However, in most of the cases, when the computed settlement values are compared with the actual settlement values after the structure is complete, a fair comparison seems to exist and it is always advisable to make this analysis and ward off any disaster that may result from differential settlement. It may also be noted here, that the calculations based upon Terzaghi's one—dimensional consolidation theory—give settlements due to primary consolidation of the compressible soil layer only. The third type of settlement, called the secondary settlement, is believed to occur due to the secondary compression of the soil during and after the pore-water dissipation. In case of organic soils, highly micaceous soils and some soft soils, the secondary compression is comparable with primary compression. In other soils, the secondary compression may be quite negligible.

A discussion of secondary compression and consequent settlement is considered outside the scope of this book. An advanced student may refer to references given at the end of this chapter.

---

1. L is the length and b is the width of the rectangle, d is the diameter circular area.

9·4. Computations for Consolidation Settlement

Theoratically, computation for the total settlement and the rate of settlement appear quite simple provided the existing pressure and the additional pressure brought about by a building on the compressible layer, along with all the data as detailed out in section 9·2, is known. In practice, however, some simplifying assumptions regarding the variation in pressure through out the depth of a compressible layer, the existing void ratios at various levels within the layer and the coefficient of consolidation are made in order to reduce the bulk of calculations. These assumptions will be described in section 9.5 followed by illustrative examples. In this article it considered sufficient to give various formulae by which the total settlement calculations may be made and later on give the method for calculating rate of settlement.

Calculations for Total Consolidation Settlement.

First Method. Let the initial void ratio of the stratum (or its portion) under consideration be e_0. Let $\triangle p$ be the additional average pressure brought about by the building on it. If H is the thickness of the stratum (or its portion) under consideration, then considering a unit area, the settlement S may be written as

$$S = \frac{\triangle e}{1+e_o} H \qquad \qquad ...(9·2)$$

where $\triangle e$ is the change in void ratio corresponding to the change in pressure $\triangle p$.

From the pressure void ratio diagram of the sample, $\triangle e$ can be read against $\triangle p$ and knowing the initial ratio e_0 of the stratum and its thickness H, the settlement S can be calculated from equation (9·2).

Equation (9.2) is analogous to equation (7·4) in Chapter 7 derived for the laboratory consolidation sample.

Second method. Let the data regarding load increment, initial void ratio and thickness of the stratum be the same as in the first method.

By definition, from equation (7·21)

$$m_v = \frac{\triangle h}{\triangle p \cdot H} \qquad \qquad ...(9·3)$$

using same notations as used in equation (7·21).

$\triangle h$ in equation (9·3) is nothing but the change in thickness H of the layer due to the increase in pressure $\triangle p$. Hence,

$$\triangle h = m_v \cdot \triangle p, H \; (=S) \qquad \qquad ...(9·4)$$

represents the required settlement, S.

Coefficient of volume compressibility m_v can be calculated from equation (7·20), knowing the initial void ratio e_0 and the coefficient of compressibility a_v for the particular load increment $\triangle p$.

Third Method. Let, again the data regarding load-increment void ratio and thickness of the stratum remain the same as in the previous two methods.

From equation (7·16), for an initial void ratio e_0 a pressure increase $\triangle p$ and the corresponding change in void ratio $\triangle e$

$$a_v = \frac{\triangle e}{\triangle p}$$

neglecting the negetive sign with a_v.

Also from equation (7·9), void ratio e attained after the load increment, in a normally loaded soil is given by

$$e = e_c - C_c \log_{10} \frac{p}{p_0}$$

$$= e_0 - C_c \log_{10} \frac{p_0 + \triangle p}{p_0}$$

where C_c is the compression index of the soil and p_0 is the initial load corresponding to e_0.

Rearranging the terms

$$(e_0 - e) = C_c \log_{10} \frac{p_0 + \triangle p}{p_0}$$

or $$\triangle e = C_c \log_{10} \frac{p_0 + \triangle p}{p_o}$$

Divide both sides by $\triangle p$

$$\frac{\triangle e}{\triangle p} = \frac{C_c}{\triangle p} \log_{10} \frac{p_0 + \triangle p}{p_0}$$

By comparison,

$$a_v = \frac{C_c}{\triangle p} \log_{10} \frac{p_0 + \triangle p}{p_0.}$$

Divide both sides again by $(1 + e_0)$

$$\therefore \qquad \frac{a_u}{1 + e_0} = \frac{C_c}{\triangle p(1 + e_0)} \log_{10} \frac{p_0 + \triangle p}{p_0}$$

From equation (3·20), we know that

$$\frac{a_u}{1 + e_o} = m_n$$

$$\therefore \qquad m_v = \frac{C_c}{\triangle p(1 + e_0)} \log_{10} \frac{p_0 + \triangle p.}{p_0}$$

Substituting this value of m_v in equation (9·4) we get another relationship for settlement in terms of the compression index $i.e.$

$$S = \frac{C_c}{(1 + e_0)} \log_{10} \frac{p_0 + \triangle p}{p_0}.H \qquad \ldots(9·5)$$

Comparison of equation (9·2), (9·4) and (9·5) shows that they are just three different ways of putting the same quantity S. However it may be mentioned that equations (9·2) and (9·4) are general in the sense that they are applicable both to the normally as well as over-consolidated soils, where as equation (9·5) is only applicable to the normally loaded soil beyond the preconsolidation range. Also C_o in equation (9 5) is a constant quantity whereas m_v in equation (9·4) is different for each load increment, since, it depends upon a_v which further depends upon $\triangle e$ for the load increment.

Time Rate of Settlement. The rate of settlement willdepend upon the boundary conditions of the compressible clay layer. In case a compossible clay layer is *sandwiched* between two layers of granular material, one at either end providing drainage at both ends, then the settlement will proceed rapidly compared to another situation where the same compressible layer has a layer of granular material only at the top and the bottom of clay layer rests on an imprevious rock (see Fig. 9·6). The two situations are shown as situation No. 3 Fig. 9·2.

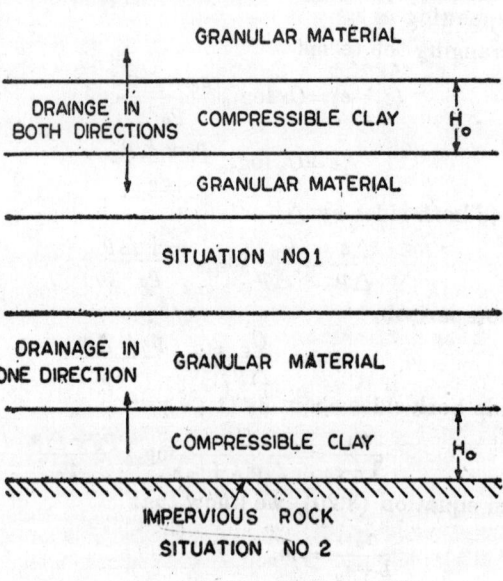

Cases of double and single drainage.
Fig 9·2.

The calculations are based upon equation (7·33) which may stated as.

$$T = \frac{C_v t}{H^2} \qquad \ldots (9\text{·}6)$$

where T = Time factor in the Terzaghi's one dimensional consolidation theory

c_v = Coefficient of consolidation relevant to the load increment $\triangle p$.

$t=$Time for consolidation of the laboratory sample or the clay layer in the field under the load increment.

$H=$Maximum length of the drainage path.

Maximum length of the drainage path H in this equation is equal to half the thickness H_0 of the compressible layer in case of double drain-age system (situation No. 1) and it is a equal to the total thickness H_0, in a case of a single drainage system (situation No. 2). c_v for the load increment is made available by the consolidation test. So T and t in equation (9.6) are related by a numerical constant c_v/H^2. If, now, it is assumed that the initial hydrostatic excess pressure at all points in the clay layer is constant[1], then curve A of Fig. 7.16 and equation (9.6) can be used to correlate, the percent consolidation U, the time factor T, the actual time of settlement t and the settlement S for the corresponding percent consolidation. In fact a table, like Table 9.3 or 9.4 is set up. In this table the time t is given some numerical values and the corres ponding values of time factor T are calculated from equation (9.6). Against the values of T percent U values are filled from curve A of Fig. 7.16. In order to fill up the 5th column, i.e, the settlement s values, corresponding total settlement S is multiplied by the percentage consolidation. In this way t is varied in steps until $U=100\%$. A number of values of t and corresponding values of s would be available from such a table. On an arithmatic or a semi-log scale settlement values are plotted against the time values t, and a time-settlement curve is obtained It is customary to plot settlement along the y-axis and the time against the x-axis if it is an arithmetic graph paper. In case a semi-log paper is used, time is plotted on the log scale and the settlement on the natural or arithmetic scale. Example 9.2 at the end of this chapter will illustrate these points.

9.5. Practical Hints

Before the various practical hints are given, it is considered essential to give a review of the actual conditions. Refer to Fig. 9.3 (a) which shows a soil profile for a building site. $ABCD$ shows a pit made into the top sand layer for placement of a footing for a column carrying a load P. Initial stress conditions in the soil directly below the centre of the proposed footing, as a function of depth before digging the excavation are shown in Fig. 9.3 (b). On excavating, material layers underneath experience a stress release as indicated in Fig. 9.3 (c). Again when the footing and the column have been placed in position, the soil layers underneath will again experience a stress increase as indicated in Fig. 9.3 (d). Following observations are made.

(1) Initial stress at the top of the clay is ab. The stress decrease due to excavation is cd and stress increase due to the

---

1. This is a gross assumption, as explained in Chapter 7. There can be other types of pressure distributions in practice. However, a discussion on all types of distributions is outside the scope of this book.

footing and the column load is *ef*. So, the *net pressure* on the top of clay layer after construction is $(ab-cd+ef)$.

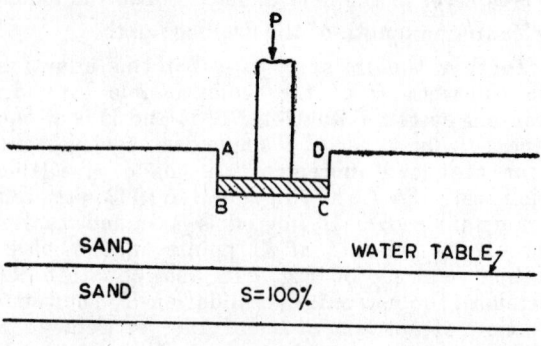

Fig. 9.3 (a)

(2) *Initial stress at the impervious rock surface is a'b'*.

The stress decrease due to excavation is c'd' and stress increase due to the footing and the column load is *e'f'*. So, the *net pressure* on the impervious rock surface is $(a'b'-c'd'+e'f')$.

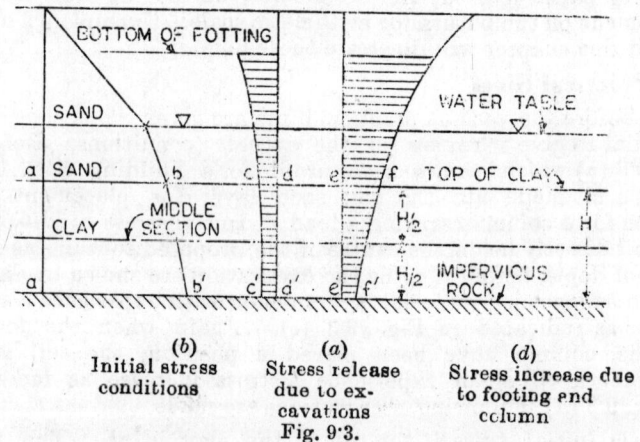

(b)	(a)	(d)
Initial stress conditions	Stress release due to excavations	Stress increase due to footing and column

Fig. 9·3.

Now, if an attempt is made to plot a stress profile through the clay thickness, with the above boundary values, one thing vould be apparent and that is that there is stress variation in the ciay layer from top to bottom. Any theoretically accurate settlement analysis must take into account this stress variation.

Also, if the water content profile for the clay layer is made available, it might appear from it that the initial void ratio of the clay layer varies from point to point.

It is rather impossible to take into account the stress variation in the clay layer from point to point. Some simplifying assumptions are very essential.

As far as the initial void ratio is concerned, it is invariably assumed that it is constant for the clay layer and an average value may be worked out for the purpose. Possibly, the value of the void ratio at the middle of the clay layer would represent a fairly approximate average from a practical point of view. Regarding the stresses, there are three ways by which they can be worked out :

    (1) Work out the initial overburden pressure at the middle of the clay layer to represent the initial conditions. Add to it the pressure exerted by the foundation at the level of the foundation, without any stress distribution, to find out the final conditions. The latter value will be too conservative.

or   (2) Take the middle section of the clay layer as representative of the whole thickness and work out the initial and final stress values at this sction. Assume that these values operate for the whole of the section.

or   (3) Work out the initial overburden pressure at the top, middle and bottom of the clay layer and take the mean value as representative of the initial conditions. Again, find out the stress decrease and increase at the three levels and add mean of the net increase to the mean initial stress to find out the final conditions.

Difference of the final and initial stress values by any of the above three methods would give the pressure increase Δp for the layer.

From the calculations point of view, the first method though conservative is the easiest since it does not involve any stress distribution. Second and third methods, of course, do take into considrration to some extent the stress variation in the clay layer, but involve more calculation work. The latter two procedures are suitable when the clay layer is relatively thin. If the soil stratum is thick and particularly if it is close to the foundation where the stress change from top to bottom is greater than 50 percent of average value at the middle, these procedures can lead to gross errors. It is, therefore, a better practice to subdivide a thick clay layer into thin slices and work out the settlement on each slice independently and add the various values so obtained to get the total settlement. Each slice in such a case should be such that the stress variation from top to bottom is not more than 50 per cent of the average value at the middle.

9·6. Settlement Correction

Consolidation theory assumes that the soil does not undergo lateral strain during the consolidation process. This condition is true for the laboratory consolidation sample, which is enclosed a ring. However, such a condition may not exist for the clay layer in the field. If a clay layer is thin and is sandwiched between sand and gravel layers which relatively do not yield or when the area under the load is large compared to the compressible layer thickness, "no lateral strain" condition may be correct assumption. However, when the compressible layer is thick and unrestrained, lateral deformation will also take place and the consolidation settlement computed from theory will not be accurate. For this condition, Skempton and Bjerrum (5) have suggested a semiempirical correction, based upon the Skempton's pore-pressure parameter[1] A. From the value of A for the soil, a correction factor C is picked up from a relevant curve for H/b ratio, where H is the thickness of the clay layer and b is the foundation width, as shown in Fig. 9·4. The corrected settlement S_c is obtained from

$$S_c = S.C$$

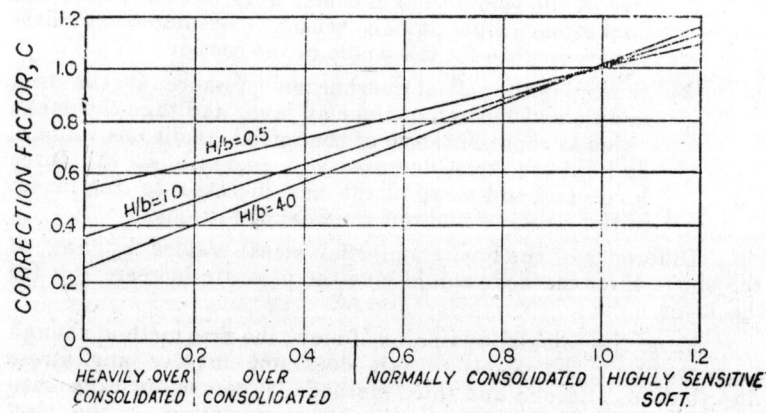

Correction Factor for Computed Settlement

(After Skempton and Bjerrum)

Fig. 9·4.

where C is the correction factor and S is the computed settlement.

---

1. See chapter 10 for discussion on pore-pressure parameters.

Some engineers are apprehensive about the use of this correction because it is extremely difficult to obtain the pore-pressure parameter A with accuracy. Where there is uncertainly about its use, it is better to present the settlement value S instead of S_o. S_o may be quoted in such cases as showing the range of uncertainty, only.

9·7. Accuracy of Computed Settlements

Settlement analysis as presented in the preceding paragraphs depends upon,

(a) The Terzaghi's Consolidation Theory.

(b) Stress distribution theories for elastic materials.

(c) Laboratory determination of the consolidation, characteristics of the soil.

As pointed out earlier, the Terzaghi's theory assumes unidirectional compression and no lateral strain, as also the flow of fluid in the vertical direction only. These two asumptions may or may not be applicable to a layer of soil in the field. The soil in the field under loading may be having lateral strain. Also there may be some drainage in the horizontal direction too. Further, a natural soil can never be homogeneous and isotropic in nature as assumed in the various stress distribution theories for elastic materials. Also, at large strain even when the stress is increasing, the true elastic stressstrain relationship does not exist in soils. So the stress distribution by elastic theories do not give the true picture of stress distribution in soils. Bouessinesq's theory, for example, gives higher stresses than the ones that actually exist in the field. Also determination of the consolidation characteristics of a soil in the laboratory involves so many variables. To quote a few, the sampling of the consolidation sample may havo been done so crudly as to destroy the natural structure of the soil mass, the sample may not be representative of the whole stratum and the investigator may not be well trained in the laboratory work.

Due to all the above reasons, it cannot be said that the settlement computations are accurate. From the data available on the existing structures, it can be said that the computed settlement for a structure may be anything between 50 per cent of the actual expected settlement. Agreement between the computed and observed values, although very poor, is still alright because of the complex nature of the problem. As such, when a designer computes the settlement for a foundation, he just reports it as the probable setttement.

9·8. Permissible Settlement

A building foundation, unless it is a raft, constitutes of a number of individual or combined footings. Settlements for the foundations are computed by taking each footing individually and considering the full dead load and the sustained[1] live load that comes on this footing.   Thus, there would be as many settlement values as are the number of footings on which the building rests.

The individual settlement values so computed are not of much consequence because if all the footings of a building settle equally, there will be no harm to the building. What is more important is the differential settlement obtained when individual settlement values are compared ; since the structure resting on these footings is rigidly connected and if the differential settlement is excessive, the structure would crack down and may eventually give way. Sometimes differential settlement between one part of a footing and other part of the same footing also leads to cracking of the structure or the failure of the foundations. As such, in case of large footings, settlements are calculated for the centre as well as for the corners or edges and then differential settlement, for the single footing itself is evaluated. The design criteria should, therefore, be the differential settlement and not the total settlement for a structure. Many building codes, therefore, place a limit on the permissible differential settlement. In a typical small framed building, the permissible differential settlement may be of the order of $0.003 L$ where L is the distance between any two adjacent columns or any two points that settle differentially. Table 9.2 gives the maximum permissible settlement values as given by some experienced soil engineers.

The maximum permissible total settlement are recommended by the Indian Standards Institution as 6.5 cms. for isolated foundations on clay, 4 cms. for isolated foundations on sand, 6.5 to 10 cms. for raft foundations on clay and 4 to 6.5 cms. for raft foundations on sand. The maximum differential settlement must not exceed 4 cms. in case of foundations on clayey soils and 2.5 cms. in case of foundation on sandy soil. The Institution also recommends that the differential settlement in public buildings such as offices, factories, flats, etc., made of concrete, should be such that the angular distortion of the frame of the building does not exceed 1/500 normally and 1/1000 where it is desired to avoid any kind of damage.

The relevant building code in the as in which the engineer works is the best guide in this respect.

9.9. Construction Period Effect

The construction of a building takes some time. As such, the soil layers under the foundations are not stressed all at once as

---

1. Sustained live load for a building can be the furniture, machinery and any other such item that is going to remain in position, all the twenty-four hours of a day and all the year round. Transient live load does not affect the settlement computations. See also Chapter 12 for further discussion.

TABLE 9·2.

Values of Permissible Settlement

S. No.	Type of structure	Maximum permissible total settlement	Maximum permissible differential settlement	Authority
1	Ordinary structures	2·54 cms (1 inch)	1·90 cms (0·75 inch)	Terzeghi and Peck (Ref. 8)
2	Framed structures	10 cms (4 inches)	0·004L[1]	Skempton and McDonald (Ref. 9)
3	Turbo generator Foundations	—	0·0002L	Sowers[2] (Ref. 10)
4	Brick Buildings	—	0·002L	Sowers[2] (Ref. 10)

assumed in the theory of consolidation and the consequent settlement analysis based upon it. As indicated in Fig, 9·3 (c) first due to excavation there is some stress release on the compressible clay layer. This stress release is naturally gradual. Again, when the building is under construction, there is gradual increase of stress on the compressible layer. Firstly, the original stress is reached and then the stress increases beyond the original stress value, to a maximum value when the building is ready. The time lapse between the regain of original stress and attainment of maximum stress value may be called the effective period of loading. For estimating the year when the ultimate settlement would be attained, it is reasonable to assume that the total load has been placed in position at the middle of the effective period of loading and then count the year of attainment of ultimate settlement.

More elaborate procedures are available to account for the construction period but they are outside the scope of this book.

9.10. Examples

Example 9·1. A saturated soil startum 5 metres thick lies above an impervious startum and below a pervious stratum. It has a compression index C_c of 0.25 and a coefficient of permeabi-

---

1. L is the distance between adjacent columns that settle different amounts or between any two points that settle differently.

2. These are just two examples. For complete data on permissible settlements, see reference 10, page 597.

lity of 3.2×10^{-8} cms/sec. Its void ratio at a stress of 1.5 kg/cm²
is 1.9. Compute :

    (*i*) the change in void ratio due to increase of stress to
2.3 kg/cm².

    (*iii*) settlement of the soil stratum due to the above in-
crease in stress.

    (*iii*) time required for 50% consolidation, for which the
time factor may be assumed as 0·20.   **(P. U. Nov. 1966)**

    Solution. In this problem,

$$C_c = 0.25 \quad \text{(Given)}$$
$$e_0 = 1.9 \quad \text{(Given)}$$

Initial pressure $p_0 = 1.5$ kg/cm²
Final pressure $p = 2.0$ kg./cm²
∴ Change in pressure

$$\triangle p = 0.5 \text{ kg/cm}^2$$

(*i*) Equation of the virgin curve is

$$e = e_0 - C_c \log_{10} \frac{P}{p_0}$$

∴ $(e_0 - e) = \triangle e = C_c \log_{10} \frac{P}{p_0}$

or $\triangle e = 0.25 \log_{10} \frac{2.0}{1.5}$

or $\triangle e = 0.0312.$ **Ans.**

(*ii*) Coefficient of compressibility a_v for the soil

$$= \frac{\triangle e}{\triangle p} = \frac{0.0312}{0.50} \text{cm}^2/\text{kg.}$$

∴ Coefficient of volume compressibility m_v for the soil

$$= \frac{a_v}{1+e_v} = \frac{0.0312}{0.5(1+1.9)} = \frac{0.3012}{0.5 \times 2.9}$$
$$= 0.0215 \text{ cm}^2/\text{kg.}$$

Settlement, $S = m_v . \triangle p . H = 0.0215 \times 0.5 \times (5 \times 100)$
$$= 5.38 \text{ cms.}$$

Coefficient of consolidation c_v for this increment

$$= \frac{k}{m_v . \gamma_w{}^1} = \frac{3.2 \times 10^{-8}}{0.0215 \times \dfrac{1}{10^3}} \text{cm}^2/\text{sec}$$

$$= \frac{3.2}{2150} \text{cm}^2/\text{sec.}$$

---

    1. Here γ_w has to be in kg/cm³ and if in the c.g.s. units, it is
taken as 1 gm/cm³, than this must be divided by 10³ to get the value
in kg/cm³.

Now

$$T = c_v \cdot \frac{t}{H^2}$$

Here

$T = 0.20$ for 50% consolidation (Given)

$$c_v = \frac{3 \cdot 2}{2150} \text{cm}^2/\text{sec}.$$

$H = 500$ cms.

$$t = \frac{0 \cdot 20 \times (500)^2}{3 \cdot 2/2150} \text{secs}.$$

$$= \frac{0 \cdot 20 \times (500)^2 \times 2150}{3 \cdot 2 \times 60 \times 60 \times 24}$$

$= 388$ days. **Ans.**

Example 9·2. In a consolidation test on an undisturbed sample of clay, following results were obtained :

Pressure kg/cm²	Void ratio	
0·20	0·953	
0·40	0·948	
0·80	0·938	
1·60	0·929	Compression values
3·20	0·878	
6·40	0·789	
12·80	0 691	
3·20	0·719	
0.80	0·754	Rebound values
0·20	0·787	

1. Plot the e-log p curve and find out the value of compression index.

2. Calculate the value of the coefficient of volume compressibility m_v for the increment 1·60 kg/cm² to 3·2 kg/cm².

3. Assume that a deep stratum of clay 5 metres thick undergoes average pressure increase corresponding to (2) above *i. e.* from 1·6 kg/cm² to 3·20 kg/cm². Find out the total settlement, assuming

(a) Double drainage conditions.
(b) Single drainage conditions.

4. Plot the time settlement curve for the two cases under (3) above. Following data is given :

(a) For the load increment 1.6 kg/cm² to 3.2 kg/cm² t_{50} for the laboratory sample is 10 minutes.

(b) Average thickness of the sample during this load increment $= 2.42$ cms.

(c) The sample was drained top and bottom.

5. Find out the coefficient of permeability for the soil.

Solution. 1. Plot the values of the void ratio on the vertical natural scale and the corresponding pressures on the horizontal logarithmic scale of a semi-log paper, as shown in Fig.9·5.

In order to find out the compression index consider two points A and B as shown :

Co-ordinate of A and B are;

A : Pressure$=4$ kg/cm^2

and $e=0.85$

B : Pressure$=8$ kg/ cm^2

and $e=0.756$

Now $e=e_o-C_c \; \log_{10} \dfrac{p}{p_o}$

or $C_c=\dfrac{e_o-e}{\log \dfrac{p}{p_o}}$

Treat A as representing the initial conditions (p_o, e_o) and B representing the final conditions (p, e).

$\therefore$ $C_c=\left(\dfrac{0.850-0.756}{\log_{10}\dfrac{8}{4}}\right)=\dfrac{0.094}{\log_{10}2}$

or $C_c=0.320$

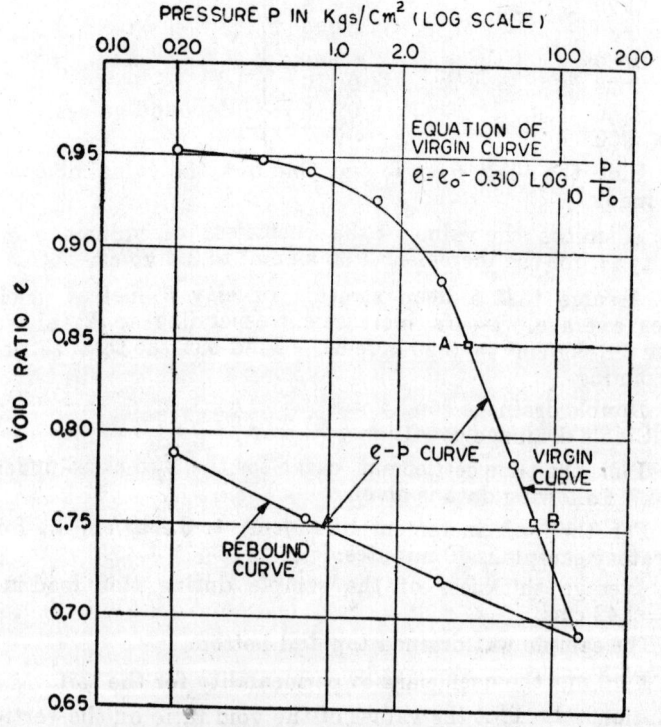

Pressure void ratio diagram
(Semi log scale)
Fig. 9.5.

(2) For the load increment 1·60 kg/cm² to 3·20 kg/cm²

$$p_o = 1·60 \text{ kg/cm}^2$$

and e_o from the curve $= 0·92$

$$p = 3·20 \text{ kg/cm}^2$$

and e from the curve $= 0·878$

Coefficient of compressibility

$$a_v = \frac{\triangle e}{\triangle p} \text{ ; neglecting sign}$$

or

$$a_v = \frac{e_o - e}{p - p_o} = \frac{0·92 - 0·878}{3·20 - 1·60}$$

$$a_v = \frac{·042}{1·6} = 0·02625 \text{ cm}^2/\text{kg}$$

∴ Coefficient of volume compressibility.

$$m_v = \frac{a_v}{1 + e_o} = \frac{0·03625}{1 + 0·92} = \frac{0·0262}{1·92}$$

$$= 0·01368 \text{ cm}^2/\text{kg. } \textbf{Ans.}$$

(8) Total settlement is given by,

$$s = m_v . \triangle p . H \qquad \text{(Equation 9·4)}$$

With the usual significance of various terms.

It is to be noted that the total settlement whether it is a double drainage or a singale drainage system remains the same. The value of H to be used in the above relation is the total thickness of the clay layer

∴
$$S = (0·01368)(1·60)(5 \times 100)$$
$$= 10·94 \text{ cms. } \textbf{Ans.}$$

(4) For the laboratory consolidation sample, average thickness during this load increment $= 2·42$ cms. (Given)

The sample in the laboratory was drained top and bottom

The longest path of drainage

$$= 1·21 \text{ cms} = H.$$

Actual time t_{50} for the sample

$$= 10 \text{ minutes} = 600 \text{ seconds.}$$

Time factor T_{50} from the curve A of Fig. 3·16 $= 0·197$

Now apply
$$T_{50} = \frac{c_v \, t_{50}}{H^2} \qquad 0·197 = c_v \frac{600}{(1·21)^2}$$

∴
$$c_v = \frac{0·197 \times (12·1)^2}{600}$$

$$= 0·481 \times 10^{-3} \text{ cm}^2./\text{sec.}$$

$$= 0·481 \times 10^{-3} \times (60 \times 60 \times 24 \times 365)$$

$$= 15,200 \text{ cm}^2/\text{year. } \textbf{Ans.}$$

Data for time settlement curve

There are two cases that have to be dealt with, one the case of double drainage and the second of single drainage.

Case I. Double Drainage. The thickness of the clay layer is 5 metres. When it is drained top and bottom, the longest path of drainage is 2·5 metres *i.e.* 250 cms. So, H in the equation (9·6) will in this case be 250 cms. and c_v will have the above value of 15,200 cm²/year.

Substituting these values in equation (9·6).

$$\therefore \qquad T = \frac{15,200}{(250)^2}\, t = \frac{1\cdot52}{6\cdot25}\, t = 0\cdot243\, t \qquad ...(9\cdot7)$$

Now prepare a table 9·3 shown below for tabulating the values of t, T, U, s. The procedure involved is as under.

Assign different values to t say zero, 0·25, 0·5, 1·0 etc. in years and get the corresponding values of time factor T from equation (9 7). Read U against each time factor, so obtained, from curve of A of Fig. 7·16. Multiply the total settlement S by this value of U to get the settlement s corresponding to the time t.

TABLE[1] 9·3

Time versus settlement values for the double drainage case

S. No·	t years	T	U%	s cms.
1	2	3	4	$5 = S \times \dfrac{(4)}{100}$
1	0	0	0	0
2	0.25	0·0607	27·5[2]	3·03
3	0.50	0·1214	39·9	4·27
4	1.00	0·2428	55·5	6·07
5	2.00	0·4856	75·5	8·27
6	4.00	0·9712	92·5	10·15
7	8.00	1·9424	8·0	10·75
8	∞	∞	100	10.94

1. Calculations in this table and Table 9·4 have been done with a slide rule and for this type of work, they are quite accurate.

2. In order to get fairly accurate values of U as shown here, the designer is advised to plot curve A of Fig. 3·16 to a larger scale and then the values.

Case II. Single Drainage. The thickness of the clay layer is 5 metres. When it is drained only at one end; the longest path of drainage is 5·0 metres *i.e.* 500 cms. So, H in the equation (9·6) will, in this case, be 500 cms, and c_v will have value of 15,200 cm^2./ year.

Substituting these values in equation (9·6)

$$T = \frac{15,200}{(500)^2}\, t = \frac{1\cdot52}{1.25}\, t = 0\cdot0607\, t \qquad ...(9\cdot8)$$

Again prepare a Table 9·4 shown below for tabulating the values of t, T, U and s. The procedure involved is the same as for Table 9·2.

TABLE 9·4

Time versus settlement values for the single drainage case.

S. No.	t years	T	U%	s cms.
1	2	3	4	$5 = S \times \dfrac{(4)}{100}$
1	0	0	0	0
2	0·25	0·0152	14·0	1·63
3	0·50	0·0304	19·50	2·13
4	1·00	0·0608	27·50	3·03
5	2·00	0·1216	39·00	4·27
6	4·00	0·2432	55·5	6·07
7	8·00	0·4864	75·5	8·2
8	∞	∞	100	10·94

The values given in columns (2) and (5) of Table 9·3 and 9·4 are plotted in Fig. 9·6. The effect of the single and double drainage on the time-rate of settlement is clearly depicted in the two curves :

(5) Equation for c_v is

$$c_v = \frac{k(1+e_o)}{\gamma_w\, a_v}$$

where the various terms have their usual significance.

Now $$m_v = \frac{a_v}{1+e_o}$$

∴ $$c_v = \frac{k}{\gamma_w\, m_v}$$

$$\therefore \qquad k = e_v . \, \gamma_w . \, m_v$$

Put $\qquad c_v = 0\cdot481 \times 10^{-3} \mathrm{cm^2./sec.}$

$$m_v = 0\cdot01368 \ \mathrm{cm^2./kg.}$$

and $\qquad \gamma_w = 1 \ \mathrm{gm./c.c.} = \dfrac{1}{10^3} \ \mathrm{kg./cm^3.}$

$$\therefore \quad k = 0\cdot481 \times 10^{-3} \times 0\cdot01368 \times \frac{1}{10^3} = 6\cdot58 \times 10^{-9} \ \mathrm{cms./sec}$$

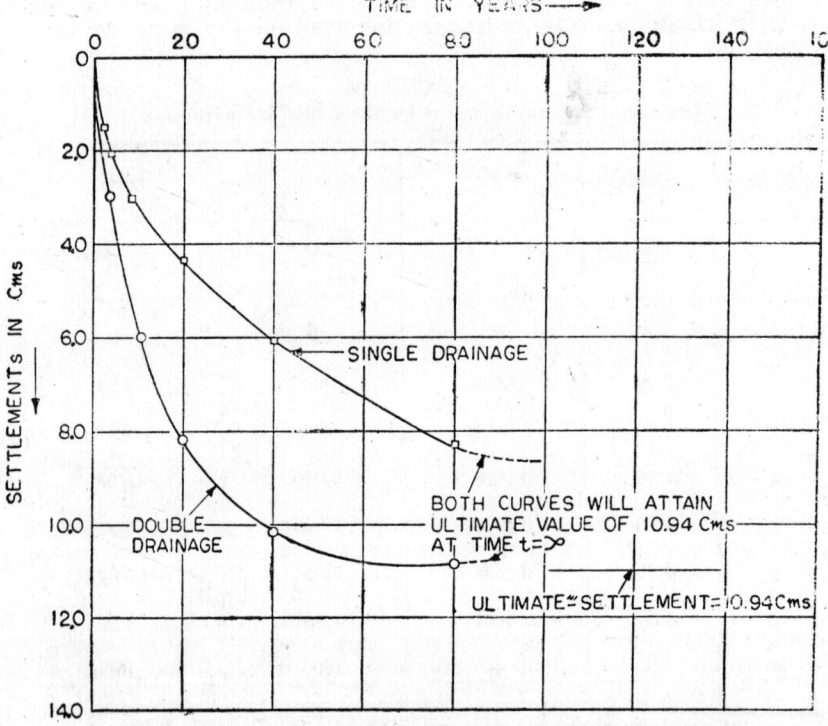

Time-settlement curves for the single and double drainage cases.
Fig. 9·6.

Example 9·3. A building 'A' shown in Fig. 9·7 (a) was constructed twenty years ago in an excavation equal to the weight of the bulding and did not show any settlement. This building rests upon individual piles. Another building 'B' again shown in Fig. 9·7 (a) which is a twelve storey structure weighing 30,000 tonnes was constructed only three years ago, on a rigid mat 30 m × 30. in size. Building 'A' has started showing some cracks. Plans of the two buildings are given in Fig. 9·7 (b)

You have been called as a consultant to asses the situation and to indicate whether any differential settlement has occurred in building 'A', due to the construction of building 'B'. The

existing pressure under '*A*' at 6·0 metres below the ground
surface *i.e.*, at the middle of the clay layer may be taken as

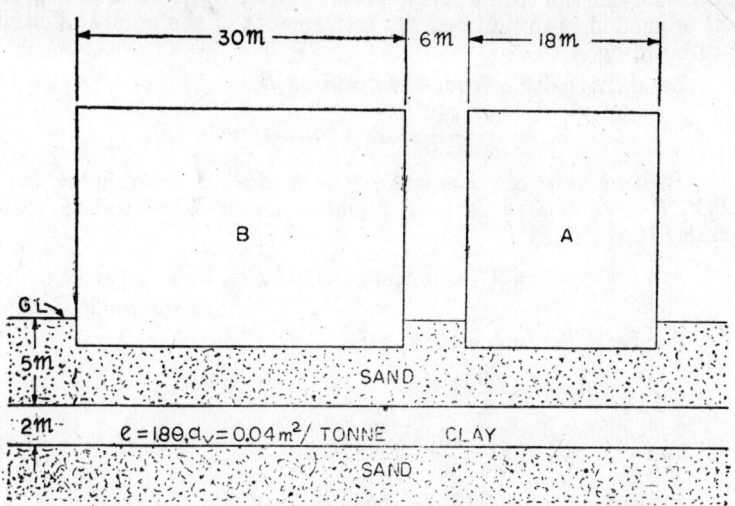

Fig. 9·7 (*a*).

5 tonnes/m². Clay properties are given in the soil profile in
Fig. 9·7 (*a*).

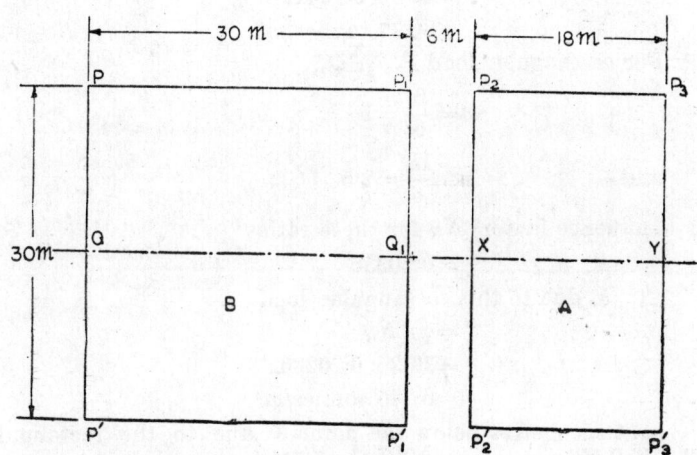

Fig. 9·7 (*b*).

Solution. The problem actually lies in finding out the additional pressure $\triangle p$ at the points six metres below points X and Y
for the building A, when the building B has been constructed.
Six metres below the ground level is the middle of the clay layer

and this section may be assumed as representing the average conditions for the 2m thick clay layer. After $\triangle p$ is known, all that is needed is to find out the settlements at the points X and Y of building A.

Load intensity q from the building B.

$$=\frac{30,000}{30\times30}=33\cdot3 \text{ tonnes/m}^2.$$

Vertical stress at six metres below X due to rectangular load $PP_1P_1'P$. Vertical stress σ_z at 6 metres below X due to the rectangular load $PP_1P_1'P$

$$=2[\sigma_z \text{ due to rectangle } PP_2XQ - \sigma_z \text{ due to}$$
$$\text{rectangle } P_1P_2XQ_1]$$

For rectangular load PP_2XQ,

$$m^1=\frac{36}{6}=6$$

and

$$n^1=\frac{15}{6}=2\cdot5$$

Influence factor $N^2{}_N$ for these ratios

$$=0\cdot24412$$

$\therefore$ σ_s due to this rectangular load

$$=q\times N_{N.}$$
$$=33\cdot3\times0\cdot24412$$
$$=8\cdot137 \text{ tonnes/m}^2.$$

For rectangular load $P_1P_2XQ_1$,

$$m=\frac{6}{6}=1$$

and

$$n=\frac{15}{6}=2\cdot5$$

Influence factor[2] N_N for these ratios

$$=0\cdot20336$$

$\therefore$ σ_z due to this rectangular load

$$=q\times N_N$$
$$=33\cdot3\times0\cdot20236$$
$$=6\cdot745 \text{ tonnes/m}^2$$

σ_z at six metres below the point X due to the rectangular load $PP_1P_1'P'$ $\qquad =2(8\cdot137-6\cdot745)$

---

1. These m and n are the symbols having the usual meanings described in the previous chapter.

2. See table 2 in the appendix for influence factors.

1. Ibid.

$$= 2\cdot783 \text{ tonnes/m}^2$$

Vertical stress at six metres below Y *due to rectangular load*
$PP_1P_1'P'$ $\quad = 2[\sigma_z$ due to rectangle $PP_3YQ - \sigma_z$ due to
rectangle $P_1P_3YQ_1]$

For rectangular load PP_3YQ,

$$m = \frac{54}{6} = 9$$

and $\quad\quad n = \frac{15}{6} = 2\cdot5$

$\therefore$ Influence factor N_N for these ratios

$$= 0\cdot24427$$

σ_z due to this rectangular load $= q \times N_N$

$$= 33\cdot3 \times 0\cdot24427$$

$$= 8\cdot142 \text{ tonnes/m}^2.$$

For rectangular load $P_1P_3YQ_1$,

$$m = \frac{24}{6} = 4\cdot0$$

and $\quad\quad n = \frac{15}{6} = 2\cdot5$

Influence factor N_N for these ratios

$$= 0\cdot24344$$

σ_z due to this rectangular load $= q \times N_N$

$$= 33\cdot3 \times 0\cdot24344$$

$$= 8\cdot114 \text{ tonnes/m}^2$$

$\therefore$ σ_z at six metres below the point Y due to rectangular
load $PP_1P_1'P'$ $\quad = 2(8\cdot142 - 8\cdot114)$

$$= 0\cdot056 \text{ tonnes/m}^2$$

Settlements at X and Y. From the data calculated above for
stress distribution, it is seen that

$$\triangle p \text{ for } X = 2\cdot784 \text{ tonnes/m}^2$$

$$\triangle p \text{ for } Y = 0\cdot056 \text{ tonnes/m}^2$$

Also m_v for the soil $= \dfrac{a_v}{1 + e_0} = \dfrac{0\cdot04}{1 + 1\cdot8} = \dfrac{0\cdot04}{2\cdot8}$

$$= 0\cdot014 \ m^2/\text{tonne}$$

Settlement at X, $S_X = m_v \cdot \triangle p \cdot H$

$$= 0\cdot014 \times 2\cdot784 \times (2 \times 100)$$

$$= 7\cdot95 \text{ cms.}$$

Settlement Y, $S_Y = m_v \cdot \triangle p \cdot H$

$$= 0 \cdot 014 \times 0 \cdot 056 \times (2 \times 100)$$
$$= 0 \cdot 16 \text{ cms.}$$

Differential settlement

$$= 7 \cdot 95 - 0 \cdot 16 = 7 \cdot 79 \text{ cms.}$$

This differential settlement has been the cause of excessive cracking of the building A.

CHAPTER-BIBLIOGRAPHY

1. Dunham, C.W., Foundation Engineering, New York, McGraw Hill Book Co., Inc., 1950.

2. Indian Standards Institution, Code of Practice for Structural Safety of Buildings, Foundations, Is 1904-1961, New Delhi.

3. Johnson, S.M., and Kavanagh T.C., "The Design of Foundation for Buildings", McGraw Hill Book Co., Inc., New York, N.Y., 1968.

4. Jumikis, A.R., Soil Mechanecs, New Delhi, Affiliated East West Press, Put Ltd., 1965.

5. Lambe, T.W., and Whitman, R.V., "Soil Mechanics", J. Wiley & Sons, Inc., N.Y., 1969.

6. Leonard, G.A., Foundation Engineering, New York, McGraw Hill Book Co., Inc., 1962.

7. Polshin, D.E., and Toker R.A., Maximum Allwable Differential Settlement of Structures, Proceedings, Fourth international Conference on Soil Mechanics and Foundation Engineering, London, Vol. 1-1957.

8. Skempton, A.W., and Bjerrum L., A Contribution to the Settlement Analysis of Foundations on Saturated Clays, Geotechnique Vol. 7, No. 4, Dec. 1957.

9. Skempton, A.W., and McDonald D.H., The Allwable Settlement of Buildings, Proceedings, Institute of Civil Engineering, Vol. 5, No. 3, 1956.

10. Terzaghi, K., Theoratical Soil Mechanics, New York, John Wesley and Sons, Inc., 1943.

11. Terzaghi, K., and Peck R.B., Soil Mechanics in Engineering Practice, New York, John Wiley and Sons, Inc., 1948.

12. Timlinson, M.J., "Foundation Design and Construction", Wiley Inter-science, John Wiley & Sons, Inc., New York, N.Y., 1969.

13. Tschebotarioff, G.P., "Soil Mechanics, Foundations, and Earth Structures", McGraw Hill Book Co., Inc., N. Y., 1951.

QUESTIONS

1. (a) What do you understand by PRECONSOSIDATION in clays and what is its significance in settlement analysis ?

   (b) A structure has been erected on a 45 feet thick layer of clay which is sandwiched between layers of free draining sands. The coefficient of consolidation of the clay sample was computed and was found to be 0·0166 cm² per minutes. Compute the time required for 50 and 90 percent settlement.

   Time factor values for 50 and 90 per cent consolidation are 0·20 and 0·85 respectively.

   (P.U. 1959)

2. A foundation slab is supported on a bed of compact sand which extends to a depth of 20 feet below the base. Under the sand, there is a stratum of clay of thickness 16 feet which in turn rests on impervious shale. The initial effective overburden pressure at the top of the clay is 0·8 ton per sq. ft. The additional pressures applied by the fundation loads are 0·80 and 0·24 tons/sq. ft. on the top and bottom of the clay stratum respectively. The clay has an average density of 120 lbs/c.ft. in saturated condition and the laboratory consolidation test on the sample of clay gave the following data :

Effective Pressure	Void Ratio
0·5	0·93
1·0	0·91
2·0	0·88
4·0	0·85

Estimate the final settlement of the foundation slab due to consolidation of clay stratum. Ground water level is situated within the sand (Assum 1 ton=20·0 lbs).   *(P.U. 1959)*

3. Same as question (3b) of Chapter 7.

4. Same as question (4) of Chapter 7.

5. Same as question (8b) of Chapter 7.

6. Subsurface exploration at the site of proposed building reveals the existence of a 2·4 metres thick layer of soft clay below a stratum of coarse sand which is 4·0 metres thick and extends from the ground surface up to the clay layer. The ground water table is at a depth of 2·5. metres below the ground surface. Laboratory tests indicate the natural water content of clay as 40%, average liquid limit 45 per cent and specific gravity of solids as 2·75. The unit weights of sand above and below the water table are respectively 1·78 and 2·10 gms/c.c.

   Estimate the probable final settlement of the building if its construction will increase the average vertical pressure on the clay by 0·71 kg/cm². *(P.U. 1963 Suppl)*

7. If the load due to a structure is uniform distributed over a certain area on the surface of a homogeneous soil mass, what will be the nature of settlement of the soil in that area and why so ?

   (P.U. Nov. 1966)

Shearing Characteristics of Soils

10.1. Introduction

As a definition, **shear strength** of a soil may be defined as the resistance of the soil to withstand shear stress. Numerically, shear strength equals the shear stress induced under loading in a material, at failure.

All stability analysis in soil mechanics, whether they relate to the foundations of buildings and bridges or to the embankment and cut slopes or to the earthen dams, etc., involve a basic knowledge of the shearing properties and shearing resistance of the soil in question. An understanding of the shearing properties and determination of the shearing resistance of a soil is not an easy proposition since the shearing resistance of a soil constitutes basically of,

1. resistance due to the interlocking of particles,
2. frictional resistance between the individual soil particles, which may be due to either sliding friction or rolling friction or both,
3. adhesion between the soil particles.

Efforts have been made to understand the shearing characteristics of soils and some theories have been evolved to describe these characteristics and are in use all over the world. The hypotheses on which these theories are based take into account almost all the factors that govern the shear resistance of soils although none of the theories can be called *perfect* in the right sense of this word. Discussion in the following pages is restricted to only two criteria—the Mohr—Coulomb and the Hvorslev, failure criteria. Section 10·3 describes the Mohr—Coulomb failure theory which finds a general application. The Hvorslev failure criterion which is applicable to remoulded cohesive soils only is described in section 10·5.

10·2. Principal Planes and Stresses : and Mohr's Stress Circle

When load is applied to a material it induces in general, a normal and a tangential stress on every plane through a point within the material, The normal or direct stress can be compressive or tensile. The tangential stress is the shear stress. The compressive stress may be treated as positive and the tensile stress as negative. The normal or direct stress may be denoted by the letter σ (Sigma) and the tangential or shear stress by τ (Tau).

Through a single point in the stressed material, infinite number of planes can pass. A student of mechanics knows that through a point, there always pass three mutually *orthogonol planes*, called the **principal planes** and the property of these principal planes is that the tangential or shear stress on each of them is zero. So, these three planes only have normal stresses. These normal stresses are consequently termed as its principal stresses. The plane with the *maximum* normal stress on it is called the **major principal plane**, the plane with the *mininum* normal stress on it is called the **mino- principal plane** and the plane with the intermediate value of stress is called the **Intermediate principal plane**. Similarly, the normal stresses on them are termed as major, minor and intermediate principal stresses and are usually designated by the letters σ_1, σ_3 σ_2 respectively. In calculations of stresses induced in soils, intermediate principal stress is of no consequence.

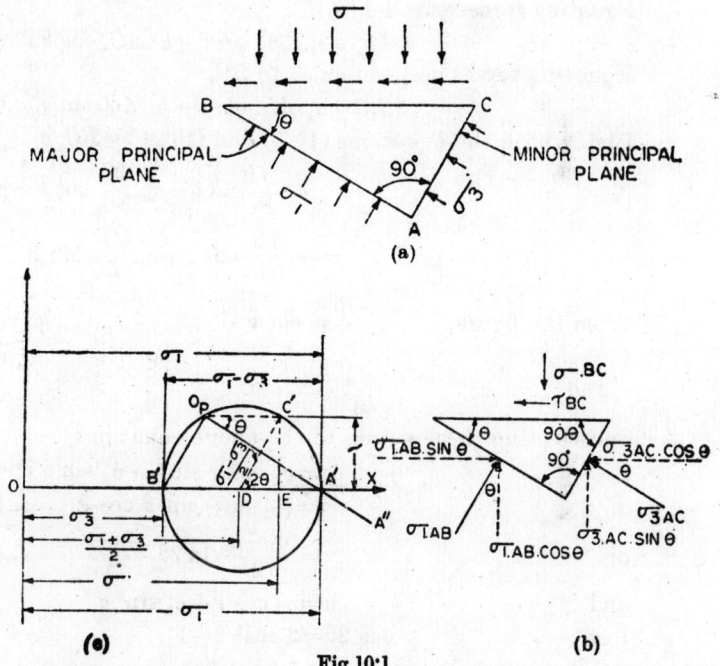

Fig 10·1.

Refer to Fig. 10·1 (a). *AB* represents the trace of a major principal plane and *AC* that of the minor principal plane at a point in a stressed soil mass. The major and minor principal stresses are designed as σ_1 and σ_3 respectively. The planes are shown mutually orthogonal (at right angles). The intermediate principal plane is the plane of the paper. Let *BC* be a plane inclined at an angle θ to the direction of the major principal plane and let this be the plane on which the normal and shear stresses are needed.

Taking unit distance perpendicular to the plane of the paper,

Normal force on $\quad AB = \sigma_1.AB$

Normal force on $\quad AC = \sigma_3.AC$

Let σ be the normal stress and τ the shear stress on the plane BC. Again considering unit distance perpendicular to the plane of the paper.

Normal force on $\quad BC = \sigma \cdot BC$ and

Shear force on $\quad BC = \tau.BC$

A force diagram can be drawn from the above value as shown in Fig. 10·1 (b). The forces $\sigma_1.AB$ and $\sigma_3 AC$ each can be resolved into components, one parallel to the plane BC and another perpendicular to BC and then the forces parallel to and perpendicular to BC can be equated for static equilibrium.

Equating forces parallel to BC.

$$\tau \cdot BC = \sigma_1.AB. \sin \theta - \sigma_3.AC. \cos \theta \quad ...(10.1)$$

Equating forces perpendicular to BC,

$$\sigma.BC = \sigma_1.AB. \cos \theta + \sigma_3.AC. \sin \theta \quad ...(10·2)$$

Divide both the equations (10·1) and (10·2) by BC

$$\therefore \qquad \tau = \sigma_1 \frac{AB}{BC} \sin \theta - \sigma_3 \cdot \frac{AC}{BC} \cos \theta$$

and

$$\sigma = \sigma_1 \frac{AB}{BC} \cos \theta + \sigma_3 \cdot \frac{AC}{BC} \sin \theta$$

From the figure, $\qquad \dfrac{AB}{BC} = \cos \theta$

and $\qquad \dfrac{AC}{BC} = \sin \theta$

Substituting these values in the above equations,

$$\tau = \sigma_1 \cos \theta \sin \theta - \sigma_3 \sin \theta \cos \theta$$

or $\qquad \tau = (\sigma_1 - \sigma_3) \sin \theta \cos \theta \qquad ...(10.3)$

or $\qquad \tau = \dfrac{\sigma_1 - \sigma_3}{2} \sin 2\theta \qquad ...(10·4)$

and $\qquad \sigma = \sigma_1 \cos^2\theta + \sigma_3 \sin^2 \theta$

But $\qquad \cos 2\theta = 2 \cos^2 \theta - 1$

$\therefore \qquad \cos^2 \theta = \dfrac{1 + \cos 2\theta}{2}$

and also $\qquad \cos 2\theta = 1 - 2 \sin^2 \theta$

$\therefore \qquad \sin^2 \theta = \dfrac{1 - \cos 2\theta}{2}$

Substitute these values of $\cos^2 \theta$ and $\sin^2 \theta$ in the equation for σ.

$$\therefore \qquad \tau = \sigma_1 \frac{1+\cos 2\theta}{2} + \sigma_3 \frac{1-\cos 2\theta}{2}$$

or $\qquad \sigma = \dfrac{\sigma_1+\sigma_3}{2} + \dfrac{\sigma_1-\sigma_3}{2}\cos 2\theta \quad$ ...(10·5)

The value of τ and σ given by equations (10·4) and (10·5) can be represented diagrammatrically also. Take a set of two axes OX and OY mutually, perpendicular to one onother as shown in Fig. 10·1 (c). Let the normal stresses be represented on the x-axis and the shearing stresses on the y axis. Set off a distance $OB' = \sigma_3$ and $OA' = \sigma_1$ on the x-axis. Points A' and B' on this axis then represent the stress conditions on the major and minor principal planes. as there are no shearing stresses on these planes and hence y-offsets are zero in each case. With $B'A'$, as the diameter, draw a circle. Through $B,'$ draw a line $B'O_p$ parallel to the minor principal plane AC, cutting the circle in O_p. Through O_p draw another line O_pA'' parallel to the major principal plane AB. Since perpendiculars to two mutually perpendicular lines (in this case major and minor principal planes) are also mutually perpendicular, therefore, angle $B'O_pA''$ is a right angle and O_pA'' passes through the point of intersection A' of the circle with the x-axis. Again through $O_p$draw a line O_pC' parallel to the plane BC. Angle AO_pC' is θ, Join C' with the centre D of the circle. Then angle $C'DA'$ would be equal to 3θ, because the angle made by the are $A'C'$ would be equal to 2θ, becaus the angle made by the arc $A'C'$ at the centre will be twice the angle made by the same arc at the circumference. The co-ordinates of point C' in this diagram represent the stresses on the plane BC under consideration. Point O_p represent the *origin of planes*. Draw $C'E$ perpendicular to the x-axis. Then OE represents the normal stresses on the plane BC and $C'E$ represents the shear stress on the plane BC. A circle of the type shown in Fig. 10.1 (c) is called **Mohr's stress circle** or simply **Mohr's circle**. Radius DC' of Mohr's circle is.

$$\frac{\sigma_1-\sigma_3}{2}$$

and the distance OD is

$$\frac{\sigma_1+\sigma_3}{2},$$

$$\therefore \qquad OE = OD + DE$$
$$= OD + DC' \cos 2\theta$$

or $\qquad OE = \dfrac{\sigma_1+\sigma_3}{2} + \dfrac{\sigma_1-\sigma_3}{2} \cos 2\theta = \sigma$, as

found analytically in the equation (10·5);

and $\qquad C'E = DC' \sin 2\theta = \dfrac{\sigma_1-\sigma_3}{2} \sin 2\theta = \tau$, as,

found analytically in equation (10·4).

In practice, what is done is this. Point B' and A' are marked (See Fig. 10·2) on the x-axis from the principal stress values. These two points, in fact, indicate the stress conditions on the minor, and major principal planes respectively, since the shear stresses on these principal planes are zero. Taking $B'A'$ as the diameter, a circle is drawn, with centre at D. Through D is drawn a line DC' at an angle 2θ to the x-axis, where θ is the angle of inclination of the plane BC where stress conditions are required, with the major principal plane AB. Then point C', so located on the circle gives the stress values on the plane BC.

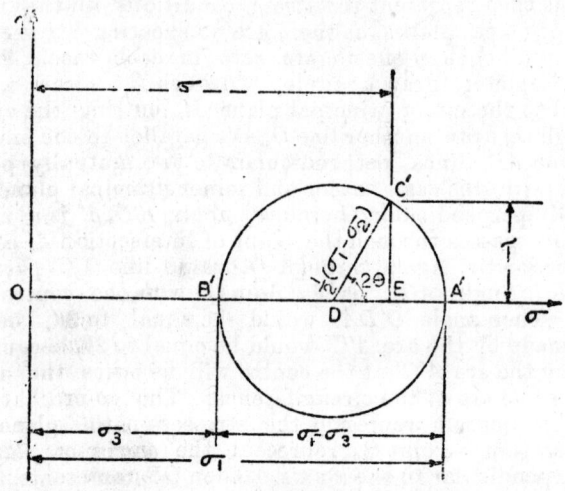

Fig. 10·2.

It is necessary to point out here that in order to get the point (in this case C') that will represent stress conditions on the plane under consideration [in this case plane BC of Fig. 10.1(a)], it is essential to draw the angle at the centre of the circle (in this case D) with the line joining the centre with the point (in this case A') representing the stress conditions on the principal plane (in this case major principal plane AB) with which the plane under consideration makes an angle θ.

It may be noted that the shear stress on BC in Fig. 10.1(a) is counter clockwise and in the Mohr's stress circle, it is shown as a positive ordinate. Thus it may be concluded that the counterclockwise shear forces may be taken as positive and clockwise forces as negative.

A number of basic relationships are evident from a Mohr's stress circle. Some of them are:

(1) Maximum shear stress can be equal to half the difference between the major and minor principal stresses $[(\sigma_1-\sigma_3)/2]$ As seen in Fig. 10.3(a), this only means that $2\theta=90°$ or $\theta=45°$ i.e.. the

maximum shear stress equal to $[(\sigma_1 - \sigma_3)/2]$ occurs on planes inclined at 45° to either the major or minor principal plane. This maximum shear stress is also called **principal shear stress.**

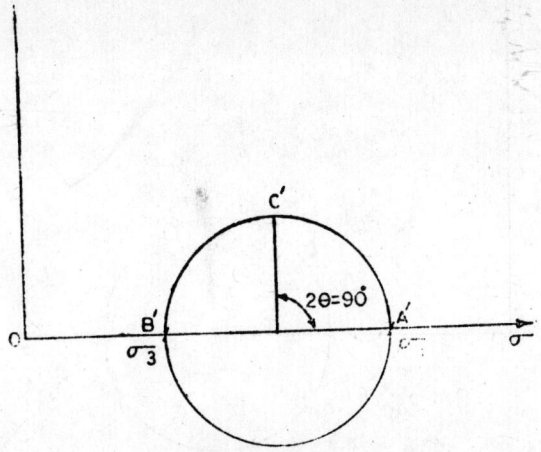

Fig. 10.3 (a).

2. Shear stresses on two mutually perpendicular planes are equal in magnitude but opposite in sign. This is seen in Fig. 10.3 (b)

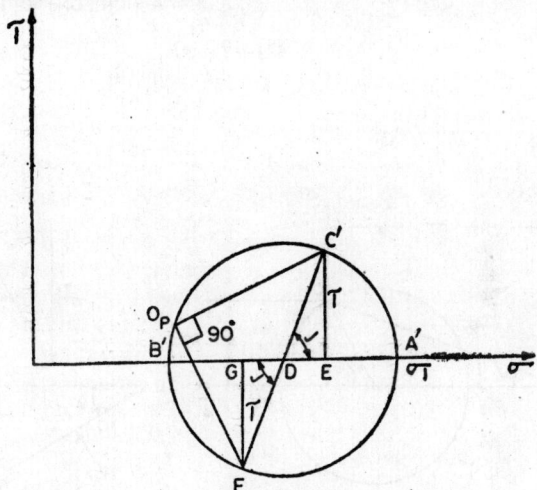

Fig. 10·3 (b)

where O_p is the origin of two mutually perpendicular planes. Points C'' and F represent the stresses on these planes. It is evident that triangles $C'_1 DE$ and FDG are identical and hence $C'E$

equals, *FG*, but the signs are opposite. These stresses are called conjugate shearing stresses.

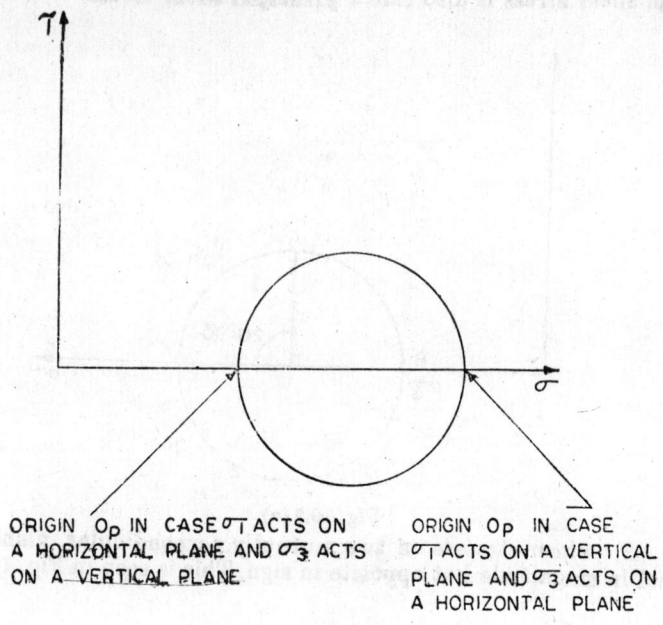

ORIGIN O$_p$ IN CASE σ_T ACTS ON A HORIZONTAL PLANE AND $\sigma_{\overline{3}}$ ACTS ON A VERTICAL PLANE

ORIGIN O$_p$ IN CASE σ_T ACTS ON A VERTICAL PLANE AND $\sigma_{\overline{3}}$ ACTS ON A HORIZONTAL PLANE

Fig. 10·3 (c)

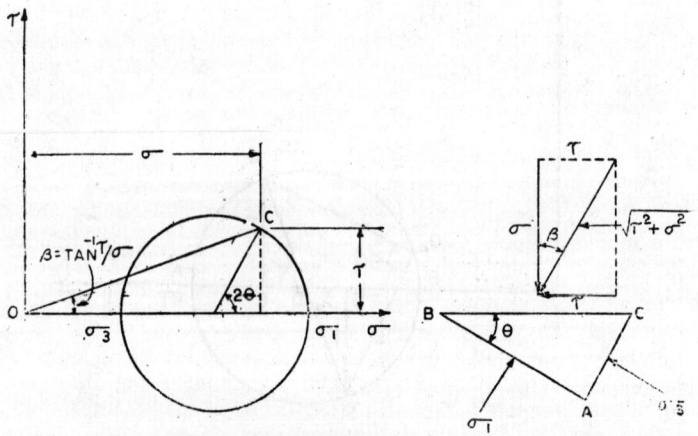

Fig. 10·3 (d)

3. If the major principal plane is horizontal and the minor principal plane is vertical or *vice versa*, the origin of planes lies on the σ-axis. This is seen in Fig. 10.3(c).

4. The resultant stress on any plane is given by $\sqrt{\tau^2 + \sigma^2}$. The angle that the resultant force on a plane makes with the normal to that plane is called angle of obliquity. This angle of obliquity in the Mohr's circle is given by β where β equals $\tan^{-1}(\tau/\sigma)$ as shown in Fig. 10 3(d) and is obtained by joining the point C', whose co-ordinates represent the state of stresses on the plane BC, with the origin 0.

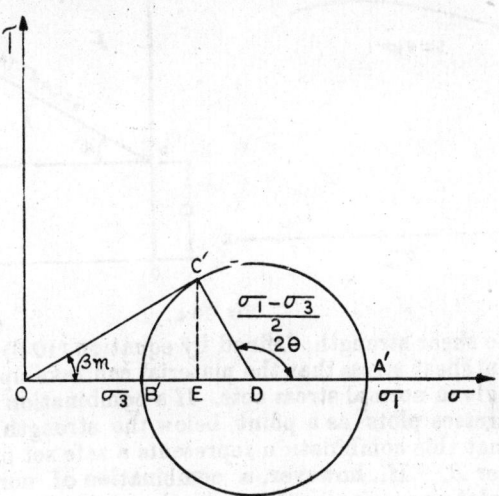

Fig. 10·3 (e)

5. Maximum of all the possible obliquity angles on the various planes is called the maximum angle of obliquity and may be designated as β_m. This maximum obliquity angle β_m is obtained by drawing through O, a tangent to the Mohr's circle. This can be seen in Fig. 10.3(e).

10.3. Mohr-Coulomb Failure Theory

Mohr-Cuolomb failure theory has been successfully applied to the failure of soils. Accord ng to this theory, the failure occurs when the shear stress on any plane equals the shear strength of the material. Moreover, the shear strength on any plane is a function of the normal stress σ on that plane, or

$$s = F(\sigma)$$

If the normal and shear stresses are plotted along the x-axis and y-axis respectively, then the above equation plots as a curve shown in Fig. 10.4(a). This plot is called the *strength envelope*.

Coulomb defined the function $F(\sigma)$ as a linear function of σ and put forth the strength equation as

$$s = c + \sigma \tan \phi \qquad \qquad ...('0.6)$$

This equation is the equation of a straight line and is shown plotted in Fig. 10.4(b). In this plot c is the intercept on the y-axis or the strength axis and $\tan \phi$ is the slope of the line. The parameters c and ϕ in the above equation are usually termed as the *cohesion*[1] and *angle of internal friction*[2] of the soil.

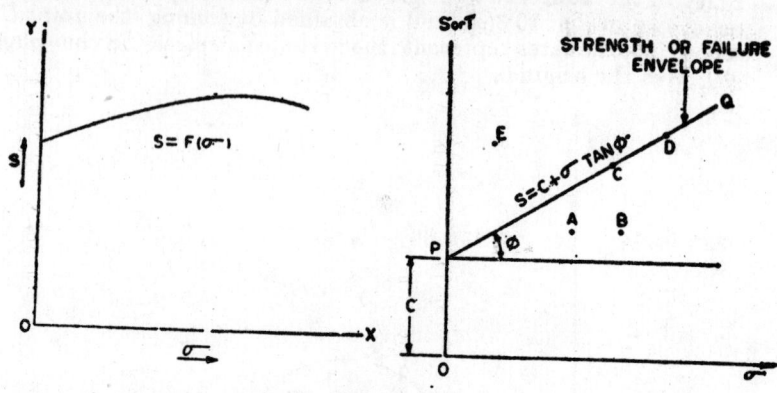

Fig. 10·4

The shear strength defined by equation (10·6) represents the maximum shear stress that the material can take up on a plane on which a given normal stress acts. If a combination of normal and shear stresses plots as a point below the strength envelope, this means that this combination represents a safe set of stresses, e. g. point A or B. If, however, a combination of normal and shear stresses plots as a point on the strength envelope e. g. point C or D, this means that the failure has occurred and the shear stress here is equal to the shear strength of the material under this normal stress. A point such as E can never exist because the stress in the material cannot exceed its strength. Hence, the strength envelope defines the limiting case of stresses and is, therefore, alternatively called the *failure envelope*. The normal and shear stresses at failure are usually designated by σ_f and τ_f respectively and for the failure conditions equation (10·6) may be written

$$\tau_f = c + \sigma_f \tan \phi \qquad \qquad \ldots (10.7)$$

If the principal stresses at failure in a stressed medium are considered, then the Mohr's circle for failure must touch this failure envelope at some point and this point of tangency represents the normal and shear stresses on the failure plane, which is inclined at an angle θ with the direction of principal plane. This situation is explained in Fig. 10.5 where for the circle A, σ_{1A} and σ_{3A} represent the principal stresses at failure in a stressed medium and the failure envelope touches this circle at the point X. Therefore, the co-ordinates of X must give the normal and shear stresses at failures on a plane inclined at an angle of θ to the major principal plane. Either the latter stresses can be measured directly from this figure to the scale to which it is drawn or equations

---

1,2. The qualities c and ϕ in this equation are not the properties of the material, as they vary with the test conditions. For details see Article 10.4.

(10.4) and (10.5) can be used to evaluate them analytically. Now, if by some physical means, conditions of loading and consequently

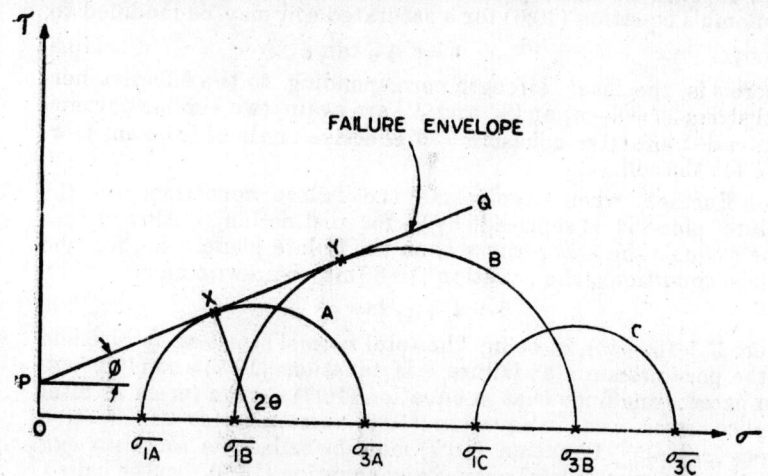

Mohr's circles and failure envelope

Fig. 10·5

the stress conditions in the medium are changed so that, at failure the principal stresses are σ_{1B} and σ_{3B} than another Mohr's circle at failure can be obtained, as shown in this figure. This circle must also touch the failure envelope and the point of tangency Y on this circle again must represent the failure conditions under the changed loading conditions. So, corresponding to each set of failure conditions, there is a point of tangency on the corresponding Mohr's circle.

If, however, the principal stresses in the medium do not correspond to failure condition, as for example the stresses σ_{1C} and σ_{3C} shown in Fig. 10.5, the corresponding Mohr's circle will not touch the failure envelope as is seen for the circle C.

In fact, the failure envelope is drawn from either a set of readings that has number of combinations of normal and corresponding shear stresses at failure which give points like C and D shown in Fig. 10.4, which are joined to get the failure envelope PQ or from a set of readings that has number of combinations of major and minor principal stresses at failure which give Mohr's stress circles like A and B shown in Fig. 10·5, from which a tangent PQ is drawn to represent the failure envelope.

How to obtain these combinations for soils is described in Article 10·7 under measurement of shear strength.

When the above concepts are to be applied to soils, it has to be kept in mind that for them, essentially two types of stresses have been defined in article 5.6, the total stresses and the effective stresses—related to each other by the pore-pressure u. Terzaghi pointed out (1) that the shearing resistance of saturated soils depends exclusively on the "effective normal stress". So, the Coulomb's equation (10.6) for a saturated soil may be modified to;

$$s = c' + \bar{\sigma} \tan \phi' \qquad \ldots(10.8)$$

where s is the shear strength corresponding to the effective normal stress $\bar{\sigma}$ ($= \sigma - u$) and c' and ϕ' are again two similar parameters called **effective cohesion** and **effective angle of internal friction** for the soil.

Further, when σ represents the failure conditions on the failure plane it is replaced by σ_f for distinction, Also in that case s equals the shear stress τ_f on the failure plane. So, for the failure conditions, the equation (10.8) may be rewritten as

$$\tau_f = c' + \bar{\sigma}_f \tan \phi' \qquad \ldots(10.9)$$

where $\bar{\sigma}_f = (\sigma_f - u_f)$, σ_f being the total normal stress at failure and u_f the pore-pressure at failure. If equations (10.7) and (10.9) are compared, one finds that in equation (10.7) σ_f is in terms of total applied stress and in equation (10.9) σ_f is in terms of effective stress ($\sigma_f - u$). Equation (10.7) may be called the *shear strength equation* in terms of *total stresses* and equation (10.9) may be called the *shear strength equation* in terms of *effective stresses*. Accordingly c and ϕ may be called total stress parameters and c' and ϕ' effective stress parameters.

10.4. Significance of c and ϕ in the Coulomb's Equation

Tests for shear strength measurements are given in Article 10.7. A glance through the direct shear test would indicate that the shear stress value τ_f at failure for a soil with a given water content and under equal test conditions but for different values of normal failure stress σ_f on a horizontal plane can be represented, in most of the cases by,

$$\tau_f = c + \sigma_f \tan \phi$$

The test conditions include the following :—

(1) type and size of the shear apparatus.

(2) duration of applied normal stress.

(3) duration during which the shear stress is increased from zero to failure point.

It has been proved that (2) these test conditions have a dominating influence on the c and ϕ values in the Coulomb's equation. Table 10.1 gives the values of c and ϕ for a silty Vienna clay. The clay before testing was remoulded at a moisture content equal to L. L. and was brought to an over-consolidated state and had an initial void ratio of 0.77 prior to shear test.

A further insight into the variation in c and ϕ is provided by Fig. 10.6(a) and (b). Fig. 10.6(a) is the $e-p$ curve of a soil on a natural scale. The corresponding shear strength curves are shown in Fig. 10.6(b). Points A, B, C on the virgin curve represent the void ratio of three samples at failure during shear testing

Table 10·1
Test Conditions and Values of c and ϕ

S. No.	Test conditions	c	ϕ	Type of test
1.	Rapid increase of both σ cand τ to prevent any drainage. Initial void ratio constant at 0·77.	0.58 Kg./cm²	0°·50′	Quick Test
2.	Complete consolidation under different values of σ followed by rapid increase of τ.	0·58 Kg.cm²	13°·30′	Consolidated quick test
3.	Complete consolidation under different values of σ, followed by an extremely slow increase of τ to failure.	0.18 Kg./cm²	24°·0′	Slow Test (No pore-pressure during shear)

and the corresponding values of normal and shear stresses are indicated by the corresponding points a, b, c on envelope I corresponding to this virgin curve. This envelope passes thr ugh the origin. Points D, E and F on the rebound curve represent the void ratios of three other samples at failure during similar testing and the corresponding values of normal and shear stresses are indicated by the corresponding points d, e and f on envelope II corresponding to this rebound curve. The second envelope does not pass through the origin, So, apparently one may conclude that a normally consolidated soil *does not possess cohesion* and its entire strength is contributed by friction only which is reflected in terms of angle of internal friction ϕ_1, whereas an over-consolidated soil, on the other hand, *possesses cohesion* and its strength is contributed both by cohesion reflected in the value of c and *friction* reflected in terms of angle of internal friction ϕ_2. Further the same soil has apparently two angles of internal friction ϕ_1 and ϕ_2. *These apparent conclusions are not true.* In fact, it is the stress history of the soil i. e., whether the soil is normally consolidated or over consolidated which forms a part of the test conditions that has lead to such conclusions.

As a conclusion, one may say that c and ϕ are not properties of the soil, since they vary with the test conditions. They are only two parameters that describe the strength of the soil under the given test conditions.

10.5. Hvorslev's Failure Criterion[1].

The Hvorslev's failure criterion finds its application only to remoulded cohesive soils. It was evolved, to start with, as an explanation to the variation in c & ϕ values as obtained fr n the

---

1. This failure criterion is sometimes also called Mohr-Coul mb-Hvorslev failure criterion, as explained in the very first paragra, under this article.

Coulomb's equation. Because of the resemblance of Hvorslev's equation with the Coulomb's equation and the fact that it can be evolved from the Mohr's stress circles also, sometimes the Hvorslev's failure criterion is called the Mohr-Coulomb-Hvorslev failure criterion.

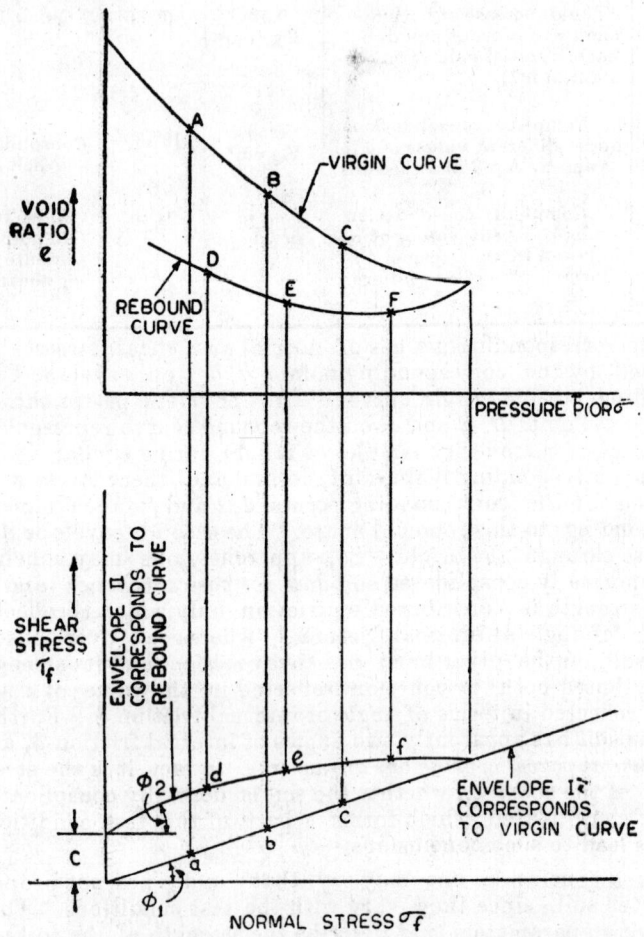

Results of Direct shear test on consolidated soil samples
Fig. 10.6

Hvorsolev (3) by his experimental work on remoulded cohesive soils found that irrespective of the stress history of a soil, its shear strength can be represented by ;

$$\tau_f = F(e, \sigma_f) \qquad (10.10)$$

where τ_f shear stress on the failure plane at failure, equal to the shear strength of the soil.

$F = (e, \sigma_f) = $ A function of the void ratio e at failure and the normal stress σ_f.

More explicitly the equation reads as

$$\tau_f = c_e + \overline{\sigma}_f \tan \phi'_e \qquad \qquad ...(10.11)$$

where $\tau_f = $ Shear stress on the failure plane at failure equal to the shear strength of the soil.

$c_e = $ Effective cohesion, generally called *"true cohesion"*.

$\overline{\sigma}_f = $ Normal stress on the failure plane at failure.

$\phi'_e = $ Effective angle of internal friction, generally called *"true angle of internal friction"*.

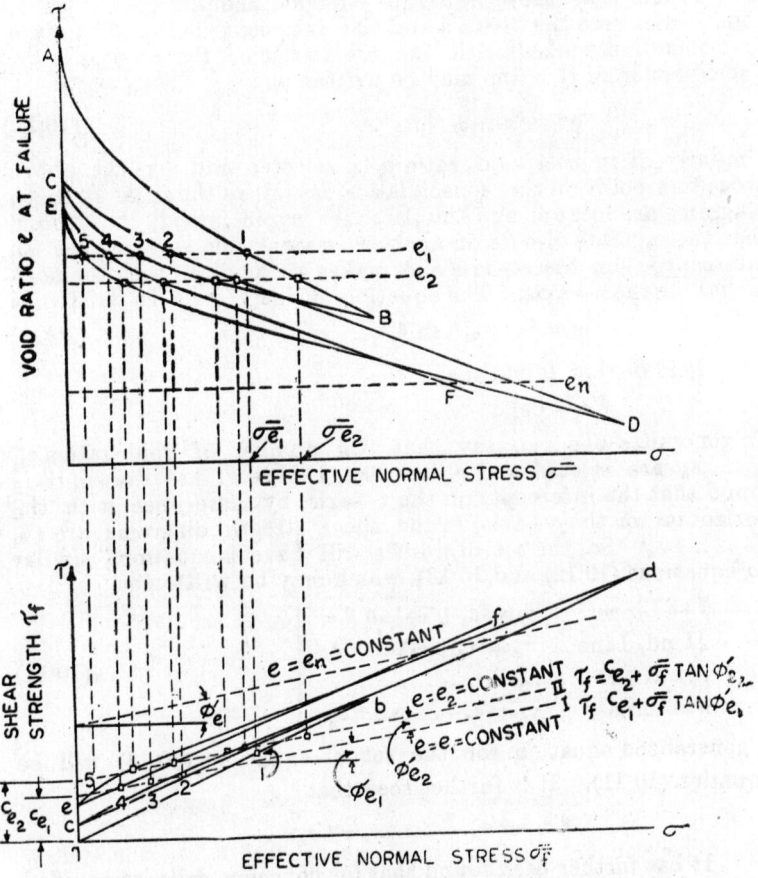

Explanation of Hvorslev's failure criterion
Fig. 10·7.

Fig 10.7 (a) and (b) explain the above equation very well. Fig. 10.7 (a) is a pressure ($\bar{\sigma}$) — void ratio (c) diagram for a soil with five legs, showing its normal and over-consolidation states. Fig. 10.7 (b) shows the corresponding shear strength envelopes drawn from direct shear tests. Now if one void ratio say e_1 is chosen and a horizontal line drawn at this void ratio as shown in Fig. 10.7 (a), it cuts the five consolidation curves at five points designated as 1, 2, 3 4, and 5, lying on different legs. Corresponding to these points, one can prepare for shear testing five laboratory sample having the same void ratio but different stress history in each case. Verticals from each of the five points on the consolidation diagram are drawn and each vertical cuts the corresponding shear strength curve in Fig. 10.7 (a). When the set of points in the shear strength diagram are joined, it is found that they lie on a straight line designated as I. Line I intersects the τ—axis and the intercept is c_{e_1}. The angle that this line makes with the horizontal or the σ—axis is $\phi'e_1$ The equation of this line may be written as :

$$\tau_f = c_{e_1} + \bar{\sigma}_f \tan \phi'e_1 \qquad \qquad ...(10.12)$$

Similarly, if another void ratio e_2 is selected and by the above procedure point on the consolidation as well as the shear strength diagrams are locat ed and the later set again joined, it is found that these points also lie on another straight line II that has an intercept c_{e_2} on the τ—axis and makes an angle ϕ_{e_2} with the horizontal or the σ—axis. The equation of line II may be written as

$$\tau_f = c_{e_2} + \bar{\sigma}_1 \tan \phi'e_2 \qquad \qquad ...(10.13)$$

It is further found that

$$\phi'e_1 = \phi'e_2$$

To generalize one can say that if n number of void ratios e_1, e_2......e_n are seleted and n lines like, I, II, etc. are drawn, it is found that the intercepts on the τ—axis, by these lines with the horizontal or the σ—axis, in the shear stregth diagrams are $\phi'e_1$ $\phi'e_2$......$\phi'e_n$. So, the set of n lines will have n equations similar to equations (10.12) and 10.13). which may be written as :

$$\left.\begin{array}{ll} \text{I st Line} & \tau_f = c_{e_1} + \bar{\sigma}_f \tan \phi'e_1 \\ \text{II nd, Line} & \tau_f = c_{e_2} + \bar{\sigma}_f \tan \phi'e_2 \\ .... \\ n\text{th Line} & \tau_f = c_{e_n} + \bar{\sigma}_f \tan \phi'e_n \end{array}\right\} \quad ...(10.14)$$

A generalized equation for the set of equations (10.14) will be equation (10.11). It is further seen that

$$\phi'e_1 = \phi'e_2.............,= \phi'e_n$$

It has further been found that for cohesive soils remoulded near the liuid limit, as in the present case,

$$\left.\begin{array}{l} \dfrac{c_{e_1}}{\overline{\sigma}_{e_1}} = k \\[2mm] \dfrac{c_{e_2}}{\overline{\sigma}_{e_2}} = k \\[2mm] \quad\cdots\cdots\cdots \\[2mm] \dfrac{c_{e_n}}{\overline{\sigma}_{e_n}} = k \end{array}\right\} \qquad \ldots(10\cdot15)$$

where k is a constant of proportionality and $\overline{\sigma}_{e_1}, \overline{\sigma}_{e_2}, \ldots\ldots\overline{\sigma}_{e_n}$ are the respective *equivalent consolidation pressures* for the void ratios $e_1, e_2\ldots\ldots e_n$. k in the above set of equations is called **cohesion factor**. It is necessary to define, here, the term *equivalent consolidation pressure*. The ebuivalent consolidation pressure for a soil sample is the pressure on the virgin curve corresponding to the void of thesample. as for example, the equivalent consolidation pressure $\overline{\sigma}_{e_1}$ for the samples, corresponding to points 1, 2, 3, 4 and 5 havinga void ratio e_1 is the pressure projected from point 1 in Fig. 10·7 (a). Similarly, equivalent consolidation pressure for any sample with a known void ratio can be pro- jected from the virgin curve. It may be noted that for one void ratio, there is only one value of equivalent consolida- tion pressure. For a normally consolidated sample, the equiva- valent consolidation pressure is the actual consolidation pres- sure but for an over- consolidate sample, it has to be projected from the virgin curve, againt the void ratio of the sample, as indi- cated above. If c_{e_1}

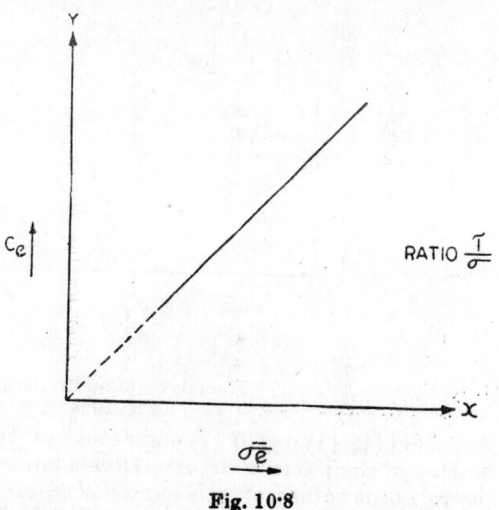

Fig. 10·8

$c_{e_2}\ldots\ldots c_{e_n}$ as obtained from Fig. 10·7 (b) are ploted on the y-axis and $\overline{\sigma}_{e_1}, \overline{\sigma}_{e_2}\ldots\overline{\sigma}_{e_n}$ also obtained from the same figure are plotted on the x axis as shown in Fig. 10·8, it can be seen that a straight line is obtained and the slope of this straight line is the constant of proportionality k. This plot establishes the relation in the set given by equation (10·15).

If, now, the values of $c_{e_1}, c_{e_2}\ldots\ldots c_{e_n}$ from the set of equations (10·15) are substituted in the corresponding equations in the set (10·14) and instead of $\phi'_{e_1}, \phi'_{e_2}\ldots\ldots\phi'_{e_n}, \phi'_e$ is substituted, then these n equations for the n lines may be writen as ;

Ist Line $\qquad$ $\tau_f = k \bar{\sigma}_{e_1} + \bar{\sigma}_f \tan \phi'_e$

2nd Line $\qquad$ $\tau_f = k\bar{\sigma}'e_2 + \bar{\sigma}_f \tan \phi'_e$ $\left.\right\}$ ...(10.15) (a)

....

nth Line $\qquad$ $\tau_f = k\bar{\sigma}e_n + \bar{\sigma}_f \tan \phi'_e$

A generalized equation for the set (10.15) (a) would be ;

$$\tau_f = k\bar{\sigma}e + \bar{\sigma}_f \tan \phi'_e \qquad \qquad ...(10.15)$$

which is the same equation as equation (10.11) except that c_e has been replaced by $k\bar{\sigma}_e$

Divide both sides of equation (10.16) by σ_e

$$\therefore \qquad \frac{\tau_f}{\bar{\sigma}e} = k + \frac{\bar{\sigma}_f}{\bar{\sigma}_e} \tan \phi'_e \qquad \qquad ...(10.17)$$

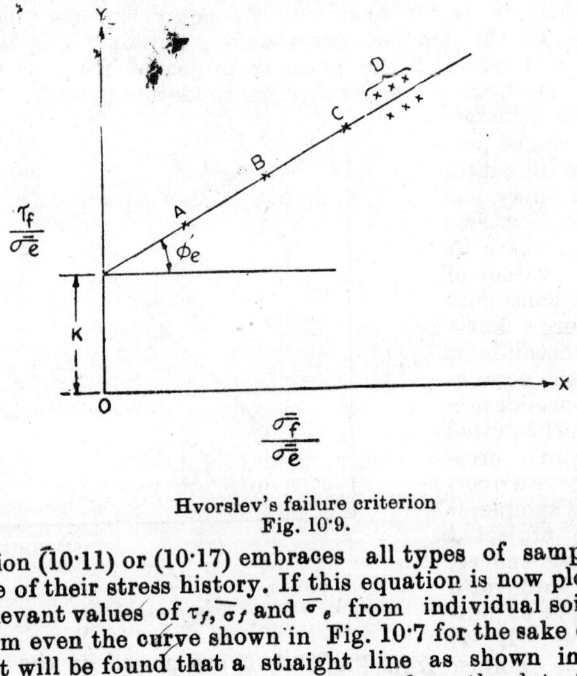

Hvorslev's failure criterion
Fig. 10.9.

Equation (10.11) or (10.17) embraces all types of samples irrespective of their stress history. If this equation is now plotted with the relevant values of $\tau_f, \bar{\sigma}_f$ and $\bar{\sigma}_e$ from individual soil samples (or from even the curve shown in Fig. 10.7 for the sake of illustration) it will be found that a straight line as shown in Fig. 10.9 results. The points A, B and C emerge from the data from over-consolidated samples and a set of points D emerge from normally consolidated samples. The slope of this line is defined by the angle ϕ'_e and the intercept on the $\frac{\tau_f}{\sigma_e}$ axis is the constant k.

It can be seen that k and ϕ'_e in Hvorslev's equation, unlike c and ϕ in Coulomb's equation are constants for the soil under consideration. That is why c_e and ϕ'_e are termed as **true cohesion** and **true angle of internal friction.** k and ϕ'_e are called Hvorslev's parameters.

The failure criterion given by equation (10·11) or by equation (10.17) is an important concept in the field of soil mechanics, although for practical problems one does not have to derive the Hvorslev's parameters. For a given set of field conditions, it is sufficient to arrive at the c and ϕ values by the Coulomb's equation and work out the practical problems based upon these values. However, most of the fundamental research on clays is evaluted in terms of the Hvorslev's parameters.

Evaluation of the k and ϕ'_e values from the data obtainable from a triaxial test has been described in Article 10·8.

10.6. Drainage Conditions in Shear Testing

Before brief descriptions of shear test are given it is felt necessary to indicate the drainge conditions before and during the test, that in-fiuence the results of a shear test. The soils specially the fine-grained soils are normally tested for shear strength when they are in a state of complete saturation. Sands, however, are sometimes tested in a dry of complete saturation. Sands, however, are sometimes tested in a dry state also. Shear tests for saturated soils are designated for three types of drainage conditions. The tests are named taking into consideration these drainage conditions.

(1) **Unconsolidated Undrained Test :** In this type of test drainage is not permitted at any stage of the test *i.e.* when it is a *dtrect shear test*[1] both during application of normal as well as shear stress, drainage is prevented and when it is a *triaxial compression test*[1], bo\ during application of chamber pressure as well as axial stress, drainage is prevented. As a consequence, the normal pressure or the chamber press exists as a hydrostatic excess pressure and no volume change takes place at any stage. This test is also called **Quick Test**.

(2) **Consolidated-Undrained Test.** In this type of test, complete drainage is permitted during the application of normal stress in a direct shear test or during the application of chamber pressure in a triaxial shear test. As such, volume change in the sample takes place and the strass gets transferred to the soil grains. However, drainage is not permitted during the application of shear stress or axial stress as the case may be. As a consequence, no volume change during shear takes place and this results in the development of pore-pressures. This test is also called **Consolidated Quick Test**.

(3) **Drained Test.** In this type of test, complete drainage is permitted under the application of shear stress. As a consequence, volume changes take place and no pore-pressure developes. The stresses are effective stresses at all times.

Each test gives different resuslts. The choice of the type of the test depends upon the field conditions. A discussion on the results obtainable from these tests and the choice of the type

---

1. For details see Article 10.7.

of the test will be given in the article 10·9, after the discussion of the triaxial compression test.

10.7. Measurement of Shear Strength

Measurement of shear strength of a soil involves essentially obtaining of test data from which the failure envelope could be plotted. Many methods are available for obtaining such data, Most widely used methods in the laboratory are :

(1) Direct Shear Test.

(2) Triaxial Compression Test.

(3) Unconfined Compression Test.

Of late, another test called the laboratory vane shear test has gained popularity as it is quick and simple. Other laboratory tests that are mostly employed for research are the *ring shear test* and the *torsion test*. Vane shear test and another type of test called the *cone test* are also employed in the field.

It is considered sufficient for the purposes of this book to discuss among the laboratory tests the direct shear test the triaixal shear test, the unconfined compression test and the vaue shear test. Other tests less commonly used are not described here. In the following sub sections brief descriptions of the actual test procedures, the reader may refer to the relevant *ISI. ASTM* or *AASHO* standards or any book[1] on laboratory t sting.

(a) **The Direct Shear Test** : The principle involved in a direct sheai test is shown in Fig. 10·10. The soil sample is placed in a metallic box called the *shear box*. The box has two pieces— the upper half and the lower half. One piece is fixed and the

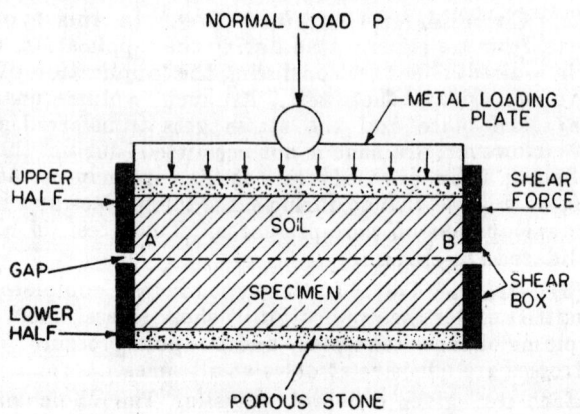

Diagrammatic sketch of principal parts of a direct test apparatus.

Fig. 10.10.

---

1. By far the best book on laboratory testing of the soils is referencee 4 at the end of this chapter.

other half can be displaced from its position by application of a horizontal force. Depeding upon whether a drained test is to be conducted or an undrained test is to be carried out, the sample is covered on either side by by porous stones or metallic gratings respectively. The two halves of the shear box can move with respect to one another. The potential plane of failure is AB. The gap between the two halves of the box depends upon the maximum particle size of the soil being tested. A rough rule is that the gap between the two halves should be sufficient so that the top half does not ride on the largest soil grain that gets between the edges Also loose soils, as explained latter, reduce in volume and this gap should be enough to allow this reduction to take place.

With the help o a yoke, a normal load P is applied to the plane AB and under this normal load. increasing shear force is applied as shown until the soil specimen fails along this horizontal plane AB. A proving ring and a dial gauge measure the shear force and shear displacement[1] respectively.

The usual size of a shear box is 6 cms $\times$ 6cms. Dispalacement between the two halves is either done manually by revolving a handle or there is a motorised system. Since this displacement can be controlled, the apparatus is calle i the *strained-controlled type unit*. However, machines are also available where shear is controlled and such units are called *stress controlled type units*. For routine testing, the former type is usually used.

It is essential that the shear displacement[1] throughout the test is kept constant. The speed of shear displecement or the rate of strain is another big variable, specially for the saturated cohesive soils. It is sufficient to mention here that the present knowledge of the subject indicates that slower the rate of strain, the lower is the shear strongth of a cohesive soil. Full discussions of the effect of the rate of strain on the shear strength of a cohesive soil is outside the scope of the book.

If drainage is permitted in a saturated soil or if the soil being tested is loose or dense sand, volume changes are measured, and the change in the thickness of the soil specimen indicates the change in volome since area of cross-section remains constant.

Data obtained from one sample include the normal load P and a set of readings consisting of the shear force F and change in thickness $\triangle$ against shear displacement δ. From the normal load P, normal stress σ* is calculated by dividing it by the area of the sample at the place AB. The shear force readings F also divided by the area of the sample at the plane AB. The shear force readings F are also divided by the area of the sample to get the values of shear stress τ against the corresponding shear displacements. Then the ratio τ/σ and the change in thickness $\triangle$ are

---

1. Realative displacement between the upper half and lower half of the box is called the shear displacement.

* Since P for a test remains constant, σ is constant for all values of ϕ in one particular test.

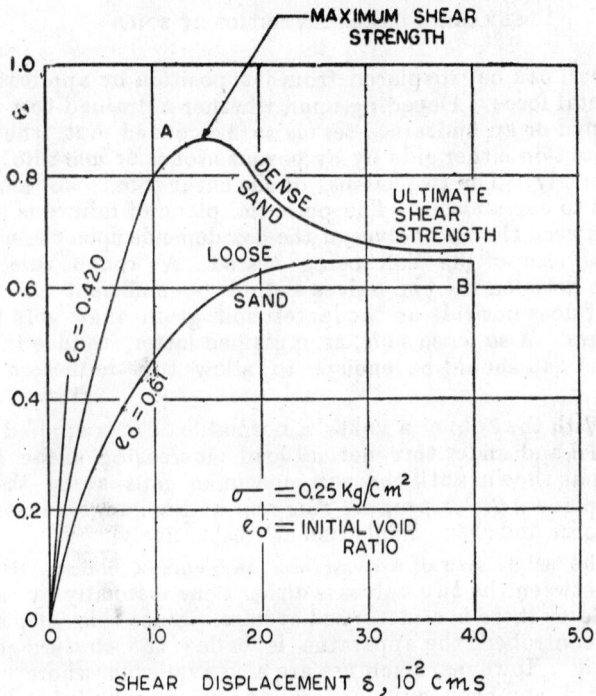

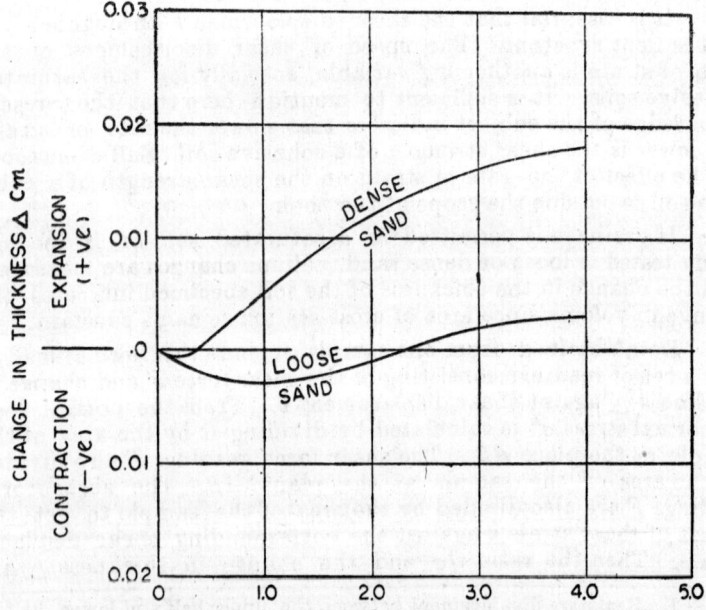

Results of a direct shear test on two samples of sand—one dense sand sample and the other loose sand sample.

Fig' 10'11

plotted against the shear displacement δ in the form of two curves shown in Fig. 10·11 (a) and (b), for either of the two types of soils selected for testing. Failure is considered to have occurred at the peak point A as in the case of dense sand.** In case ratio τ/σ attains a constant value for a large range of deformation, as for the loose sand, then the point B with sufficiently large strain defines the failure. It may be noted that the dense sand after attaining a maximum shear strength at the point A attains an ultimate value lower than the maximum value but closer to the value of shear strenth for loose sand, thus indicating that the dense sand has two values of shear strength—the maximum and the ultimate. Fig. 10·11 (b) shows that dense sand on application of shear stress expands and loose sand contracts for most of the range of shear displacement.

In order to obtain a failure envelope, several samples have to be tested, each under a different normal load P. From the results of such tests, failure shear values (τ_f) are plotted against the corresponding failure normal stress values (σ_f) to obtain the desired envelope

(b) **The Triaxial Compression Test.** Before the triaxial testing equipment or the test procedure is described, it is considered necessary to explain the *principle* involved.

Let a small element of soil ABCD shown in Fig. 10·12 be acted upon by the stresses σ_1 and σ_3. AB, BC, CD and DA in

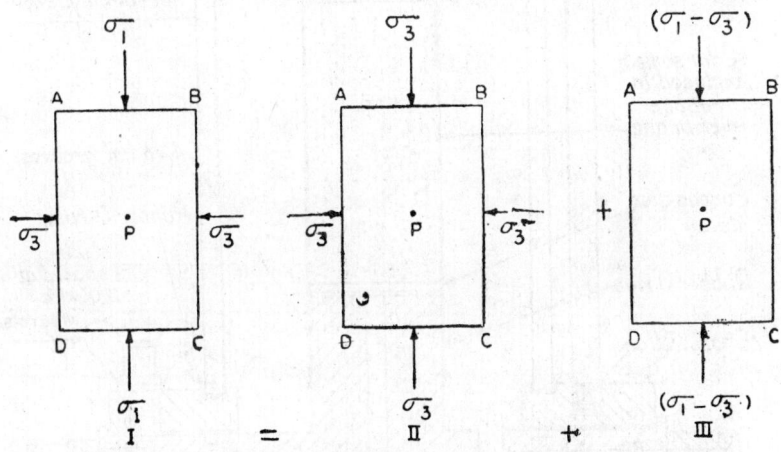

Principal behind triaxial testing.
Fig 10·12.

fact, represent four planes around a point, the depth of the diagram having not been shown. Further, let the planes on which

---

** These curves for the sand are only meant as an illustration. Similar curves for cohesive soils can also be drawn. The shape of the latter type of curves will differ. A discussion on shear characteristics of cohesive soils given in Fig. 10·11 will illustrate this point.

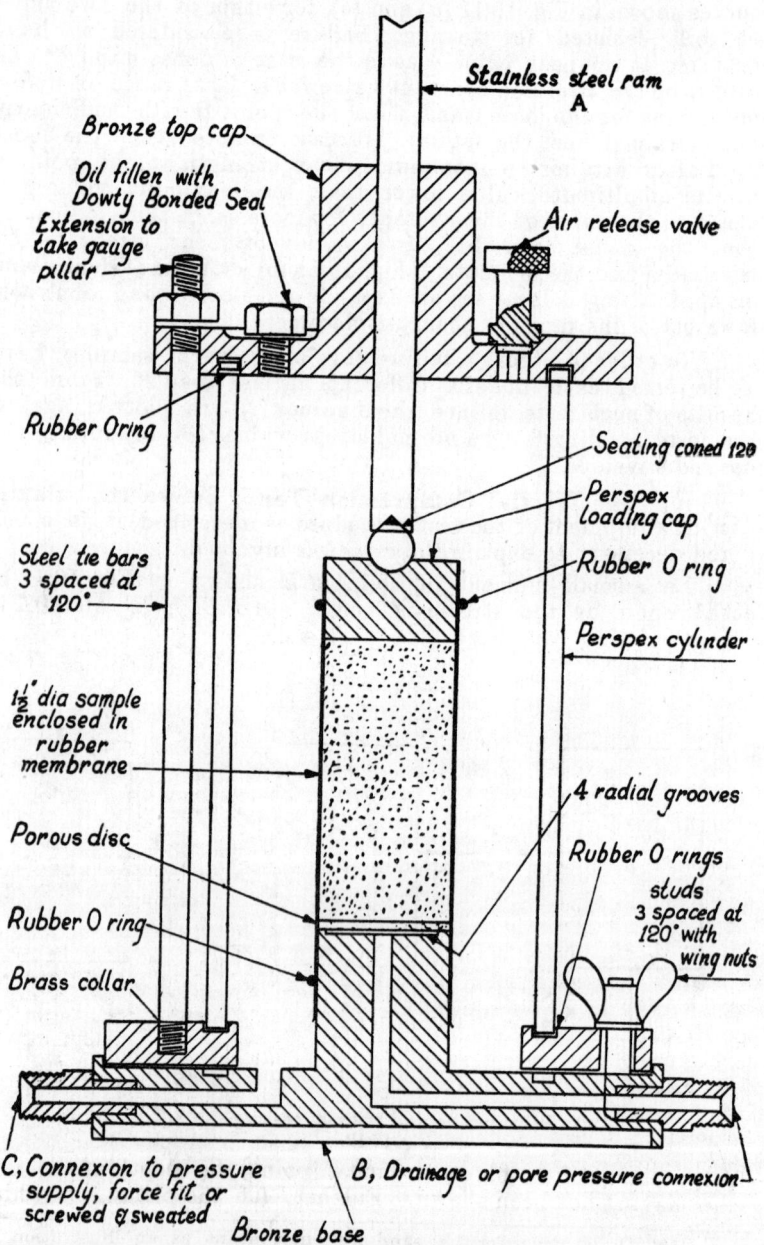

Triaxial testing equipment,

Fig. 10·13

these stresses act be assumed as the principal planes. The horizontal planes AB and CD, therefore, become the major principal planes and the vertical planes AD and BC, the minor principal planes as σ_1 represents the major principal stress and σ_3 represents the minor principal stress. This general stress system shown on the left hand side of the sign of equality can be split up into two systems—an isotropic stress system consisting of an alround stress σ_3 and a uniaxial stress system consisting of a stress[1] $(\sigma_1 - \sigma_3)$ acting on the major principal planes in the vertical direction—shown on the right hand side of the sign of equality. If a soil sample is made to fail under the above two stress system, one can assume that the failure has taken place under the general stress system shown on the left hand side. This fact has been made use of in the design of triaxial testing equipment.

The equipment shown in Fig. 10·13 usually called a *triaxial cell* is meant for a cylinderical soil sample of 3·8 cms (1.5 inches) diameter and a height to diameter ratio ranging between 2 to 2.5, the usual height to diameter ratio[2] being 2. Bigger units in which samples of 7.6 cms diameter can be tested are also available. For routine testing of soils free from stones, equipment shown in Fig. 10.13 is normally employed, since it is more convenient to handle than a bigger unit. The principal features of a unit are :

1. The base.
2. The removable cylinder and top cap.
3. The loading piston.
4. Loading cap and porous stone.
5. Rubber membranes and rubber 0-rings.

The base is a single piece usually machined from brone forg-and has a padestal 3.8 cms. in diameter on which the sample is placed. Provision exists in the base padestal for drainage of the sample. Four pressure connecctions can be made at the base. One connection serves for filling of the pressure cell with some fluid used in the cylinder.[3] Two connections from the padestal on which the sample is placed can be used either to connect two burettes for drainage during consolidation in a consolidated undrained test and also during a drained test, or they can be used to connect the base of the sample to a pressure measuring unit, when the pore-pressure is required to be measured in an undrained test. The fourth connection usually provided in the top cap is used when it is reguired to pass water through the sample for saturation. This connection is also sometimes used to measure pore-pressures at the top of a sample.

---

1. This uniaxial stress $(\sigma_1 - \sigma_3)$ is called the deviator stress.
2. To be more precise the practice is to use a height to diameter ratio of two for clays and three for sands.
3. This cylinder is also called pressure chamber.

The removable cylinder is made from transparent perspex, which helps in setting up the test and enables the exeperimenter to observe the mode of failure. The material of the cylinder can take up a pressure[1] of as much as 40 kg/cm². The top cap to which the cylinder is permanently fixed is a bronze casting. In the centre of this cap is fitted a bush through which the loading piston can pass and rest on the loading cap placed on the sample top. An air release valve is also fitted to this top cap for releasing the entrapped air and there is also an oil filler hole for adding oil before long duration tests.

The loading piston carries the axial load that is applied to the sample and is usually made from stainless steel. The main problem that one faces in the design of the piston and the bushing through which it goes, is the inbetween clearance. The ideal clearance should not allow any leakage, and should at the same time not present any problem of friction between the piston and the bushing. Most of the cells designed and manufactured by the British firms carry a clearance of 0·00762 cms. The Norwegian designed cell popularly called the "Norwegian Cell" carries a further improvement i.e. it has a revolving bush that eliminates most of the friction and the clearance is very small so that there is no leakage in the system.

The loading cap on the sample top through which the axial load is transmitted to the sample, is either made of brass or of transparent perspex. Many designs of the cap are available. The cap has an exact diameter of 3·8 cms and usually carries a hole through it for circulation of water, if needed. Sometimes this top connection is also used to connect a pore pressure measuring unit, in saturated samples. When an undrained test is to be conducted, a plain perspex disc may be used. Whatever the type of design, the upper surface of the cap must carry a conical recess for putting a steel ball to ensure axial loading of the sample.

On the base pedestal, for drainage during consolidation or during drained tests or for measuring pore-pressures in undrained tests, porous disc is placed between the bottom of the sample and the pedestal. The porous disc is usually 0·32 cms. thick and is made from vitrified material.

The sample is enclosed in a thin rubber memberane to seal it from the chamber fluid. The rubber membrane may be made from self-vulcanizing latex or may be purchased commercially. The essential qualities of the membrane should be that it should exert the minimnm restraint to the sample and also should prevent any leakage both from the chamber into the sample and from the sample into the chamber.

The test procedure is fairly simple but needs careful attention on the part of the worker. The prepared sample is placed on the base pedestal. Porous stone is used in case drainage is to be

---

1. Units with higher chamber pressure are also available.

permitted. A rubber membrane covers the sample. Rubber O-rings are used at the two ends to seal off the sample. The perspex cylinder is placed in position. Loading piston is then placed in postion on the loading cap. The chamber is filled with a fluid, usally water or glycerine. Any air within is removed. The required lateral pressure σ_3 is then applied. This pressure is kept constant throughont the loading when the uni-axial stress is under application. It may be noted here that the intermediate principal stress is equal to the minor principal stress, since the fluid exerts pressure equally in all directions. The application of an alround pressure like this corresponds to the stress system designated as II in Fig. 10·12. Uni-axial load from a loading machine is applied to the piston till the sample fails. Application of uniaxial load corresponds to the stress system designated as III in Fig. 10.12. There are two types of loading machines available—one is the strain controlled type and the other is the stress controlled type. For routine testing the strain-controlled type loading unit is normally used. In such a unit the rate of strain is controlled and must be kept constant throughout the test as is done for the direct shear test. The load is transmitted as a reacion. Either the plunger of the loading machine moves down at a constant rate and records the reaction electrically when the position of the cell is fixed or the loading piston rests against a fixed proving ring and the cell is moved upwards at a constant rate.

Typical data obtained from a sample inculde a set of readings consisting of applied axial load against corresponding deformation induced in the sample, when the chamber pressure is kept constant.[1] Where pore pressure are also measured as in

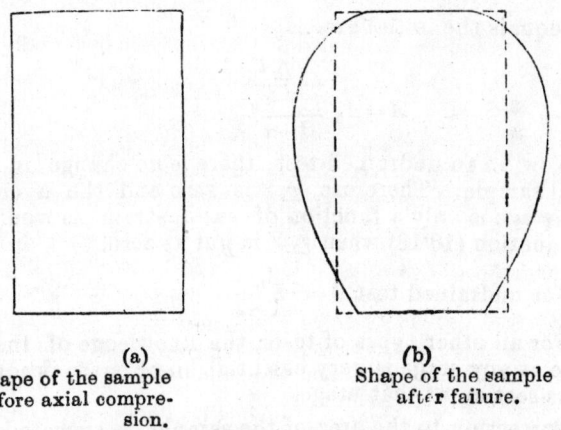

<table>
<tr><td>(a)
Shape of the sample
before axial compre-
sion.</td><td>(b)
Shape of the sample
after failure.</td></tr>
</table>

Fig. 10·14.

1. This data is typical from an axial compression test. Many other types of tests are possible in the triaxial testing equipment and data will accordingly vary. But the description of all types of tests is beyond the scope of this book.

the case of undrained tests, the data include pore pressure rad-diugs also.

It may be noted that initially the soil sample is of uniform area of cross-section. As the deformation proceeds, there is late-ral bulging of the sample and ultimately at failures, the sample assumes a shape like that of a barrel. The typical shapes in eleva-tion of the sample before and after the test are shown iu Fig. 10·14(a) and (b). As seen in Fig. 10·14(b) the cross-sectional area of the sample does not remain uniform. The axial stress or the deviator stress ($\sigma_1 - \sigma_3$) under these circumstances is calculated on the basis of *average cross-sectional area*, obtained by assuming cylindrical shape after failure as shown by the dotted lines in Fig. 10·14(b). The calculations involved in determing average cross-sectional area are as under :

Let A_0 denote the initial cross-section area, l_0 be the initial length and V_0 the initial volume of the sample. If A denotes the average crosssectional area after a change δ in length and $\triangle V$ in volume, then $A (l_0 + \delta)$ must equal $(V_0 + \triangle V)$

i.e.
$$A(l_0 + \delta) = (V_0 + \triangle V)$$

or
$$A = \frac{V_0 + \triangle V}{l_0 + \delta}$$

Divide the numerator by V_0 and the denominator by l_0, then

$$A = \frac{1 + \dfrac{\triangle V}{V_0}}{1 + \dfrac{\delta}{l_0}} \cdot \frac{V_0}{l_0}$$

In axial compression test $\left(-\dfrac{\delta}{l_0} \right)$ denotes the axial strain ϵ and $\dfrac{V_0}{l_0}$ equals the initial aree A_0

$$\therefore \qquad A = A_0 \frac{1 + \dfrac{\triangle V}{V_0}}{1 - \epsilon} \qquad \qquad ...(10·18)$$

Now, in an undrained test, there is no change in volume of the soil sample. Therefore, $\triangle V$ is zero and the average area A at any stage is only a function of axial strain as would be clear from equation (10·18) when $\triangle V$ is put as zero.

For undrained test $A = \dfrac{A_0}{1 - \epsilon}$ \qquad \qquad ...(10·19)

For all other types of tests, the knowledge of the change in volume at any stage is very essential in order to know the area of cross-section at that stage.

Correction to the area of the sample as suggested in equati-ons (10·18) and (10·19) is called **area correction**.

If the axial load P at any deformation δ is known and if A is the cross-sectional area at that stage, then

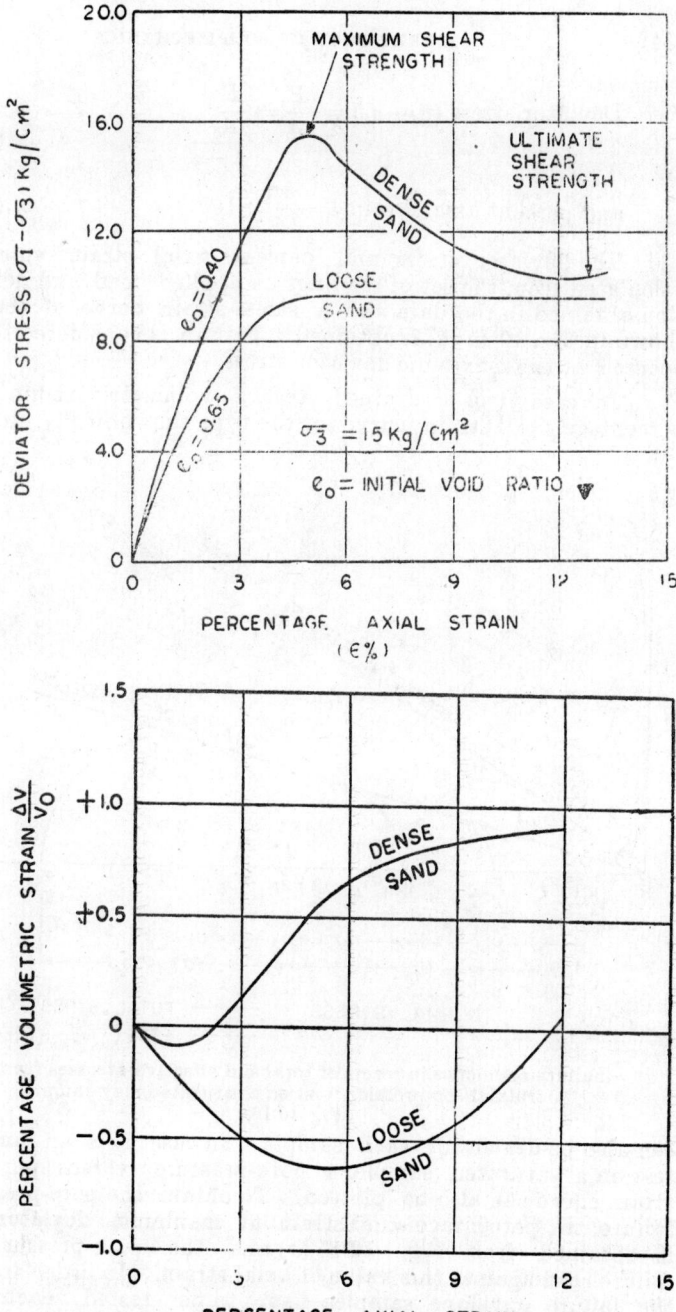

Results of a triaxial compression test on two samples of sand
one dense sand sample and the other loose sand sample
Fig 10·15.

Deviator stress $(\sigma_1 - \sigma_3) = \dfrac{P}{A} = \dfrac{P}{\dfrac{A_0}{1-\epsilon}}$ ...(10·20)

and percent axial strain $\epsilon = \dfrac{\delta}{l_0} \times 100$ ...(10·21)

The deviator stress and percent axial strain values are calculated from the set of readings on applied load and deformation obtained in the data and a stress-strain curve of the type shown in Fig. 10·15 (a) is obtained. Failure is considered to have occured at the maximum deviator stress.

In case it is a drained test, a volumetric strain versus percentage axial strain curve of the type shown in Fig. 10·15 (b)

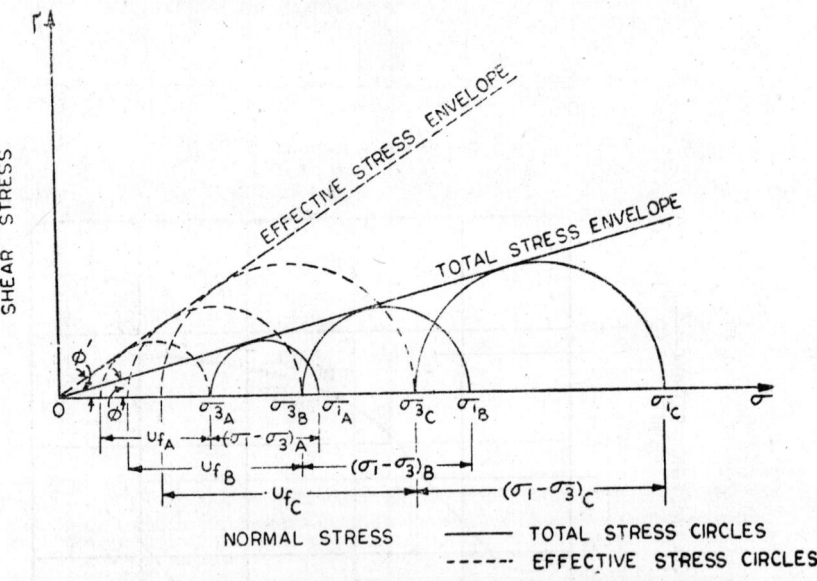

Failure envelopes in terms of total and effective stresses from a triaxial compression test on a saturated clay sample
Fig. 10·16.

can also be drawn for each sample. In case it is an undrained test on a saturated sample, a pore-pressure versus percentage strain curve can also be plotted. To obtain the pore-pressure at failure, the percentage axial strain at maximum deviator stress is obtained from Fig. 10·15 (a) and the pore-pressure u_f at failure read against this value of axial strain. In order to obtain the failure envelope samples have to be tested, each under different lateral pressure. Results can be obtained both in terms of total and effective stresses and from them total and effective stress envelopes can be plotted. Fig. 10·16 shows[1] the results o an undrained triaxial test on three samples, *A*,*B*, and *C* of a soil, plotted in terms of total and effective stresses, The two stress envelopes are also shown. These stress envelopes. represent the results of a triaxial test in terms of Coulomb's equation. Interpretation of the data from a triaxial compression test on cohesive soils in terms of Hvorslev's equation is given in article 10·8.

The triaxial test is a much better test compared to the direct shear test. The relative merits are as under :—

In any types of shear test, it is essential that the stresses and strains are distributed uniformally in the soil so that the results could be plotted as if the stresses and strains are being applied to a point. From the stress distribution point of view, a triaxial test is a much better test compared to a direct shear test, as in a direct shear test the failure is progressive whereas in a triaxial test the strength is mobilized all at once. In a direct shear test the soil near the edges fails earlier than the soil at the centre and, therefore, the material is not strained uniformally. In a triaxial test also the material near the ends does not get strained as much, due to the end restraint, as the material at the centre. However, the failure zone lies at the centre of the sample where the conditions are much more uniform compared to the the conditions at the ends. So, from the stress-strain point of view, conditions are much better in a triaxial test.

2. In a direct test the plane of failure is pre-determined. In the case of a triaxial test, the failure plane is determined by the applied stresses and it is the weakest plane.

3. Volume changes in a direct test are measured from the changes in the height of the sample. They are not accurate since the area of failure surface, although assumed constant, is not constant. The triaxial test is conducted usually on saturated samples and the volume changes are measured by the burettes attached to the base of the cell and are much more accurate compared to those measured in the direct test.

4. State of stresses on the principal planes at all deformations are known in a triaxial test and as such the Mohr's stress circles and failure envelopes can be plotted for any deformation or strain.[2] This is not true in a direct shear test. In a direct shear test the stresses is known only on the horizontal failure plane and one can draw Mohr's circle and determine the orientation of planes after obtaining the failure stresses only.

5. The triaxial testing equipment can be used for performing a number of different types of tests according to the drainage conditions that suft the field conditions, whereas it is not so with the direct shear test apparatus.

---

1. These results are typical of the results in an undrained triaxial test. In case of a drained test, only one envelope will be drawn since the stresses in the data obtained are effective stresses.

2. For details of a comprehensive work along these lines, the reader may refer to reference 5.

There are, however, many limitations of the triaxial test. Among these limitations the more important ones are the influence of the intermediate principal stress and change in principal stress directions in practical problems and the influence of the end restraint of a labortary sample. A discussion on these limitations is outside the scope of this book.

(c) **Unconfined Compression Test**. Uncofined compression test is a limiting case of a triaxial shear test. The sample in this test is unsupported i.e. there is no lateral pressure applied to the sample, the test is limited to the clayey and silty soils only since a sandy material will not stand without any support. Also there is no drainage from the sample and hence it is esssntially a quick test. The size of the test sample is the same as in the case of a triaxial test. One type of machine that can be used in this type of test is shown in Fig, 10·17. It consists of a moveable lower platen A which is connected to gears through a threaded screw which can move vertically up and down at a fixed speed. To the upper fixed yoke is fitted a proving ring with a dial gauge C. The bottom portion of the proving ring carries the upper platen B. A strain gauge is fixed to a vertical rod and its pin rests against the top platen. The *test procedure* is

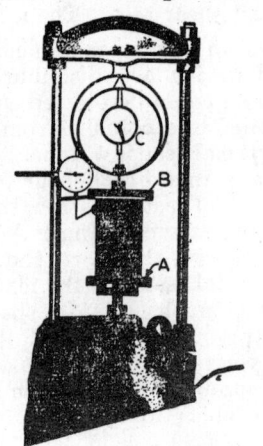

Unconfined Compression Machines.
Fig. 10.17

fairly simple. The surfaces of the top and bottom platens are lightly oiled and the soil sample is placed between the top andbottom plates. The light oiling of the platens reduces the friction between the soil and the surface of the platens. This reduction in friction reduces, to some extent, the tendency of the specimen to became barrel-shaped. The lower plate is then moved upwards by either the handle attached to the gears or electrically with the help of a motor. The rate of upwards movement is kept constant. The sample gets compressed between the two platens. As the lower platen moves up, there is a reaction on the fixed proving ring. The load can be read on the proving ring dial gauge and the corresponding deformation on the strain gauge. A set of readings consisting of loads and corresponding deforma tions is taken. The deformation is continued till the sample fails. The failure point is reached when the load readings start decreasing instead of increasing. The deformation is discontinued at this point.

Calculations for the compressive stresses are made by ssuming the volume of specimen as constant and the shape as a cylinder as in the case ot a triaxial test. The values of strains corresponding to various deformation readlngs are calculated first.

Then the corrected cross-sectionl areas corresponding to these strain values are calculated. Stress at any strain is given by the quotient of the load at that stage divided by the corrected cross-sectional area. A stress versus percentage strain curve is plotted from these calculations and from th's maximum compressive stress given by the peak of the curve is noted, and this represents the failure conditions. In case the curve becomes flat at large strains and there is no well-defined peak then the stress at an arbitrary strain (usually 15 to 20%) represents the failure conditions. The unconfined compressive strength of a soil is usually denoted as q_u.

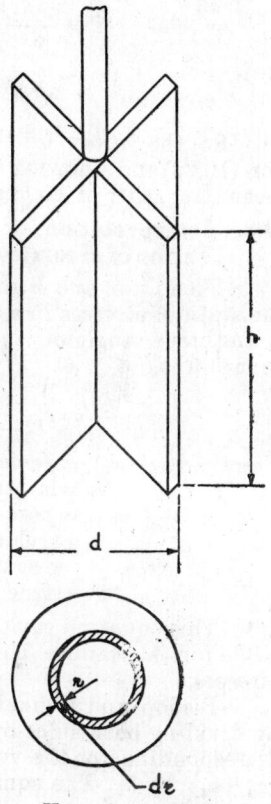

(*d*) **Laboratory Vane Shear Test** Laboratory Vane shear test is the quickest test to determine the shear strength of cohesive soils. The vane tester consists of a pair of thin steel blades connected to a vertical shaft as shown in Fig. 10·18 (*a*) to which is attached a torque measuring arrangement (not shown). The tester is pushed into the soil sample and a torque is applied on the shaft. Rotation of the vane inside the soil sample like this only means shearing off a cylindrical piece of the soil from the sample. The value of the torque applied to rotate the blades in the soil is read on a scale. The shear strength may be calculated from this torque as under.

Let τ be the shear stress acting along the surface, top and bottom of the sheared cylinder of soil, cross-section of which is shown in Fig. 10.18 (*b*). Let h represent the height and d the diameter of the sheared cylinder.

Vane shear tester.
Fig. 10.18.

Consider a small cylindrical ring of width dr at a distance r from the centre as shown. Then the shear stress across the area of this ring is $(2\pi r\, dr)\, \tau$.

Moment of this force about the axis of the cylinder is $(2\pi r\, dr\, \tau)r$.

Also the stress along the peripheral surface of the cylinder is $(\pi.d\, h).\tau$.

Total torque applied is given by the moment of the resistance offered by the soil about the axis of the shaft.

$$\therefore \text{Torque } T = 2 \int_{0}^{d/2} (2\pi r \, dr \, \tau) + (\pi.d.h.\tau)\frac{d}{2}$$

or
$$T = 4\pi.\tau. \left[\frac{r^3}{3}\right]_{0}^{d/2} + \pi \frac{d^2.h.\tau}{2}$$

or
$$T = \frac{\pi\tau d^3}{6} + \frac{\pi\tau d^2.h}{2}$$

$$= \tau.\pi.d^2\left(\frac{d}{6} + \frac{h}{2}\right) \qquad \qquad ...(10.22)$$

So, the values of the torque T may be substituted in equation (10.22) and knowing the dimensions of the vane tester, the shear strength τ of soil can be found.

10.8. Interpretation of Triaxial Compression Test Data in Terms of Hvorslev's Parameters

Skemption and Bishop (7) have given an expression stating the state of stresses in a triaxial test from which true cohesion c_e and true angle of internal friction ϕ'_e can be found. The equation is.

$$\frac{(\sigma_1 - \sigma_3)_f}{2} = c_e \frac{\cos\phi_e'}{1 - \sin\phi_e'} + (\sigma_3 - u_f).\frac{\sin\phi_e'}{1 - \sin\phi_e'}$$

where
$(\sigma_1 - \sigma_3)_f$ = deviator stress at failure

σ_3 = lateral chamber pressure

u_f = pore-pressure at failure

c_e = true cohesion corresponding to the void ratio test sample

ϕ'_e = true angle of internal friction for the soil.

This equation can be derived[1] by the considering a Mohr's circle for the failure conditions corresponding to the effective stresses.

Bishop and Henkel (8) have further expressed this equation by dividing both sides by $\bar{\sigma}_e$ (the equivalent consolidation pressure corresponding to the void ratio of the test sample) and replacing $(\sigma_3 - u_f)$ by $\bar{\sigma}_3$. The equation will read as ;

$$\frac{(\sigma_1 - \sigma_3)_f}{2\bar{\sigma}_e} = \frac{c_e}{\bar{\sigma}_e} \frac{\cos\phi'_e}{1 - \sin\phi'_e} + \frac{\bar{\sigma}_3}{\bar{\sigma}_e} \frac{\sin\phi'_e}{1 - \sin\phi'_e}$$

Evidently this is the equation of a straight line. If now $\dfrac{(\sigma_1 - \sigma_3)_f}{2\bar{\sigma}_e}$ is plotted against $\dfrac{\bar{\sigma}_3}{\bar{\sigma}_e}$—from data obtained from a number of samples, a straight line can be passed through these plotted points.

Let the equation of this plot be written

$$\frac{(\sigma_1 - \sigma_3)_f}{2\bar{\sigma}_e} = c_1 + \frac{\bar{\sigma}_3}{\bar{\sigma}_e} \tan\lambda \qquad \qquad ...(10.24)$$

---

1. The derivation of this equation is left to the reader as it is fairly simple.

where c_1 is the intercept on the $\dfrac{(\sigma_1-\sigma_3)_f}{2\sigma_e}$ axis and λ is the angle made by this line with the horizontal.

A comparisons of equations (10.23) and (10.24) shows that

$$c_1 = \frac{c_e}{\sigma_e} \cdot \frac{\cos\phi_e{}'}{1-\sin\phi_e{}'}$$

and $\quad \tan\lambda = \dfrac{\sin\phi'_e}{1-\sin\phi'_e}$

From the first of these two equations, one can work out the value of c_e correspoding to the void ratio e and from the second, one can work out $\phi_e{}'$. It may be noted that $\dfrac{c_e}{\sigma_e}$ gives the cohesion factor k.

10·9 Pore Pressure Parameters A and B

In order to understand the shear strength of a saturated soil, it is essential to understand the development of pore-pressure in a soil. Skempton (6) has explained the relationship between the applied stresses and the pore-pressure induced by these stresses with the help of two co-efficients, that have come to be known as the Skempton's pore-pre-pressure parameters A and B. These co-efficients have been universally recognised and they find their utiliy in many practical problems. As such, they are presented below.

Let a saturated sample of soil brought from the field have an allround pressure p in the field as shown in Fig. 10·19 and let $\triangle\sigma_1$ and $\triangle\sigma_3$ (again shown) be the changes brought during shear in the major and minor principal stresses in a triaxial test in the laboratory. This set of changes in stresses can be split up into two systems—one the all-round pressure chauge $\triangle\sigma_3$ and the second a uniaxial stress change $(\triangle\sigma_1-\triangle\sigma_3)$ which as already stated may be called deviator stress.

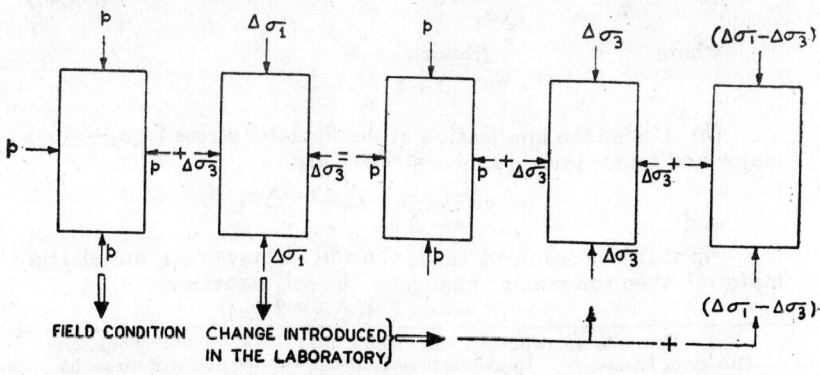

Fig. 10.19.

Under the all-round pressure $\triangle \sigma_3$ applied in the laboratory, let $\triangle u_a$ be the change in the pore-pressure of the sample and again under the application of the deviator stress $(\triangle \sigma_1 - \triangle \sigma_3)$, let $\triangle u_d$ be the change in the pore-pressure of the sample.

Assume that the sample has a volume of V and a porosity of η so that ηV represants the volume of voids in the sample and since the voids are fully saturated, ηV represents the volume of water in the sample. If c_c and c_v are assumed as the volume compressibilities of the soil structure and the pore-water respectively, then calculations follow as under :

(a) Under the application of the all-round pressure alone, the intergrandlar stress in the sample is
$$\triangle \bar{\sigma}_3 = (\triangle \sigma_3 - \triangle u_a)$$
∴Change in volume of the structure
$$= -c_c V.(\triangle \sigma_3 - \triangle u_a)$$
Now, considering the volume of water in the voids, change in this volnme
$$= -c_v \eta V \triangle u_a ;$$
since the change in the pore-pressure is $\triangle u_a$
The two volumes calculated above are eqnal.
$$\therefore \quad c_v. \eta V. \triangle u_a = c_c. V(\triangle \sigma_3 - \triangle u_a).$$
Volume V cancels throughout.
$$\therefore \quad \triangle u_a(c_c + \eta c_v) = c_c \triangle \sigma_3$$
or
$$\frac{\triangle u_a}{\triangle \sigma_3} = \frac{c_c}{c_c + \eta c_v}$$
or
$$\frac{\triangle u_a}{\triangle \sigma_3} = \frac{1}{1 + \eta \dfrac{c_v}{c_c}}$$
Put the   R.H.S.$=B$
$$\therefore \qquad \frac{\triangle u_a}{\triangle \sigma_3} = B \qquad\qquad \ldots(10\cdot25)$$
where
$$B = \frac{1}{1 + \eta \dfrac{c_v}{c_c}}$$

(b) Under the application of the deviator stress $(\triangle \sigma_1 - \triangle \sigma_3)$ major and minor intergranular stresses are
$$\triangle \bar{\sigma}_1 = (\triangle \sigma_1 - \triangle \sigma_3') - \triangle u_d$$
and $\triangle \bar{\sigma}_3 = -\triangle u_d$
Naw, if it is assumed that the soil behaves as an elastic material, then the volume change of the soil structure[1]
$$= -c_c. V.\tfrac{1}{3}(\triangle \bar{\sigma}_1 + 2 \triangle \bar{\sigma}_3)$$

1. For this relationship the reader may lock up any book on Theory of Elasticity. In elastic materials the volume changes brought up by application of three principal stress $\triangle \sigma_1$, $\triangle \sigma_2$ and $\triangle \sigma$, is the same as by the mean of the stresses. In this case $\triangle \bar{\sigma}_2 = \triangle \bar{\sigma}_3$.

Again, considering the volume of water in the voids, change in this volume $\qquad = -c_v.\eta V.\triangle u_d$;
since change in pore-pressure is $\triangle u_d$.

The two volumes calculated above are again equal.

$\therefore \qquad c_v.\eta V.\triangle u_d = c_o.V.\tfrac{1}{3}(\triangle \bar\sigma_1 + 2\triangle \bar\sigma_3)$

Cancel V throughout and substitute the values of $\triangle \bar\sigma_1$ and $\triangle \bar\sigma_3$

$\therefore \qquad c_v.\eta\triangle u_d = c_o.\tfrac{1}{3}[(\triangle \sigma_1 - \triangle \sigma_3) - \triangle u_d + 2(-\triangle u_d)]$

or $\qquad c_v.\eta.\triangle u_d = c_o.\tfrac{1}{3}[\triangle \phi_1 - \triangle \sigma_3 - 3\triangle u_d]$

or $\qquad \triangle u_d[c_v.\eta + c_o] = c_o.\tfrac{1}{3}(\triangle \sigma_1 - \triangle \sigma_3)$

The soil, in fact, does not behave as a truly elastic material. Therefore, replace the numerical coefficient 1/3 by an arbitrarily chosen coefficient A.

$\therefore \qquad \triangle u_d[c_v.\eta + c_o] = c_o.A(\triangle \sigma_1 - \triangle \sigma_3)$

or $\qquad \dfrac{\triangle u_d}{(\triangle \sigma_1 - \triangle \sigma_3)} = A\,\dfrac{c_o}{c_v.\eta + c_o} = A\,\dfrac{1}{1 + \eta\,\dfrac{c_v}{c_o}}$

But $\qquad \dfrac{1}{\eta + \dfrac{c_v}{c_o}} = B$

$\therefore \qquad \dfrac{\triangle u_d}{\triangle \sigma_1 - \triangle \sigma_3} = A.B. \qquad\qquad \text{...(10.26a)}$

If the total pore-pressure induced by the application of $\triangle \sigma_1$ and $\triangle \sigma_3$ on the sample is $\triangle u$, then

$$\triangle u = \triangle u_a + \triangle u_d$$

$\therefore \qquad \triangle u = B.\triangle \sigma_3 + A.B(\triangle \sigma_1 - \triangle \sigma_3)$

or $\qquad \triangle u = B[\triangle \sigma_3 + A(\triangle \sigma_1 - \triangle \sigma_3)]$

The coefficients A and B in the above equation are called the **Skempton's pore-pressure coefficients**, after the name of Prof. A.W. Skempton, who derived there equations.

When the soil is saturated, the volume compresibiiity of the soil structure is much more than the volume compressibility of water, i.e., $c_o >> c_v$.

$\therefore$ Ratio $\dfrac{c_v}{c_o}$ may be assumed as zero for all practical purposes. This makes the coefficient B eqnal to one for the saturated soils.

When the soil is dry, the volume compressibility of air in the voids is much more than the volume compressibility of the soil structure, i. e., $c_v >> c_o$.

$\therefore$ Ratio $\dfrac{c_v}{c_o}$ may be assumed as infinity for all practical purposes. This makes the coefficient B equal to zero for dry soils.

For partially saturated soils, value of B as found experimentally lies between one and zero. A curve of the variation of

B with the degree of saturation of the soil is shown in Fig. 10.20. So, for saturated soils, equation (10.26) may be written as

$$\triangle u = \triangle \sigma_3 + A(\triangle \sigma_1 - \triangle \sigma_3)$$

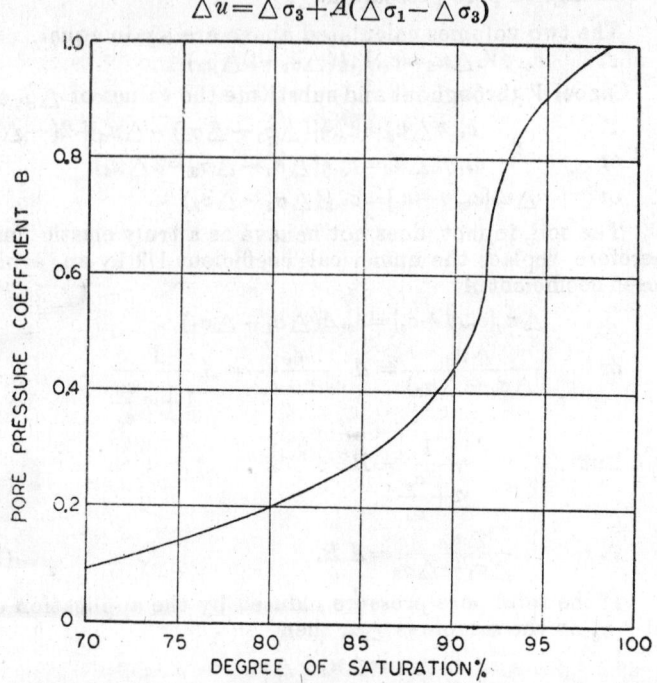

Pore-pressure coefficient B versus degree of saturation for saturated cohesive soils.

Fig. 10.20.

Further for an undrained test under constant lateral pressure, $\triangle \sigma_3$ equals zero.

$$\therefore \qquad \triangle u = A(\triangle \sigma_1 - \triangle \sigma_3)$$

In this type of test $(\triangle \sigma_1 - \triangle \sigma_3)$ is nothing but the deviator stress designated previously in artical 10-7(b) as $(\sigma_1 - \sigma_3)$. Also the change in pore-pressure was previously designated as u. At failure if $(\sigma_1 - \sigma_3)$ and u are known than the pore-pressure coefficient A at failure is given by,

$$A_f = \frac{u_f}{(\sigma_1 - \sigma_3)_f} \qquad \qquad ...(10.27)$$

where the subscripts just indicate that the values of the quantities are the ones that are obtained on failure of a soil sample.

It may be noted from equations (10.25) and (10.26a) that A and B are just dimensionless numbers.

10·9A. Discussion on Results Obtainable from Triaxal Tests on Saturated Cohesive Soils

How to conduct various types of triaxial tests involving different drainage conditions on saturated cohesive soils and what results to expect from such tests is very important for a soil engineer. A brief account of the various steps involved in conducting each type of test and the results to be expected is the subject-matter of this article.

(a) **Unconsolinated Undrained Test** (or Quick Test). The sample of the saturated soil from the field or the remoulded sample prepared in the laboratory is placed in the triaxial cham-

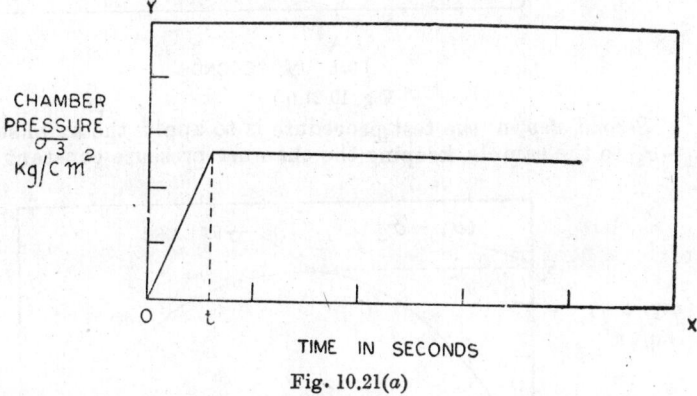

TIME IN SECONDS

Fig. 10.21(a)

ber with the rubber sleeves and '0' rings in position. *First step* is to apply chamber pressure σ_3 to the sample. No drainage is permitted in the sample. In case the initial pore-pressure in the sample is zero, the full[1] chamber pressure will be reflected on the pressnre measuring device (pressure guage or a pressure transducer). In ease it is a field sample, it may have some negative pore-pressure in it due the release of the overburden pressure. As such this negative pore-pressare plus the pore-pressure recorded on the pore-pressure measuring device, in this case, must equal the applied pressure. The time for appreciation of the full pore-pressure depends upon the permeability of the soil and its elastic properties. The application of the chamber pressure and the development of the pore pressure, if plotted as ordinates against time as abscissa, will give curves of the type as shown in Fig. 10.21(a) and (b). Theoretically the time t to introduce the pressure σ_3 in the chamber and the time t_1 required by the sample for-pressure development equal to $\triangle u_a(=\sigma_3)$ should be the same. However, in practice there is little lag between application of the all-round pressure in the chamber and its appreciation in the sample. This

---

1. Due to elastic compression of the sample, it is just possible that the pore-pressure developed may not be 100% of the applied chamber pressure. But the statement, in general, is true.

time lag depends upon many factors and discussion on these factors is not within the scope of this book.

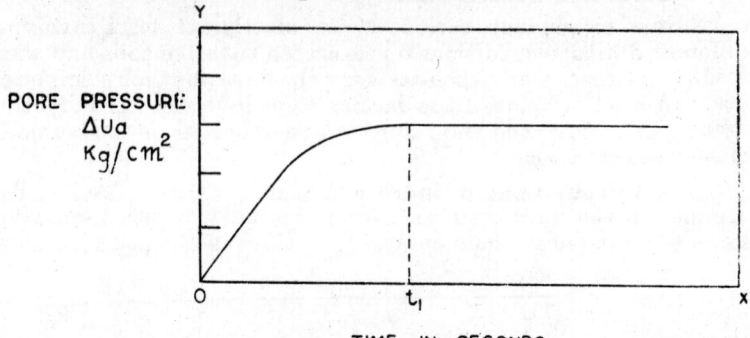

TIME IN SECONDS
Fig. 10.21.(b)

Second step in the test procedure is to apply the **axial stress** $(\sigma_1 - \sigma_3)$ to the sample, keeping the chamber pressure constant and

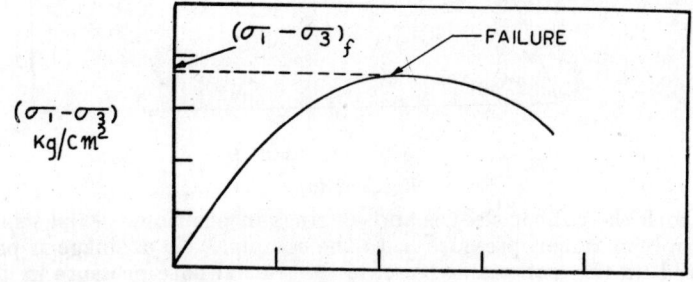

PERCENTAGE STRAIN (ϵ %)
Fig. 10.21(c)

under this stress the sample is allowed to shear off, without any drainage being allowed, again. This application of the deviator stress will further develop pore-pressure in the sample. A typical deviator stress versus percentage strain curve for a normally consolidated soil sample is shown in Fig. 10.21(c) and the corresponding pore-pressure *versus* percentage strain curve is shown in Fig. 10.21(d).

If u_f denotes the total pore-pressure at failure under the all-round chamber pressure and the deviator stress, then

$$u_f = \triangle u_a + \triangle u^1_d$$

where $\triangle u_d$ is the pore-pressure at failure under deviator stress. The pore-pressure $\triangle u_a$ has not been allowed to dissipate at any stage and it only exists as an hydrostatic excess pressure in the water, so that the failure effective stresses $\bar{\sigma}_1 = (\sigma_1 - u_f)$ and

---

1. See Fig. 10·21 (d). On this graph of $\triangle u_d$ *versus* ϵ percentage $\triangle u_d$ at failure is shown as $\triangle u_f$.

$\sigma_3 = (\sigma_1 - u_s)$ are independent of the chamber pressure. The effective deviator stress $(\bar{\sigma}_1 - \bar{\sigma}_3)$ equals the total deviator stress $(\sigma_1 - \sigma_3)$ as

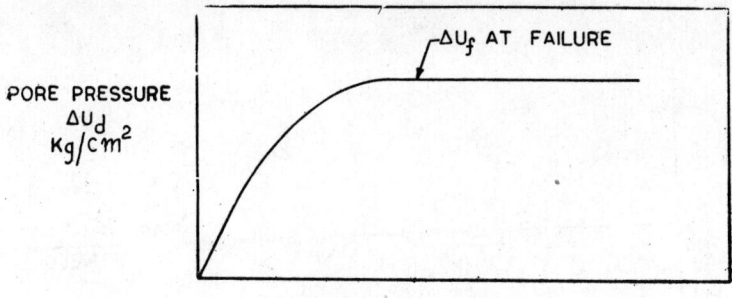

PORE PRESSURE Δu_d Kg/cm^2

PERCENTAGE STRAIN $(\epsilon \%)$

Fig.10.21(d)

may be varified from the difference of $\bar{\sigma}_1$ and $\bar{\sigma}_3$. Variation in the chamber pressure only changes the pore-pressure u_f and the effective stress in the sample remains constant. As a consequence, if a number of such tests are conducted with a different chamber pressure in each case the stress circles in all cases will have the same diameter equal to $(\sigma_1 - \sigma_3)$. The only difference would be that the minor principal stress in each case will vary[1] and stress circle relating to one sample will get displaced with respect to the stress circle relating to another sample by an amount equal to the difference between the chamber pressures in the two cases. This applies both for the total as well as the effective stresses. The effective stress circle of one sample will get displaced from its total stress circle by an amount equal to the pore-pressure Δu_d at failure under deviator stress. The stress envelopes for the total and effective stresses will coincide. A horizontal line tangential to all the circles will represent the failure envelope. The shear strength of the soil in this test will be represented by an intercept made by this envelope with the τ-axis. The angle ϕ (and ϕ') in this case is zero.

The above observations are presented in Fig. 10.22 for three samples A, B and C.

(b) **Consolidated-Undrained Test.** Let it be assumed that the sample is at a void ratio e_0 when it is placed inside the chamber of the triaxial apparatus. Also assume that it has pressure more than the preconsolidation load of the sample, so that at void ratio e_0 it is in a normally consolidated state. *As a first step*, fluid is pumped into the chamber and *pressure* applied. The sample is allowed to consolidate under this pressure by allowing drainage from the sample. A burette connected to the base of the sample measures the volume change. When the sample has been completely drained, there will be no rise of water in the burette any more. The soil sample would have reached a void

---

1. See Fig. 10·21 (d). On this graph of Δu_d *versus* ϵ percentage Δu_d at failure is shown as Δu_f.

ratio e_1 corresponding to this chamber pressure and e_1 would be smaller than e_o.

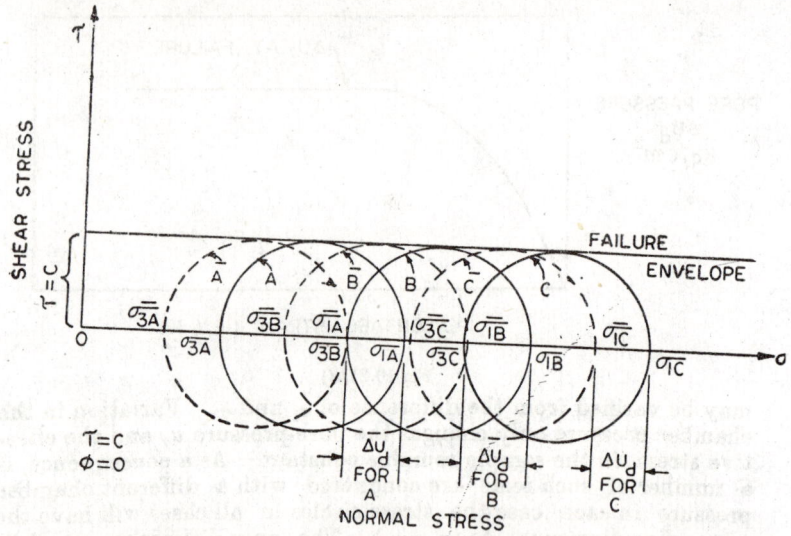

Fig. 10.22

Second step in the test procedure is to apply axial stress $(\sigma_1 - \sigma_3)$ to the sample. Keeping the chamber pressure constant, under this deviator stress the sample is allowed to shear off, without any drainage being allowed. Stress-strain and pore-pressure data for the sample is collected. On the stress-strain curve for this data, the failure point is located and the deviator stress at failure $(\sigma_1 - \sigma_3)_f$ is read. Then, another sample of the soil is taken and it is consolidated first to some other void ratio say e_2 corresponding to a different chamber pressure and then it is sheared. The second sample will give a different value of deviator stress at failure. If a number of soil samples are consolidated to various void ratios under a number of different pressures, and then sheared off as indicated above, then a number of values of deviator stresses and pore-pressures at failure corresponding to the various chamber pressures used for consolidation will be available. Each set of data from a sample will give a different total and effective stress circle as shown in Fig. 10.16 for three different samples A, B and C tested at three void ratios e_1, e_2 and e_3. A total and an effective stress envelope tangential to the respective stress circles can be drawn as shown. The values of c and ϕ can be picked up from the total stress envelope and those of c' and ϕ' from the effective stress envelope. In Fig. 10.16 it is seen that c and c' are zero. This is expected since the soil samples were normally consolidated and for normally consolidated soils[1] the Mohr-Coulomb

---

1. For sands also, the Mohr-Coulomb envelope passes through the origin.

envelope passes through the origin of the set of axes. For over-consolidated soils, the envelopes give intercepts c and c' on the τ axis.

(c) **Drained Test (or Slow Test).** In this type of test complete drainage is permitted during *both the steps* explained in detail in (a) and (b) *i. e.*, drainage is permitted both during consolidation under the chamber pressure as also during the application of the deviator stress. This directly leads to the effective principal stresses at failure. Stress circle can be drawn for each sample for the effective stresses so obtained and the failure envelope is drawn tangential to these stress circles. The values of c' and ϕ' obtained from the drained tests are *fairly*[1] identical with those obtained from an undrained test.

The effective stress envelope for a normally consolidated soil passes through the origin and for an over-consolidated soil, it cuts the axis and gives the intercept c.

Many other types of these tests are possible on the triaxial equipment. However, they are outside the scope of this book. A comprehensive treatment on the various tests is available in reference 9 given at the end of this chapter.

These tests find their practical utility also. The quick test finds its application in the design of foundations. If a foundation load is transmitted to an unconsolidated clay, it is essential to find out the factor of safety of the clayey soil mass against shear immediately after placing the foundation in position. The consolidated undrained or the slow test values may be used in the stability calculations against failure in shear of consolidated dams and slopes of embankments constructed from cohesive soils, under conditions of draw down of water.

10.10. Shear Characteristics of Granular Materials

As can be seen in Fig. 10.10(a), the shear strength developed by a sand sample depends upon whether the sample is *dense* or *loose*. This is true for gravels and gravel-sand mixtures also. A dense sand has two values of shear strength—the maximum and the ultimate value—whereas a loose sand has only the ultimate shear strength. The two types of materials have different volume change characteristics. The dense sands expand and the loose sands contract on application of shear stresses. These characteristics are represented slightly differently in Fig. 10.23.

Two samples of soil are taken at void ratios represented by the points A and B as shown Fig. 10·23 (b) and they are sheared in a direct shear testing apparatus. The normal load[2] in these tests is σ_1. The sample with the void ration corresponding to

---

1. Data also exist that indicate that ϕ' from a drained test is more than the one from a consolidated undrained test. For further discussion on this subject, see article 10·11 also.

2. In this discussion by load is meant the unit load.

the point A expands and on failure reaches a void ratio represented by the point C and the sample with the void ratio corres-

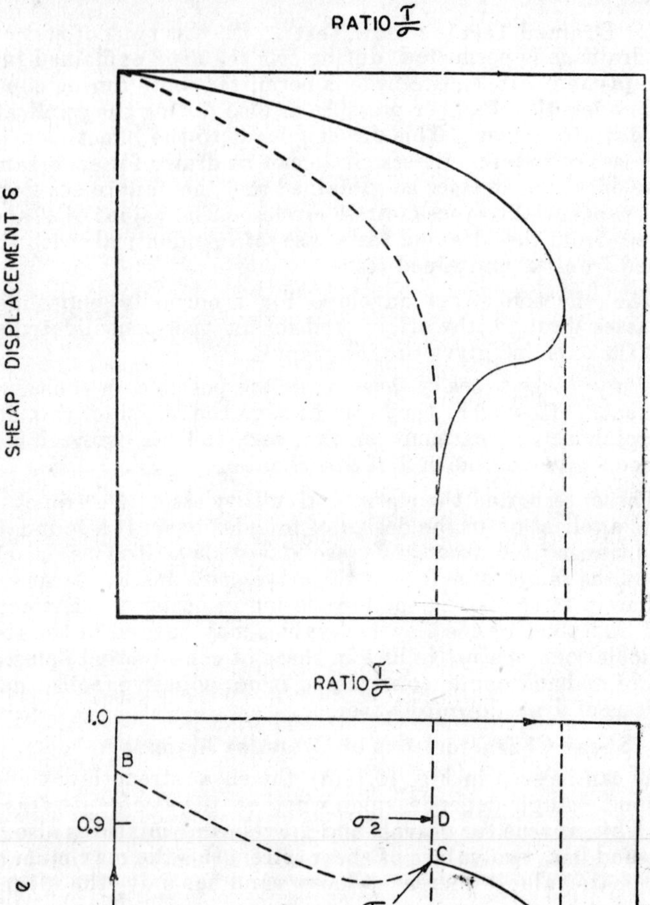

RATIO $\dfrac{\tau}{\sigma}$

SHEAP DISPLACEMENT $\mathfrak{S}$

RATIO $\dfrac{\tau}{\sigma}$

VOID RATIO $\mathscr{C}$

1.0

B

0.9 $\sigma_2 \rightarrow$ D

C

$\sigma \nearrow$

0.8

0.7 A

0.6

Fig. 10·23.

ponding to the point B, contracts and it also reaches a failure void ratio corresponding to point C. The two curves showing the ratio (τ/σ) *versus* the shear displacement δ corresponding to these two tests are shown in Fig. 10·23 (*a*). Now, if two similar samples are again sheared under a normal load $\sigma_2 < \sigma_1$, the two samples are observed to ultimately have a void ratio corresponding to point *D* in Fig. 10·23 (*b*). The void ratio corresponding to point *C* is evidently lower than the void ratio corresponding to point *D*. So the sample which had the initial void ratio corresponding to point *B* contracts more under a normal load of σ_1 than another sample with the same initial void ratio but under a normal load σ_2, when shear stresses are applied to the two samples. On the contrary, a sample with an initial void ratio corresponding to the point *A* expands less under a normal load σ_1 than another sample with the same initial void ratio but under a normal load σ_2 when shear stresses are applied to the two samples. If percentage contractions and expansions are calculated, these results can be plotted on the two legs of the horizontal axes, against their respective initial void ratios as shown in Fig. 10·24. Join *DD* and *CC*. These two lines represent the influence of the normal loads σ_1 and σ_2 respectively on the expansion or contraction characteristics of a soil at a particular void ratio. For example, consider a sample with a void ratio 0·85 and let it be supposed that it has to be sheared under a normal load of σ_2. Then the material will expand, since a horizontal from this void ratio cuts the line *DD* on the expansion side. Now suppose that the same sample has to be sheared under a normal

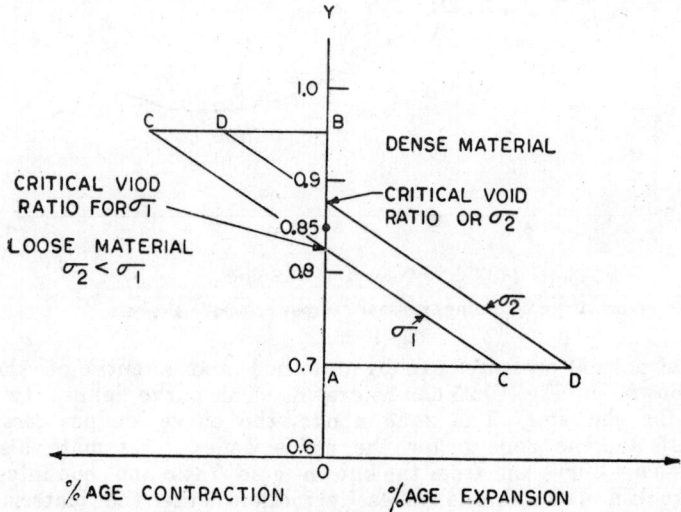

Fig. 10·24.

load of σ_1. Then it can be seen from the figure that the sample under this higher pressure will contract.

As a conclusion, one may say *that there is no such thing as a loose or a dense material unless the void ratio of the material and its confining pressure are also mentioned.*

Another conclusion from the figure is that there is one void ratio for each normal pressure at which a sample neither expands nor contracts. As an example, if σ_2 pressure is considered, then the line DD cuts the void ratio axis at a void ratio of 0·87. So, if a soil sample is taken at a void ratio of 0·87, the sample on application of shear stresses will neither expand nor it will contract. Such a void ratio where there is no expansion or contraction of the material is called the **critical void ratio**. It is, again, pointed out that the critical void ratio and the confining pressure are closely related. If a number of soil samples at different void ratios are tested under different normal (or confining) pressures, and their expansion or contraction is noted, a set of lines like CC and DD can be drawn. From this set, critical void ratios for

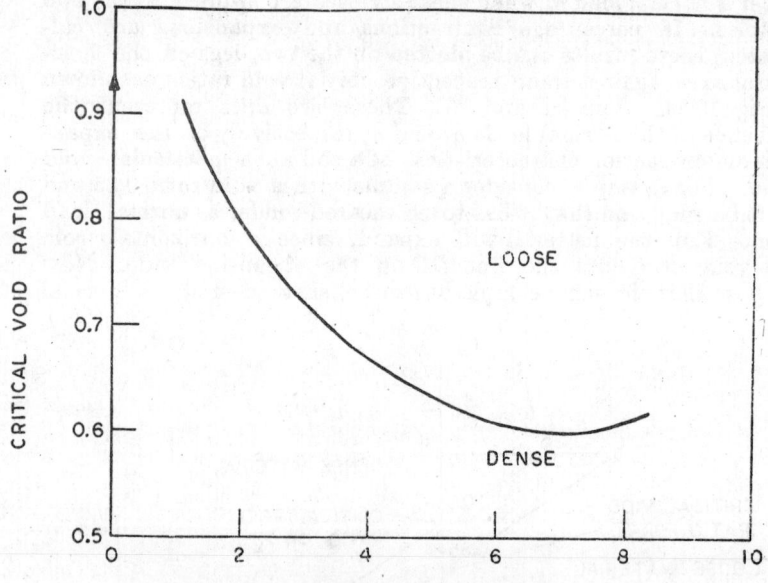

Normal (or confining) pressure *versus* critical void ratio.

Fig. 10·25.

different normal pressures can be obtained and a curve of the type shown in Fig. 10·25 can be drawn. This curve defines two zones for the soil. The zone above the curve defines *loose* materials and the zone below the curve defies *dense* materials. From such a curve and from the known void ratio and confining pressure of a material, one can easily judge whether the material is loose and will contract on application of shear stresses or the material is dense and will expand on application of shear stresses. All that one has to do is to plot the known void ratio and the confining pressure and see where the point falls *i.e.* whether it falls in the zone above the curve or in the zone below the curve.

The tendency of the dense materials to expand on application of shear stresses is termed as **dilatancy**. Property of dilatancy is helpful in some practical cases. When a soil is saturated and if drainage is prevented during application of shear stresses, it will expand and will throw negative pore-pressure in the soil. As seen from equation (10·29) below, this simply means development of larger shear strength, everything else remaining constant. This concept is helpful in reasoning out as to why it is essential to place the material in an earthern dam for example, in a compacted state. If an earthquake happens to occur in the area where the earthern dam is located, the soil will undergo temporary increase of shear stresses with development of negative pore-pressure (since the pore-pressure shall have no time to dissipate). This development of negative pore-pressure in fact increases the shear resistance of the material and would save the dam from failure.

Loose saturated material on the other hand may show an opposite tendency *i.e.* it may exhibit what is called *liquefaction*[1] and result in a shear failure.

When the failure envelope is drawn for cohesionless material, the envelope is nearly a straight line passing through the origin and the strength of the soil can be represented as

$$s = \sigma \tan \phi \qquad \qquad ...(10·28)$$

in terms of total stresses and as,

$$s = \bar{\sigma} \tan \phi' = (\sigma - u) \tan \phi' \qquad ...(10·29)$$

in terms of effective stresses.

The value of ϕ' in these soils ranges from 28 to 42 degrees. In general, this value increases with increasing density.[2] Extermely loose sands with unstable structure may have as low a value ϕ' as 10 (11).

At the same relative density, value of ϕ' depends upon the particle size distribution and the shape of particles. A well-graded soil may have a ϕ' several degrees more than a soil with single size particles. Similarly, a soil composed of angular grains has larger value of ϕ' compared to a soil composed of rounded particles. Larger values of ϕ' in the above cases are possibly due to better interlocking of a well-graded soil and frictional resistance in a soil composed of angular particles. Besides the above factors, the minerological composition of the soil is also believed to affect the value of ϕ'. The effect of moisture on ϕ' is small and is of the order of 1 to 2 degrees.

---

1. Liquefaction is nothing but the temporary transformation of the granular material into a thick slurry suspension.

2. See the chapter on B.C. of soils, where ϕ' and the penetration resistance in terms of the number of blows in a standard penetration test are related.

10·11. Shear Characteristics of Saturated Cohesive Soils

A soil that contains substantial amount of clay is cohesive in nature. The shear characteristics of such a soil depends upon the stress history of the soil. In article 7·4, definitions of normally consolidated and over-consolidated clays were presented. It may be recalled that a normally consolidated soil is one in which the stress has never increased beyond the present overburden pressure and an over consolidated soil is one that has experienced a stress higher than its present pressure. If a consolidated undrained triaxial test is considered, a normally consolidated soil is one in which the chamber pressure σ_3 prior* to the application of the deviation stress $(\sigma_1 - \sigma_3)$ is equal to or greater than the preconsolidation pressure in the field or equal to or greater than a pressure σ_c under which the soil has been consolidated in the laboratory and an over-consolidated soil is one in which the chamber pressure σ_3^1 prior to the application of the deviator stress $(\sigma_1 - \sigma_3)$ is less than a pressure σ_c under which the soil had previously been consolidated. The normally consolidated and the over consolidated clays exhibit different shearing characteristics, as mentioned earlier and hence are presented separately.

Normally Consolidated Clays. Laboratory investigations into the shear characteristics of the normally consolidated clays indicate the following general characteristics :

1. The shape of the stress-strain curve for a *normally* consolidated clay resembles *fairly* with the stress-strain curve of a dense sand as shown in Fig. 10·15.

2. A normally consolidated soil contracts in a drained test. In respect of volume changes on application of shear stresses, it behaves like loose sand.

3. Because of the tendency of a normally consolidated clay to contract in an undrained test with pore-pressure measurements, the pore-pressure is positive and it increases with increasing strain as shown in Fig. 10·21 (d).

4. Skempton's pore-pressure coefficient A at failure *i.e.* A_f for typical normally consolidated soils is found to be approximately unity.

5. If the consolidated undrained triaxial compression test results on normally consolidated clays are plotted in the form of Mohr's diagram, it is seen that both the total as well as the effective stress envelopes[1] pass through the origin, as seen in Fig. 10.16

---

*. In the undrained tests the chamber pressure prior to the application of deviator stress represents the intergranular stress on the material.

1. An envelope from a drained test will also pass through the origin and in that case also.

This indicates that in the Coulomb's equation in terms of total or effective stresses, cohesion c or c' is zero.

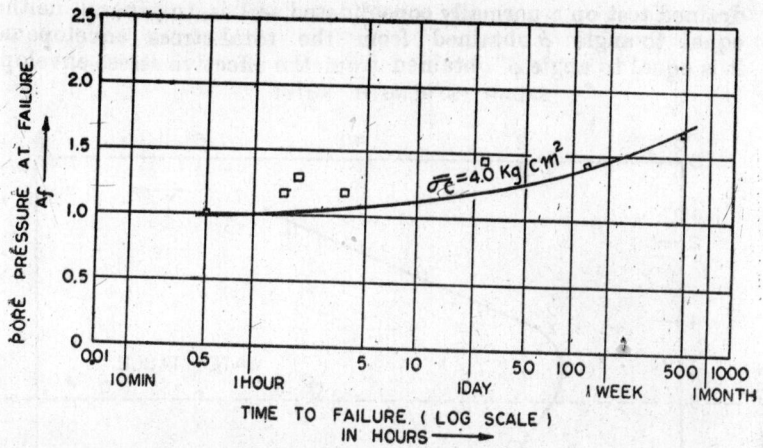

Undrained shear strength $\frac{(\sigma_1 - \sigma_3)}{2}$ versus time to failure (After Bjerrum and co-workers)
Fig. 10·2?.

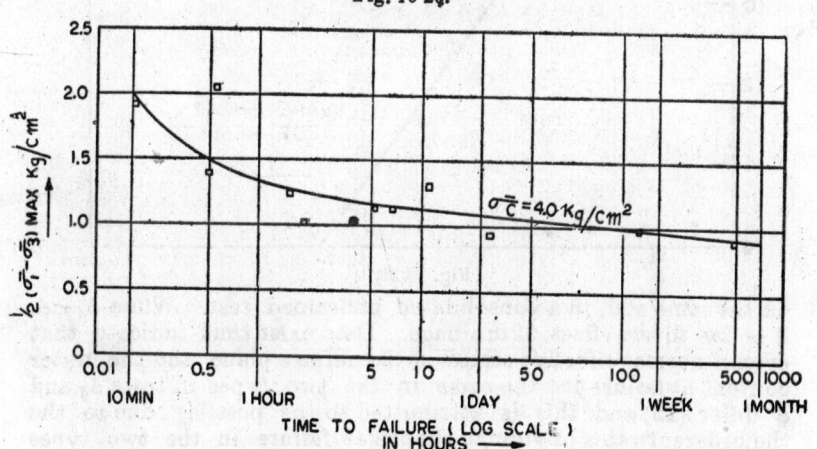

Pore pressure at failure A_f versus time to failure
(After Bjerrum and co-workers).
Fig. 10·26

6. The rate of application of shear stress has a tremendous effect on the strength of a cohesive soil. Results on a normally consolidated marine clay from Fornebu, Oslo (13) indicate that the undrained shear strength of the soil decreases with the increase in time to failure and the pore-pressure parameter A_f increases as the time to failure increases. These results are shown in

Fig. 10·26 and 10·27 respectively. These characteristics have been verified on many other soils also.

7. The angle of internal friction say ϕ_d obtained from a drained test on a normally consolidated soil is, in general, neither equal to angle ϕ obtained from the total stress envelope nor it is equal to angle ϕ' obtained from the effective stress envelope[1]

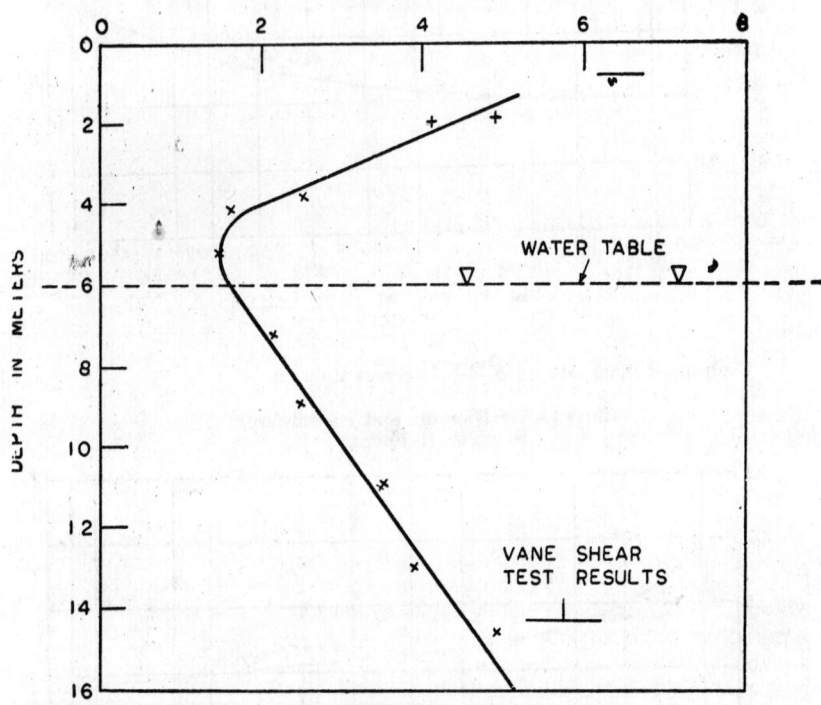

Fig. 10·28.

on the same soil, in a consolidated undrained test. When $\phi_d \neq \phi$. it is due to the effect of drainage. Data exist that indicate that even if normal effective stress on the failure plane and the water content at failure are the same in the two types of tests ϕ_d and ϕ' differ (12) and this is attributed to be possibly due to the the different rates of volume change at failure in the two types of tests.

In the field an important feature of a saturated normally consolidated clay is the increase of shear strength with depth, below the water table, as shown in Fig. 10·28. Possible reason for

---

1. This statement may not be strictly true, since in the case of weald clay the failure stresses of the two series of tests fall on the same envelope if the effective stresses are used, thus indicating that $\phi_d = \phi'$

this increase in shear strength is the decreasing void ratio due to increasing overburden pressure, with depth. Above the water table, the strength is usually found increasing towards the ground surface, possibly due to the capillary pressure or dessication. Variation above the water table is also indicated in this figure.

Over-consolidated Clays. Almost all clays met with in the field are over-consolidated. Hence an understanding of their strength charactistics is very essential.

In the laboratory, an over-consolidated soil sample for triaxial testing can be prepared by first consolidating the saturated soil to a pressure $\bar{\sigma}_0$ larger then the preconsolidation pressure and then removing some pressure which will permit the sample to rebound to a smaller pressure $\bar{\sigma}_3$. Then, in order to know its strength under this allround pressure $\bar{\sigma}_3$, the deviator stress ($\sigma_1 - \sigma_2$) may be applied. Laboratory investigations into the shear characteristics of over-consolidated clays indicate the following general characteristics :

1. The shape of the stress-strain curve compared to that of a normally consolidated soil sample is not very much different. At lower strains the former curve is steeper than the latter one and at higher strains, after a maximum value, it becomes flatter than the one for normally consolidated clays.

2. In a drained test, the over consolidated soil expands like dense sand.

3. Because of the tendency of an over-consolidated material to expand in an undrained test with pore pressure measurements, the pore-pressure at failure depends upon the degree of over-consolidation reflected by the **over-consolidation ratio**, which is defined as the ratio of the maximum allround consolidation pressure σ_0 to which the sample has been subjected before allowing expansion to the present effective allround pressure $\bar{\sigma}_3$ i.e. ratio $\dfrac{\bar{\sigma}_0}{\sigma_3}$—gives the consolidation ratio[1]. From the existing data on various clays, published so far, it can be concluded that the pore-pressure at failure for over-consolidation ratios less than 4 is positive and for over consolidation ratio more than 4 is negative. For an over-consolidation ratio of about 4, the pore-pressure is approximately zero. This means that at this over consolidation ratio the sample neither expands nor contracts i.e. at this over-consolidation ratio, there is no volume change in the clay, if a drained test were to be conducted.

4. Unlike normally consolidated soils in which Skempton's pore-pressure coefficient A at failure i.e., A_f is unity, in the over-consolidated soils A_f depends upon the degree of over-consolidation ratio. Henkel (14) has shown that in the case of *weald clay* the plot of A_f versus the over-consolidation ratio is as indicated

---

1. An over-consolidation ratio of unity means that the soil is normally consolidated.

in Fig. 10.29. This result is typical of the over-consolidated clays as verified on clays from many other parts of the world.

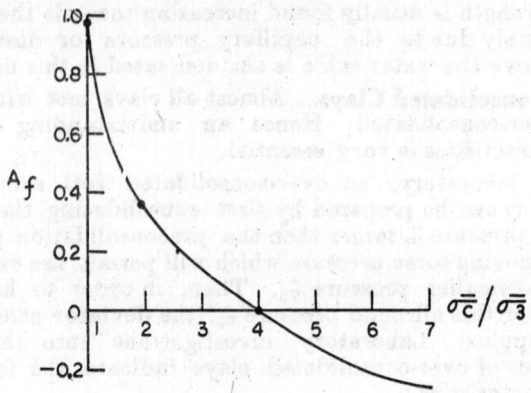

A_f versus over-consolidation ratio for Weald Clay (After Henkel)
Fig. 10·29.

5. If the consolidated undrained triaxial compression test results on over-consolidated clays are plotted in the form of a

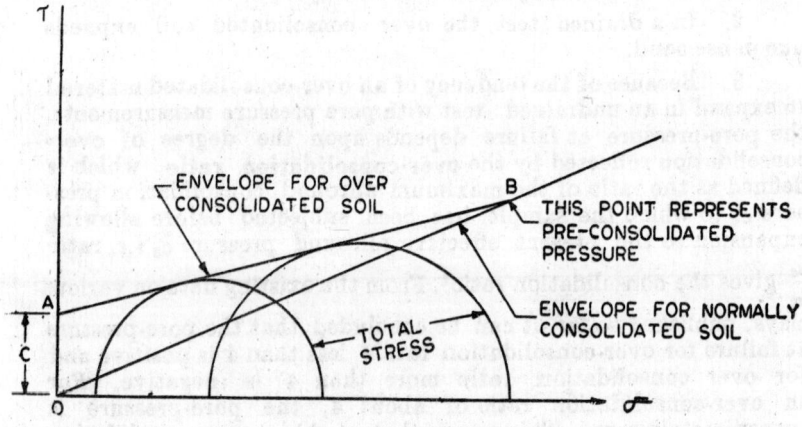

Envelopes for over-consolidated and normally consolidated samples.
Fig. 10·30.

Mohr's diagram, in terms of total stresses, as shown in Fig. 10·30, it is seen that the envelope does not pass through the origin. This indicates that in the Coulomb's equation for total strsses c is not zero. If now, the diagram is drawn in terms of effective stresses, again it will be found that c is also not zero (This is not shown). The envelope on normally consolidated soil samples of the same clay, is also shown in Fig. 10·30. The two envelopes meet at the point B. This point B represents the pre-consolidation pressure for the soil. One may say that for the over-consolidated soil the envelope is only *AB* and for the normally consolidated soil, it is *OB* and its extension.

6· Like normally consolidated soils, the over-consoildated soils also show decrease in strength with the increase in time to failure. The effect of the rate of application of shear stresses or the time to failure on A_f values is not well understood for over-consolidated clays.

It is considered essential, here, to point out how the total and effective stress circles will plot for the normally and over-consolidated soils, in an undrained triaxial test with pore-pressure measurements. The solid circles in Fig. 10·31 are the total stress circles and the dotted circles are the effective stress circles for three conditions : —

(1) When pore-pressure u is positive i.e. $u>0$

(2) When pore-pressure u is zero i.e. $u=0$

(3) When pore-pressure u is negative i.e. $u<0$

The first condition above pertains to a normally consolidated soil and the last two conditions to lightly and heavily over-consolidated soils respectively.

——— TOTAL STRESSES

------ EFFECTIVE STRESSES

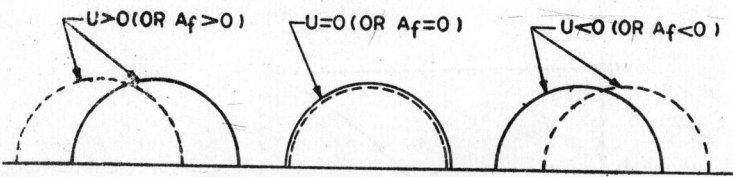

Total and effective stress circles for undrained triaxial tests
with pore-pressure measurements.
Fig. 10·31.

10.12. Shear Strength of Compacted Soils

When soils are compacted for construction of embankments or earthen dams, etc., they are in a state of partial saturation and hence represent a three phase system i.e., a specimen would contain solid soil particles, water and air or gas. The behaviour of such a soil depends upon the amount of gas and air in the voids. Measurement of shear strength of a partially saturated soil becomes complicated since a specimen under stress will have both pore-water and pore-air pressures. Under a low stress, the air occupies larger void space and under high stress, a part of it gets dissolved and the remaining air gets compressed. As a consequence, even if undrained tests are conducted, the soil undergoes volume changes.

The principle of effective stress $\bar{\sigma}^* = (\sigma - u)$ as advocated by Terzaghi is not applicable here, since this principle can only be applied to a two-phase system. Bishop (14) was the first to suggest a general principle of effective stress, which he stated as,

$$\bar{\sigma} = \sigma - [u_a - x(u_a - u_w)] \qquad (\dots 10.28)$$

where σ and $\bar{\sigma}$ are the total and effective stresses, u_a is the pore-air pressure, u_w is the pore-water pressure and x is a parameter related to the degree of saturation of the soil. For fully saturated soils, x is unity, which reduces equation (10.28) to the equation given by Terzaghi.

Equation (10.28) is applicable to partially saturated soils. Wu (15) indicates that x and degree of saturations can be co-related and for a typical soil the relationship is as indicated in Fig. 10.32.

The process of compaction employs very high dynamic or static loading, so that a compacted clay, in some respects, resembles a heavily over-consolidated clay. The strength envelope of a compacted clay resembles the strength envelope AB of a heavily consolidated clay, shown in Fig. 10.30. The evaluation of the effective stresses in a partially saturated soil will involve the measurement of both the pore-air and the pore-water pressures, a detailed account of which is outside the scope of this book.

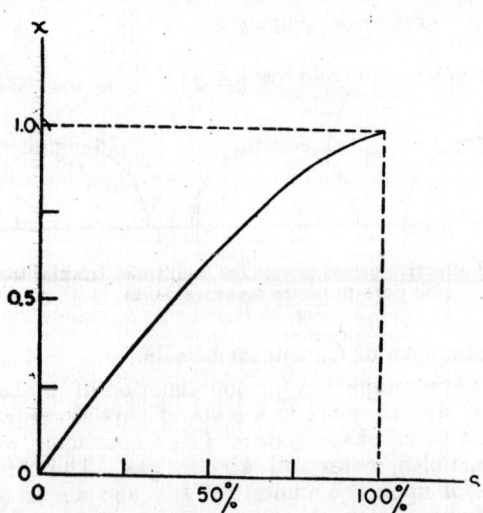

Parameter x versus degree of saturation S.

Fig. 10.32.

10. 13. Field Measurments of Shear Strength

Field measurement of the soil strength is very often required and is helpful specially on large engineering projects, Use is of-

---

*See equation 5.9 page 108 of this text.

ten made either of the soil penetrometers or the standard penetration test equipment or the field vane shear equipment.

Many types of soil penetrometers such as the proving ring penetrometers[1] and the TVA penetrometer[1], etc. are available. These penetrometers are allowed to penetrate the soil at a constant rate and the penetration resistance offered by the ground is a measure of the shear strength of the soil.

The standard penetration test and testing equipment is described in detail in Chapters 12 and 16. It is sufficient to indicate here that this equipment gives the density of the soil upon which depends the angle of internal friction of the soil. From the angle of internal friction so determined, one can find out the bearing capacity of the soil.

Vane shear testing equipment with the help of which a vane shear test on cohesive soils can be conducted upto a depth of 100 feet is available. The size of the blades in the field vane is larger (makes usually $2\frac{1}{2}$ inches or 6.25 cms. dia. cylinder) than the size of the blades in the laboratory vane (makes usually $\frac{3}{4}$ inch or 1·9 cms. dia. cylinder). This test normally forms a part of the sub-surface exploration for large engineering projects and has gained popularity because of the simplicity of its operation.

10.14. Examples.

Examples 10.1. Two undrained triaxial tests were performed on a soil. In the first test, the confining pressure was 1·50 kg/cm² and the failure occurred at a deviator stress of 5·0 kg/cm². In the second test, the confining pressure was 3.0 kg/cm² and the failure occured at a deviator stress of 8·0 kg/cm². Estimate the values of shear parameter c_u and ϕ_u.   **(P.U. April, 1965)**

Note. c_u and ϕ_u as stated in this question are nothing but the c and ϕ parameters in the Coulomb's equation.

Solution. The solution lies in drawing Mohr's stress circles for the two test samples. The data is as under :—

$$\text{1st sample} \qquad \sigma_3 = 1·50 \text{ kg/cm}^2$$
$$\text{and } (\sigma_1 - \sigma_3) = 5·0 \text{ kg/cm}^2$$
$$\therefore \qquad \sigma_1 = 6·50 \text{ kg/cm}^2$$
$$\text{2nd sample} \qquad \sigma_3 = 3·0 \text{ kg/cm}^2$$
$$\text{and } (\sigma_1 - \sigma_3) = 8·0 \text{ kg/cm}^2$$
$$\therefore \qquad \sigma_1 = 11·0 \text{ kg/cm}^2$$

With these principal stresses draw the Mohr's stress circles as shown in Fig. 10·33. Draw an envelope of failure AB. This cuts the Y-axis in A. The values of c and ϕ read from this envelope are :

$$c = 0·6 \text{ kg/cm}^2$$
$$\phi = 29·5° \qquad \textbf{Ans.}$$

1. See Soil Test Incorporated Catalog, edition III, page 43 for details and use of the equipment.

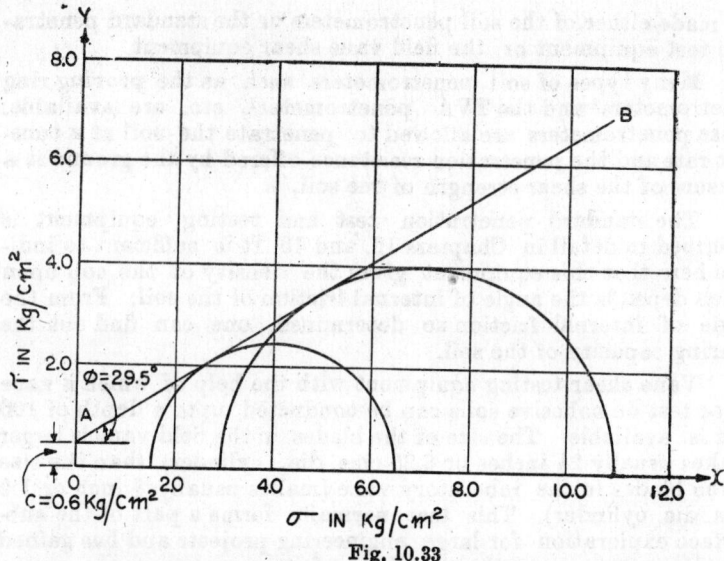

Fig. 10.33

Example 10.2. In a direct shear test on clean sand failure occurs at a stress of 0.5 kg/cm². The normal stress intensity is 0.7 kg/cm². Plot the Mohr's stress circles and determine the angle of internal friction for the soil. **(P.U. 1966)**

Solution. Take a set of two axes and let the horizontal axis represent the normal stresses and the vertical axis the shear stresses. In this particular case

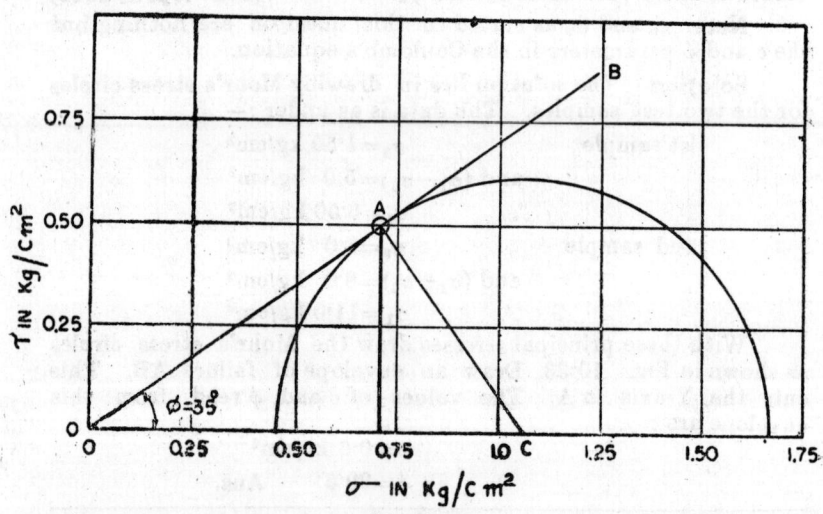

Fig. 10.34

Normal stress $\sigma = 0.7$ kg/cm^2
Shear stress $\tau = 0.5$ kg/cm^2

both at failure. Thes two stresses when plotted on the chosen axes give the point A. as shown in Fig. 10.34. This point represents the failure plane (horizontal plane in this case). Also point A must lie on the failure envelope. Since the material tested is sand, the envelope must also pass through the origin O. Assuming that the envelope is a straight line, OAB represents the envelope. The Mohr's circle must be tangential to this envelope. So draw a circle tangential to OAB, as shown. This is the stress circle for the material.

Angle of internal friction for the soil, as read from this diagram is 35° **Ans**.

Example 10.3. When an unconfined compression test is conducted on a cylinder of soil, it fails under axial stress of 1·4 kg/cm^2. The failure plane makes an angle of 54° with the horizontal. Determine the cohesion and angle of shearing resistance of the soil sample.   **(P.U. Nov. 1964)**

Solution. Since it is unconfined compression test.

the axial stress under which the sample fails is 1·4 kg/cm^2 ; so that if it is treated as the deviator stress.

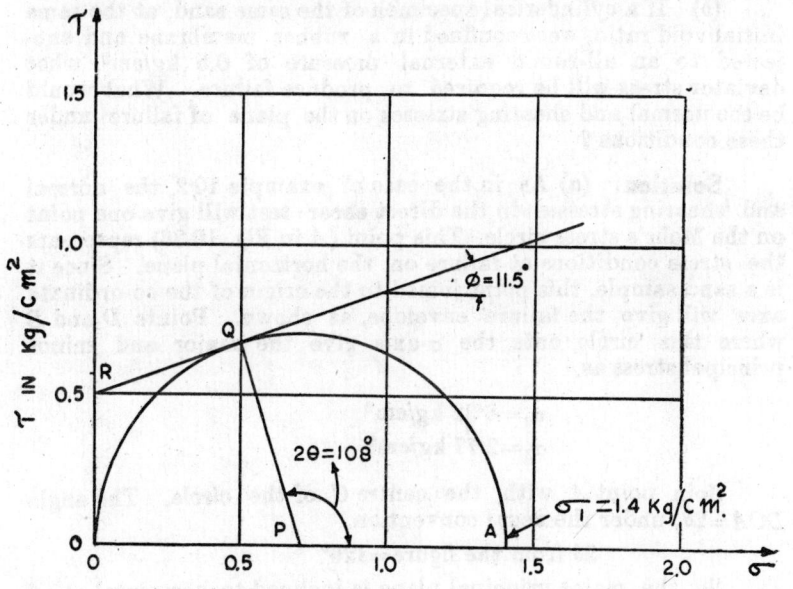

Fig. 10·35.

$$(\sigma_1 - \sigma_3) = 1.4 \text{ kg/cm}^2$$

But $\sigma_3 = 0$

$\therefore$ $\sigma_1 = 1.4 \text{kg/cm}^2$

With these principal stresses, draw Mohr's stress circle as shown in Fig. 10.35. At the centre P of this circle, draw a line PQ which makes an angle $2\theta = 108^\circ$, with PA as shown. Let this line cut the circle in the point Q. Then Q represents the stress conditions on the failure plane. The failure envelope, therefore, must touch the circle at this point Q. So at Q, draw a line perpendicular to PQ and let this line cut the τ-axis in R. Then OR represents the cohesion and the angle made by RQ with the horizontal represents the angle of shearing resistance or angle of internal friction. The two values obtained from this diagram are,

Cohesion $c = 0.525 \text{ kg/cm}^2$ and

$\phi = 11.5^\circ$ **Ans.**

Example 10.4. (a) A specimen of sand tested with a direct shear device, failed under a shear stress of 2.40 kg/cm^2. The normal stress was 4.16 kg/cm^2. Determine the magnitude and directions of principal stresses in the zone of failure.

(b) If a cylinderical specimen of the same sand, at the same initial void ratio, were confined in a rubber membrane and subjected to an all-round external pressure of 0.5 kg/cm^2, what deviator stress will be required to produce failure. What would be the normal and shearing stresses on the plane of failure under these conditions ?

Solution. (a) As in the case of example 10·2, the normal and shearing stresses in the direct shear test will give one point on the Mohr's stress circle. This point (A in Fig. 10.36) represents the stress conditions at failure on the horizontal plane. Since it is a sand sample, this point joined to the origin of the co-ordinate axes will give the failure envelope, as shown. Points D and B where this circle cuts the σ-axis give the major and minor principal stress as.

$$\sigma_1 = 8.32 \text{ kg/cm}^2$$
$$\sigma_3 = 2.77 \text{ kg/cm}^2$$

Join point A with the centre C of the circle. The angle $DCA = 2\theta$, under the usual convention.

2θ from the figure $= 120^\circ$

So, the major principal plane is inclined to horizontal at an angle $\theta = 60^\circ$ and the minor principal plane is inclined to the horizontal at an angle $(90 - \theta) = 30^\circ$. This is shown at the top of Fig. 10.33.

The above results can be arrived at *analytically* also, as shown below,

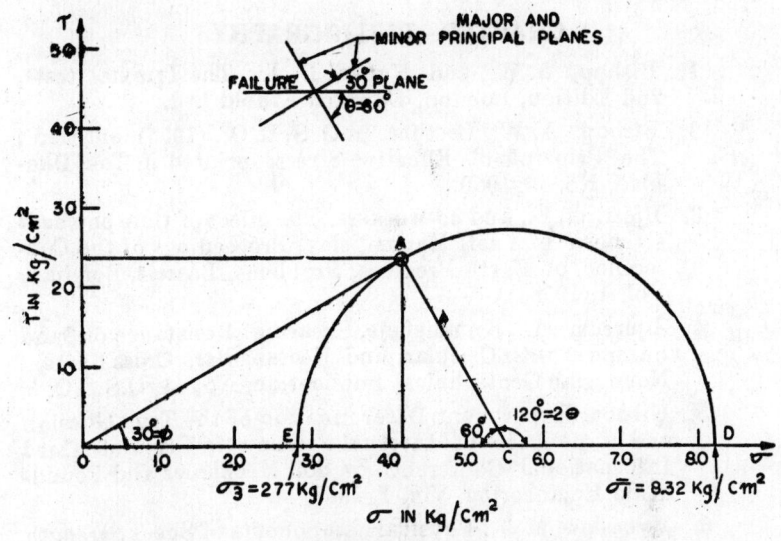

Fig. 10·36.

For sands $\tau_f = \bar{\sigma}_f \tan \phi'$

$\therefore$ $\phi' = \tan^{-1} \dfrac{\tau_f}{\bar{\sigma}_f} = \tan^{-1} \dfrac{2.40}{4.16} = \tan^{-1} 0.577 = 30°$

$$\bar{\sigma}_3 = \frac{\bar{\sigma}_f}{1 + \sin \phi'} = \frac{4.16}{1 + 0.5} = 2.77 \text{ kg/cm}^2$$

and $\sigma_1 = \dfrac{\bar{\sigma}_f}{1 - \sin \phi'} = \dfrac{4.16}{1 - 0.5} = 8.32 \text{kg/cm}^2$

Also $\theta = 40° + \dfrac{\phi'}{2} = 45° + 15 = 60°$

(*b*) In this case $\bar{\sigma}_3 = 0.5 \text{ kg/cm}^2$.

$\therefore$ $\bar{\sigma}_1 = \bar{\sigma}_3 \tan^2 \left(45° + \dfrac{\phi'}{2} \right) = 0.5 \tan^2 60°$

$$= 0.5 \times 3.0 = 1·5 \text{ kg/cm}^2$$

Deviator stress required to fail the sample
$$= (\bar{\sigma}_1 - \bar{\sigma}_3) = 1.5 - .5 = 1.0 \text{ kg/cm}^2$$

Failure stresses in this case would be :—
$$\bar{\sigma}_f = \bar{\sigma}_3 (1 + \sin \phi') = 0.5 (1 + 0.5) = 0.75 \text{ kg/cm}^2$$

and $\tau_f = \dfrac{\bar{\sigma}_1 - \bar{\sigma}_3}{2} \cos \phi' = \dfrac{1}{2} \times 0.866$

$$= 0.433 \text{ kg/cm}^2$$

Note : The above results can be obtained from Mohr's circle also.

CHAPTER—BIBLIOGRAPHY

1. Bishop, A. W., and Hankel D. J., The Triaxial tests, 2nd Edition, London, Edwards Arnold Ltd.

2. Bishop, A. W., Lecture in O. S. L. O., (1955) entitled ; The Principle of Effective Stress, printed in Tek Ukeblad, No. 39, 1959.

3. Bjerrum, L., and co-workers, The effect of time on shear strength of a soft Marine clay, Proceedings of the Conference on Earth Pressure Problems, Bassels, Belgium, Vol. 1, 1958.

4. Bjerrum, L., Kummeneje, Shearing Resistance of Sand Samples with Circular and Rectangular, Cross Section Norwegian Geotechnical Publication, No. 44, O.S.L.O.

5. Gibson, Experiment Determination of the True Ghesion and True Angle of Internal Friction in Clays of Third International Conference on Soil Mechanics and Foundation Engineering, Vol. 1, 1953.

6. Hvorslew, M. J., Physical Components of Shear Strength ASCE Research Conference on Shear Strength of Cohesive Soils, Boulder, Colorado, 1960.

7. Lambe, T. W., Testing of Soils for Engineers.

8. Lambe, T. W., and Whitman, R. V., "Soil Mechanics", J. Willey & Sons, Inc., N. Y., 1969.

9. Means, R. E., Parcher J. V., "Physical Properties of Soils", Charles F. Merild Books Inc., Columbus, Ohio,

10. Schmmertman, J. H., and Osterberg O. J., Experimental Investigation of the Development of Cohesion and Friction with Axial Strain in Saturated Cohesive Soil, ASCE Research, Conference on Shear Strength of Cohesive Soils, Boulder, Colorado, 1960.

11. Scott, R. F., Principles of Soil Mechanics, Mass., Additional-Wesley Publishing Co., Inc., 1963.

12. Skempton, A. W., The pure pressure Coefficient A and B Gestechnique, Vol. IV.

13. Skempton, A. W., and Bishop B. W., Soils, chapter X in Building Materials, their Elasticity and inelasticity, by Reines, Amsterdam (Holland), North Holland Co.

14. Taylor, D. W., Fundamentals of Soil Mechanics, New York, John Wiley and Sons, Inc., 1948.

15. Terzaghi. K., The Shearing Resistance of Saturated Soils and the angle between the planes of Shear, Proceedings of First International Conference on Soil Mechanics and Foundation Engineering, Vol. 1, 1936.

16. **Terzaghi, K., The Coulumb's Equation for the Shear Strength of Cohesive Soils, translated by L. Bjerrum in Form Theory to Practice in Soil Mechanics, New York, John Wiley and Sons, 1960,**

17. **Wu., T. H., Soil Mechanics, Boston, Allyn and Bacon, Inc,, 1966.**

QUESTIONS

1. (a) What are the advantages of triaxial shear test over the direct shear test ? Describe briefly the triaxial shear test with a simple sketch.

(b) The result of a series of triaial tests on a soil have been expressed by means of the values c=1·2 ton per sq. foot and ϕ=20°. On the assumption that the equation $s = c + p \tan \phi$ is valid, what vertical pressure would be required to produce a failure of the mass of soil if the lateral pressure were equal to 0.3 tons per sq. ft. ? (*P. U. 1959 Annual*)

2. (a) What is unconfined compression test ?

Describe the test procedure briefly and mention how it helps in designing footings on cohesive soils.

(b) The sand in a deep deposit has an angle of internal friction of 40°, a dry density of 110 lbs. per cu. ft. and saturated unit weight of 131 lb/cft. If the water table is at a depth of 5 ft., what is the shearing resistance of the material to sliding along a horizontal plane at a depth of 10 ft. (*P. U. 1959 Suppl.*)

3. (a) What is a triaxial shear test ? Describe the test procedure and explain the practical significance of the test.

(b) A sample of sand subjected to triaxial shear test failed when the minor principal stress was 3200 lbs per square ft. and the major principal stress was 11,500 lbs/sq. ft. Draw Mohr's circle and find out the angle of internal friction and the angle which the plane of rupture makes with the minor principal plane. (*P. U. 1960 Annual*)

4. (a) Under ordinary conditions, how will you determine when failure occurs in a soil specimen during a direct shear test ? Define unit shearing strength of a soil.

(b) In a thick deposit of sand, the water table is 7 feet below the top and the sand above the water table is moist having a unit weight of 118 lbs./cft. The saturated sand has a unit weight of 128 lbs./cft. Assuming that the angle of internal friction for the damp sand is 32° and it is not altered due to submergence, compute the shearing resistance against sliding along a horizontal plane at a depth of 20 feet from the top. (*P. U. 1960 Suppl.*)

5. (a) Give the expression for the Coulomb's Law of shear strength for a cohesive soil and discuss briefly the factors on which cohesion and angle of internal friction depend.

(b) Describe briefly three types of tests (depending upon drainage conditions) that are performed on a cohesive soil in a triaxial machine. Give one field example of each test. (*P. U. 1961 Annual*)

6. (a) What is Mohr's Strength Theory for soils ? Sketch typical strength envelopes for a soft clay, a clear sand and a silty clay ?

(b) In an undrained triaxial test on a sample of saturated clay the all round pressure is maintained at 1·00 kg./cm². The unconfined compressive strength is 3·52 kg./cm². At what vertical applied stress in addition to all-round pressure should the sample fail ? (*P. U. 1962 Annual*)

7. How is the unconfined compressive strength of a saturated clay determined in the laboratory ? Sketch the apparatus used. How is unconfined compressive strength related to bearing capacity of a saturated clay ? (*P. U. 1963 Suppl.*)

8. From the geometry of Mohr's Circle, prove that for any material for which Mohr's Strength theory is valid

$$\sigma_1 = \sigma_3 N\phi + 2c\sqrt{N\phi},$$

where σ_1 and σ_2 are the limiting major principal stresses c and ϕ the strength parameters and

$$N\phi = \tan^2\left(45° + \frac{\phi}{2}\right) \qquad (P. U. April 1964)$$

9. Define slow, quick and consolidated quick triaxial shear tests, illustrating their use by at least one field example. (P. U. Nov. 1965)

10. What is Mohr's Strength Theory for soil ? Sketch typical strength envelopes for a clean sand, silty clay and a soft clay.

(P. U. April, 1966)

11. What is Mohr's Stress circle ? The stresses during a triaxial test on a sample of soil are as under :—

Major principal stress=5 kg/cm²
Minor principal stress=3 kg/cm².

Draw the Mohr's stress circle and determine the state of stress on a plane inclined at 55° with the major plane. (P. U. Nov. 1965)

12. State Coulomb's Law of Shear Strength of Soils for $c - \phi$ soil. Establish the relationship between principal stresses at failure for a cohesive soil. (P. U. Nov. 1965)

13. (a) Describe Mohr's Strength Theory as applicable to soils.

(b) Describe briefly the different types of shear tests that can be performed in the laboratory by controlling the drainage conditions in samples. (P. U. April, 1966)

14. (a) What are the various types of shear tests of soils from the view point of drainage of the soil ? Give one example of each of field conditions; which the different types of tests simulate.

(b) Describe the procedure of carrying out the above types of tests in a triaxial compression test apparatus.

(c) A soil sample 3.25 inches long and 1.5 inches in diameter fails in an unconfined compression test under an axial load of 25 lbs. The axial deformation of the sample at failure was observed to be 0·25 inch. What is the shear strength of the soil sample ? (P. U. Nov. 1966)

15. (a) Distinguish between drained shear parameters and undrained shear parameters,

(b) Tests conducted on a new engineering material indicate that the Mohr's envelope is a parabola symmetrial about the σ axis with its focus at the origin of the $z - \sigma$ co-ordinates,

An unconfined compression test indicates q_a=4000 PSF (2 kg/cm²) with a failure plane 60° from the major principle plane. Find.
(i) Shear & normal stress on the plane of failure.
(ii) Draw & mark on Mohr's circle accurately. (Mysore 1965)

16. A consolidated undrained triaxial test is performed on a specimen of saturated clay with a value of 63=2.0 kg/cm². At failure $(\sigma_1 - \sigma_3)$=2.8 Kg/cm² and u=1.8 kg/cm². The failure plane in the test makes an angle of 57° with the horizental.

1. Calculate the normal & shear stresses on the failure surface and the maximum shear stress in the specimen.
2. If the clay specimen had a ϕ=24°, and c of 0.8 kg/cm², show why failure occurs on a plane of 57° instead of the plane of maximum shear.
3. If the specimen in (2) is loaded slowly to failure in a drained test with 63=2.0 kg/cm², what will be the major principle stress at failure.

Earth Pressures, Earth Pressure Theories and Similar Problems

11.1. Introduction

Like a liquid the soil too, exerts a lateral pressure on any structure with which it is in contact. A study of this property of the soil is, therefore, very essential as it is applicable in the design of retaining walls, sheet piles bulk heads, revetments, underground structures like conduits, and tunnels and in fact in many other structures. In this chapter, it is intended to discuss the fundamentals regarding earth pressures and earth pressure theories and give indications about some of the applications, based upon these fundamentals side by side.

11.2. Kinds of Lateral Earth Pressure

Essenttally there are two kinds of earth pressure—the active earth pressure and the passive earth pressure. The active and passive earth pressures develop as a result of soil structure interaction and they are attributed to two distinct and different physical conditions. Consider a retaini g wall, retaining a mass of clean, dry and cohesionless sand. Now, suppose, that due to some reason, the retaining wall moves outward, as shown in Fig. 11·1 (a). The sand will have a tendency to move alongwith the wall and shall like to attain its natural slope depending upon its angle of repose. Due to this tendency, it will exert some pressure on the wall. This *pressure* that the soil developes due to the *outward movement* of the wall is called the *active earth pressure*[1]. For stability, the wall should be able to with stand this pressure, Now, suppose that due to some other reason, the wall moves towards the soil, as shown in

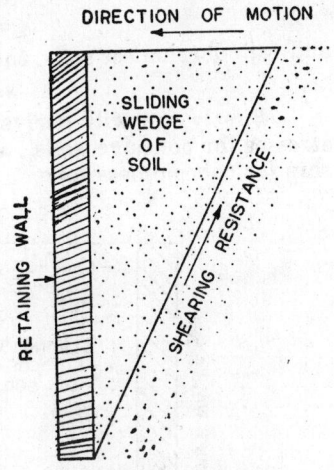

DIRECTION OF MOTION

SLIDING WEDGE OF SOIL

RETAINING WALL

SHEARING RESISTANCE

Development of active earth pressure

Fig. 11·1 (a)

---

1. The best way to remember it is this : The soil stretches, and chases the wall, so it is active and hence the name *active earth pressure*.

Fig. 11.1 (b). In order to maintain stability, the soil will offer resistance and in this process of offering resistance, it will develop some pressure. This *pressure* that the soil developes due to the *inward movement* of the wall is called the *passive earth pressure*[1]. The passive earth pres sure may be much more than the active earth pressure.

The active and the passive earth pressures are associated with the movement of the structure in contact with the soil and they represent two extreme physical conditions. There can, therefore, be an *intermediate* value of pressure that should be associated with the stationary condition of the structure of soil. This pressure that the soil developes at *zero movement* may be called the **earth pressure at rest.**

11.3 Relationship Between Principal Stresses at Failure

Before the fundamental concepts and theories about earth pressures are detailed out, it is felt necessary to give the relationship between principal stresses both for cohesive as well cohesionless soils. It may be recalled here that for cohesive soils, the Mohr-Coulumb criterion for failure gives the equation for failure/strength envelope as

$$s = c + \sigma \tan \phi$$

and for cohesionless soils, the equation reads as

$$s = \sigma \tan \phi$$

Firstly, we will derive the relationship between principal stresses for cohesive soils, and then will deduce the relationship for cohesionless soils.

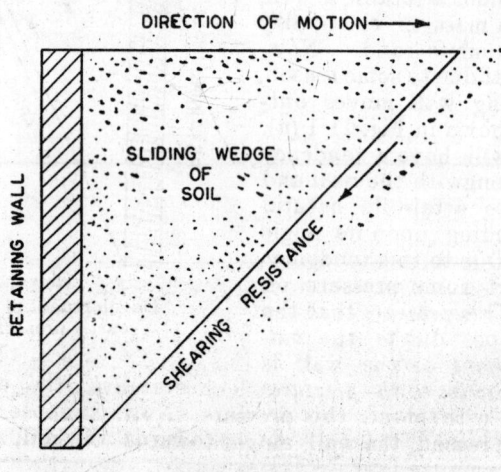

DIRECTION OF MOTION →

SLIDING WEDGE OF SOIL

RETAINING WALL

SHEARING RESISTANCE

Fig. 11·1 (b)
Development of passive earth pressure.

---

1. The best way to remember it is this : The wall chases the soil ; so th soil is passive and hence the name *passing earth pressure.*

Let AB be the envelope to the Mohr's circle drawn with principal pressure σ_1 and σ_3, as shown in Fig. 11·2. Equation of this envelope is

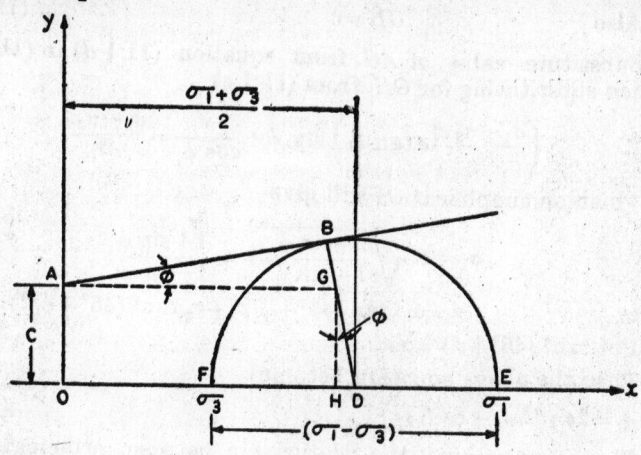

Fig. 11·2.

$$s = c + \sigma \tan \phi$$

BD represents the radios circle and equals $\dfrac{\sigma_1 - \sigma_3}{2}$

Draw a line AG parallel to the x-axis meeting BD in G. Drop a perpendicular GH on the x-axis.

$$\angle GAB = \angle DGH$$

and $\qquad\qquad GH = c$

$\therefore \qquad\qquad \dfrac{GH}{GD} = \cos \phi \qquad\qquad$ (From $\triangle DGH$)

$\therefore \qquad\qquad GD = GH/\cos \phi \qquad\qquad$ ...(11·1 a)

Also $\qquad\qquad \dfrac{BG}{AG} = \sin \phi \qquad\qquad$ (From $\triangle ABG$)

$\therefore \qquad\qquad BG = AG \sin \phi \qquad\qquad$ ...(11·1 b)

But $\qquad BG + GD = BD = \dfrac{\sigma_1 - \sigma_3}{2}$

$$AG \sin \phi + \frac{GH}{\cos \phi} = \frac{\sigma_1 - \sigma_3}{2} \qquad\qquad \text{...(11·1 c)}$$

Now AG, from the figure $= OD - HD$

Here $\qquad\qquad OD = \dfrac{\sigma_1 + \sigma_3}{2}$

and $\qquad\qquad \dfrac{HD}{GH} = \tan \phi$

or $\qquad\qquad HD = GH \tan \phi$

Substituting these values for OD and HD

∴ $AG = \dfrac{\sigma_1 + \sigma_3}{2} - GH \tan \phi$ ...(11.1 d)

Also $GH = c$ ...(11.1 e)

Substitute value of AG from equation (11.1 d) in (11.1 c) and then substituting for GH from (11.1 e)

∴ $\left(\dfrac{\sigma_1 + \sigma_3}{2} - c \tan \phi \right) \sin \phi + \dfrac{c}{\cos \phi} = \dfrac{\sigma_1 - \sigma_3}{2}$

which on simplification will give

$$\sigma_1 = 2c \sqrt{\dfrac{1 + \sin \phi}{1 - \sin \phi}} + \sigma_3 \dfrac{1 + \sin \phi}{1 - \sin \phi}$$

or $\sigma_1 = 2c \tan (45° + \phi/2) + \sigma_3 \tan^2 (45° + \phi/2)$

Put $\tan^2 (45° + \phi) = N\phi$

Then the above equation becomes

$\sigma_1 = 2c \sqrt{N\phi} + \sigma_3 N\phi$ ...(11.1)

This equation gives the relationship between principal stresses in case of cohesive soils.

For the cohesionless soils, $c = 0$ and the envelope passes through the origion.

∴   Substituting $c = 0$ in equation (11.1), we get

$\sigma_1 = \sigma_3 N$ ...(11.2)

which gives the relationship between principal stresses for cohesionless soils.

It may be noted that $N\phi = \tan^2 (45° + \phi/2)$ is also equal to $\dfrac{1 + \sin \phi}{1 - \sin \phi}$. This can be deduced trignometrically. $N\phi$ in the equations (11.1) and (11.2) is called the *flow value*.

The relationships expressed by the equations (11.1) and (11.2) are very useful in earth pressure problems.

11.4. Coefficient of Earth Pressure at Rest

The coefficient of earth pressure at rest (defined subsequently) is obtained either from the considerations based upon theory of elasticity or from the available field data.

Consider a soil deposit and assume that it is elastic, homogeneous, isotropic and semi-infinite in extent. Let a retaining wall retain this deposit as shown in Fig. 11.3. A prism of material at a depth z below the surface is acted upon by the vertical stress σ_v and radial a (horizontal) stress σ_h. Let E and μ represent the modulus of elasticity and Poisson's ratio for this material. Then,

Vertical strain $\epsilon_v = \dfrac{\sigma_v}{E}$

and lateral strain in the horizontal direction.[1]

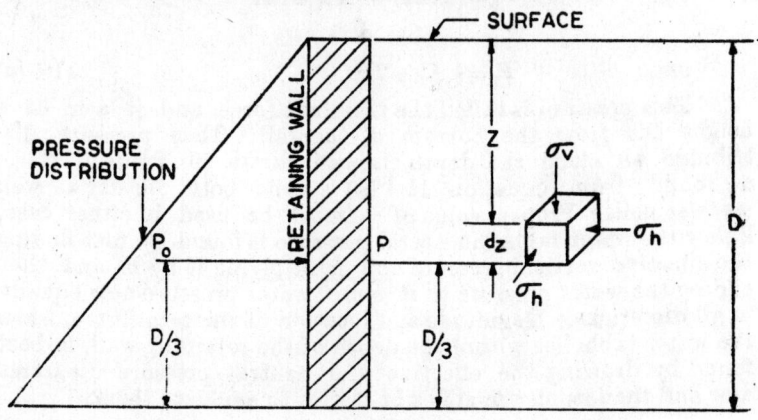

Fig 11·3.

$$\epsilon_h = \frac{\sigma_h}{E} - \mu\left[\frac{\sigma_h}{E} + \frac{\sigma_v}{E}\right].$$

Since the material is of infinite extent in the horizonatal direction, ϵ_h must equal zero, as it cannot deform horizontally.

$$\therefore \qquad \frac{\sigma_h}{E} - \mu\left[\frac{\sigma_h}{E} + \frac{\sigma_v}{E}\right] = 0$$

$$\therefore \qquad \frac{\sigma_h}{\sigma_v} = \frac{\mu}{1-\mu} \qquad \qquad \ldots(11\text{·}3)$$

Vertical stress σ_v at depth z equals γz, where γ is the unit weight of the material. So, the horizontal stress $\sigma_h = \left(\dfrac{\mu}{1-\mu}\right)(\gamma z)$.

The ratio $\left(\dfrac{\mu}{1-\mu}\right)$ is designated as K_o and is called the *coefficient of earth pressure at rest.*

This means that,

$$\sigma_h = K_o\,(\gamma z) \qquad \qquad \ldots(11\text{·}4)$$

where $K_o = \dfrac{\mu}{1-\mu} =$ Coefficient of earth pressure at rest. The horizontal pressure σ_h determined in equation (11·4) is the *earth pressure at rest.*

If the height of retaining wall is D as shown in Fig. 11·3, the total horizontal pressure P_0 per unit width of the wall exerted by

---

1. Applying generalized Hook's Law.

the soil on the wall can be found out by intergrating equation (11·4) between the limits $z=0$ to $z=D$

$$\therefore \quad P_0 = (\sigma_h \, dz) = \int_0^D K_0(\gamma z) \, dz$$

or $\qquad P_0 = \frac{1}{2} K_0 \gamma D^2 \qquad \qquad ...11\cdot4 \ (a)$

This pressure is called the resultant force and it acts at a height $D/3$ from the bottom of the wall. Then pressure distributed all along the depth is also shown in figure 11·3 σ_h as found from equation 11·4 (a) is valid both for dry as well as wet soils. Proper value of γ must be used in either case. Below the water table the lateral pressure is found by first finding the effective vertical pressure and multiplying it by K_0 and then adding the water pressure to it, since water pressure acts equally in all directions. Magnitude and location of the total force, when the water table lies within the depth of the retaining wall, is best found by drawing the effective and neutral pressure diagrams first and then using principles of statics to evaluate them.

Theoretically K_0 equals $\left(\dfrac{\mu}{1-\mu} \right)$

where a proper choice of the Poisson's ratio for the soil may be made. Choice of μ for a soil is not easy since in no way a soil can be considered elastic, homogeneous and isotropic material, as already pointed out in Article 8·1.

Incidentally, it may be mentioned that we have assumed that there are no shearing stresses on the planes on which σ_v and σ_h act i.e. we have assumed that σ_v and σ_h are principal stresses.

Some values of K_0 are available from experience for various types of soils and these values may be made use of in any practical problem, when needed. Table 11·1 gives these values.

TABLE 11·1

Values of K_0 for differing types of soils

S. No.	Soil Type	Coefficient of Earth Pressure at Rest K_0
1	Loose Sand	0·4
2	Dense Sand	0·6
3	Sand compacted in Layers	0.8
4	Soft Clay	0·6
5	Hard Clay	0·5

11·5 Plastic Equilibrium in Soils

A mass of soil is said to be in a *state of plastic equilibrium* if failure is about to occur simultaneously at all points within this mass. Such a state of failure in soils can only be brought about by the tectonic forces. Normally, the yielding of a rigid structure in contact with a soil mass produces failure only in a portion of the whole mass. As a consequence, a state of the soil mass in which failure is imminent at all points within the mass is generally referred as *a general state of plastic equilibrium* and a state of the soil mass in which a part of the mass undergoes failure is referred to as *a local state of plastic equilibrium.*

Rankine was the first to indicate the stress conditions associated with the states of plastic equilibrium in a semi-infinite mass of gravity alone. States of plastic equilibrium identical with those that Rankine investigated came to be known as the Rankine's state of plastic equilibrium. These states of plastic equilibrium in semi-infinite masses only serve as an introduction to the practical problems in which more complicated and generally local states of plastic equilibrium exist.

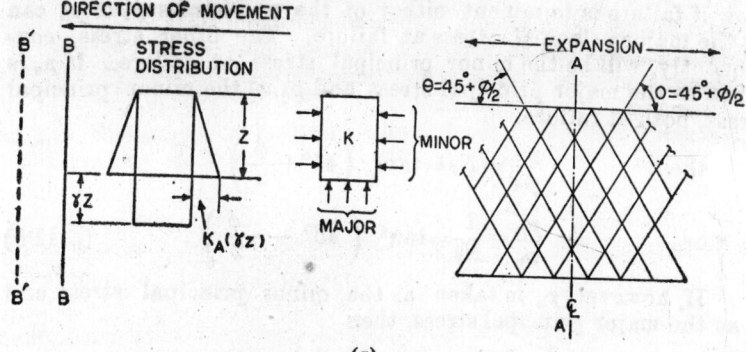

(a)

Rankine's Active State

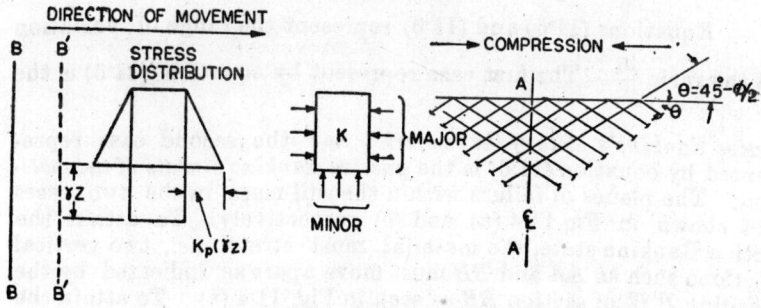

(b)

Rankine's Passive State

Fig. 11·14.

Rankine's concept of plastic equilibrium in a semi-infinite soil mass is presented in Fig. 11·4. The mass is considered elastic, homogeneous and isotropic : For simplicity, discussion in this article is ristricted to a cohesionless soil only. Consider an element K of unit area within this mass. This element is symmetrical with respect to a vertical plane. The vertical stress p_v on the base at a depth z below the surface is given by,

$$p_v = \gamma z$$

where γ is the unit weight of the soil. Due to symmetry p_v is a principal stress. Therefore, a normal stress p_h on the vertical sides of the element at a depth z, will be a principal stress.

According to equations (11·2) the relationship between the major and the minor principal stresses in cohesionless soils, at failure is given by

$$\frac{\sigma_1}{\sigma_3} = N_\phi = \tan^2 (45° + \phi/2)$$

Now if failure is imminent either of the two stress p_v or p_h can be the major principal stress at failure. The other stress, consequently, will be the minor principal stress at failure. If p_v is taken as the major principal stress and p_h as the minor principal stress, both at failure.

then,
$$\frac{p_v}{p_h} = N_\phi = \tan^2 \left(45° + \frac{\phi}{2} \right)$$

or
$$\frac{p_h}{p_v} = \frac{1}{N_\phi} = \tan^2 \left(45° - \frac{\phi}{2} \right) \qquad (\ldots 11\cdot5)$$

If, however, p_v is taken as the minor principal stress and p_h as the major principal stress, then

$$\frac{p_h}{p_v} N_\phi = \tan^2 \left(45° + \frac{\phi}{2} \right) \qquad \ldots(11\cdot6)$$

Equations (11·5) and (11·6) represent the range of variation of the rario $\dfrac{p_h}{p_v}$. The first case represent by equation (11·5) is the active Rankine's state of the material and the second case represented by equation (11·6) is the passive Rankine's state of the material. The planes of failure within the soil mass in the two cases are shown in Fig 11·4 (a) and (b) respectively. To attain the active Rankine state, the material must stretch i.e., two vertical sections such as AA and BB must move apart as indicated by the position B'B' of section BB as seen in Fig. 11·4 (b). To attain the passive Rankine state, the material must compress i.e. two vertical sections such as AA and BB must move closer as indicated by the position B'B' of section BB as seen in Fig. 11·4 (b). The planes of failure are inclined to the horizontal at an angle θ equal to $\left(45°+\dfrac{\phi}{2}\right)^1$ in the active case and equal to $\left(45°-\phi\right)^1$ in the passive case. The pattern of failure represented in these figures by these planes crossing each other is called *shear pattern*.

The ratio of horizontal to vertical stress *i.e.*, ratio $\dfrac{p_h}{p_v}$ in the active case is always designated by $K_A\left(=\dfrac{1}{N_\phi}\right)$ and in the passive case by K_P $(=N_\phi)$, so that

$$p_h=K_Ap_v=\dfrac{1}{N_\phi}\ p_v=\tan^2\left(45°-\dfrac{\phi}{2}\right)p_v \qquad \qquad ...(11\cdot7)$$

in the active case and

$$p_h=K_Pp_v=N_\phi\ p_v=\tan^2\left(45°+\dfrac{\phi}{2}\right)p_v \qquad \qquad ...(11\cdot8)$$

in the passive case.

The coefficient K_A and K_P in the above equations are called *Rankine's coefficient of active earth pressure* and *Rankine's coefficient of passive earth pressure* respectively. Most of the times, the word Rankine is dropped while expressing these coefficients.

Numerical values of K_A and K_P depend upon the value of the angle of internal friction ϕ. These two numerical values represent two extreme limits for the coefficient of earth pressure. The *coefficient of earth pressure at rest* is an intermediate[1] value between these limits. Also K_A will be smaller than K_P. Equations (11·7) and (11·8) clearly show that the horizontal pressure in the active case will be smaller than the one in the passive case *i.e.*, if the material stretches, the lateral pressure is smaller than if it compresses.

The stress distributions in the two cases are also indicated in Fig. 11·4 (a) and (b) respectively.

The active and the passive states of equilibrium as defined by Rankine are the extreme states of equilibrium for the material and any state inbetween these extremes (including the state of rest) is called an *state of elastic equilibrium.*

The above states of a soil mass investigated by Rankine are general states of plastic equilibrium. In actual physical problems, normally, local states of plastic equilibrium are encountrend. The states of stresses in the plastic zone as also the shape and extent of the zone depend upon the *deformation* and *boundary conditions*

---

1. If σ_1 and σ_3 are the major and minor principal stresses at failure, the plane of failure, is inclined at an angle of $\left(45°+\dfrac{\phi}{2}\right)$ to the direction of the major principal plane. So the above angles emerge from this principle.

associated with the structure under consideration. For a general state of plastic equilibrium to occur, it is essential that the sec-

tion BB moves to the position $B'B'$, which constitutes the deformation condition (see Fig. 11·4). Also the base and sides of the section must be smooth so as to ensure zero shear stresses on the horizontal and vertical planes and this condition of smoothness constitutes the necessary boundary condition. In an actual structure, these simple deformations and boundary conditions do not exist. Following are given as mere illustrations:

(a) Refer to Fig. 11·5. It shows a vertical section through a prismatic box, in which sand gets deposited, in the same state as in the field. It is assumed that the sand is dry. Assume that the wall BB of the box is smooth and the base BA is rough. Now, if the wall BB yields about the bottom point B to the position BB', the sand mass contained in the wedge BBC is free to deform and in this zone the state of stresses correspond to the active Rankine case, whereas the sand in this rest of the box due to the friction at the base cannot be expected to attain the active Rankine state in the zone BBC is mobilized, if the ratio $\dfrac{BB'}{BC}$ in

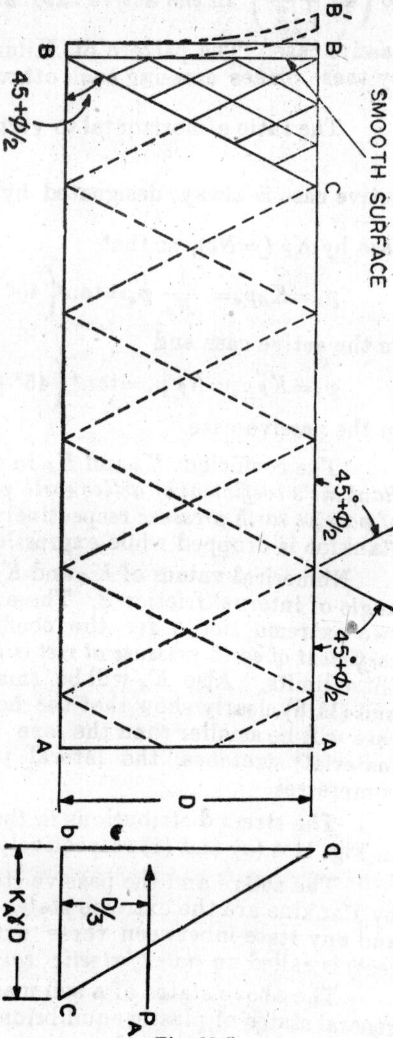

Fig. 11·5 equals the ratio $\dfrac{BB'}{AB}$ in Fig. 11·4 (a). The shear pattern in the zone is also indicated in the Figure. Beyond this condition, the material will slide along the surface CB which makes an angle of $45°+\phi/2$ with the horizontal. The pressure distribution in this case is shown on the right.

Fig. 11·5.

(b) Another situation is illustrated in Fig. 11·6, in which a smooth vertical retaining wall yields about the top point B. Both

theoretical and experimental investigations show that in such a case, the surface of sliding starts at the bottom point B at an angle of $45°+\phi/2$ with the horizontal and after becoming steeper as it goes upwards, it intersects the ground surface at

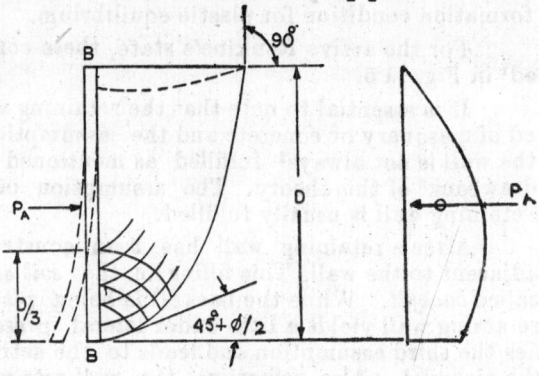

Fig. 11·6.

90 degrees. The plastic equilibrium starts at the bottom, and progresses towards the top. The distribution of pressure is roughly parabolic as against triangular in the previous case.

This case also corresponds to the Rankine's active case. Similar cases where a supporting wall moves inwards and yields either about the top or about the bottom point can be quoted to show the development of passive Rankine state only in a portion of the soil mass retained by the wall.

These cases only go to show that the general state of plastic equilibrium investigated by Rankine is only a starting point in the study of earth pressures and that the actual cases are much more complicated.

11·6. Rankine's Earth Pressure Theory

The question of finding out the lateral earth pressure against retaining walls is one of the oldest in the civil engineering field. A lot of work both theoretical as well as experimental, has been done in the area and many hypotheses and theories have been proposed. Of all of them, Rankine's and Coulomb's theories have been well accepted by the profession and since they have stood the test of the time, they have been often called the *classical earth pressure theories.* The Rankine's theory shall be discussed in this article and the Coulomb's theory shall follow in a subsequent one. As originally proposed, the Rankine's theory covered the uniform cohesionless soils only, although latter on it was extended to stratified, partially immersed cohesionless masses and cohesive soils too.

There are three assumptions that are associated with the Rankine's theory. They are,

1. The retaining wall face is smooth.
2. The wall is vertical.

3. The wall yields about the base and thus satisfies the deformation condition for plastic equilibrium.

For the active Rankine's state, these conditions are indicated[1] in Fig. 11·5.

It is essential to note that the retaining walls are constructed of masonary or concrete and the assumption of smoothness of the wall is not always[2] fulfilled as mentioned above. This is a drawback[2] of this theory. The assumption of verticality of the retaining wall is usually fulfilled.

After a retaining wall has been construsted, soil is filled adjacent to the wall. This filling i.e. the *soil adjacent to the wall* is called *backfill*. Whiie the backfill is being placed in position, the retaining wall yields a litte under lateral p essure and thus satisfies the third assumption and leads to the active Rankine state of the material. Also, sometimes the wall gets pulled towards the soil and again satisfies the third assumption and leads to the passive Rankine state of the material. Thus the magnitude and distribution of lateral earth pressures on retaining walls are obtained based upon the above conditions duly taken into consideration by Rankine. Following cases shall be discussed :

(*a*) Active earth pressure on dry or moist cohesionless soils without surcharge.

(*b*) Active earth pressure on partially immersed cohesionless soils, without surcharge.

(*c*) Active earth peessure on dry or moist and partially submerged cohesionless soils with surcharge.

(*d*) Active earth pressure on cohesive soils.

(*e*) Passive earth pressure on cohesionless soils.

---

1. The side of the box in Fig. 11·5 is comparable to a retaining wall.

2. Roughness of the back of the wall commonly leads to reduced active and passive earth pressures and hence the assumption is on the safer side. The assumption of a smooth vertical surface is almost fullfilled in the case of a cantilever retaining wall. If this wall yields outwards the failure of the material is along the surfaces AB and BC as shown in Fig. 11.7 These surfaces rise at an angle of $45° + \phi/2$ with the horizontal Within the zone ABC, the ma-

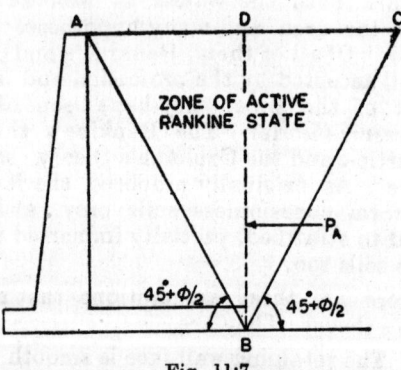

ZONE OF ACTIVE RANKINE STATE

P_A

$45 + \Phi/2$ $45 + \Phi/2$

Fig. 11·7.

terial is the active Rankine state and there are no shearing stresses on the plane BD. So this plane BD represents what maybe called smooth vertical wall.

(*f*) **Passive earth pressure on cohesive soils.**

In the above cases, it will be assumed that the surface of the material is horizontal and coincides with the top of the retaining wall. The effect of the inclined surface will be indicated in a subsequent paragraph. Effect of stratification will also be discussed in a subsequent note.

Case (a) *Active earth pressure on dry or moist cohesionless soils, without surcharge.*

The magnitude and distribution of earth pressure in this case is comparable to the case (*a*) of local state of plastic equilibrium discussed in article 11·5. and illustrated in Fig. 11·5. The vertical pressure p_v and depth z is

$$p_v = \gamma z$$

where γ is the unit weight of the dry[1] or moist sand. The horizontal pressure p_h at this depth is

$$p_h = K_A p_v = K_A(\gamma z) \qquad \qquad ...(11·9)$$

If the height of the wall is designated as D then the total (resultant) horizontal pressure P_A can be obtained by integrating the above equation between the limits

$$z = 0$$

and $\qquad z = D$

$$P_A = \int p_h dz = \int_0^D K_A\, \gamma z dz = \tfrac{1}{2} K_A \gamma D^2 \qquad ...(11·10)$$

The pressure distribution is triangular as can be seen from equation 11·9 since z varies from zero at the top to D at the bottom. This triangular distribution is shown in Fig. 11·8 ; with the total (resultant) horizontal pressure P_A acting at one-third the height from the bottom.

The value of K_A in equation (11·10) is

$$K_A = \frac{1}{N_\phi} = \tan^2\left(45° - \frac{\phi}{2}\right)$$

as seen in equation (11·7).

Case (b) *Active earth pressure on partially immersed cohesionless soils, without surcharge.*

Fig. 11.9 (*a*) shows a soils profile in which the water table stands at a depth D_1 below the ground surface. If a cut is made in this profile and a retaining wall of height $D > D_1$ is constructed in this material to retain the soil on one side of the cut, the unit weight γ of the sand above the water table is more than the submerged unit weight γ_b of the sand below the water table. So, the

----

1. If the soil is dry, then in the calculations for the unit weight γ, the degree of saturation S may be taken zero. If it is a moist soil, the relevant degree of saturation may be taken.

effective unit vertical stress above and below the water table will be different. If z denotes any depth below the ground surface, then the

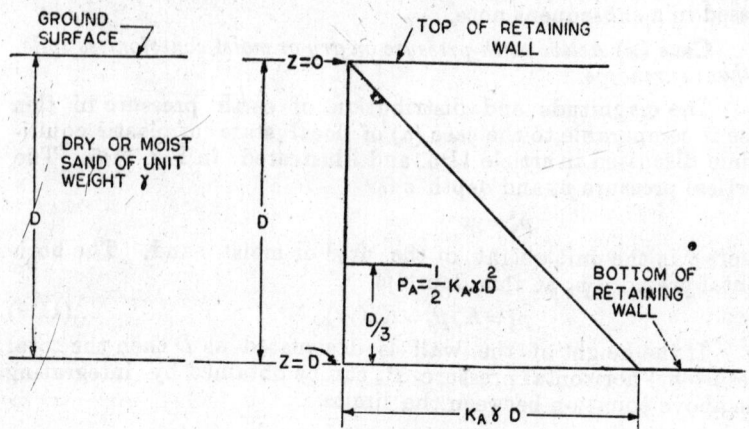

Active earth pressure distribution in dry or moist cohesionless soil.
Fig. 11·8.

effective stress above the water table

$$=\gamma z, (z < D_1)$$

and effective stress below the water table

$$=\gamma D_1 + \gamma_b(z - D_1)(z > D_1)$$

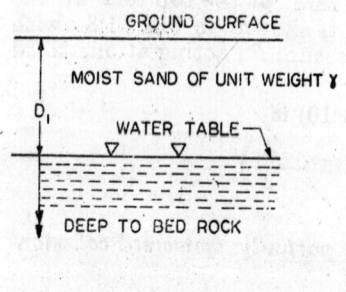

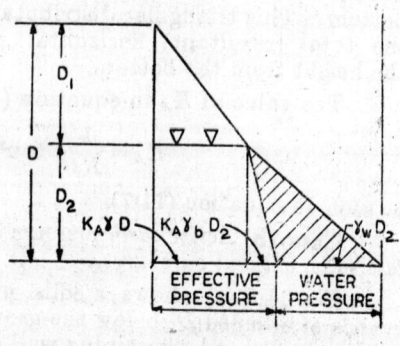

(a)   (b)

Active earth pressure distribution in partially immersed sand.
Fig. 11·9.

Therefore, the effective unit horizontal stress at a depth $z > D_1$ will be given by

$$p_h = [\gamma D_1 + \gamma_b(z - D_1)]K_A$$

In addition the water pressure equal to $\gamma_w\,(z-D_1)$ at a depth $z>D_1$ below the ground surface, acts equally in all directions, so that the total unit horizontal stress at $z>D_1$ is given by,

$$p_h=[\gamma D_1+\gamma_b(z-D_1)]K_A+\gamma_w(z-D_1)$$

If the total height of the retaining wall is D out of which D_1 is above the water table and D_2 is below it, then in the above equation

$$z-D_1=D_2,$$

so that unit horizontal pressure at the base is given by

$$p_h=K_A[\gamma D_1+\gamma_b D_2]+\gamma_w D_2 \qquad ...(11.11)$$

The pressure distribution is shown in Fig. 11.9 (b). The effective and water pressure components are shown very distinctly in the diagram. The total pressure and its point of application can be worked out from the pressure diagram using principles of statics.

Case (c). *Active earth pressure on dry or moist and partially immersed cohesionless soils with surcharge.*

If the backfill carries a uniform load of intensity q per unit area, the effective vertical stress p_v increases by an amount q at any depth[1] z. Therefore, the increase in the active Rankine pressure in the horizontal direction is K_A times q_2 i.e.,

$$\triangle p_h=K_A q \qquad ...(11.12)$$

This applies both to the dry or moist backfill as wall as to the partially immersed backfill. The pressure diagram in either case is increased by this amount $K_A q$ throughout the depth, as shown in Fig. 11.10 and 11.11.

Case (d) *Active earth pressure on cohesive soils.*

The shear strength of cohesive soils is governed by equation

$$s=c+\sigma \tan \phi$$

and it was established in articles 11.3, that the ratio of the major to minor principal stresses in a cohesive soil is given by

$$\sigma_1=2c\sqrt{N_\phi}+\sigma_3 N\phi \qquad \text{(equation 11.2)}$$

where all the symbols have their usual significance. As the wall is smooth, the vertical pressure p_v at any depth z below the ground surface is γz where γ is the unit weight of the material. In the active earth pressure case, this forms the major principal stress equal to σ_1 and the horizontal pressure p_h forms the minor principal stress σ_3.

$$\therefore \qquad p_v=2c\sqrt{N_\phi}+p_h N\phi$$

Divide both sides by N_ϕ and simplify to get p_h.

$$\therefore \qquad p_h=p_v\cdot\frac{1}{N_\phi}-2c\ \frac{1}{\sqrt{N_\phi}}$$

---

1. No stress distribution is to be taken into account here.

or
$$p_h = \gamma z \frac{1}{N_\phi} - 2c . \frac{1}{\sqrt{N_\phi}} \qquad \qquad ...(11.13)$$

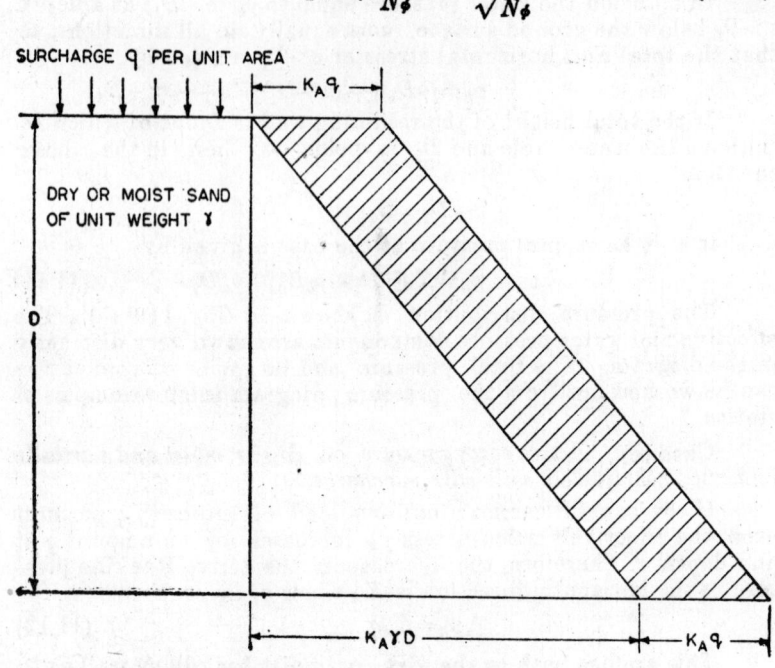

Effect of uniform surcharge on active earth pressure
distribution in dry or moist sand.
Fig. 11.10.

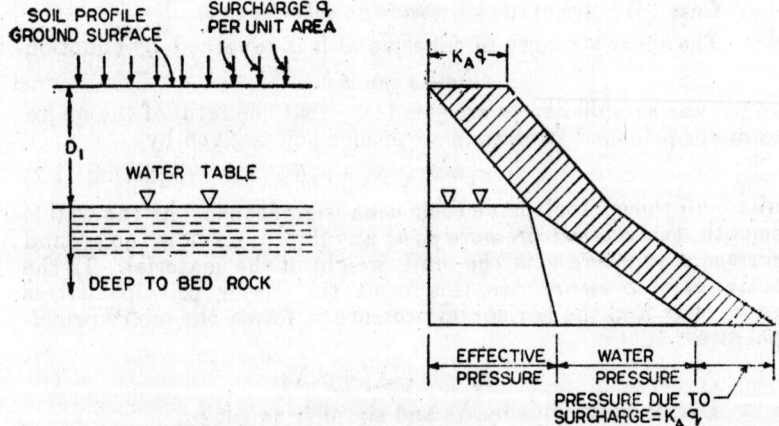

Effect of uniform surcharge on active earth pressure
distribution in partially submerged sand.
Fig. 11.11.

When $z=0$,

$$p_h = -2c \frac{1}{\sqrt{N_\phi}}$$

which is a negative value.

When $p_h = 0$,

$$z = \frac{2c}{\gamma} \sqrt{N_\phi}$$

This particular value of z when the horizontal pressure is zero is designated as z_0

$$\therefore \qquad z_0 = \frac{2c}{\gamma} \sqrt{N_\phi} \qquad \qquad ...(11.14)$$

So, for $z > z_0$. p_h has a positive value. Hence for the height D of a retaining wall in a cohesive material, the pressure upto a depth

$$z_0 = \frac{2c}{\gamma} \sqrt{N_\phi}$$

from the surface is negative and below this depth it becomes positive.

The pressure distribution is shown in Fig. 11.12.

In order to calculate the total (resultant) pressure P_A, equation (11.13) may be intergrated between the limits,

$$z = 0$$

and $$z = D.$$

$$\therefore \qquad P_A = \int_0^D p_h\, dz = \frac{1}{2}\gamma D^2 . \frac{1}{N_\phi} - 2c \frac{D}{\sqrt{N_\phi}} \qquad ...(11.15)$$

If this resultant P_A is zero, then from equation (11.15), D_c works out as,

$$D = \frac{4c}{\gamma} \sqrt{N_\phi}$$

This particular value of D when the resultant pressure is zero is designated as, D_c

$$\therefore \qquad D_c = \frac{4c}{\gamma} \sqrt{N_\phi} \qquad \qquad ...(11.16)$$

A comparison of equations (11.14) and (11.16) shows that

$$D_c = 2z_0$$

This means that a vertical bank of cohesive soil smaller than D' should be able to stand without any lateral support. This

condition is not strictly true in practice. since the pressure against
the wall increases from a value of $-\dfrac{2c}{\sqrt{N_\phi}}$ at $z = 0$ to $+\dfrac{2c}{\sqrt{N_\phi}}$ at $z = D_c$
of as seen in Fig. 11·12, whereas on a vertical unsupported face
a bank the normal stress is zero at every point. As a result,
the greatest depth upto which a vertical cut can stand unsupport-
ted is a little less than D_c. For a soft clay angle of internal
friction is zero $i.e.$

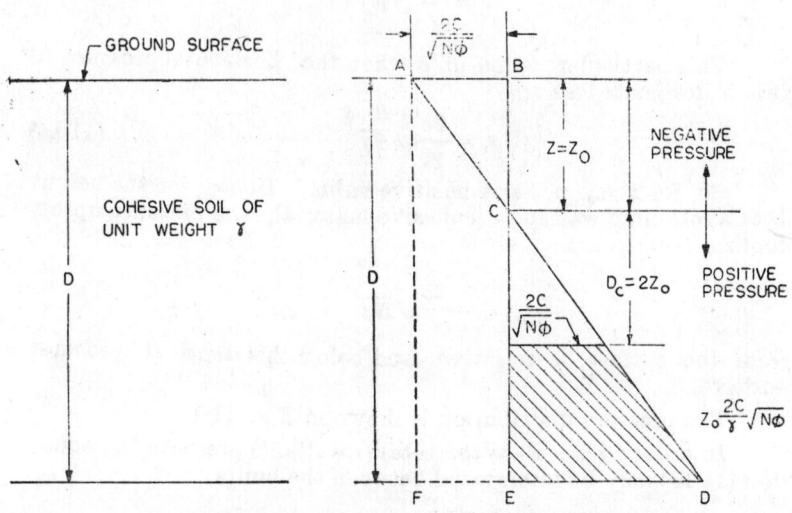

Active earth pressure distribution in a cohesive soil
Fig. 11·12

$$\phi = 0$$
$$\therefore \quad N_\phi = \tan^2 (45° + 0)$$
$$= \tan^2 45° = 1$$

So, for such soils,

$$P_A = \frac{1}{2}\gamma D^2 - 2cD \qquad \qquad ...(11\cdot17)$$

and
$$D_c = \frac{4c}{\gamma} \qquad \qquad ...(11\cdot18)$$

Usually due to the tension in the soil upto a depth z_0 below
the surface, the soil loses contact with the retaining wall. As a
consequence the triangle ABC representing the total negative
pressure should be excluded while calculating P_A. The area of
this triangle is,

$$\frac{1}{2}\left(\frac{2c}{\sqrt{N_\phi}}\right)\left(\frac{2c}{\gamma}\sqrt{N_\phi}\right) = \frac{2c^2}{\gamma}$$

The modified total (resultant) pressure[1] on the retaining wall will be.

$$P_A = \frac{1}{2}\gamma D^2 \cdot \frac{1}{N_\phi} - 2c \frac{D}{\sqrt{N_\phi}} + \frac{2c^2}{\gamma} \qquad ...(11.19)$$

when $\phi = 0$,

$N_\phi = 1$

and therefore for a soft clay,

$$P_A = \frac{1}{2}\gamma D^2 - 2cD + \frac{2c^2}{\gamma} \qquad ...(11\cdot20)$$

If the surface of the soil carries a uniform surcharge per unit area, the horizontal pressure throughout the depth will increase by an amount $K_A q$

where $K_A = \dfrac{1}{N_\phi}$

and the total increase in pressure for the depth D will be, $K_A q D$

or $\dfrac{qD}{N_\phi}$

For the soil carrying a uniform surcharge,

$$P_A = \frac{1}{2}\gamma D^2 \frac{1}{N_\phi} - 2c \frac{D}{\sqrt{N_\phi}} + \frac{2c^2}{\gamma} + \frac{qD}{N_\phi}$$

Case (e). *Passive earth pressure on cohesionless soils.*

For the passive earth pressure to develop, the retaining wall must get pushed towards the soil as indicated in Fig. 11.13 (a). A shear pattern of the type shown is developed due to this deformation. The planes of shear rise at an angle of $45° - \dfrac{\phi}{2}$ to the horizontal. at any depth z below the surface, the vertical pressure p_v equals γz where γ is the unit weight of the dry or moist soil. In the passive earth pressure case, p_v becomes the minor principal stress and p_h, becomes the major principal stress.

From equation (11·6), for the cohesionless soils, the ratio of p_h to p_v in the passive earth pressure equals

$$N_\phi (=K_P).$$

$\therefore$ $$p_h = N\phi p_v$$
$$= K_P(\gamma z) \qquad ...(11.21)$$

So, the total (resultant) pressure for the height D of the wall will be given by integrating equation (11·21) between the limits

$$z = 0$$

and $$z = D$$

---

1. The modified resultant pressure can also be found out by considering the area of triangle ADF and subtracting from it the area $ACEF$ in Fig. 11·12.

$$\therefore \qquad P_P = \int_0^D K_P(\gamma z) dz$$

$$= \frac{1}{2} K_P \gamma D^2 \qquad \qquad ...(11\cdot22)$$

The pressure distribution is triangular as shown in Fig. 11.13 (*b*).

In case the soil is partially immersed in water and the water table lies above the base of the retaining wall, then the passive earth pressure at any depth D is found out by finding out the effective and neutral stresses separately as done in sub-section (*b*) above.

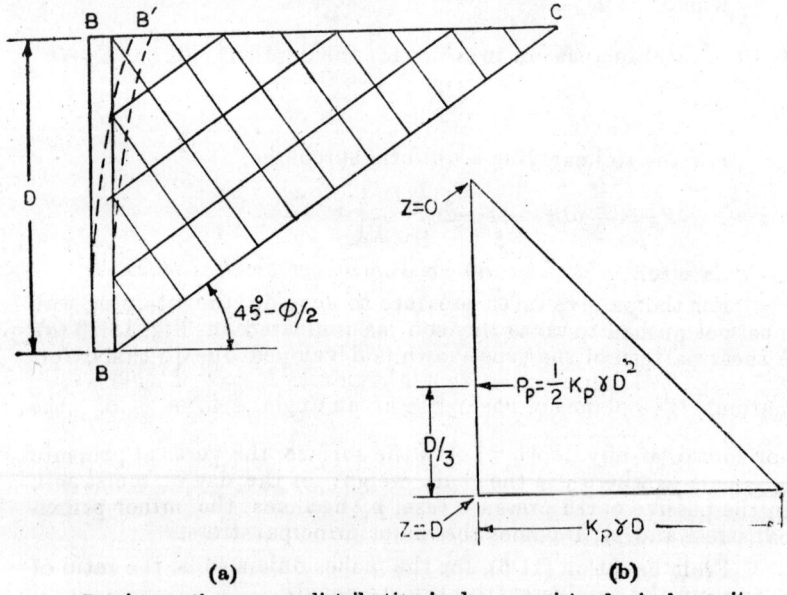

(a)   (b)

Passive earth pressure distribution in dry or moist cohesionless soil.

Fig. 11.13

If there is a uniform surcharge q per unit area on the back-fill, then the increase in horizontal pressure equals $K_P q$ and this increase is uniform throughout the depth of the soil behind the wall.

Pressure distribution diagram for the above two cases are similar in shape to the diagram shown in Fig. 11.10 and 11.11.

Case (f). *Passive earth pressure on cohesive soils.*

Equation (11·1) gives

$$\sigma_1 = 2c \sqrt{N_\phi} + \sigma_3 N_\phi$$

where all the quantities have their usual significance. In the passive earth pressure, σ_1 equals p_h and σ_3 equals p_v.

$$\therefore \qquad p_h = 2c\sqrt{N_\phi} + p_v N_\phi.$$

At any depth z below the surface, vertical stress p_v equals γz so that the above equation becomes.

$$p_h = 2c\sqrt{N_\phi} + N_\phi \gamma z \qquad \ldots (11 \cdot 23)$$

which gives the horizontal pressure at any depth. The total (resultant) pressure for the depth D of a retaining wall, retaining this material, can be obtained by integrating equation (11·23) between the limits

$$z = 0$$

$$\text{to } z = D$$

$$\therefore \qquad P_P = \int_0^D p_h$$

$$= 2c\sqrt{N_\phi}D + \frac{1}{2}\gamma N_\phi D^2 \qquad \ldots (11 \cdot 24)$$

If the backfill carries a uniform surcharge of q per unit area, then the horizontal pressure p_h at any depth z is increased by an amount $K_p q$

where $\qquad K_p = N_\phi$ i.e. p_h

becomes,

$$p_p = 2c\sqrt{N_\phi} + \gamma z N_\phi + qN \qquad \ldots (11 \cdot 25)$$

Total increase in pressure for the depth D of the retaining will be $K_P q\, D$ or $N_\phi qD$.

$\therefore$ For the soil carrying a uniform surcharge,

$$P_P = 2c\sqrt{N_\phi}D + \frac{1}{2}\gamma N_\phi D^2 + N_\phi qD \qquad \ldots (11 \cdot 26)$$

In equation (11·26) let the quantity $2c\sqrt{N_\phi}D$ equal X, $\frac{1}{2}\gamma N_\phi D^2$ equal Y and $N_\phi qD$ equal Z. Then in Fig. 11·14, the portion X is represented by the diagram $ABCD$, Y is represented by the triangle EFG and Z is represented by the rectangle $DCFE$. Point of application of X and Z, therefore, lies at $D/2$ from the bottom and that of Y lies at $D/3$ from the bottom.

Since X and Z have the same point of application, they can be taken together.

$\therefore$ Designate the quantity $(X+Z)$ as P_P' and the quantity Y as P_P''.

$$\therefore \qquad P_P'' = 2c\sqrt{N_\phi}D + N_\phi qD \qquad \ldots (11 \cdot 27)$$

$$P_P'' = \tfrac{1}{2}\gamma N_\phi D^2 \qquad \ldots (11 \cdot 28)$$

P_P' has its point of application at $D/2$ from the bottom and P_P' 1 as it point of application at $D/3$ from the bottom. Also

$$P_P' + P_P'' = P_P.$$

If equations (11·27) and (11.28) are now compared with equation (11·26), it can be seen that equation 11·27 is obtainable from equation (11·26) if γ in the latter is substituted equal to zero and equation (11 28) is obtainable from equation (11·26) if c and q in the latter are substituted as zero. So, the total passive earth pressure can be obtained in two steps.

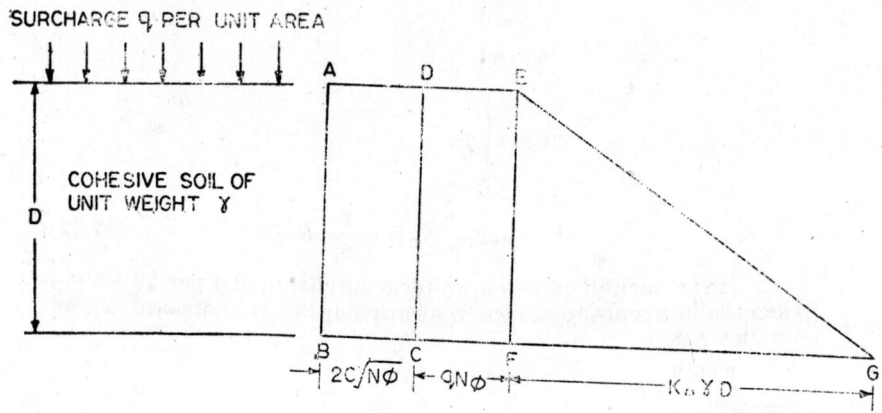

SURCHARGE q PER UNIT AREA

COHESIVE SOIL OF UNIT WEIGHT γ

Fig. 11.14.

1. Assume the unit weight γ of the soil as zero. The portion of passive earth pressure so obtained has its point of application at one-half the height of the wall from the bottom.

2. Assume both the cohesion c and surcharge q as zero. The portion of passive earth pressure so obtained has its point of application at one-third the weight of the wall from the bottom.

Obtaining the passive earth pressure in cohesive soils by the above procedure in two steps, simplifies calculations by more elaborate procedures like the *General Wedge Theory* and is also useful in the computation of bearing capacity of shallow foundations by Terzaghi's theory.

The sub-division of P_P into P_P' ann P_P'' is only true if the wall is vertical and perfectly smooth. In any other case, this will only be an approximate procedure.

Effect of Inclination of the ground surface on active earth pressure of dry or moist cohesionless soils :

In order to understand the effect of inclination of the ground surface on the active (or passive) earth pressure, consider an infinite slope in a cohesionless and homogeneous material as shown in Fig. 11·15 (a). Now consider a vertical slice $ABCD$ of the material having unit depth perpendicular to the plane of paper enclosed between the ground slope XX and a plane YY drawn parallel to the ground surface. The forces acting on this slice are, the weight W of the slice, the reaction at the level of YY

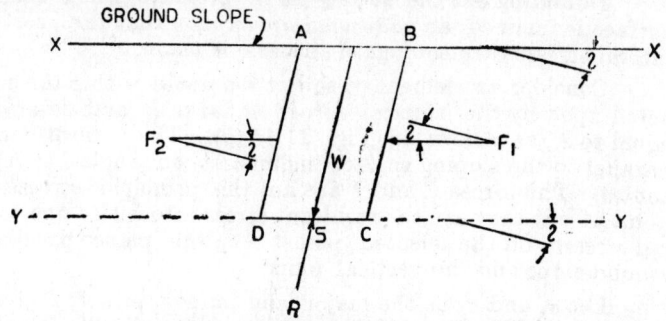

Fig. 11.15 (a)

and two forces F_1 and F_2 acting on the sides. For equilibrium of the slice, the weight W must equal the reaction R and they must have a common line of action, with directions of action opposite to each other. Also the forces F_1 and F_2 acting on the

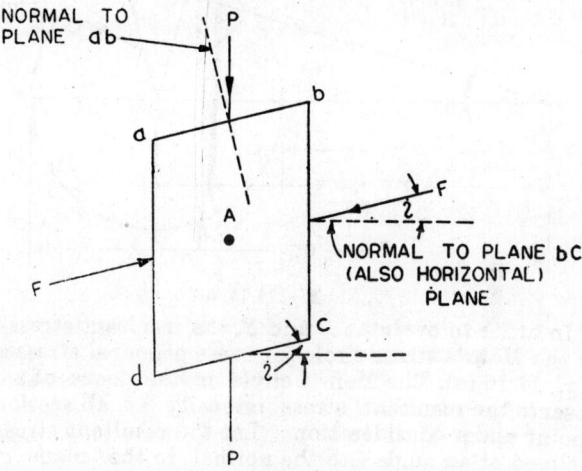

Fig. 11.15 (b)

sides must be equal and opposite and their line of action must be parallel to the slope XX. Otherwise, there would be a residual moment with respect to the point E.

It can be seen that the forces on the slice are either vertical or are parallel to the sloping surface. In theory of elasticity it has been proved that if the stress on a given plane at a given point is parallel to another plane, the stress on the latter plane at the same point must be parallel to the first plane. Such planes are called *conjugate planes* with respect to stresses on them and the stresses are called *conjugate stresses*. The forces considered in Fig. 11.15 (a) are the conjugate stresses.

In finding out the active earth pressure when the ground surface is inclined, an *additional assumption* that the vertical and lateral stresses are conjugate stresses, is made.

Consider an element of soil at a point A within the soil mass acted upon by the vertical forces equal to P and lateral forces equal to F, as drawn in Fig. 11·15 (b). The lateral forces act parallel to the sloping surface inclined at an angle i to the horizontal. The forces P and F are not the principal stresses at the point A but they are the resultant stress intensities due to principal stresses on the selected planes *i.e.* the plane parallel to the ground slope and the vertical plane.

Let σ_1 and σ_3 be the major and minor principal stresses at the point A. Then for stability, from equation (11·2) for cohesionless materials, one can deduce that

$$\frac{\sigma_1 - \sigma_3}{\sigma_1 + \sigma_3} = \sin \phi \qquad \qquad ...(11\text{-}29)$$

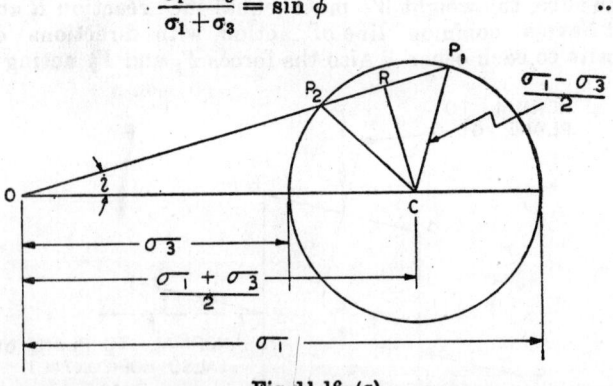

Fig. 11.16. (a)

In order to evalutae P and F, the resultant stress-intensities, draw the Mohr's stress circle for these principal stresses, as shown in Fig. 11·16 (a). The Mohr's circle is the locus of a point that represents the resultant stress intensity at all sections through the point under consideration. Let the resultant stress intensity be inclined at an angle i to the normal to the plane on which it acts. Through the origin O, draw a line inclined at an angle i to the line OC, where C is the centre of the circle. This line cuts the circle in two points P_1 and P_3. There are, therefore. two planes on which the resultant stress intensity is inclined at an angle i to the normal, the resultant stress intensities[1] P and F

---

1. If $P = OP_1$ and $F = OP_2$, this simply means that the lateral pressure F is less the vertical pressure P. So, this stress system implies the active Rankine case.

being represented by OP_1 and OP_2 respectively. Through C drop a perpendicular CR on the line OP_2P_1. Join CP_1 and CP_2. From Fig. 11·16 (a)

$$OR = OC \cos i = \frac{\sigma_1 + \sigma_3}{2} \cos i$$

and

$$CR = OC \sin i = \frac{\sigma_1 + \sigma_3}{2} \sin i$$

Also

$$P_2R = RP_1 = \sqrt{CP_2^2 - CR^2}$$

$$= \sqrt{\left(\frac{\sigma_1 - \sigma_3}{2}\right)^2 - \left(\frac{\sigma_1 + \sigma_3}{2}\right)^2 \sin^2 i}$$

Substitute from equation (11·29), $(\sigma_1 - \sigma_3) = (\sigma_1 + \sigma_3) \sin \phi$.

$$\therefore \qquad P_2R = RP_1 = \frac{\sigma_1 + \sigma_3}{2} \sqrt{\sin^2 \phi - \sin^2 i}$$

Now stress $P = OP_1 = OR + RP_1$

or

$$P = \frac{\sigma_1 + \sigma_3}{2} \cos i + \frac{\sigma_1 + \sigma_3}{2} \sqrt{\sin^2 \phi - \sin^2 i}$$

or

$$P = \frac{\sigma_1 + \sigma_3}{2} \left[\cos i + \sqrt{\sin^2 \phi - \sin^2 i} \right]$$

or

$$P = \frac{\sigma_1 + \sigma_3}{2} \left[\cos i + \sqrt{\cos^2 i - \cos^2 \phi} \right]$$

Also stress $F = OP_2 = OR - P_2R$

or

$$F = \frac{\sigma_1 + \sigma_3}{2} \cos i - \frac{\sigma_1 + \sigma_3}{2} \sqrt{\sin^2 \phi - \sin^2 i}$$

or

$$F = \frac{\sigma_1 + \sigma_3}{2} \left[\cos i - \sqrt{\cos^2 i - \cos^2 \phi} \right] \dots (11·31)$$

The stress ratio F/P designated as K from equations (11·30) and (11·31) is,

$$\frac{F}{P} = K = \frac{\cos i - \sqrt{\cos^2 i - \cos^2 \phi}}{\cos i + \sqrt{\cos^2 i - \cos^2 \phi}} \qquad \dots (11·32)$$

This ratio is called the **conjugate ratio** or **Rankine's lateral pressure ratio**.

The lateral pressure in this case is *incidentially* the Rankine active earth pressure.

$$\therefore \qquad p_h (= F) = K p_v (= P) \qquad \dots (11·33)$$

Let the point A being considered be at a depth z below the inclined surface in a material whose unit weight is γ [See Fig. 11·16 (b)]. Then if the width of the element in the plane of the figure is x, then the vertical pressure on the plane cd parallel to the ground surface is

$$(\gamma z)(x \cos i)/x = \gamma z \cos i.$$

∴ From equation (11·33), horizontal pressure p_h is given by,

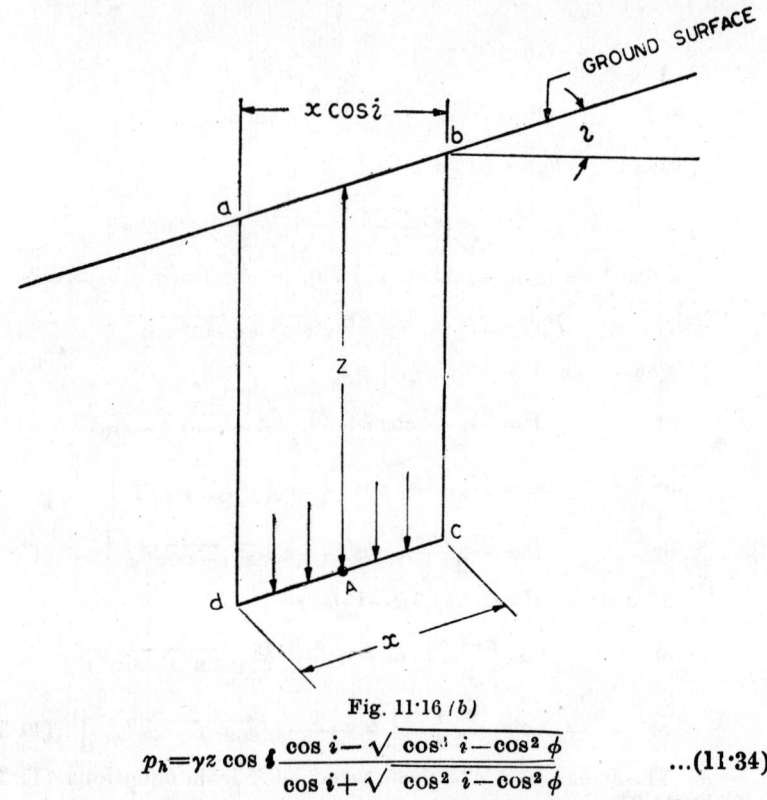

Fig. 11·16 (b)

$$p_h = \gamma z \cos i \cdot \frac{\cos i - \sqrt{\cos^1 i - \cos^2 \phi}}{\cos i + \sqrt{\cos^2 i - \cos^2 \phi}} \qquad ...(11·34)$$

By a similar reasoning, the Rankine's passive earth pressure is given by,

$$p_h = \gamma z \cos i \cdot \frac{\cos i + \sqrt{\cos^2 i - \cos^2 \phi}}{\cos i - \sqrt{\cos^2 i - \cos^2 \phi}} \qquad ...(11·35)$$

Effect of stratification on the active earth pressure of dry or moist cohesionless soil. Fig. 11·17 shows two layers of homogenous cohesionless soils (dry or moist) with thicknesses D_1 and D_2; unit weights γ_1 and γ_2 and angles of internal friction ϕ_1 and ϕ_2, respectively. The shearing stresses on the horizontal planes are zero and hence the stresses on the horizontal and vertical planes are the principal stresses. At any depth $z < D_1$, the vertical principal stress σ_{v_1} is $\gamma_1 z$ and, therefore, the horizontal pressure σ_{h_1} at that depth is given by,

$$\sigma_{h_1} = K_{A_1} \sigma_{v_1} = K_{A_1} (\gamma_1 z) \qquad (...11·36)$$

where K_{A_1} is the coefficient of active earth pressure for the first layer and equals $\dfrac{1}{N_{\phi_1}}=\tan^2(45°-\phi_1/2)$. In the figure this pressure is represented by the line AB_1 and at the depth D_1, its value is

$$K_{A_1}\,\gamma_1 D_1=BB_1.$$

At another depth $z>D_1$ the vertical principal stress $^\sigma v_2$ is given by,

$$^\sigma v_2 =\gamma_1 D_1+\gamma_2(z-D_2)$$

and therefore, the horizontal pressure $^\sigma h_2$ at that depth is given by,

$$^\sigma h_2 = K_{A_2}\ ^\sigma v_2 = K_{A_2}\,[\gamma_1 D_1+(z-D_1)] \qquad ...(11\cdot37)$$

where K_{A_2} is the coefficient of active earth pressure for the second layer and equals $\dfrac{1}{N_{\phi_1}}=\tan^2\left(45°-\dfrac{\phi_2}{2}\right)$. In the figure, this pressure is represented by the line B_2C_2 and at the depth $z=D_1+D_2$ its value becomes,

$$^\sigma h_2=K_{A_2}\,[\gamma_1 D_1+\gamma_2(D_1+D_2-D_1)]$$

or

$$^\sigma h_2=K_{A_2}[\gamma_1 D_1+\gamma_2 D_2]$$

or

$$^\sigma h_2=K_{A_2}\,\gamma_2\left[D_2+\dfrac{\gamma_1}{\gamma_2}D_1\right]$$

$$=CC_2\ \text{(in the figure)}.$$

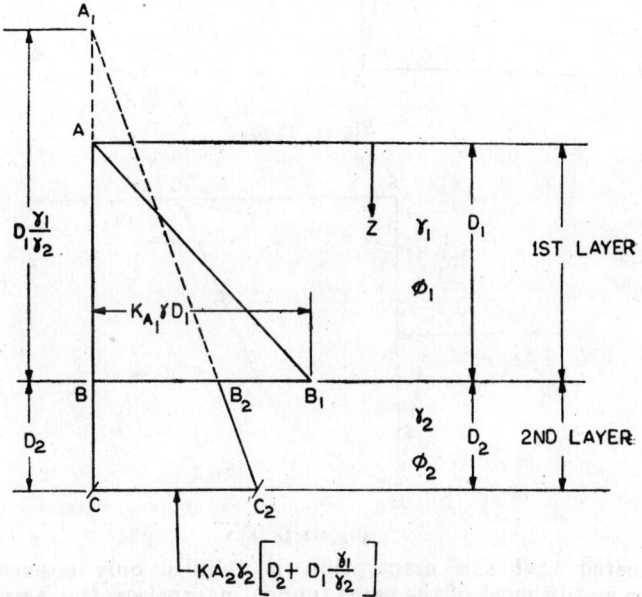

Active earth pressure in dry or moist two layered system.
Fig. 11·17.

The line C_2B_2 when projected back cuts the reference line ABC

in A_1. The distance A_1B equals $D_1\,\dfrac{\gamma_1}{\gamma_2}$ and it represents the thick-

ness of a layer of soil with a unit weight γ_2 which exerts on the surface of the second layer the same pressure as the first layer with a unit weight γ_1. actually resting on this surface.

The centre of pressure can be determined from the prinici-ples of statics, using the pressure distribution diagram in Fig. 11·17.

If the number of layers is more than two, the above analy-sis may be extended to cover the particular system.

11·7.   Coulomb's Theory of Active Earth Pressure

A reference to article 11·6 will indicate that it was assumed in the Rankine theory, that the wall face is smooth. It was also

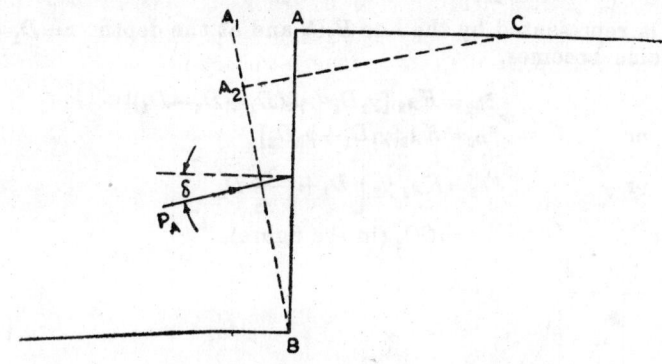

Fig. 11. 18 (a

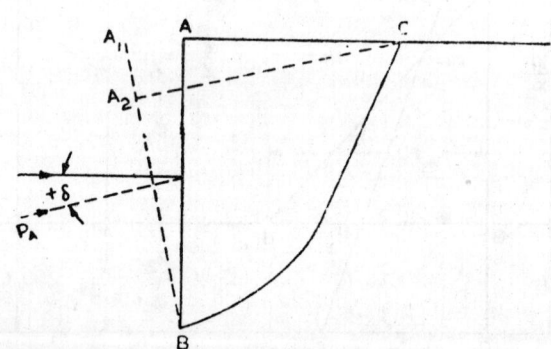

Fig. 11·18 (b)

indicated that this assumption is fulfilled only in exceptional cases and in most of the cases found in practice, the assumption is not correct. Footnote on page 331 indicates that the roughness of the wall leads to reduced earth pressures and hence the above assumption of smoothness in the Rankine theory is on the safer side. However, the error involved in making this assumption is not small.

The Columb's theory assumes that the wall surface is rough. The effect of the roughness of the wall surface on the active earth pressure is shown in Fig. 11·18 (a). It represents a section through a cohesionless material, supported along a vertical face AB. When the the wall AB deforms the surface CA of the wedge, CBA also deforms as shown by its position CA_2 after the deformation. The downward movement of the material along the rough surface AB changes the direction of the earth pressure from its original horizontal position to an inclined position. The new direction is inclined at an angle δ to the normal to the wall. The angle δ is called the **angle of wall friction**. For active earth pressure δ is considered positive when it is measured downward from the normal to the wall as shown in this figure.

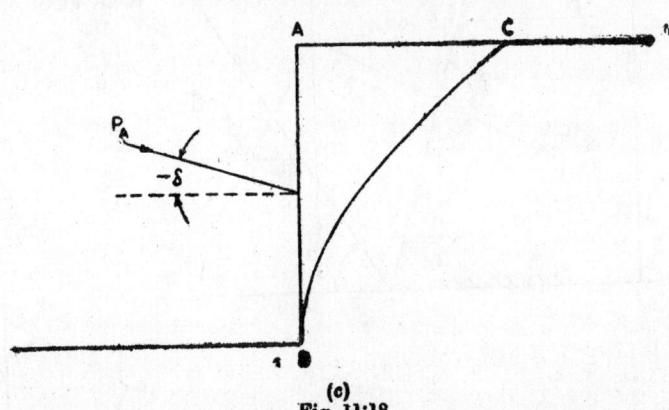

(c)
Fig. 11·18,

When the angle δ is not equal to zero, the shape of the sliding wedge ABC is not triangular as shown in Fig. 11·18 (a), for simplicity. Investigations have shown that where the wall surface is a vertical plane and the backfill carries no surcharge, the sliding surface of the wedge ABC is curvilinear and of the type shown in Fig. 11·18 (b) for positive value of δ and it is curvilinear and of the type shown in Fig. 11·18 (c) for negative values of δ. Methods to find out the active earth pressure P_A for such cases are available. However, the final equations in all the methods are very complicated. Sufficiently accurate results can be obtained by assuming that the surface BC is a plane. This assumption has been made by Coulomb in his theory. Also it is not necessary in this theory, that the wall surface must be vertical or that the surface of the fill must have a uniform shape. The theory is equally applicable to inclined walls, a wall with a broken surface, a curved or a broken fill surface and also to distributed and concentrated surcharge loads on the fill.

The important assumptions, therefore, in this theory are,

1. The retaining wall surface is rough.

2. The wall surface need not be vertical.

3. The sliding surface BC of the sliding wedge ABC is a plane (Refer to Fig. 11·18).

4. The wall yields about the base and thus satisfies the ceformation condition for plastic equilibrium.

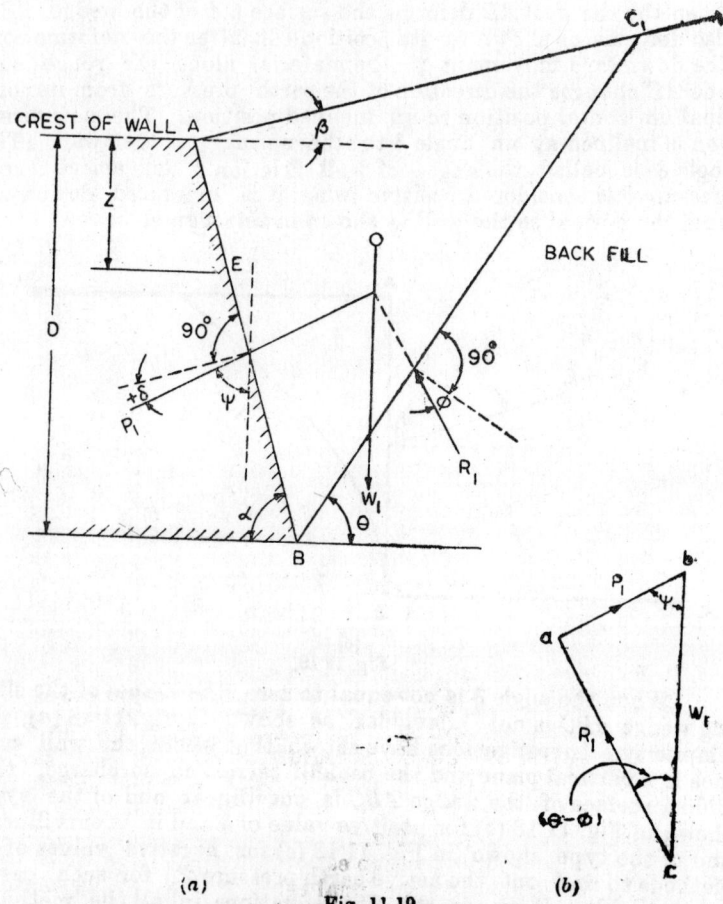

Fig. 11.19
Coulomb's theory of active earth pressure.

A simple case of active earth pressure on an inclined wall with a uniformly sloping fill, as shown in Fig. 11·19 (a) will be discussed for illustration. The material being retained is cohesionless and homogenous. The depth perpendicular to the plane of the paper is taken as unity.

In the figure BC_1 represents an arbitrarily chosen plane through the material passing through the base B. ABC represents the failure wedge. This wedge is acted upon by the following forces :

(a) The weight W_1 of the wedge.

(b) The reaction R_1 acting on the face BC_1 at angle ϕ to the normal to the face.

(c) The reaction P_1 along the back of the wall, at an angle δ to the normal to the wall face AB.

Let,

a=angle that the back of the wall makes with the horizontal.

β=angle that the surface of the back fill makes with the horizontal.

θ=angle that the right hand side BC_1 of the wedge ABC_1 makes with the horizontal.

ϕ=angle of internal friction of the soil.

These angles are shown in Fig. 11·19 (a). Since the wedge is in equilibrium under the action of these three forces W_1, R_1 and P_1 in the first place these forces meet in a point and secondly the polygon of forces shown in Fig. 11·19 (b) is a closed polygon. The direction of all the three forces are known and the weight W_1 of the wedge ABC_1 can be calculated provided the unit weight γ of the material is known. So, with these known parameters, P_1 can be found from the force polygon.

The value of P_1 depends upon the slope angle θ of the surface BC_1. For $\theta=(180-a)$, P_1 equals zero. With decreasing value of θ, P_1 increases and passes through a maximum. Then it decreases and for $\theta=\phi$, it again becomes zero. The wall should be sufficiently strong to withstand the greatest lateral pressure, P_{max} equal to the active earth pressure P_A. Hence what is needed is to find out the maximum value of P. Coulomb found out the value of P_A[1] is given by

$$P_A=\tfrac{1}{2}\gamma D^2 \frac{K_A}{\sin a \cos \delta} \qquad \ldots\ldots\ldots(11·38)$$

where
$$K_A=\frac{\sin^2(a+\phi)\cos \delta}{\sin a \sin (a-\delta)\left[1+\sqrt{\dfrac{\sin (\phi+\delta)\,\sin\,(\phi-\beta)}{\sin (a-\delta)\,\sin\,(a+\beta)}}\right]^2}$$

1. The passive earth pressure P_P is given by,
$$P_P=\tfrac{1}{2}\gamma D^2 \frac{K_P}{\sin a \cos \delta}$$

where
$$K_P=\frac{\sin^2 (a+\phi)\cos \delta}{\sin a \sin (a-\delta)\left[1-\sqrt{\dfrac{\sin (\phi+\delta)\sin (\phi-\beta)}{\sin (a-\delta)\sin (a+\beta)}}\right]^2}$$

When from above, value of K_A depends exclusively on the values of a, β, δ and ϕ. For $a=90$, $\beta=0$ and $\phi=\delta=30°$, the difference between the exact value of the earth pressure corresponding to the failure wedge of the type shown in Fig. 11·18 (b) and the Coulomb's value is less than five percent. With decreasing values of δ, the error decreases further and for $\delta=0$ (and $a=90$ and $\beta=0$, of eourse) the Coulomb's value obtained from equation (11.38) equals the Rankine value obtainable from equation (11.10).

The analytical solution given in equation (11.38) for even this simple case is quite complicated. So, graphical solutions designed to evaluate P_A are normally used and they are simple. One of the methods is the "trial and error method" in which a number of trial failure wedges of the type ABC_1 are chosen and the force polygons are drawn for all such wedges, from which values of P_1, P_2 etc. are found. The maximum value of P obtained from any one of the force polygons is the value of P_A, Culmann (1866), Rebhann (1871), Poncelet (1840) and Engesser (1880) have given better

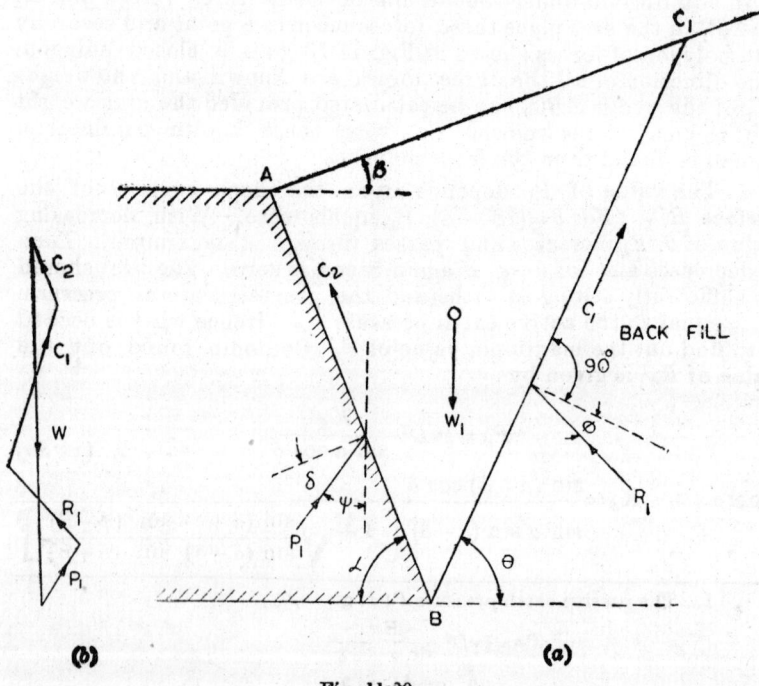

Fig. 11·20.

graphical procedures to determine the value of P_A. Culmann's and Rebhann's methods will be described in the subsequent articles.

Using the principles described above, the active earth pressure in a cohesive material can also be determined. Fig. 11·20 (a) represents the forces acting on the wedge ABC_1. They are,

(a) The weight W_1 of the wedge.

(b) The reaction R_1 acting on the face BC_1 at an angle ϕ to the normal to the-face.

(c) The reaction P_1 along the back of the wall, at an angle δ to the normal to the wall face AB.

(d) Cohesion $c_1 = c \times (BC_1 \times 1)$ along the plane BC_1 and in the direction B to C_1.

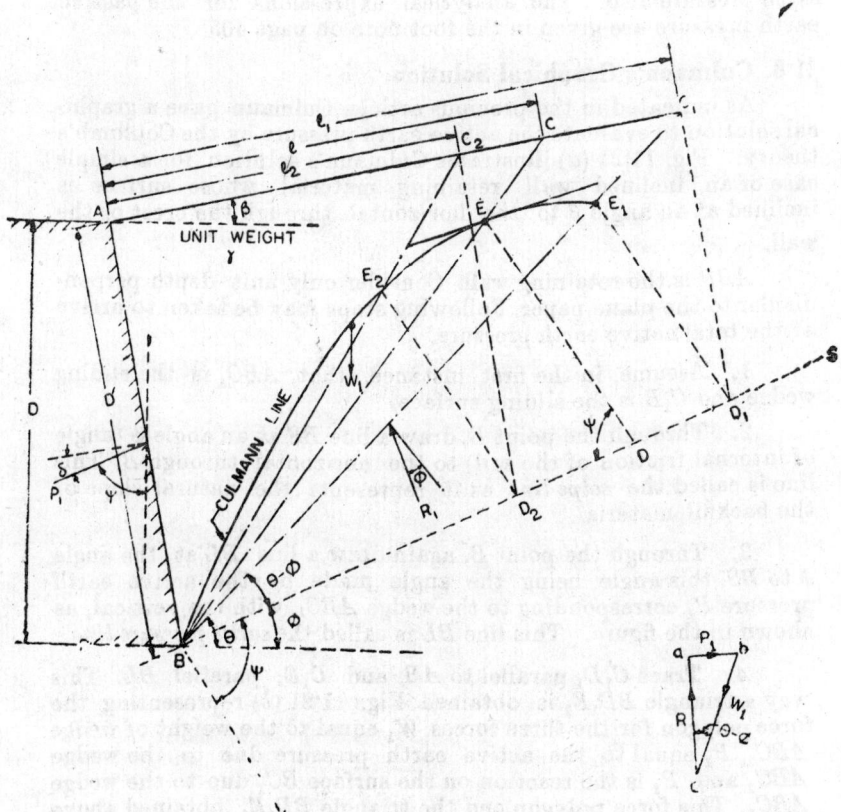

Fig. 11·21.

(e) Cohesion $c_2 = c' \times (AB \times 1)$ along the wall face AB and in the direction B to A.

c and c' in (d) above are respectively the cohesion of the soil and the cohesive force between the soil and the wall face AB.

Magnitudes of W_1, c_1 and c_2 are known and the directions of all the above five forces are also known. So, the force polygon of the type shown in Fig. 11.20 (b) can be drawn and this force polygon will give the value of P_1.

The final active earth pressure is the resultant of the force P_1 and the force c_2 along the wall. Similar force polygons can be drawn for other assumed failure wedges and the maximum value of the active earth pressure P_A determined by trial and error.

The above principles can be extended to find out the passive earth pressure also. The analytical expressions for the passive earth pressure are given in the foot note on page 405

11·8. Culmann's Graphical Solution.

As indicated in the previous article, Culmann gave a graphical solution to evaluate the active earth pressure by the Coulomb's theory. Fig. 11·21 (a) illustrates Culmann's solution for a simple ease of an inclined wall retaining material whose surface is inclined at an angle β to the horizontal through the crest of the wall.

ABP is the retaining wall. Consider only unit depth perpendicular to the plane paper. Following steps may be taken to arrive at the total active earth pressure.

1. Assume, in the first instance, that ABC_1 is the sliding wedge and C_1B is the sliding surface.

2. Through the point B, draw a line BS at an angle ϕ (angle of internal friction of the soil) to the horizontal through B. This line is called the *solpe line* as it represents the natural slope of the backfill material.

3. Through the point B, again draw a line BL at the angle ψ to BS, this angle being the angle made by the active earth pressure P_1 corresponding to the wedge ABC_1, with the vertical, as shown in the figure. This line BL is called the *earth pressure line*.

4. Trace C_1D_1 parallel to AB and D_1E_1 parallel BL. This way a triangle BD_1E_1 is obtained Fig. 11·21 (b) representing the force polygon for the three forces W_1 equal to the weight of wedge ABC_1, P_1 equal to the active earth pressure due to the wedge ABC_1 and P_1 is the reaction on the surface BC_1 due to the wedge ABC_1. This force polygon and the triangle BD_1E_1 obtained above are identical.

$$\therefore \qquad \frac{P_1}{W_1} = \frac{E_1D_1}{BD_1}$$

or
$$P_1 = W_1 \frac{E_1 D_1}{BD_1}$$

But
$$W_1 = \frac{1}{2}\gamma D'.l_1 \qquad \text{(From the figure).}$$

or
$$P_1 = \frac{1}{2}.\gamma..D'l_1 \frac{E_1 D_1}{BD_1} \qquad \qquad ...(11·39)$$

5. Now assume that ABC_2 is the sliding wedge and C_2B is the sliding surface. In the manner laid down in steps 1 to 4, the value of P_2, the earth pressure corresponding to wedge ABC_2 can be evaluated from a comparison of triangle BD_2E_2 with another force polygon (not shown) from the wedge ABC_2. From the similar considerations as above, P_2 will work out as

$$P_2 = \frac{1}{2}\gamma.D'.l_2 \frac{E_2 B_2}{BD_2} \qquad \qquad ...(11·40)$$

A number of sliding wedges can arbitrarily be chosen likewise and values of $P_1 P_2 ......$determined.

6. Since in the figure C_1D_1 is parallel to C_2D_2 etc. the ratio

$$\frac{AC_1}{BD_1} = \frac{AC_2}{BD_2} ; \text{ etc.}$$

i.e.
$$\frac{l_1}{BD_1} = \frac{l_2}{BD_2} = ...$$

let this ratio be denoted by the letter n, so that $\frac{l_1}{BD_1}$ in equation

(11.39) and $\frac{l_2}{BD_2}$ in equation (11·40), etc. can be replaced by n.

Substitution gives

$$P_1 = \frac{1}{2}.\gamma.D'.nE_1 D_1$$

and
$$P_2 = \frac{1}{2}\gamma.D'.n.E_2 D_2$$

Similarly, values of earth pressures can be obtained for other arbitrarily chosen sliding wedges. The factor

$$\frac{1}{2}.\gamma.D'.n = M \text{ (say)}$$

will be common in all the equations and is independent of the slope of the assumed surface of sliding, so that

$$P_1 \propto E_1 D_1$$
and
$$P_2 \propto E_2 D_2, \text{ etc.}$$

In fact, to some scale in the figure, $P_1 P_2$, etc. are represented by distances $E_1 D_1$, $E_2 D_2$. etc.

7. Points E_1, E_2 etc, so obtained lie on a curve called the *Culmann line*. In order to find out the maximum active earth pressure P_A, a line tangential to the Culmann line and parallel to the slope line BS is drawn. This tangential line touches the Culmann line at the point E. Through E a line ED parallel to the earth pressure line BL is drawn. Then, the distance ED to the figure represents the active earth pressure P_A whose value is given by,

$$P_A = M.ED$$

$$= \frac{1}{2} \gamma . D' . n . ED \qquad ...(11.41)$$

The above procedure can be applied to a curved or broken fill surface as also to broken retaining wall surface. In such cases, the weights W_1, W_2 etc. of the assumed failure wedges are calculated and plotted as distances, BD_1, BD_2, etc. along the line BS to a convenient scale. Then through the points D_1, D_2 etc., D_1E_1, D_2E_2 etc. are drawn pareallel to the earth pressure line BL. The rest of the procedure is the same as detailed out for the above case.

Pressure Distribution. In case the surface of the backfill and the back of the wall are plane, the pressure distribution on the wall is hydrostatic. This simply means that the active earth pressure P_A will act, in such cases, at one-third of the height of the wall from the base. This contention can be proved as under.

Fig. 11.19 shows the conditions indicated above and equation (11.38) gives the value of active earth pressure P_A for the soil. The normal component of this pressure, say P_{An} acting in a direction normal to the wall is

$$P_{An} = P_A \cos \delta$$

$$= \frac{1}{2} \gamma D^2 \frac{K_A}{\sin \alpha} \qquad ...(11.42)$$

If, now, a point at a depth z below the crest is considered then the normal component of active earth pressure upto this depth i.e. on the section AE shown in Fig. 11.19 is given by substituting z in place of D in equation (11.42). Let this be denoted by P'_{An}.

$$P'_{An} = \frac{1}{2} \gamma z^2 \frac{K_A}{\sin \alpha} \qquad ...(11.43)$$

The intensity of the normal to component of active earth pressure at the point E is given by differentiating the equation (11.43) with respect to

$$\frac{dP'_{An}}{dz} = \gamma . z \frac{K_A}{\sin \alpha} \qquad ...(11.44)$$

Equation (11.44) shows that the normal component of active earth pressure has a hydrostatic distribution since on the right hand side of this equation z is the only variable. For a wall with a vertical face

$$\alpha = 90°$$

Hence equation (11.44) gives the intensity at z equal $\gamma.z.K_A$ which is simpler to comprehend.

In case the surface of the backfill and the back of the wall are not plane, the pressure distribution is not hydrostatic. For such cases a graphical procedure is followed to locate the point of application of P_A. For details of this procedure, see reference 4, as such details are considered outside the scope of this book.

The graphical procedure laid down in this article for determination of the active earth pressure can be extended to find out the passive earth pressure also. The only difference in the procedure for the latter case is, that the line BS in Fig. 11.21 is drawn below the horizontal at an angle ϕ.

11.9. Use of Charts.

It may be noted that the values of ϕ and δ are at best, rough estimates. As such for design purposes, probably a rough but sufficiently accurate estimate of earth pressures would be enough. Use of certain charts is suggested for ready and sufficiently accurate calculations. Earth pressure tables and graphs for the values of K_A and K_P are available for different values of ϕ, δ and β.

Figs. 11·22 (a) and (b) show two graphs for the values of active earth pressure K_A for different values of ϕ and δ in case of a cohesionless backfill These graphs are applicable to the following boundary conditions :

1. The back of the wall is a vertical plane.

2. The surface of the backfill is a horizontal plane.

i.e. $\beta = 0$.

Most of the field problems on retaining walls meet the above boundary conditions. Even if the back of the wall is slightly inclined to the horizontal, these graphs could be used for rough estimation of the value of K_A.

Fig. 11·23 shows another graph which gives the value of passive earth pressure K_P for different values of ϕ and δ for a cohesionless backfill and applicable under the above boundary conditions.

Fig. 11·22 (a) and (b) indicate that K_A decreases with the increase in the value of δ, when δ is positive, i.e., $\delta > 0$. This simply means that taking the active earth pressure P_A from the Rankine's theory where δ is assumed zero compared to Coulomb's theory when δ is taken > zero is on the safer side. However, when $\delta < 0$ i.e. when it is negative, the value of P_A if obtained from the Rankine's theory would be erroneous because it will be less than the one obtained from the Coulomb's theory.

Fig. 11·23 indicates that K_P increases with the increase in the value of δ. So, obtaining passive earth pressure P_P from the Rankine theory where δ is assumed zero compared to the Coulomb's theory, when δ is more than zero, is not on the safer side.

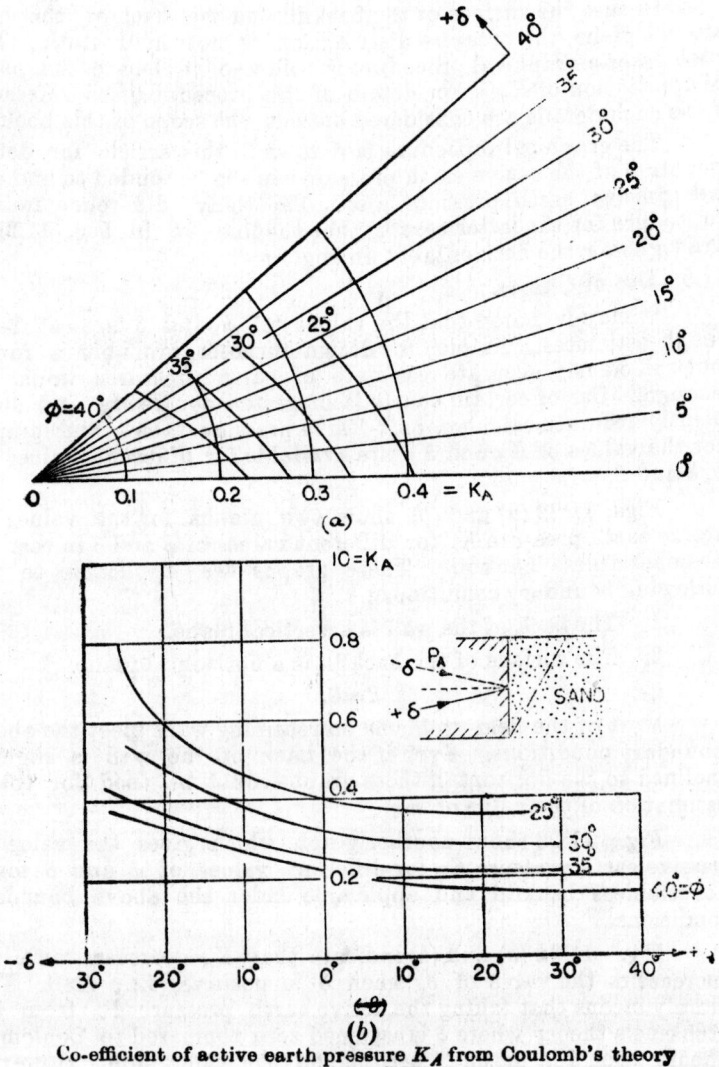

Co-efficient of active earth pressure K_A from Coulomb's theory
(After Syffert)
Fig. 11·22

It may be recalled that it is assumed in the Coulomb's theory that the surface of sliding is a plane (assumption 3 page

348). It was indicated, however, in Fig. 11·18 (b) and (c) that the actual surface of sliding is not a plane but is curved surface. This is true both for the active as well as for the passive case. Hwo much error is involved in assuming the surface of sliding as

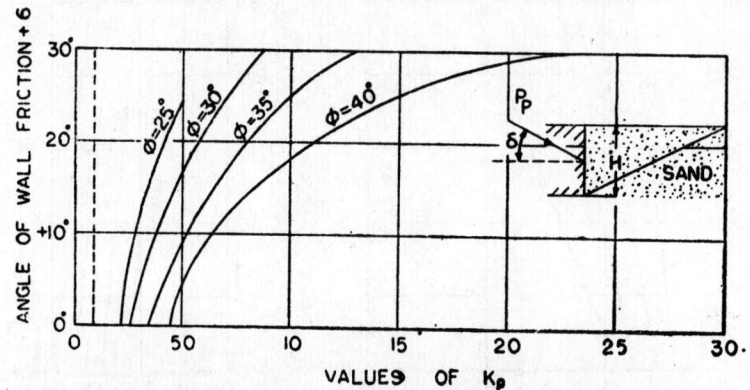

Co-efficient of passive earth pressure K_P from Coulomb's theory.

Fig. 11.23

a plane instead of a curved surface is clear from Fig. 11·24 (a) and (b). The solid lines in the figure show the values of K_A and K_P calculated from the Coulomb's theory for a vertical wall with a horizontal surface of backfill, as a function of ∂ and ϕ. If these values are calculated by assuming the surface of failure as a curved surface, the results are different. The difference is not appreciable for the active earth pressure K_A, but it is quite significant for the passive earth pressure K_P. The corrected K_P values are shown by dotted lines.

It is also clear from Fig. 11·24 (a) that the effect of the third assumption on K_P is not as much when δ is small. However, as δ becomes larger and larger, the effect of this assumption is quite pronounced. As such, as a rule, if $\delta^1 < \phi/3$, the third assumption of a plane surface of sliding may be considered valid otherwise the curvature of the surface must be taken into account.

11·10. Rehbann's Construction

As indicated in article 11·8, Rehbann gave a simple graphical method to find out the active earth pressure behind a retaining wall in case of granular backfill. Fig. 11·25 shows his graphical construction. ABM represents the retaining wall. AC_1 represents the inclined surface of the backfill.

---

1   When $\partial = \phi$, the error involved may be of the order of 30%.

Step 1. At B, draw a line BC_1 inclined at an angle ϕ to the horizontal, cutting the backfill surface in C_1.

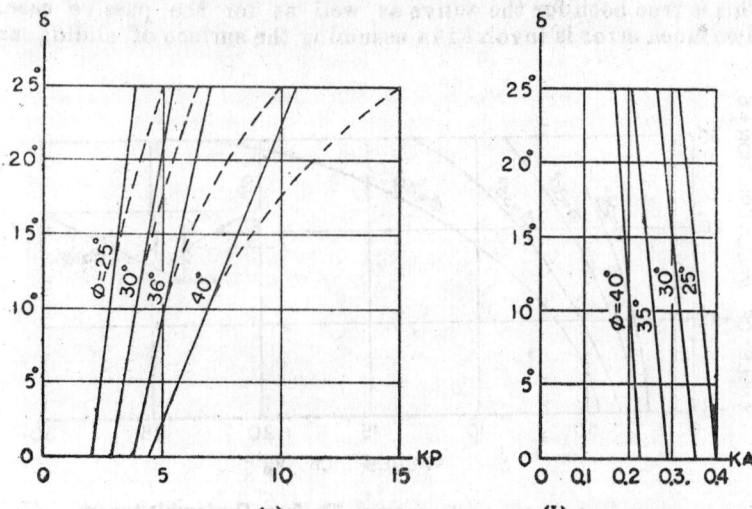

(a)

(b)

Effect of assuming plane surface of sliding on the calculations for K_A and K_P

Fig. 11·24.

a plane instead of a curved one is shown from Figs. 11·24 (a) and (b). The solid lines in the figure show the values of K_A calculated from the Coulomb's theory for a vertical wall with inclined surfaces of backfill as a function of β and ϕ. If these values are plotted, K_P assuming the actual failure as a curve-surface, the dotted are obtained. The differences are not appreciable for the active earth pressure K_A, but it is quite appreciable in the passive earth pressure K_P, the converted K_P values are shown by dotted lines.

It is also clear from Fig. 11·24 (a) that the value of the third arc-circle, K_A, is not affected when a vertical wall, however, in presence of a wall; the effect of the assumption is quite pronounced. It means at $\phi = 37.5^\circ$ $\delta = 0$, the assumption of a plane surface of sliding may be accepted valid; otherwise the true surface of sliding surface can never give a smooth.

11·30. Rebhann's Construction.

As detailed in article 11·13, Rebhann developed a simple graphical method to find out the active earth pressure behind a retaining wall on cohesionless granular backfill. This is however, but graphical construction, which represents are represented in Fig. 11·25 showing the inclined surface of the backfill.

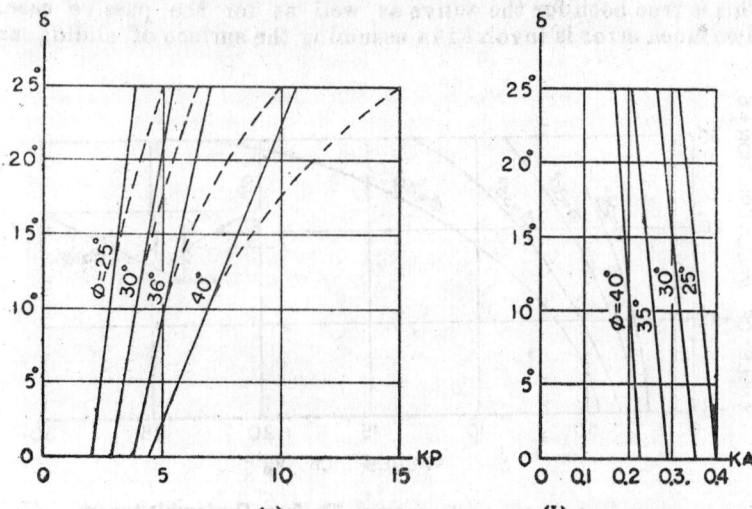

Fig. 11·25.

Step 2.   At B, again draw BL inclined at an angle to ψ to the line BC_1.

Step. 3.   Draw a semi-circle on the line BC_1.

Step 4.   Through A, draw a line AE parallel to BL, meeting BC_1 in a point E.

Step 5.   At E, draw a perpendicular to BC_1 meeting the semi circle in the point F.

Step 6.   With B as the centre and BF as the radius. draw an arc FG, meeting BC_1 in the point G.

Step 7.   Through G, draw a line GC parallel to BL or AE, meeting the backfill surface AC_1 in C.

Step 8.   Join BC. Then BC represents the plane of rupture. Earth pressure P_A is given by

$$P_A = \frac{1}{2} \gamma . x^2 . \sin \psi \qquad \qquad ...(11.45)$$

where $\qquad\qquad x = CG$

For proof of this construction, the student may refer to reference 8.

11·11.   Arching in Ideal Soils

If one part of a support retaining soil mass yields, while the rest of it remains in position, the soil near the yielding part of

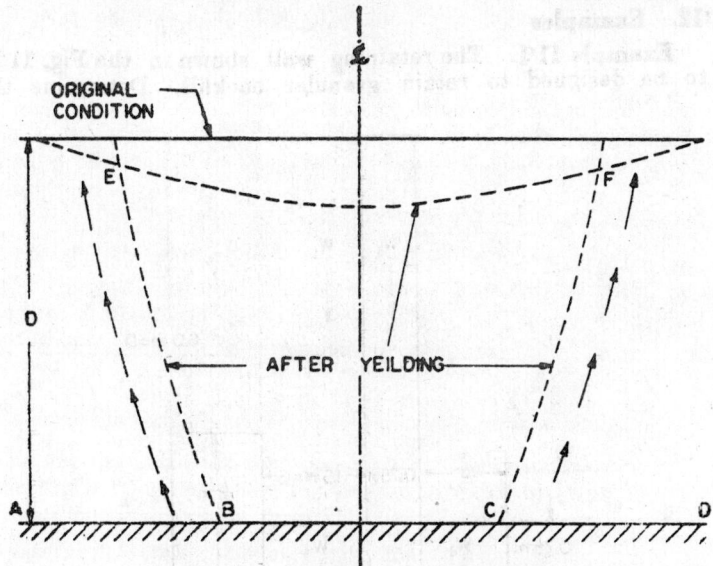

Fig, 11·26.

the support moves from its original position. The relative movement within the soil is opposed by the shearing resistance that acts along the surface (or surfaces) of contact between the yielding and stationary positions of the soils. The situation is explained in Fig. 11·26 in which a mass of soil is retained on a horizontal platform. If portion BC of the platform yields and the portions AB and CD remain stationary, the mass of soil affected by this yield is shown by the dotted lines. Shearing resistance opposing the movement of the soil acts along the faces of contact BE and CF between the yielding and stationary soil mass. This shearing resistance, since it opposes the movement of the soil, reduces the total pressure on the yielding position BC by an amount equal to its vertical component. Since the total pressure on the entire support is the same whether BC yields or not and equals the weight of the soil above, the reduction of pressure on BC is accompanied by an equal increase of pressure on the unyielding parts AB and CD. This transfer of pressure from a yielding soil mass to stationary parts of the supports is called the **arching effect** (4) and the soil is said to **arch** over the yielding **part of the support.**

The above definition forms the basis of calculations of earth pressures behind retaining walls that yield about the lower and rather than the upper end, as assumed in the preceding articles. The theories of arching are beyond the scope of this book. The reader, however, may refer to reference 4 for any detailed discussion on this topic.

11·12. Examples

Example 11·1. The retaining wall shown in the Fig. 11·27 is to be designed to retain granular backfill. Determine the

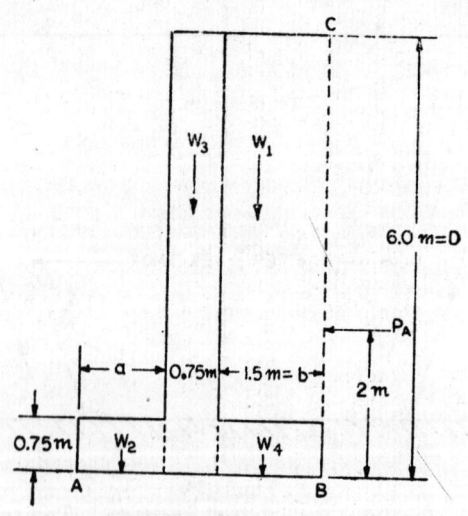

Fig. 11·27.

width of the base in front of the wall i.e. dimension 'a' in order to provide a factor of safety of 20 against overturning alone, under the following two cases :—

(a) The backfill is well drained and slightly moist. Assume the degree of saturation as 20%.

(b) The drainage system gets clogged and the backfill becomes submerged below a depth of one meter the surface of the backfill. Above this depth, it is fully saturated.

The soil and concrete properties may be taken as under :

For the Soil : Specific gravity of solids

$$G = 2.65$$

Void ratio $= 0.594$

Angle of internal friction $= 30°$

For concrete. Weight of concrete

$$= 2300 \text{ Kg./m.}^3$$

Solution. (a) *When the backfill is well drained and degree of saturation S is 20%.*

Unit weight of the soil

$$\gamma = \frac{G + Se}{1 + e} \cdot \gamma_w$$

$$= \frac{2.65 + 0.20 \times 0.594}{1 + 0.594} \times 1$$

or $\gamma = \dfrac{2.7688}{1.594}$

$$= 1.740 \text{ gms./cc}$$
$$= 1740 \text{ kg./m.}^3$$

Co-efficient of active earth pressure

$$K_A = \tan^2 (45° - \phi/2)$$

$$= \tan^2 30° = \frac{1}{3}.$$

Now consider only one metre depth perpendicular to the plane of the paper.

∴ $$P_A = \frac{1}{2} \gamma. K_A. D^2$$

$$= \frac{1}{2} \times 1740 \times \frac{1}{3} \times 6^2$$

or $$P_A = 10,440 \text{ kgs.}$$

Designate the weight of the soil on the diamension "b" of the retaining wall as W_1 and divide the retaining wall structure into three portions with weights W_2, W_3 and W_4 as shown. Then,

$$W_1 = (5.25 \times 1.5 \times 1) \times 1740 = 13,700 \text{ kgs.}$$

$$W_2=(a \times 0.75 \times 1) \times 2300 \quad =1725 \ a \text{ kgs.}$$

$$W_3=(6.0 \times 0.75 \times 1) \times 2300= \ 10,350/\text{- kgs.}$$

$$W_4=(1.5 \times 0.75 \times 1) \times 2300=2587 \text{ kgs.}$$

The earth pressure P_A induces the overturning moment and the weights W_1, W_2, W_3 and W_4 offer the resisting moment to the structure.

Take the moments about the point A.

$$Overturning \ moment = P_A \times \frac{D}{3}=10,440 \times \frac{6}{3} =20,880 \text{ kg. metres}$$

Resisting moments :

Due to $W_1=(a+0.75+0.75)=W_1 \ (a+1.5)$

$\qquad =(13,700a+20,600) \text{ kg.metres}$

Due to $W_2=W_2 \left(\dfrac{a}{2} \right)=1725a \ . \ \dfrac{a}{2} \ 862.5a^2 \text{ kg-metres}$

Due to $W_3=W_3 \left(\ a+ \dfrac{0.75}{2} \right)=10,350 \ (a+0.375)$

$\qquad =(10,350a+3880) \text{ kg-metres}$

Due to $W_4=W_4 \ (a+0.75+0.75)=W_4 \ (a+1.5)$

$\qquad =2587 \ (a+1.5)=(2587a+3880) \text{ kgs-metres}$

Total resisting moment$=(862.5a^2+26,637a+28,360) \text{ kg-metres.}$

If a factor of safety of two is desired, this resisting moment must be equal to two times the overturning moment.

$\therefore \quad 862.5a^2+26,637a+28,360=2 \times 20,880.$

This quadratic can be solved to get the value of a. Solving it,

$$a=0.50 \text{ metres.}$$

(b) *When the drainage system gets clogged, etc.*

If degree of saturation S is taken as unity, then,

$$\gamma_{sat}=\frac{G+Se}{1+e} \ \gamma_w= \frac{2.65+1 \times 0.514}{1+0.594} \times 1$$

or $\qquad \gamma_{sat}=2.04 \text{ gms/c.c.}=2040 \text{ kgs/m}^3.$

$\therefore \qquad \gamma_{sub}=2.04-1.00=1.04 \text{ gms/c.c.}=1040 \text{ kgs/m}^3$

Let z denote the depth below the crest level of the retaining wall.

Upto $z=1$ metre, the soil is fully saturated

The vertical pressure at $z=1 \text{ metre}=\gamma_{sat} \times 1$

$\qquad\qquad =2040 \text{ kgs/m}^2$

So, the horizontal pressure at this depth$=2040 \ K_A$

$\qquad\qquad\qquad =2040 \times \tfrac{1}{3}=680 \text{ kgs/m}^2.$

Below $z=1$ metre, the soil is submerged. So, the horizontal pressure at any point for $z>1$ metre will depend upon the effective vertical pressure at that depth times the coefficient of active earth pressure plus the water pressure.

The effective vertical pressure at $z > 1$

$$= \gamma_{sat} \times 1 + \gamma_{sub} (z-1)$$

At $z = 6$ metres, effective vertical pressure

$$= 2040 + 1040 (6-1)$$
$$= 2040 + 1040 \times 5 = 2040 + 5200$$
$$= 7240 \text{ kg/m}^2$$

Component of horizontal pressure due to this effective pressure at

$$= 6\text{metres}, \quad = 7240 \times K_A$$
$$= 7240 \times 1/3 = 2413 \text{ kg/m}^2$$

Also the water pressure acts equally on all directions.

Water pressure at $z = 6$ metres $= 1000 \times 5$
$$= 5000 \text{ kg/m}$$

Total horizontal pressure at $z = 6$ metres
$$= 2413 + 5000 = 7413 \text{ kg/m}^2$$

From the above calculations, a pressure diagram shown in Fig. 11.28 (a) can be drawn.

Let the total pressure represented by bcd be P_1, that represented by $cdag$ be P_2 and the one represented by $cg f$ be P_3. These pres-

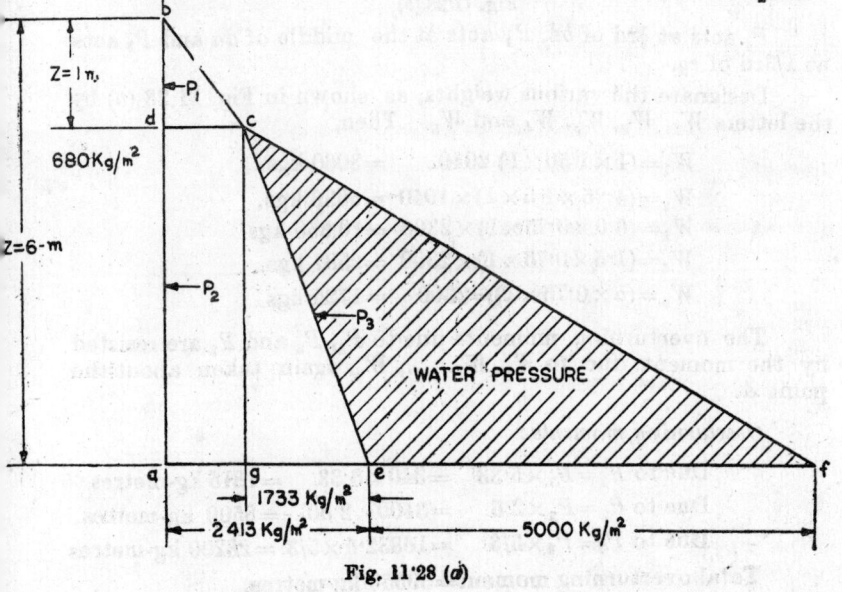

Fig. 11·28 (a)

ssures are to be calculated due to one metre depth of the back-full along the length of the wall. These pressure act horizontally at the section BC and are responsible for the overturning moment about the point A.

$$P_1 = \tfrac{1}{2} \times (1 \times 1) \times 680 \qquad = 340 \text{ kgs.}$$
$$P_2 = (5 \times 1) \times 680 \qquad = 3400 \text{ kgs}$$
$$P_3 = 1/2 \times (5 \times 1) \times 6733 \quad = 16832 \cdot 5 \text{ kgs}$$

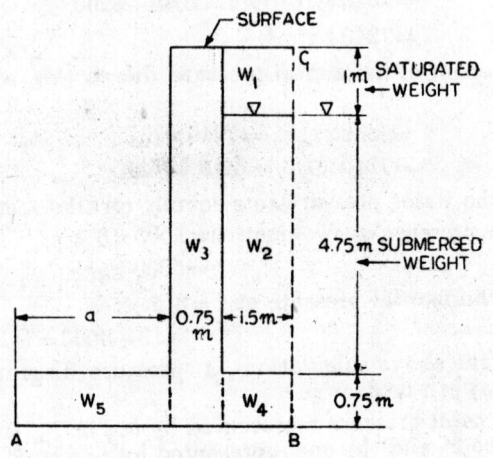

Fig. 11·28 (b)

P_1 acts at $\tfrac{1}{3}$rd of bd, P_2 acts at the middle of da and P_3 acts at 1/3rd of cg.

Designate the various weights, as shown in Fig. 11.28 (b) by the letters W_1, W_2, W_3, W_4 and W_5. Then,

$$W_1 = (1 \times 1\cdot 50 \times 1)\ 2040 \qquad = 3060 \text{ kgs.}$$
$$W_2 = (4\cdot 25 \times 1\cdot 5 \times 1) \times 1040 \ = 6630 \text{ kgs.}$$
$$W_2 = (6\cdot 0 \times 0\cdot 75 \times 1) \times 2300 \ = 10{,}350 \text{ kgs.}$$
$$W_4 = (1\cdot 5 \times 0\cdot 75 \times 1) \times 2300 \ = 2587 \text{ kgs.}$$
$$W_5 = (a \times 0\cdot 75 \times 1) = 2300 \qquad = 1725 \text{ kgs.}$$

The overturning moments due to P_1, P_2 and P_3 are resisted by the moment due to W_1, W_2......, W_5. again taken about the point A.

Overturning moments :

$$\text{Due to } P_1 = P_1 \times 5\cdot 33 \quad = 340 \times 5\cdot 33 \quad = 1815 \text{ kg-metres}$$
$$\text{Due to } P_2 = P_2 \times 2\cdot 5 \quad = 3400 \times 2\cdot 50 \quad = 8500 \text{ kg-metres}$$
$$\text{Due to } P_3 = P_3 \times 5/3 \quad = 16832\cdot 5 \times 5/3 = 28200 \text{ kg-metres}$$

Total overturning moment = 39550 kg-metres.

Resisting moments :

Due to $W_1 = W_1 (a + 0\cdot75 + 0\cdot75) = 3060 (a + 1\cdot5)$
$\qquad = (3060a + 4590)$ kg-metres

Due to $W_2 = W_2 (a + 0\cdot75 + 0\cdot75) = 6630 (a + 1\cdot5)$
$\qquad = (6630a + 9945)$ kg. metres

Due to $W_3 = W_3 (a + 0\cdot375) = 10{,}350 (a + 0\cdot375)$
$\qquad = (10{,}350a + 3880)$ kg-metres

Due to $W_4 = W_4 (a + 0\cdot75 + 0\cdot75) = 2587 (a + 1\cdot5)$
$\qquad = (2587a + 3880)$ kg-metres

Due to $W_5 = W_5 \left(\dfrac{a}{2} \right) = 1725a.$ $\dfrac{a}{2} = 862\cdot5a^2$

Total resisting moment $= 862\cdot5a^2 + 22{,}627a + 22{,}295$ kg-metres

Now, if a factor of safety of two is desired, the total resisting moment should be equal to twice the total overturning moment.

$\therefore\ 862\cdot5a^2 + 22{,}627a + 22{,}295 = 2 \times 39{,}550 = 79{,}100$

This is quadritic in a and can be solved to get the value of a. Solving for a,

$$a = 2\cdot25 \text{ metres}$$

Comments. 1. The results in part (a) and (b) indicate that the clogging of drainage system can lead to failure in case the wall is not designed for this eventual condition.

2. We could have neglected the first-factor in the resisting moment accompanied by the term a^2 to find out "a" dimension since this is quite negligible compared to the other factors.

Example 2. Find out the active earth pressure P_A for a granular soil under the following conditions :

(a) $\phi = 30°$ and $\delta = 0$ $\qquad$ (Rankine's case)

(b) $\phi = 30°$ and $\delta = 25°$ $\qquad$ (Coulomb's case)

Assume that the back of the wall is vertical and the surface of the backfill is horizontal, in both cases. Comment on the results obtained.

Solution. (a) From Fig. 11·23 (b), when $\phi = 30°$ and $\delta = 0$

$\qquad K_A = 0.33$

$\therefore \qquad P_A = \tfrac{1}{2}\gamma . K_A . D^2 = \tfrac{1}{2}\gamma . D^2 \times 0\cdot33$

or $\qquad P_A = 0\cdot165\ \gamma\ D^2,$

where γ and D are the unit weight and depth of the soil respectively.

(b) From the same figure, when $\phi = 30°$ and $\delta = 25°$

$\qquad K_A = 0\cdot28$

$$\therefore \qquad P_A = \frac{1}{2}\gamma . K_A . D^2 . \frac{1}{\cos\delta} = \frac{1}{2} \times . \gamma . D^2 \frac{0.28}{0.93}$$

or $\qquad P_A = 0.15\gamma D^2,$

where again γ and D are the unit weight and depth of the soil, respectively.

Comments. The earth pressure in the Coulomb's case is reduced by an amount

$$= \frac{0.165 - 0.15}{0.165} \times 100$$

$$= \frac{0.015}{0.165} \times 100 = 9.1\%. \text{ Ans.}$$

CHAPTER — BIBLIOGRAPHY

1. Huntington, W.C., Earth Pressure and Retaining Walls, New York, John Wiley & Sons, Inc.

2. Johnson, S.M., and Kavanagh T.C., "The Design of Foundation for Buildings" McGraw Hill Book Co., New York, N.Y., 1968.

3. Junarkar, S.B., Mechanics of structures, Vol. II, Anand Charotar Book Stall, 1961.

4. Lambe, T.W., and Whitman, R.V., "Soil Mechanics," J. Wiley & Sons, Inc., New York, 1969.

5. Leonard, G.A., "Foundation Engineering, "McGraw Hill Book Co., Inc., New York, N.Y., 1962.

6. Rankine, W.J.M., on the Stability of Loose Earth, Phil., Trans, Roy. Soc. London, Vol. 147.

7. Rebhann, G., Theorie des Erddrukes under Futtermauern, Wien.

8. Terzaghi, K., General Wedge Theory of Earth Pressures, Transactions, ASCE Vol. 106.

9. Terzaghi, K., Theoretical Soil Mechanics, New York, John Wiley & Sons, Inc., 1943.

10. Terzaghi, K., Anchored Balkheads, Transactions ASCE, Vol. 199.

11. Tomlinson, M.J., "Foundation Design & Construction, Wiley Interscience, John Wiley & Sons, Inc., New York, N.Y., 1969.

12. Wu. T.H., Soil Mechanics, Boston, Allyn and Bacon, Inc., 1966.

QUESTIONS

1. (a) Explain the Coulomb's theory of active earth pressure against retaining walls giving the assumptions made, if any.

(b) A wall with a smooth vertical back 10 feet high retains a mass of purely cohesive soil having a cohesion $c = 200$ lbs/sq. ft. and a unit weight of 110/c. ft. The value of ϕ may be taken as zero. What is the total active Rankine pressure against the wall ? At what distance above the base is the centre of pressure ? At what depth is the intensity of pressure zero ?

(P.U. 1959)

2. (a) Give an explanation of active and passive earth pressure ? What is meant by earth pressure at rest ?

(b) Tho backfill of a retaining wall 30 feet high is of cohesionless soil for which the coefficient of active pressure is 0·28. The surface is level with the top of the wall and carries a uniformity distributed surcharge load of 224 lbs per sq. ft. The density of top six feet of fill is 112 lbs/c. ft. and below this level, the density is 125 lbs/c. ft. Neglecting wall friction, find the magnitude and point of application of the resultant thrust per foot run of the wall.   *(P.U. 1959 Suppl.)*

3. A vertical retaining wall 20 feet high supports a cohesionless soil that weighs 115 lbs per c. ft. The upper surface of the soil rises from the crest of the wall at an angle of 20° with the horizontal. The angle of internal friction of the soil is 28° and the angle of wall friction is 20°. By Culmann's method, compute the total active earth psessure against the wall.   *(P.U. 1960)*

4. (a) What is meant by active, passive and 'at rest' earth pressure ? Give their approximate values for a sand with $\phi = 30.°$

(b) Derive the expression for active earth pressure for a cohesive soil by Rankine's method.

(c) If the wall is rough, will the actual values of active earth pressure be more or less than Rankine's values and to what extent ?   *(P.U. 1961)*

5. (a) What are the assumptions in the Coulomb's Theory of active earth pressure ? Discuss the merits of Coulomb's Theory over the Rankine's Theory.

(b) A retaining wall 20 feet high supports a cohesionless fill that weighs 115 lbs/c. ft. The upper surface of the fill rises from the crest of the wall at an angle of 20° with the horizontal. The angle of internal friction is 28° and the angle of wall friction is 20°. By Culmann's method or otherwise find the total active earth pressure against the wall and determine its point of application and direction.   *(P.U. 1963)*

(6) (a) There are two retaining walls of the same height, one with a perfectly smooth back and the other with a rough back. The backfill has the same properties for both the walls. Which of the walls will experience a greater lateral pressure and why ?

(b) A retaining wall 5·0 metres high with a sloping back at 10° to the vertical as shown in the figure below, supports a cohesionless backfill rising from the crest at an angle of 5°, with the horizontal. The backfill weighs 1·90 gms/c.c. and carries a uniformly distributed load of 0·· tonne per square metre. The angle of shearing resistance of the backfill is 30 and the angle of wall friction is 20. Find, by a graphical method, the total active pressure per metre length of the wall. Also mark the direction and point of application of the resultant pressure at the back of the wall.

(P.U. 1963 Suppl.)

7. (a) Discuss and compare the salient features of Rankine's and Coulomb's theories of Earth Pressure.

(b) A cohesionless backfill with horizontal surface has a weight of 1·82 gms. per c.c. and carries a uniformly distributed surcharge of 2·50 tonnes per sq. metre. The active lateral earth pressure at a point 6 metres down the top of the retaining wall 12 metres high is measured to be 3·50 tonnes per sq. metre. What is the magnitude, direction and point of the total Rankine active lateral force against the wall ?   *(P.U. 1964)*

8. Give a brief explanation of active lateral pressure and passive resistance pressure. State the basic assumptions underlying the Rankine and Coulomb theories of lateral pressure.

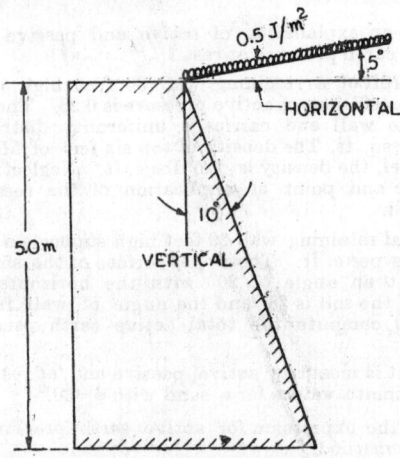

Fig. 11·29.

(b) A smooth vertical wall 8 metres high is pushed against a mass of soil having a horizontal surface and the shearing resistance given by Coulomb's equation in which $c=1,500$ kg/sq m. and $\phi=18°$. The unit weight of the soil is 1·95 gm/cm³. Its surface carries a uniform load of 900 kg/sq. m. What is the distance from the base of the wall to the centre of pressure? What is the intensity of lateral pressure at the base of the wall? *(P.U. Nov. 1964)*

9. A 10 m. high retaining wall has a batter of 1 in 12. The backfill is of clean dry sand with $\phi=33°$ and the unit weight 1·60 gm/c.c. The angle of wall friction may be assumed 24°. Using Culmann's graphical

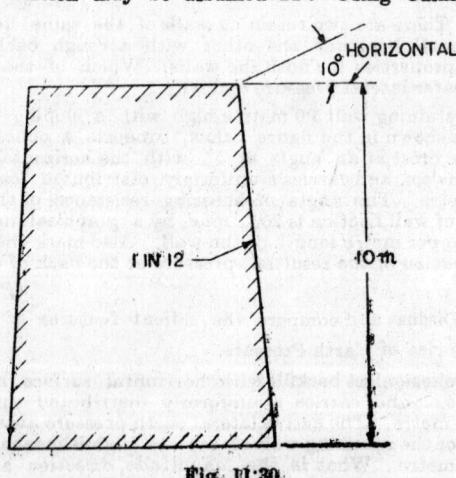

Fig. 11·30.

method, estimate the value of active earth pressure against the wall in tonnes per metre length of wall. Show its magnitude and point of action and direction in the figure.   (*P.U. April 1965*)

10. (*a*) State the assumption of Rankine's earth pressure theory.

(*b*) List the salient points of difference between the Coulomb's and Rankine's earth pressure theories.   (*P.U Nov. 1965*)

11. (*a*) State the assumptions made in Rankine theory of earth pressure ?

(*b*) What are the short comings of Rankine theory of earth pressure.

(*c*) Describe the Rebhann's construction for the computation of earth pressure behind a retaining wall.   (*P.U. April 1966*)

12. (*a*) Explain the terms : active earth pressure, passive earth pressure and earth pressure at rest.

(*b*) Describe with a neat sketch, the procedure of graphical solution, based on wedge theory, for active earth pressure, of a cohesive soil on a retaining wall.

(*c*) What is the depth of tension cracks in a cohesive soil having a unit cohesion of 2000 kg/m and an angle of internal friction of 15° ? The unit weight of the soil is 1750 kg/cubic metre.   (*P.U. Nov. 1966*)

13. Establish the relationship between principal stresses at failure in a noncohesive soil, with the help of Mohr s concept of strength of soil.
                                                        (*P.U. April, 1967*)

14. A vertical retaining wall 10 m. high supports a cohesionless fill with $\gamma = 2$ gms/c.c. The upper surface of the fill rises from the crest of the wall at an angle of 20° with the horizontal. Assuming $\phi = 30°$ and $\delta = 20°$, indicate the position of failure plane on the figure and determine the total active earth pressure *either* by Culmann's or Rehbann's construction.
                                                        (*P.U. April, 1967*)

15. (*a*) With the help of active and passive pressure wedge analysis, derive an equation for the ultimate bearing capacity of a C–ϕ soil.

(*b*) A smooth vertical wall 6 metres high is pushed against a mass of soil having a horizontal surface carries a uniform load of 900 kg. per square metre. The soil is at a moisture content of 32.10%. The properties of this soil are : Cohesion=0·20 kg./cm2, angle of internal friction=15°, void ratio=0.85 and specific gravity of soil solids=2.66.

1. Find the total passive-Rankine's pressure and its point of application.

2. Determine the intensity of lateral pressure at the base of wall.
                                                        (*Baroda, 1966*)

16. (*a*) List the assumptions and limitations of Rankine's theory of earth pressure.

(*b*) A retaining wall 15′ height supports a cohesive back fill. Calculate the active earth pressure against the wall if shear strength of soil is represented by $S = 1000$ lbs/sq. ft. $+ \sigma \tan 19°$ Using Mohr's strength theory locate the plane of critical failure.

                                                        (*Mysore 1966*)

12

Bearing Capacity of Soils

12·1. Introduction

The foundation of a structure transmits the weight of the structure and the live loads to the soil underneath. These loads have to be transferred in such a manner that the soil is not stressed beyond its capacity to support the loads. The best thing to do would be, then, to place the foundations on the hard *bedrock*, since the capacity of rock in natural beds is relatively very high compared to the loads brought down by a structure on its foundations. This would ensure a minimum contact area between the foundation and the *foundation bed*. However, it is not always possible to place the foundations on rock, since usually the rock is very much far away from the ground surface and it would be invariably uneconomical to place the foundations at such large depths. As a consequence, it is invariably required to place the foundations on the soil strata available at relatively shallow depths. A scientific design of the foundation based upon the *bearing capacity* of the soil, then, becomes very essential.

This chapter deals with the fundamentals governing the bearing capacity of soils and its importance in foundation design.

12·2. Definitions

A few definitions are presented first.

Foundation. The lowest part of a structure is called a foundation.

Foundation Bed. The material on which a foundation rests is called a foundation bed.

Ultimate Bearing Capacity. Maximum unit pressure that a soil can stand *without rupture in shear or without excessive settlement* of the structure is called the **ultimate bearing capacity** of the soil. This is also sometimes called *ultimate bearing value* or *ultimate soil pressure*.

Safe Bearing Capacity. Ultimate bearing capacity divided by a reasonable factor of safety is called the safe bearing capacity. It is for this bearing capacity that the design of the foundation is done. Factors of safety in foundation design range from 3 to 5 or even more. The value in a particular case will depend upon the importance of the structure, differential settlement it can stand and the soil strata underneath as exhibited by the soil profile at the site.

12.3. Factors Affecting Bearing Capacity

Bearing capacity for foundation design depends upon one or more of the following factors :

1. *Type of soil.* As indicated in the first chapter, the soils can be subdivided into two categories—the coarse grained soil and the cohesive soils. In general, the coarse grained soils have better bearing capacities compared to cohesive soils.

2. *Physical features of the foundation.* Some physical features of the foundation affect the bearing capacity. They are

(a) Type of the foundation.

(b) Size of the foundation.

(c) Depth of the foundation below the ground surface.

(d) Shape of the foundation.

(e) Rigidity of the structure.

3. Amount of *total and differential settlement* that the structure can stand :

4. *Physical properties* of the soil, such as density, shear strength etc.

5. *Water conditions* in the ground, e.g., the position of the water table has a considerable influence on the bearing capacity values.

6. *Original stresses.*

12.4. Gross and Net Bearing Capacity

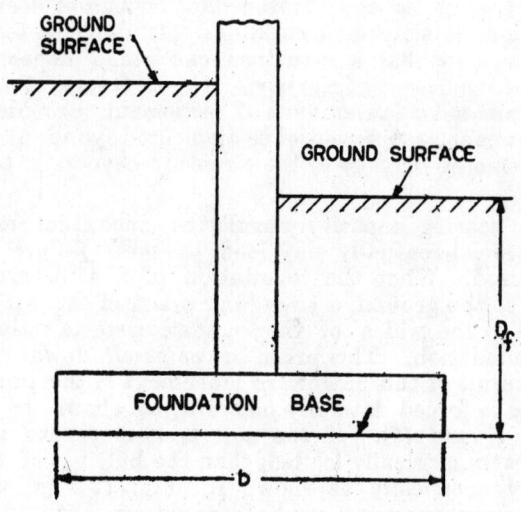

Fig. 12.

Sometimes it becomes essential to differentiate between what is called the *gross value* of bearing capacity and the *net value* of bearing capacity. **Gross bearing capacity** is the total unit pressure which the soil can take up. **Net bearing capacity** is the gross bearing capacity minus the vertical pressure that is produced on the horizontal plane at the level of the base of the foundation by an adjascent surcharge, *i.e.* the net bearing capacity is equal to the total unit pressure minus the unit weight of the soil between the ground surface and the bottom of the foundation. If the ground surface on one side of the foundation is higher than on the other side, as shown in Fig. 12·1, then the net bearing capacity is found out by taking the minimum depth of surcharge. *If* the foundation rests on the ground surface, there is no difference between the gross and the net bearing capacity values.

The ratio of the depth of the soil surcharge (D_f) and the width of the foundation (b) is called the **surcharge ratio**. In case, a pavement or a concrete floor rests upon the soil surcharge, it is appropriate to find out an equivalent surcharge which is the height of a column of soil whose unit weight is equal to the combined unit weight of the pavement and the soil above the base of the foundation.

12·5. Two Requirements

Before design or safe bearing capacity of soil is chosen, it it essential to take into consideration two types of action by the soil when subjected to load.

1. The bearing capacity must be low enough to ensure that the compression settlement caused is not detrimental. The amount of settlement that a structure can stand depends to a large extent on the degree of rigidity of the structure. Furtheron, a structure can stand a fair amount of settlement provided it is uniform. However, a differential settlement beyond a certain amount may cause damage to the structure beyond a tolerable limit.

2. The bearing capacity should be such that excessive shear strain which eventually may lead to *shear failure*[1] of the soil, is not caused. When the foundation of a structure transmits the load to the ground, a triangular prism of the soil whose width is equal to the width of the foundation starts acting as a part[2] of the foundation. This prism forces itself down into the soil and as a result of this downward movement of the prism, the soil at its sides is forced laterally outward, as shown by arrows in Fig. 12·2 (*a*) and (*b*). If the soil is homogenous and the foundation is symmetrically loaded, than the bulging of the soil may occur symmetrically as shown in Fig. 12·2 (*a*) and the

---

1. This shear failure is usually called the **bearing capacity failure**.
2. Terzaghi's theory on bearing capacity of soil—see article 12·13

foundation would settle uniformly. If, however, the soil is non-homogeneous or the foundation is eccentrically loaded, then, in that case the upheaving of soil on one side, as for example side *A* in Fig. 12·2 (*b*), would be more compared to the upheaving on the

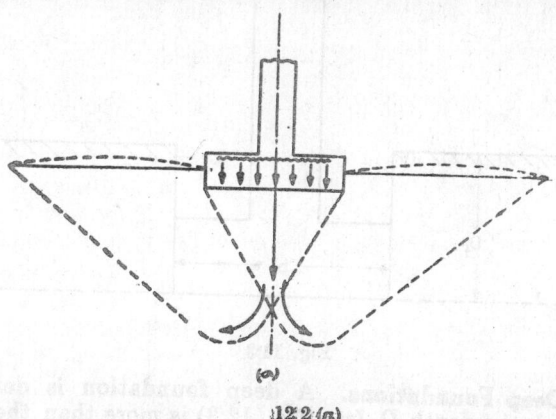

(*a*)

Fig. 12·2 (*a*)

other side, say *B*. This simply means that the foundation has settled more at the corner *A* than at the corner *B*. This results in tipping of the structure to the left, which eventually leads to the shear failure of the foundation.

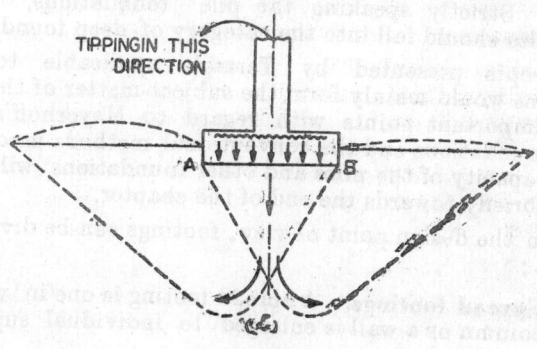

TIPPING IN THIS
DIRECTION

(*b*)

Fig. 12·2 (*b*)

Compression settlement and shearing failure are two independent things and as such need independent investigation too. The design or safe bearing capacity, evidentally, would be the least of the two values, obtainable from the above two independent investigations.

12 6 Classification of Foundations

For purposes of discussion, foundations can be classified into two distinct categories :

1. **Shallow Foundations.** A shallow foundation is defined as one in which the depth D_f (see Fig. 12·3) is equal to or less than the width b of the foundation.

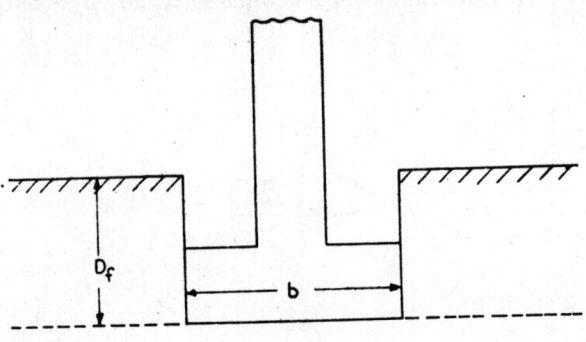

Fig. 12·3

2. **Deep Foundations.** A deep foundation is defined as one in which the depth D_f (see Fig. 12·3) is more than the width b of the foundation.

The above two classifications can be attributed to the classical works of Terzaghi (1) and Meyerhoff (2) respectively. The treatment given by these gentlemen is mainly restricted to footings. Strictly speaking the pile foundations, wells and cassions also should fall into the category of deep foundations.

Concepts presented by Terzaghi applicable to shallow foundations would mainly form the subject-matter of this chapter although important points with regard to Meyerhoff's concept for deep foundations and the conventional methods used for load carrying capacity of the piles and other foundations will also be discussed briefly towards the end of the chapter.

From the design point of view, footings can be divided into four types :

1. **Spread footings.** A spread footing is one in which the base of a column or a wall is enlarged to individual support for the load.

2. **Strap footings.** A strap footing constitutes of two or more individual footings connected by a beam called a *strap*. It is also sometimes called *cantilever footing* or *pump handle foundation*.

3. **Combined footing.** A combined footing supports two or more columns in a row.

4. **Mat or raft foundation.** A mat or a raft foundation is a large footing that supports a number of columns in two or more rows.

Fig. 12·4 shows the above types of footings.   The individual
canhave a different arrangement than the one shown in the figure

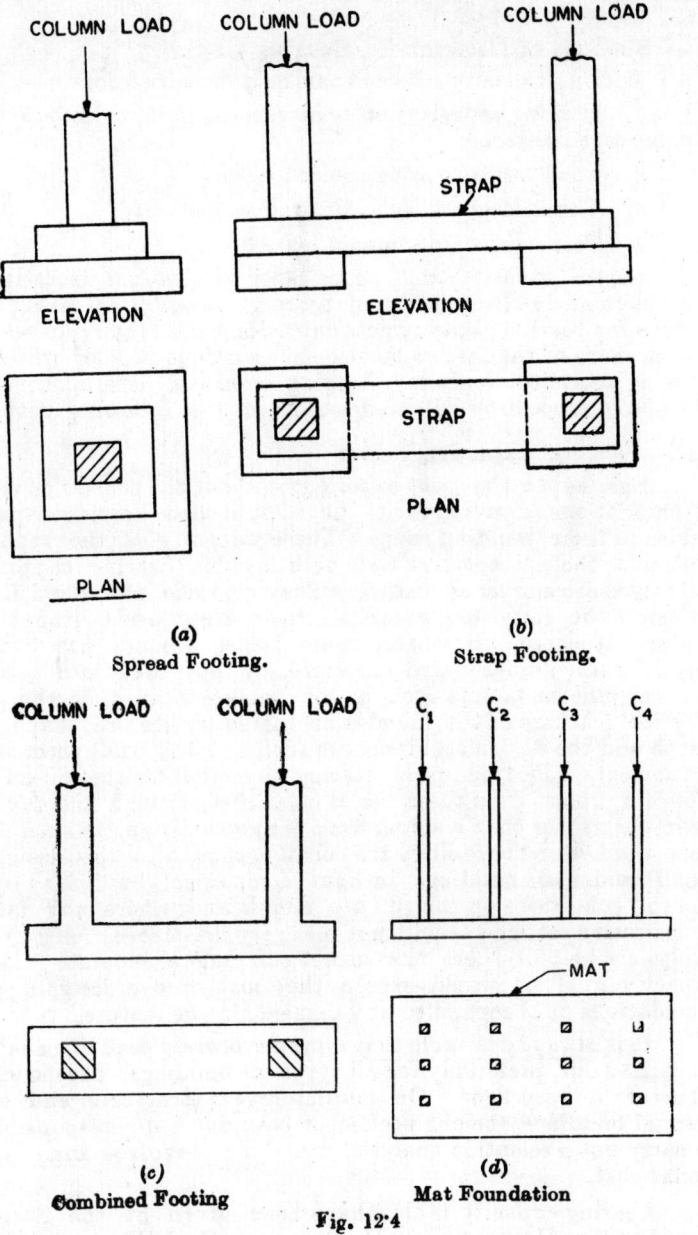

(a)
Spread Footing.

(b)
Strap Footing.

(c)
Combined Footing

(d)
Mat Foundation

Fig. 12·4

for spreading load. Similarly the strap between the two footings can be designed with a slightly different arrangement, than the one shown in the figure.

12.7 Methods of Determining Bearing Capacity

Bearing capacity of a soil can be determined from :—

1. Bearing capacity tables in various building codes, based upon past experience.

2. Load test on a construction site.

3. Theoretical and semi-theoretical methods.

4. Test results from model testing.

All the investigating agencies or individual investigators who have evolved any tables of bearing capacity values or procedure for bearing capacity determination have taken into account one or more of the factors mentioned in article 12·3 on which the bearing capacity depends. Bearing capacity determination by the above procedures will be discussed in the following articles.

12.8 Bearing Capacity Tables

Based upon the past experiences about the success or failure of foundations, many agencies have published bearing capacity tables in their building codes. These tables give the types of soils and their respective safe or allowable bearing capacities. All tables are similar in nature. They presume that the soil can support the indicated pressure with safety and without any undue settlement. However, these tables are not based upon any scientific analysis, and therefore, do not take into account some significant factors such as the compressibility of the soil, physical features of the foundation including its size, shape and depth and the design requirements such as loads and permissible settlements. Picking up the bearing capacity for the soil on the site of a project from these tables is, at best, a rough and a crude estimate, rather than a sound basis for good design. As such these values of bearing capacities are usually applied to the design of small residential buildings or light commercial buildings where the soil conditions on the site are simple and where any failure or excessive settlement will not be very disastrous. Also these tables are helpful where the cost of carrying a separate bearing capacity analysis would exceed the cost of overdesigning the foundations or of repairs to any damage, in the feature.

It is always desirable that a proper bearing capacity analysis be carried out, preferably for all types of buildings, except where structure is very light. On multistoreyed structures and commercial buildings, the engineer must consider it his responsibility to carry out a scientific analysis, even if it involves some additional cost.

Bearing capacity tables have been given by the National Board of Fire Underwriters (3), American Civil Engineers Hand

Book (5). American Standards Association, New York City, Atlanta and Chicago Building Codes. These values are also available in reference 4, at the end of this chapter. The Indian Standards Institution has also given a table of bearing capacity values in their standard IS—1904, 1961—Code of practice for the structural safety of Building Foundations and this table has been printed as table III in the appendix at the end of this volume.

12.9 Load and Settlement.

When a load such as footing is placed on a soil, deformation takes place and the footing settles. If the load is increased in stages and under each load increment final settlement is observed, the settlement increases and ultimately a stage is reached when failure under increased load occurs. The behaviour of the soil under such a loading to failure as observed from model testing and by excavations adjascent to the fullsized foundations indicates three or four stages before failure, as described below :

Firstly, on application of load, the soil distorts. This results in lateral bulging of the soil column directly below the foundation and the settlement of the ground surface in the immediate vicinity of the foundation. As a consequence, in saturated clays, sometimes, a bulging of the surface distances away from the foundation takes place.

Secondly, around the periphery of the foundation, a local cracking may be observed.

Thirdly, a triangular prism of the soil immediately below the foundation starts acting as a part of the foundation and forces the soil downwards, at its apex and outwards at its faces, as already explained in Fig. 12·2 (*a*).

Finally, the zone of shear rupture spreads to larger distances away from the foundation. The surface of shear failure is usually curvilinear. It is the ground surface above this zone of failure that shows upward bulging and cracking.

As a consequence of the above changes, the structure gets damaged and this failure is termed as the **bearing Capacity failure.**

If a load settlement curve is drawn from the load *versus* final settlement readings in such a model test or a full scale test, the results would generally be of the type shown in Fig. 12·5. Curve *A* in this failure is typical of a cohesionless material, curve *B* is typical of a mixture of sand and clay (cohesionless and cohesive soils) and curve *C* is typical of a cohesive soil. It is seen from all these curves that, in general for small loads, the settlement is proportional to the load. This characteristic is specially true for the curve *A*. This curve after a certain load adopts a sharp downward slope, indicating that the failure has occurred. During the early part of this curve upto point *E* when the settlement is proportional to the load, elastic distortion and compression of the soil takes place. During the transition zone, *EF*, from milder to steeper slope, it indicates local cracking of the soil. The last steeper leg of the curve indicates that the settlement is disproportionate to the load and that failure has occurred. The theoretical failure point can be determined by drawing tangents to the initial and final legs of the curve and the point of intersection of these tangents gives the theoretical point of failure *i.e.*, the ultimate load that the soil can carry.

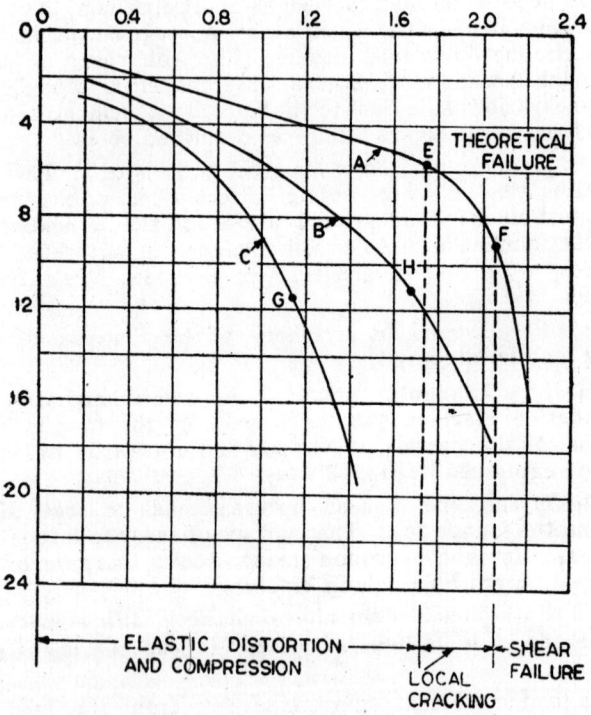

LOAD INTENSITY Kg/Cm^2

Curve A—Cohesionless soil *e.g.*, sand.
Curve B—Mixture of cohesionless and cohesive soil.
Curve C—Purely cohesive soil, *e.g.*, clay.
Load—Settlement curves—Load test.

Fig. 12·5.

It may not be that easy to determine the theoretical point of failure for curves *B* and *C*. In such a case, the ultimate load carrying capacity may be taken as corresponding to a point where the settlement curve becomes fairly steep and approaches a straight line, *e.g.*, point *G* or *H* in this figure.

12.10. Bearing Capacity from Load Tests.

Perhaps the best procedure to find out the safe bearing capacity of a soil is to test a full sized foundation under its design load long enough to observe its complete settlement and then to increase the load to failure. This procedure suffers from these defects;

1. The length of time required to achieve full consolidation settlement may be too much.

2. The load required to carry out the test may be too large.

3. The cost of carrying out a full-scale test is enormous.

As a consequence, a short-term model test is normally used to evaluate the safe bearing capacity. The test is called a **plate loading test**. In this test, a test plate of an appropriate size is placed inside a pit. The width of the pit is at least five times the width of the test plate. The test plate can be circular or square in shape. Usually square plates of sizes ranging from 30 to 75 cms. are used. For clayey, silty and sandy soils normally met with in the field, the size of the plate should be 60 cms. square. For gravelly and dense sandy soils, a 30 cms. square plate may be used. In fact, the size of the plate should be as large as possible, consistent, of course, with the capacity of the loading device. At the centre of the pit a hole of the size of the plate is dug and the bottom level of this hole should correspond to the *proposed* bottom level of foundation. The depth of the hole below the bottom of the test pit shall be such that the ratio of the width to depth of the loaded area in the pit equals the ratio of the width to depth of the actual foundation. With reference

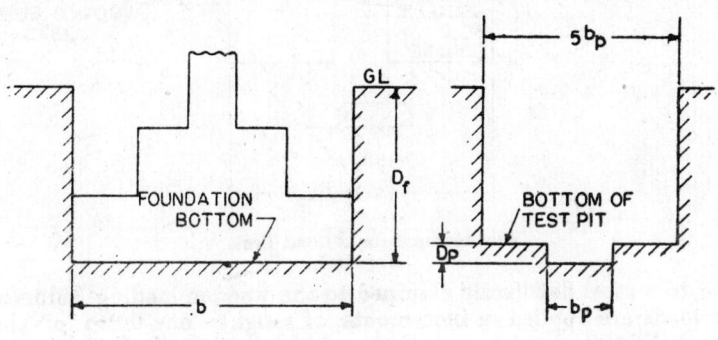

Actual Foundation

Fig. 12·6.

Test Pit

to the Fig. 12·6, this means that

$$\frac{b_p}{D_p} = \frac{b}{D_f} \qquad (12 \cdot 1)$$

where

b_p = width of the hole or width of the plate.

D_p = Depth of the hole below the bottom of the test pit.

b = width of the foundation.

D_f=Depth of the foundation below the ground level.
The plate is placed on the carefully levelled bottom of this hole.
Usually a thin sand layer of plaster of Paris is placed in between to
take care of the irregularities between the bottom of the pit and
the plate. A load is placed on the plate through a loading frame
as shown in Fig. 12·7. The platform of the frame is loaded by
sand bags, cast iron (C.I.) bars or concrete blocks. Sometimes,
reaction loading with the help of a jack is also used. Settlement
of the load is measured with the help of micrometer dial gauges

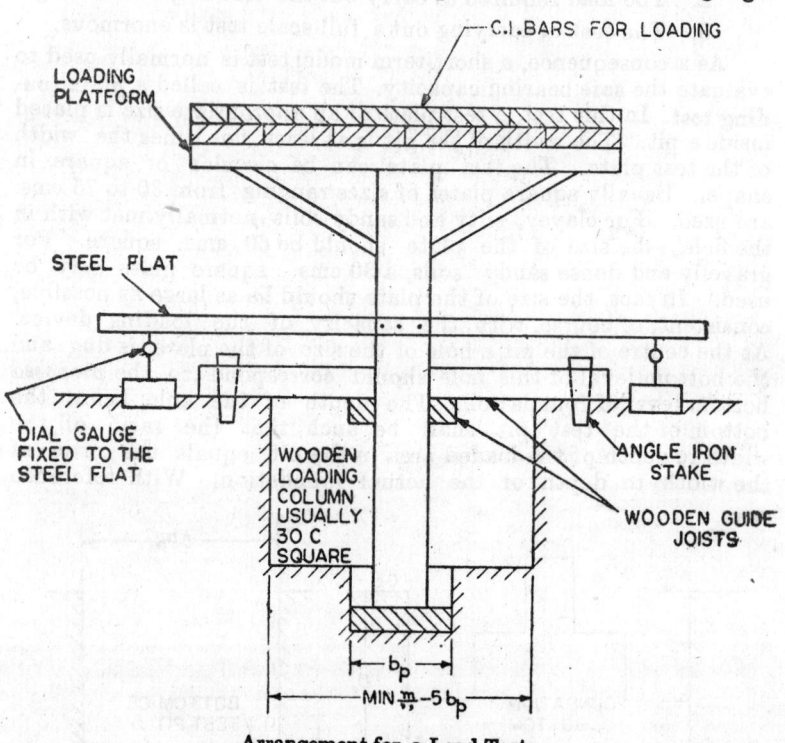

Arrangement for a Load Test.
Fig. 12·7.

fixed to a steel flat firmly clamped to the wooden loading column.
The loads are applied in increments of roughly one-tenth of the
expected failure load or one-fifth of the proposed design load.
Each increment is kept constant and readings taken at regular
intervals of 1, 2, 5, 10, 20, 40 and 80 minutes, etc. till any
further settlement ceases or the rate of settlement becomes less
than 0·002 cms : (0·001 inch) per hour. This is checked
from a plot of the settlement *versus* time curve on a semi-logarithmic scale where the settlement is plotted on the natural
scale and the time on the log-scale. This curve is similar to the
dial readings *versus* time curve on a semi-log scale for a consolidation test. As soon as the settlement ceases or the rate of

settlement goes below the above quoted figure, the load is increased. Total settlement curve of the types shown in Fig. 12.25 is then platted. How to obtain the point of failure has already been explained in article 12·9.

The plate is only a sort of model of the actual foundation and, therefore, the load-test results only express the short-term loading characteristics of the model and not necessarily the long-term loading characteristics of a full-sized foundation.

As has been pointed out in article 12·5, safe bearing pressure for a soil is determined either from settlement considerations or from shear failure point of view. It may be recalled here (and it was pointed out in Chapter 9) that, from theoretical considerations the distortion settlement of both rigid and uniformly loaded areas increases in direct proportion to the foundation pressure and the foundation width. In a load test, the size of the plate is much smaller than the size of the actual foundation. Consequently, for the same pressure the settlement of the bearing plate will be different from the settlement of the foundation. As such, it is essential to co-relate the settlement of the bearing plate with the settlement of the foundation.

The distortion settlement of a square footing on homogeneous elastic soil is given by

$$S = \frac{0.6qb}{E} \qquad ...(12.2)$$

where S is the settlement, q the pressure intensity, b the width (or length) of the foundation and E the modulus of elasticity of the soil. For saturated and homogeneous clays, E is a constant quantity so that the settlement of the plate or the actual foundation for a given pressure intensity will be proportional to the respective widths. If S_f denotes the settlement of the foundation of width b and S_p denotes the settlement of the plate of width b_p, then the following approximate[1] relationship between these quantities follows, for clayey soils.

$$\frac{S_f}{S_p} = \frac{b}{b_p} \qquad ...(12.3)$$

Modulus of elasticity E for sands, increases with the foundation width and such a simple relationship between the settlements and widths is not possible. Also the sand undergoes some compression settlement or reduction in void ratio during a test. An empirical relationship between foundation settlement S_f and plate settlement S_p (distortion plus compression) based upon their respective widths for a cohesionless sand is as under :

$$\frac{S_f}{S_p} = \left[\frac{b_f (b_p + 1)}{b_p (b_f + 1)} \right]^2 \qquad ...(12.4)$$

---

1. The settlement of the foundation (S_f) for clayey soil is normally determined from the consolidation characteristics of the soil and not from the plate-load test.

where the various symbols have the meaning as in equation (12·3). In equation 12·4, the foundation width b_f and the plate width b_p are expressed in feet. This equation is based upon the work of Terzaghi and Peck (6) and Mazanti and Sowers (7).

This expression has also been incorporated in the Indian Standard IS-1888, 1962. If the widths of the foundation and the plate are expressed in cms, the above expression will read as,

$$\frac{S_f}{S_p} = \left[\frac{b_f\,(b_p + 30\cdot48)}{b_p\,(b_f + 30\ 48)}\right]^2 \qquad ...(12\cdot5)$$

Equation (12·4) or (12·5) can be used to find out the load-intensity for a footing of known size for the permissible settlement i.e., b_f, b_p and S_f being known, S_p can be found out and then the load-intensity can be read from the corressponding load-settlement curve. (See curve A, Fig. 12·5 for a cohesionless soil).

Method of Plotting. One method to plot the load-settlement curve from the load test results is shown in Fig. 12·5, where both the load intensities and corresponding settlements are plotted to a natural scale. An improved method of plotting the results and evaluating the ultimate bearing capacity has been suggested by Abbet (8). This method consists of two steps.

1. The results of the load settlement test are first plotted to a natural scale as shown in Fig. 12·5. The early straight line

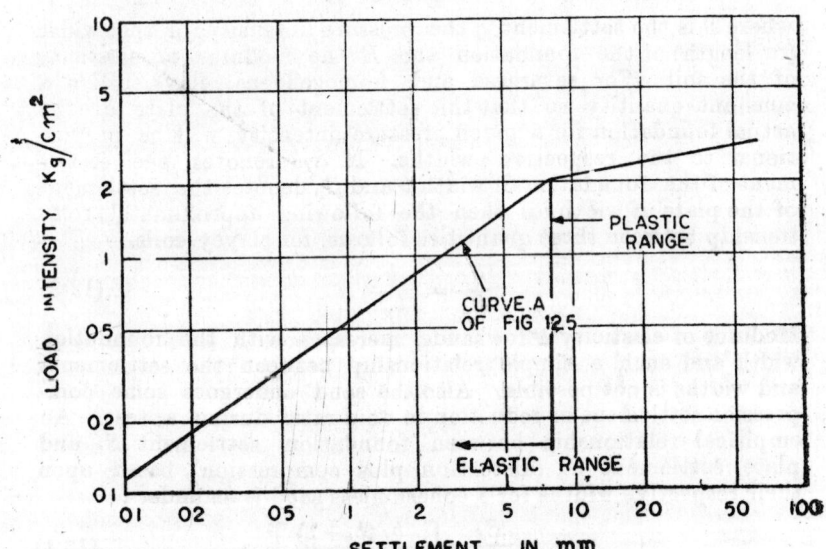

Load-settlement curve, Log-Log scale.
Fig. 12·8.

portion of the curve is projected backward to cut the settlement axis. Any settlement at zero load-intensity so indicated is due to the uneven seating of the test plate. This is called *zero correction*. This zero correction is applied to all settlement readings. The corrected settlement values and the corresponding load intensities are then plotted on a log-log scale, as shown in Fig. 12·8. If a proper zero correction is made, this logarithmic plotting will give two straight lines that deviate from each other at a point. The point of intersection of the two straight lines gives the ultimate bearing capacity.

This method has also been recommended by the Indian Standard Institution in their standard IS : 1888-1962.

Housel has also suggested a method for determining ultimate load carrying capacity and the student may refer to reference 9 for this method.

Limitation of load-test. Load test results unless properly interpreted may lead to serious misgivings. Limitations of this test are as under.

(*i*) The settlement of a full-sized foundation differs from that of a small sized plate. As such unless *size effects*[1] are taken into consideration by the use of the equations (12.3) and (12.4) or (12·5), the settlement of the plate does not reflect the settlement of the full-sized foundation although the pressure may be the same. Even the use of these equations for pure sands or clays may not give the correct settlement for the full-sized foundation because a soil, however, homogeneous it may be, is usually never purely a sand or a clay. Some agencies[2] recommend the repetition of the load test by using plates of two or more sizes and then by extrapolation the settlement for the full size of the foundation may be found from the settlements for these plates.

(*ii*) The load test results reflect the characteristics of the soil located only within a depth of nearly twice the width of the plate. As such, the test may not reflect the characteristics of the soil stressed by the full-sized foundation. This can be visualized by considering the pressure bulb for the two cases. It seems necessary here, first to indice as to what a pressure bulb is.

Stress distribution theories were presented in Chapter 8. By any one of these theories, say applicable to isotropic, homogeneous, elastic and semi-infinite masses, the stress intensity at any point in the soil mass under the loaded area can be found out. If the stress intensities at a number of points in the soil mass are known, contours of equal stress intensity under the

1. It is shown in Art. 12·15 that the bearing capacity in sandy soils increases with the width of the footing. Hence the importance of size effects from this point of view too.
2. See IS. 1888, 1962, page 11.

loaded area can, then, be drawn. Fig. 12·9 shows such contours of equal intensity, in terms of the load intensity q for infinitely long and a square foundation. These stress intensity contours are based upon the Bouessinesq's analysis, Similar diagrams can be drawn from the Westerguard's analysis also. Contours of the type shown in Fig. 12.9 are called a **pressure bulb.**

A pressure bulb of 0·1 times the load intensity q is usually taken to indicate the zone of increased stress that may cause settlement of the foundation. As seen in the figure this zone extends to about twice the width of the foundation. This is the condition for a homogeneous elastic and isotropic medium. In an actual soil, this zone may not even go to this depth.

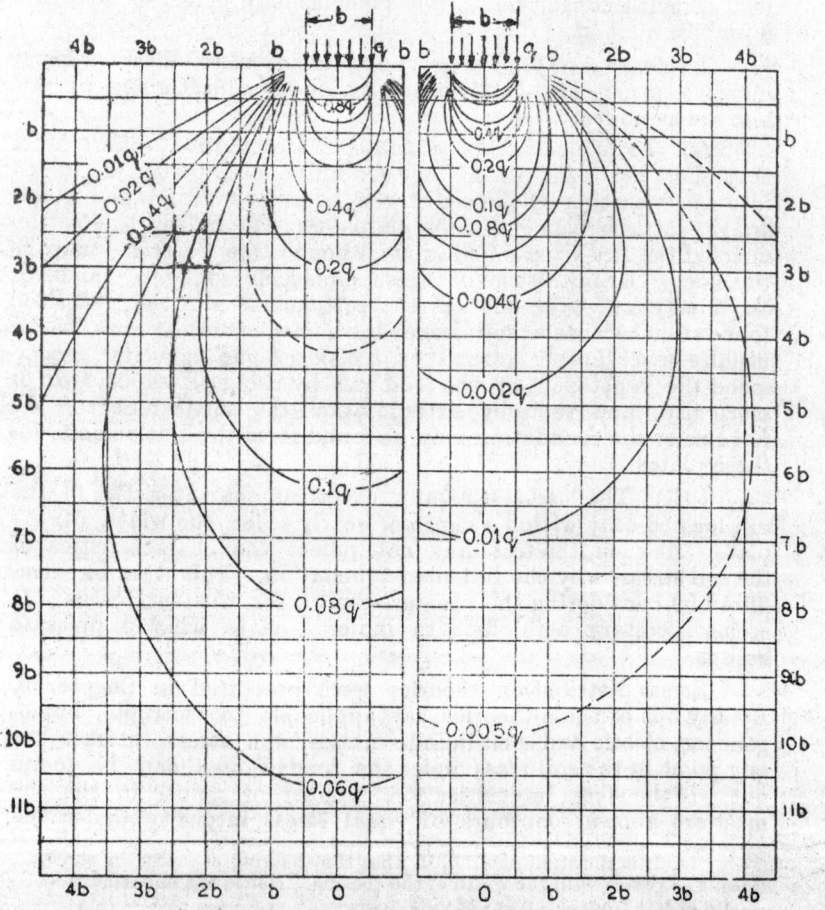

Contours of equal vertical stress intensity beneath a foundation on an isotropic, homogeneous, elastic and semi-infinite mass (Boussinesq's case)
Fig. 12·9.

If, now, the pressure bulb of 0·1q intensity be drawn for the actual foundation, as shown in Fig. 12·10, then it is seen that the depth and extent of the soil enclosed by this contour is much larger in the case of actual foundation than in the case of the plate. Now, suppose that the soil underneath the foundation constitutes of two layers—a layer of granular material underlain by a layer of clayey soil. The granular layer, naturally, would have higher capacity compared to the clayey layer underneath it. If the pressure bulb of the plate extends only upto the granular soil as shown in this figure, the bearing capacity from the plate-loading test does not reflect the characteristics of the clay layer. As such, the results from a plate-loading test do not give a realistic value of the bearing capacity, in case the soil within the pressure bulb of the plate and the pressure bulb of the foundation is not the same.

(*iii*) In case of clayey soils, the consolidation settlement is not reflected in the settlement measurements, since the consolidation settlement may take years to develop, whereas a plate loading test is usually over in a few days.

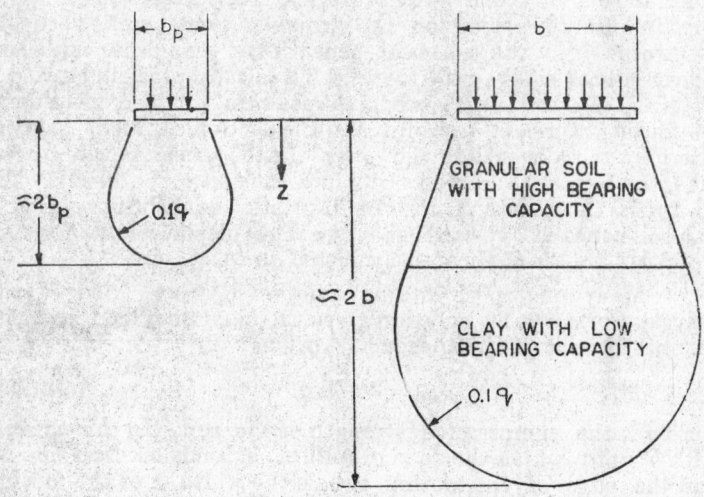

Pressure bulbs for the plate and the actual foundation.
Fig. 12 10.

(*iv*) The use of the bearing capacity data from a plate loading test is not suitable for the design of a continuous footing, since the bearing capacity of a square or a circular footing is usually 30 per cent more than that of a continuous footing whose width is equal to the width of the square footing or the diameter of the circular footing.

12·11. Bearing Capacity from Theoretical and Semi-theoretical Considerations.

Many theoretical and semi-theoretical methods of bearing capacity determination are available in the soil mechanics literature. It is outside the scope of this book to discuss all these methods. Also the scope of the book limits a very detailed discussion on the methods presented hereafter. As such, important and basic concepts behind only few theoretical and semi-theoretical analyses will be given in the subsequent articles. The starting point in the analytical methods was provided by Prandtl (11) who gave a theoretical solution for the plastic failures of metals and whose solution for a horizontal surface (a special case of his general solution) is applicable to foundations. Extension and modification of these concepts were done by Terzaghi (1). Both these methods are applicable to shallow foundations ($D_f \leqslant b$; see article 12·6). Myerhoff (2) gave the solution for deep foundations ($D_f > b$; see article 12·6). Following articles give the salient points of all these three analyses and any related work.

12·12. Prandtl's Analysis

Prandtl was mainly concerned with the penetration of punches into metals. Movement of these punches was guided. A *basic* assumption of his solution for foundations is that an axially loaded footing of width b and a large length, L where L/b approaches infinitely[1] placed on the ground surface, sinks vertically downwords into the material beneath it, thus producing shear failure on both sides of the footing. The situation is illustrated in Fig. 12·11. Material in the wedgeshaped zone I immediately under the footing spreads laterally and the section through this zone undergoes the distortion shown by dotted lines. As the footing sinks, the soil zone I exerts pressure on zones II and III. The soil zones II are assumed to be in plastic equilibrium and they push soil zones III upwards as units. The displacements for zone II and III are also indicated by dotted lines.

Many angles of internal friction ϕ were tried in this analysis. For $\phi = 0$, his solution gives the maximum load intensity (*i.e.* unit ultimate bearing capacity) q_{ult} as

$$q_{ult} = 2·571 \ q \qquad\qquad ...(12·6)$$

where q is the compressive strength of the soil. In this case the width b' upto which the zone of failure extends on each side beyond the edge of the footing equals the width b of the footing. For all other values of $\phi > 0$, ratio b'/b is more than unity and q_{ult} increases rapidly with the value of ϕ. Table 12·1 below gives the values of the ratio b'/b and q_{ult}/q for different values of ϕ.

The above derivations do not take into account the weights of the zones II and III.

Experience shows that the complete shear failures of the soil under loaded footings occur in plastic clays only. Angle of internal

---

1. By definition, such a footing would be called a continuous footing. By assuming a continuous footing, the problem becomes two dimensional and hence easier to solve.

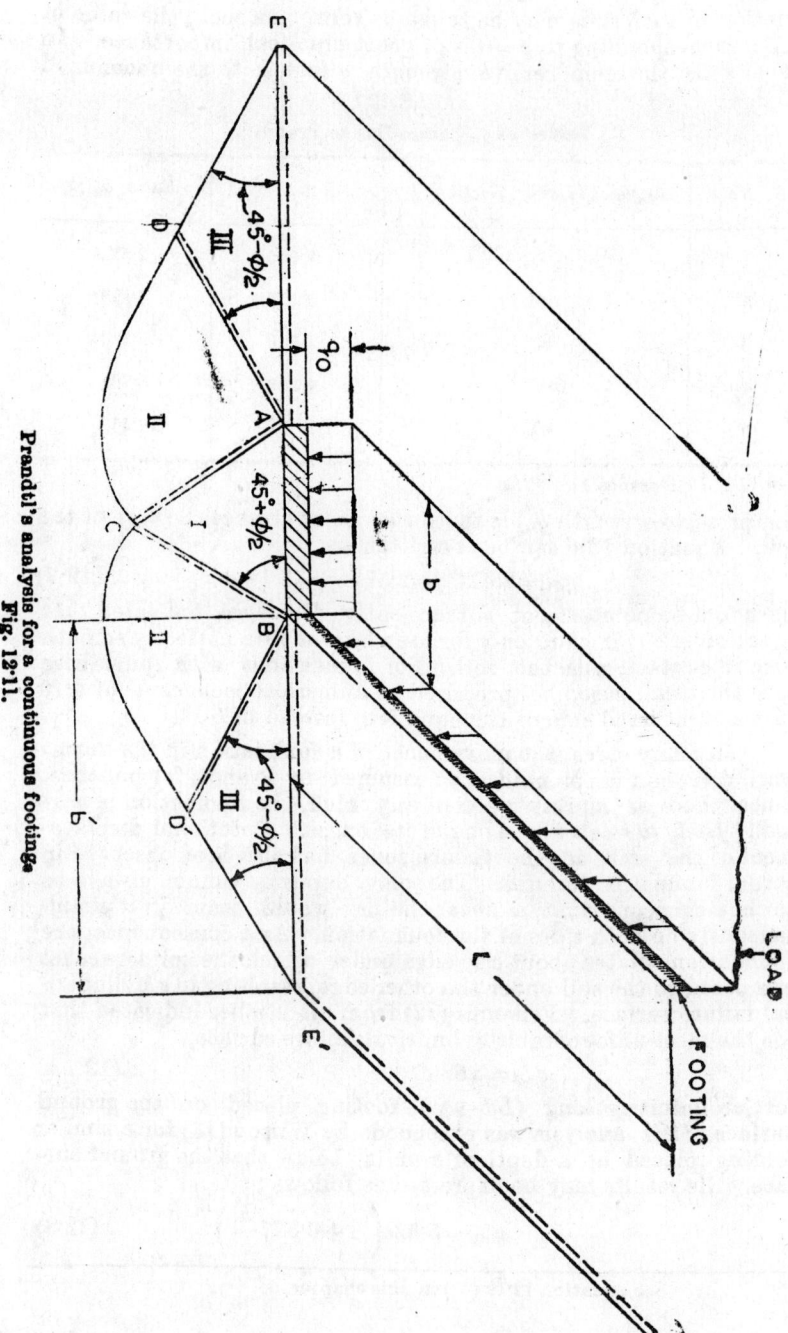

Prandtl's analysis for a continuous footings.
Fig. 12·11.

friction in such soils may be taken as zero. As such, the value of q_{ult}/q corresponding to $\phi=0$ is of great practical importance. In such soils the compressive strength q (equal to the unconfined

TABLE 12·1
Values of q_{ult} according to Prandtl[1]

S. No.	Angle of internal friction, ϕ	Ratio, $b'b$	Ratio, q_{ult}/q
1	0	1·000	2·571
2	10°	1·572	3·499
3	20°	2·530	5·194
4	30°	4·290	8·701
5	40°	8 462	17·560

1. Reference 11

compressive strength q_u) is twice the shear strength s $(=c)$ of the soil. Equation 12·6 can be re-written as

$$q_{ult}=2.571\ q=2·571\ q_u=5·14\ c \qquad ...(12·7)$$

As would be pointed out subsequently, Terzaghi indicated that equation (12·7) is valid only for foundations with perfectly smooth base in contact with the soil. For foundations with rough base as is the usual case, he proposed a numerical coefficient[1] of 5·70 on the right hand side of the equation, instead of 5·14.

In many cases, the movement of a foundation in the downward direction is *not guided*, as assumed by Prandtl for punches. Since there is no restriction of any kind, the foundation is normally *freely to rotate* about one of its edges. Rotational displacement of the soils in the failure zones has also been observed in actual foundation failures. The clay deposits cannot always be so homogeneous that a shear failure would occur in it simultaneously on both sides of the foundation. As a consequence, the foundation rotates about one edge under which the soil is weaker compared to the soil under the other edge, resulting in a cylinderical failure surface. Fellenius (12) from his studies indicated that for the most unfavourable cylindrical failure surface,

$$q_{ult}=5·52\ c \qquad ...(12·7\ a)$$

for an infinitely long $(L/b\to\infty)$ footing placed on the ground surface. His analysis was extended by Wilson (13) for a similar footing placed at a depth of h units below that the ground surface. His results may be expressed as follows :

$$q_{ult}=5·52c\left(1+0·377\frac{h}{b}\right) \qquad ...(12·8)$$

1. See equation 12·19 (   ) in this chapter.

In case the footing is not infinitely long *i.e.*, L/b does not approach infinity, then the resistance to rotation gets increased by the shearing strength of the soil on vertical planes beneath the two ends of the footing. The increased ultimate bearing capacity of clays in case of a footing of width b and length L, is given by,

$$q_{ult} = 5.52\, c \left(1 + 0.38\, \frac{h}{b} + 0.44\, \frac{b}{L}\right) \qquad \dots (12.9)$$

For a square footing where $b = L$, placed on the ground surface (*i.e.*, $h = 0$), the ultimate bearing capacity from the above equation is as under :

$$q_{ult} = 7.95\, c \qquad \qquad \dots (12.10)$$

In order to find out the safe bearing capacity q_0 of the soil from the above equations for use in design, a suitable factor of safty F_s is required. Factors of safety for use in bearing capacity analysis are discussed in article 12.19.

12.13. Terzaghi's Analysis

Tarzaghi analysed first the conditions for the failure of soil support under shallow continuous footings ($L/b \to \infty$) and then suggested modifications for isolated square, rectangular and circular footings. The conditions of failure for a continuous footing, as considered by him, are shown in Fig. 12.12 to 12.14. Three kinds of conditions will be discussed :

(a) Perfectly smooth base of a footing resting on surface of an ideal soil.

(b) Rough base of a footing resting on the surface of an ideal soil.

(c) Rough base of a footing resting at a level below the ground surface.

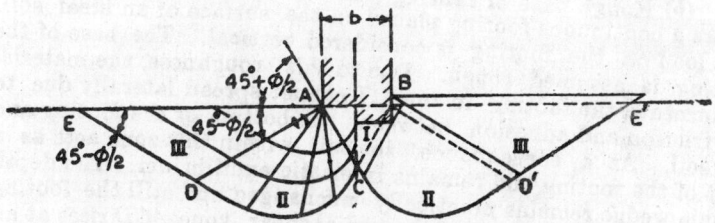

A footing with a smooth base resting on the surface of an ideal soil.
Fig. 12.12.

(a) **Perfectly smooth base of footing (surface loading case).** Fig. 12.12 shows a continuous footing placed on the surface of an ideal soil[1]. The load on the footing is considered vertical. The base is assumed to be perfectly smooth. The state of plastic failure in such a case corresponds to that of a semi-infinite weightless solid and is indicated on the left hand side. This case

---

1. By ideal soil is meant that it is elastic, homogeneous and isotropic.

is identical with that analysed by Prandtl. The failure region can be divided, again, into five zones designated as I, II and III. Zone I is a wedge-shaped zone and is located beneath the foundation. In this zone the major principal, stresses are vertical and Terzaghi has called it the *active Rankine zone*, because the shear pattern in this zone is identical with the shear pattern for an active Rankine state. Zones designed as III are identical with the passive Rankine state and have been called as the *passive Rankine zones*. The boundaries of the active zone rise at an angle of $(45° \times \phi/2)$ and those of passive zones at an angle of $(45° - \phi/2)$, with the horizontal. Zones designated as II are located between the zone I and zone III are called the *zones* of *radial*-shear. The dotted lines on the right hand side of the central line indicate the boundaries of the zones I and III at the instant of failure and the solid lines represent these boundaries when the foundation sinks into the material. The material in the wedge-shaped zone I immediately under the footing spreads laterally and the section through this zone deforms as shown.

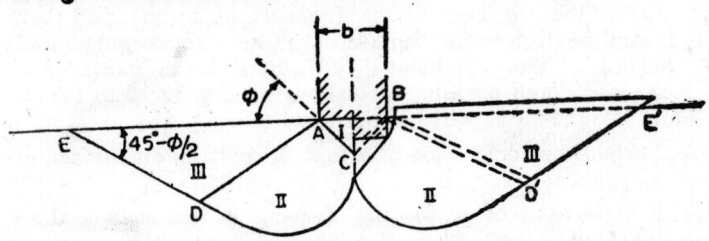

A footing with a rough base resting on the surface of an ideal soil.
Fig. 12.13.

(b) Rough base of footing (surface loading case). Fig. 12·13 shows a continuous footing placed on the surface of an ideal soil. The load on the footing is considered vertical. The base of the footing is assumed rough. Due to this roughness, the material underneath the footing in zone I cannot spread laterally due to the friction and adhesion between the bottom of the footing and the soil. As a consequence, the soil within this zone acts as a part of the footing and remains in elastic equilibrium. The depth of this wedge remains practically unchanged and still the footing sinks. The boundary CA of the radial shear zone ACD rises at an angle of ϕ to the horizontal, provided the adhesion and friction at the base of the footing are sufficient to prevent any sliding motion. At C, the radial shear zone is tangential to the vertical.

(c) Rough base of the footing. (loading below the surface) Fig. 12·14 shows a continuous footing placed at a certain level below the surface, in an ideal soil. The load on the footing is considered vertical. The depth D_f below the ground surface at which the footing is placed is less than the width b of the footing *i.e.* it is a shallow footing. The base of the footing is assumed rough. In this case, the shearing resistance ef the soil located above the base of the footing is neglected. Instead, if γ is the unit weight of this soil, then the soil above the bottom level of the footing is replaced by an equivalent surcharge $q=\gamma D_f$, per unit of area. The failure zones are deemed to be located only

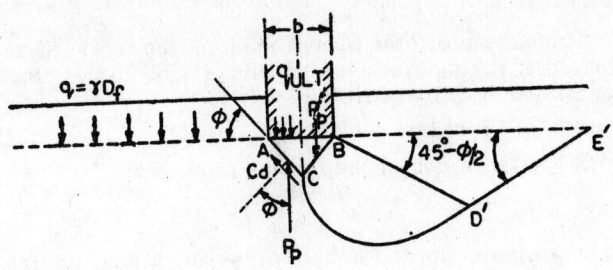

A footing with a rough base resting at a level below the surface on an ideal soil $(D_f < b)$

Fig. 12·14

below the botton level of the footing and are considered identical with those shown in Fig. 12·13.

In brief, the assumptions in the Terzaghi's bearing capacity analyis in the above case are ;

1. The footing is continuous *i.e.*, ratio L/b approaches infinity.

2. The footing is shallow *i.e.*, $D_f < b$.

3. The base of the footing is rough.

4. The shearing resistance of the soil above the base of the footing is neglected. The soil above this level is replaced by an equivalent surcharge $q=\gamma.D_f$ acting at that level. Consequently, the failure regic.. is considered not to extend above the base of the footing and this surcharge $q=\gamma D_f$ only provides the pressure on the failure region.

5. Load on the footing is vertical.

In any one of the three cases described above, the failure occurs only when the pressure exerted by the load on to the soil adjoining the inclined boundaries of the wedge-shaped zone equals the passive earth pressure. The ultimate bearing capacity can be determined by algebrically summing the vertical components of the forces that act on the soil located within the wedge-shaped zone I and equating the sum to zero.

Let Q_{ult} be the total ultimate load per unit length of the footing ($=q_{ult} \times b$) that the footing can support. The wedge ABC will be in equilibrium, at failure, under the following forces :

1. The total ultimate load Q_{ult}, acting vertically downwards and representing the total ultimate bearing capacity of the footing.

2. The weight of the footing equal to $\gamma.\frac{1}{2}(b)(b/2 \tan \phi)=\frac{1}{4}\gamma b^2 \tan \phi$ acting downwards. (This weight is small and hence it is neglected).

3. The resultant passive earth pressure P_P acting on the faces AC and BC and inclined at an angle ϕ to the normals to these faces. The direction of P_P in the figure is vertical.

4. The cohesive force C_d acting along the faces AC and BC.

Therefore, taking into account the above forces, the equation of equilibrium can be written as :

$$Q_{ult}+\tfrac{1}{4}\gamma b^2 \tan \phi=2(P_P+C_d \sin \phi) \qquad ...(12.11)$$

Since b is the width of the footing as shown,

$$C_d=c\left(\frac{b/2}{\cos \phi}\right)$$

where c represents the soil cohesion[1] acting along the faces AC and BC.

Equation (12·11) becomes

$$Q_{ult}+\tfrac{1}{4}\gamma b^2 \tan \phi=2(P_P+c \; b/2 \tan \phi) \qquad ...(12.12)$$

If the weight of the soil on the left hand side is neglected, then the equation (12.12) becomes,

$$Q_{ult}=2(P_P+c \; b/2 \tan \phi) \qquad ...(12.13)$$

The resultant passive earth pressure P_P in the above equations constiutes of three components as under :

1. A component P_{P_γ} which is produced by the weight of the shear zone $E'D'CB$. This component is calculated by assuming the cohesion c and the surcharge q equal to zero (*See* Page 124 of reference 1, for details).

2. A component P_{P_c} which is produced by the cohesion c of the soil. This component is computed by assuming the unit weight γ of the soil and the surcharge q equal to zero, (*See* Page 125 of reference 1, for details).

3. A component P_{P_q} which is produced by surcharge q. This component is computed by assuming the weight γ and the cohesion c of the soil equal to zero (*See* Page 126 of reference 1, for details).

The values of P_{P_γ} P_{P_c}, and P_{P_q} are obtained for different surfaces of failure and then they are added to obtain P_P. As such, evaluation of P_P is approximate, although not much error is involved in assuming different failure surfaces. This value of P_P equal to $(P_{P_\gamma}+P_{P_c}+P_{P_q})$ can be substituted in equation (12.12) or (12.13). Substituting this value in equation[2] (12.12), one would get,

---

1. This soil cohesion is obtained from the Coulomb's equation.
2. If value of P_P is substituted in equation (12·13), the only change ultimately comes in the value N_γ in equation (12·16a).

$$Q_{ult} + \frac{1}{4}\gamma b^2 \tan\phi = 2\left[P_{P_\gamma} + P_{P_c} + P_{P_q} + c.\frac{b}{2}\tan\phi\right]$$

$$\text{or } Q_{ult} = b\left\{\left[2\frac{P_{P_\gamma}}{b} - \frac{\gamma b}{4}\tan\phi\right] + \frac{2P_{P_q}}{b} + \left[\frac{2P_{P_c}}{b} + \frac{c}{2}\tan\phi\right]\right\}$$

$$...(12.14)$$

If an analysis[1] is made for the values of, P_{P_γ}, P_{P_c} and P_{P_q},

it is possible to write down from equation (12.14), the following equation, for the total ultimate load.

$$Q_{ult} = b\left[\frac{\gamma b}{2}N_\gamma + \gamma D_f N_q + cN_c\right] \qquad ...(12.15)$$

The ultimate bearing capacity, q_{ult} per unit area from equation (12.15) will be,

$$q_{ult} = \frac{\gamma b}{2}N_\gamma + \gamma D_f N_q + cN_c \qquad ...(12.16)$$

Equation (12.16) is sometimes called the **general bearing capacity equation.** In equations (12.15) and (12.16). N_γ, N_q and N_c are the Terzaghi's dimensionless bearing capacity factors due to the soil weight, surcharge and cohesion respectively. Their numerical values depend upon the value of the angle of internal friction ϕ of the soil, as given below :

$$N_\gamma = \frac{1}{2}\tan\phi\left[\frac{K_P}{2\cos^2\phi} - 1\right] \qquad ...(12.16a)$$

$$N_q = \frac{a^2}{2\cos^2(45^0 + \phi/2)} \qquad ...(12.16b)$$

and $$N_c = \cot\phi\left[\frac{a^2}{2\cos^2(45° + \phi/2)} - 1\right] \qquad ...(12.16c)$$

where K^2_P = co-efficient of passive earth pressure for a cohesionless soil.

ϕ = angle of internal friction for the soil

and $a = \epsilon^{(3\pi/4 - \phi/2)\tan\phi}$

Terzaghi has plotted these bearing capacity factors as functions of the angle of internal function ϕ. They are shown in Fig. 12.15, by solid lines.

---

1. This analysis is outside the scope of this book.
2. It may be noted that K_P is also a function of ϕ

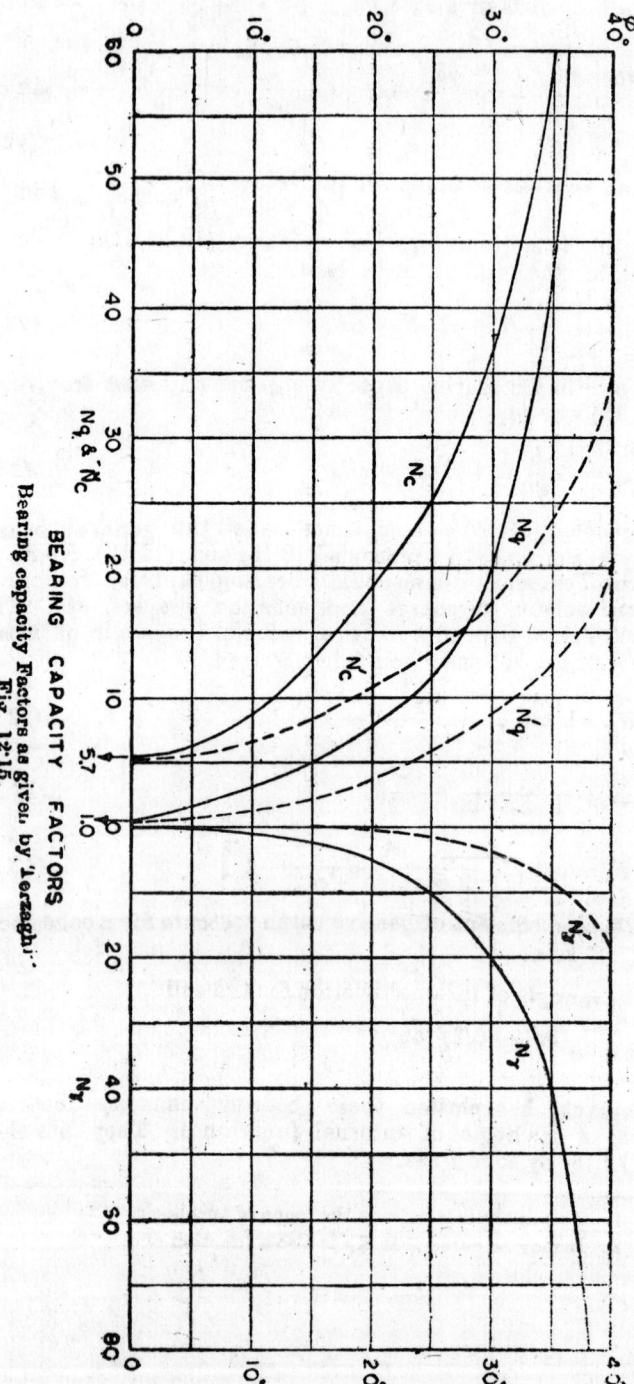

BEARING CAPACITY FACTORS

Bearing capacity Factors as given by 'Terzaghi'.

Fig. 12·15.

General and Local Shear Failures. When there is no load on the soil surface, the soil is in a state of elastic equilibrium. When the load on the soil is increased beyond a critical value, the soil gradually passes into a state of plastic equilibrium. For this transition from an elastic to a plastic state, there may exist either of the two possible conditions given below :

1. The physical properties of the soil may be such that the strain that takes place before the plastic failure occurs, is very small. As such, the footing does not sink into the ground unless the plastic failure has actually occurred. The corresponding load-settlement curve is of the type designated as A in Fig. 12·5. The failure of this type is called a **general shear failure.**

2. The physical properties of the soil may be such that large strain takes place before the plastic failure occurs. The corresponding load-settlement curve in this case is of the type designated as C in Fig. 12·5. The failure of this is called **local shear failure.**

General shear failure occurs normally in dense or stiff soils and the planes of failure are well defined as shown in Fig. 12·13 (say). Local shear failure occurs in lose soils and the planes of failure in such a material are not as well defined as shown in Fig. 12·13.

Equation (12·16) given in the preceding paragraph is applicable to a case of general shear failure. Terzaghi has suggested that the bearing capacity for footings resting on loose materials where failure amounts to a local shear case may be calculated approximately by using the equivalent parameters c' and ϕ' for an equivalent ideal plastic material,

where $\qquad\qquad c'=2/3\ c$

and $\qquad\qquad \tan\phi'=\tfrac{2}{3}\tan\phi$

The above suggestions are based upon the available stress-strain relations for such soils. The bearing capacity equation in that case reads as,

$$q_{ult}=\frac{\gamma b}{2}\,N'_\gamma+\gamma D_f N'_q+c'N'_c$$

or $\qquad q_{ult}=\frac{\gamma b}{2}\,N'_\gamma+\gamma D_f N'_q+\tfrac{2}{3}cN'_c \qquad\qquad ...(12·17)$

where N'_γ, N'_q and N'_c are the bearing capacity factors for local shear case which are obtained by replacing c and ϕ by c' and ϕ' in equations 12·16 (a) to (c). These factors are also plotted against ϕ values in Fig. 12·15. After picking up these factors against the angle of internal friction ϕ of the soil equation (12·17) may be used to find out the bearing capacity q_{ult}.

If the stress-strain curve of a soil is intermediate between the curves A and C, the ultimate bearing capacity will lie between the two values given by equations (12·16) and (12 17).

12.14. Meyerhof's Analysis

Meyerhof (2) extended the work of Terzaghi. His analysis differs from that of Terzaghi in a few respects.

1.   Terzaghi neglected the shear resistance of the soil above the bottom level of the footing and replaced this soil as surcharge. In other words the failure region has not been considered in this analysis to extend above the base. Meyerhof assumed that the shear zones extend above the base of the footing.

2.   Terzaghi assumed that the central wedge surface AC and BC make an angle ϕ with the horizontal. Meyerhof's analysis assumes this angle made by the surface AC and BC as ψ where $\psi > \phi$.

As a result of the above differences, the bearing capacity factors by Meyerhof for the same angle of internal friction are less than the ones given by Terzaghi. This is shown in Fig. 12 16.

A detailed discussion on Meyerhof's analysis is outside the scope of this book.

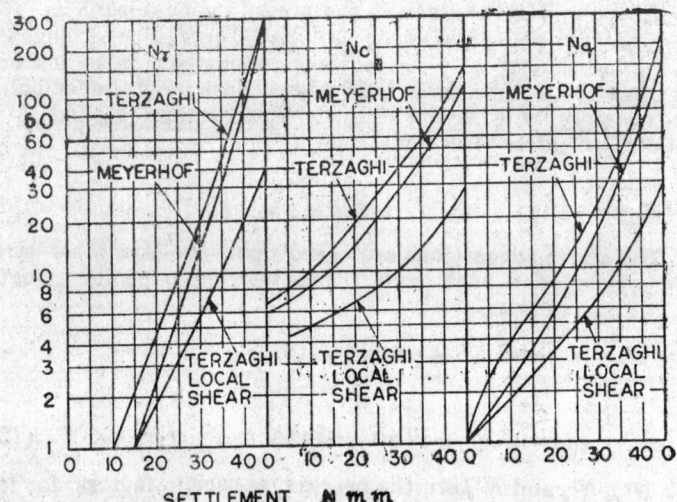

SETTLEMENT $N\ m\ m$

Fig. 12·16.

12 15. Discussion on General Bearing Capacity Equation

The general bearing capacity equation is,

$$q_{ult} = \frac{\gamma b}{2} N_\gamma + \gamma D_f N_q + c N_c \qquad \qquad ...(12\cdot16d)$$

This equation and the graphs for N_γ, N_q and N_c reflect many important factors on which the bearing capacity of a soil depends. These factors will be discussed in the light of whether the material is cohesionless or cohesive.

Cohesionless Material. If the soil on which the footing is to rest is cohesionless, then the cohesion c is zero. Substituting this in the above equation, q_{ult} for a cohesionless soils becomes

$$q_{ult} = \frac{\gamma b}{8} N_\gamma + \gamma D_f N_q \qquad \qquad ...(12 \cdot 18)$$

This equation shows that the ultimate bearing capacity of a cohesionless soil depends directly upon the

1. *Unit weight of the soil.* The unit weight of a soil with a degree of saturation S is given by

$$\gamma = \frac{G + Se}{1+e} \gamma_w$$

In case the soil is submerged under water, the unit weight is given by

$$\gamma_b = \frac{G - 1}{1 + e} \gamma_w$$

The submerged unit weight γ_b is less than the unit weight γ of the soil by an amount equal to the unit weight of water. As such, an estimation *of the location of water table is* very essential. If the water table is too close to the ground surface, the unit weight γ_b will be effective. Therefore, the bearing capacity will be less than what it would be, if the water table were far away.

In fact, the ultimate bearing capacity should be found out for the worst conditions *i.e.* by taking the unit weight of the soil as γ_b so that even if in a moist soil at a later date the water table rises, the bearing capacity is not lowered, unless, of course, it is very sure that the water table is not likely to rise during the life of the structure.

2. *Width of the footing.* As seen in the equation (12·18), the unit ultimate bearing capacity increases with the increase in the width which occurs in the first factor on the right hand side. Also the total ultimate bearing capacity per unit length of the footing will also increase with the increase in width since in order to get Q_{ult} (the total ultimate bearing capacity per unit length) q_{ult} is multiplied by the width b.

3. *Depth of the footing.* The depth of the footing below the ground surface, D_f occurs in the second factor on the right hand side of equation (12·18). As such, the unit ultimate bearing capacity q_{ult} increases with the increase in the depth of the footing or the height of the surcharge, above the base of the footing.

Graphs of N_γ and N_q *versus* ϕ in Fig. 12·15 show that the values of N_ϕ and N_q increase with the increase in the angle of internal friction. It is also known that the angle of internal friction depends upon the relative compactness or relative density of a cohesionless mater.al. More compact a soil is more is its angle of internal friction. As such, the bearing capacity of the cohesionless soil also depends indirectly on the *relative density* of the soil.

Cohesive Material. If the soil on which the footing is to rest is cohesive and $\phi=0$, the values of N_γ, N_q and N_c can either be calculated from the set of equations (12·16a) to (12·16c) or they can be read directly from the graphs in Fig. (12·15).

$\therefore$ For $\qquad \phi=0;\ N_\gamma=0,\ N_q=1$ and $N_c=5\cdot7$.

Substitution of these values in the general bearing capacity equation for a continuous footing gives

$$q_{ult}=\gamma.\,D_f+5.7c \qquad\qquad \text{...(12·19)}$$

This equation shows that the bearing capacity of such a soil depends upon the unit weight γ of the soil. The observations with regard to the location of water table, as given for cohesionless materials are equally applicable here also. In addition this equation shows that,

1. The ultimate bearing capacity of the soil depends upon the *cohesion* of the soil. If cohesion is taken as one-half of the unconfined compressive strength q_u, the above equation will reduce to,

   $$q_{ult}=\gamma.D_f+2.85\,q_u \qquad\qquad \text{...(12·20)}$$

   One may say that the bearing capacity in this case depends upon the unconfined compressive strength of the soil.

2. The ultimate bearing capacity of the soil depends upon the depth of the footing below the ground level or the height of the surcharge above the base of the footing.

   If in equations (12·19) and (12·20), D_f is substituted as zero, the equations reduce to

   $$q_{ult}=5\cdot7c \qquad\qquad \text{...(12·19a)}$$

   and $\qquad q_{ult}=2\cdot85c \qquad\qquad \text{...(12·20a)}$

   Equation (12·19a) may be compared with Prandtl's equation (12·7). Prandtl's equation is meant for a continuous footing with a smooth base whereas Terzaghi's equation is meant for a continuous footing with a rough base, both resting on the ground surface.

3. The ultimate bearing capacity of the soil is independent of the width of the footing, since the width b does not appear on the right hand side in any of the equations (12·19), (12·20), (12·19a) or (12·20a).

12·16. Bearing Capacity of Isolated Footings. (*Effect of the shape of the footing.*)

There it no rigorous theoretical method that can give the bearing capacity of an isolated square, circular or rectangular footing. Based upon his experience and the experimental results he had, Terzaghi has given the following empirical equations for the approximate bearing capacity of isolated square and circular footings loaded axially.

For circular footings,

$$q_{ult} = 1·3c \ N_c + \gamma D_f \ N_q + 0·6 \ \gamma R N_\gamma \qquad ...(12·21)$$

For square footings,

$$q_{ult} = 1·3c \ N_c + \gamma D_f \ N_q + 0·4 \ \gamma b D_\gamma \qquad ...(12·22)$$

where $\qquad q_{ult}, c, \gamma, D_f, b, N_c, N_q$ and N_γ

have their usual significance (b is the side of square footing) and R is the radius of the circular footing.

For a granular soil, cohesion c is zero. Therefore, for a footing placed on the surface (*i.e.,* $D_f = 0$) of such a material,

$$q_{ult} = 0·6\gamma \ R N_\gamma \qquad ...(12·21 \ a)$$

for the circular footing and

$$q_{ult} = 0·4 \ \gamma b N_\gamma \qquad ...(12·22 \ a)$$

for the square footing.

The corresponding value for the continuous fooing for these conditions will, from equation (12·18), work out as

$$q_{ult} = 0·5\gamma b N_\gamma \qquad ...(12·23)$$

For a cohesive soil, angle of internal friction ϕ may be taken as zero[1]. Therefore, for a footing placed on the surface (*i.e.,* $D_f = 0$) of such a material,

$$q_{ult} = 1·3 \times 5·7c = 7·4c \qquad ...(12·24)$$

both for a circular footing as well as a square footing.

The corresponding value for the continuous footing for these conditions as given by equation (12·19 *a*) is

$$q_{ult} = 5·7c \qquad ...(12·25)$$

Equations (12·22 a) onwards to (12·25) as given above give a fair idea of the effect of the shape of the footing on the bearing capacity values.

Hansen (16) extended the work of Terzaghi to include the effect of the shape and depth of the footing as well as inclination of loads. His equations for the granular and cohesive soils are as under :

For granular soils, where c=0.

---

1. When $\phi = 0$, $N_\gamma = 0$, $N_q = 1$ and $N_c = 5·7$.

$$q_{ult} = \frac{1}{2}\, \gamma N_\gamma b \left(1 - 0.3\, \frac{b}{L}\right)\left(1 - 1.5\, \frac{H}{V}\right)^2$$
$$+ \gamma\, D_f N_c \left(1 + 0.2\, \frac{b}{L}\right)\left(1 + 0.1\, \frac{D_f}{b}\right)\left(1 - 1.5\, \frac{H}{V}\right)$$

...(12·26)

with limitations, $\quad b \leqslant L;\ D_f \leqslant 15b;\ H \leqslant V \tan \phi.$

For cohesize soils, where $\phi = 0$

$$q_{ult} = 5c\left(1 + 0.2\, \frac{b}{L}\right)\left(1 + 0.2\, \frac{D_f}{b}\right)\left(1 - 1.3\, \frac{H}{V}\right) + \gamma D_f$$

...(12·27)

with limitations, $\qquad b \leqslant L\ ;\ D_f \leqslant 2.5b$ and $H \leqslant 0.4V.$

where

$\qquad q_{ult} =$ Unit ultimate bearing capacity.

$\qquad \gamma =$ Unit weight of the soil.

$\qquad b =$ Width of the footing.

$\qquad L =$ Length of the footing.

$\qquad H, V =$ Horizontal and vertical components of the load acting on the footing.

$\qquad D_f =$ Depth of the base of the footing below the ground surface.

$\qquad c =$ Cohesion of the soil $= \frac{1}{2}\, q_u$ where q_u is the unconfined compressive strength.

$\qquad \phi =$ angle of internal friction of the soil.

$\qquad N_\gamma, N_c =$ Terzaghi's dimensionless bearing capacity factors as given in Fig. 12.15.

Article 12·4 defines what are called the gross and the net ultimate bearing capacities. Skempton has given the net ultimate bearing capacity for a footing with a width b and a length L, placed on a cohesive material. His equation read as :

$$q'_{ult} = cN_c$$

...(12.28)

where

$$N_c = 5\left(1 + 0.2\, \frac{b}{L}\right)\left(1 + 0.2\, \frac{D_f}{b}\right)$$

when the ratio $\dfrac{D_f}{b} \leqslant 2.5$

and

$$N_c = 7.5\left(1 + 0.2\, \frac{b}{L}\right)$$

when $\dfrac{D_f}{b} > 2.5$

q'_{ult} in these equations is the net ultimate bearing capacity. Rest of the symbols have the meanings already stated.

For surface loading $D_f = 0,$

from equation (12.28), for a square footing $(b=L)$,

$$q'^1{}_{ult}=6c \qquad \ldots(12.29)$$

as against Terzaghi's equation (12.24) which gives

$$q_{ult}{}^1=7\cdot 4c$$

It may be noted that the Terzaghi's equation (12.24) is applicable to shallow foundations only $(D_f < b)$ whereas there is no such limitation on the Skempton's equations (12.27) and (12.28).

12 17. Footings on Sands.

Article 12·15 gives an insight into the various factors on which the bearing capacity of cohesionless soils depends. The subject needs further treatment due to its importance.

In determining the bearing area needed for a footing, the computations are based upon the net ultimate bearing capacity, as defined in article 12.4. According to the Terzaghi's theory, the net ultimate bearing capacity q'_{ult} for a continuous footing resting on a sand deposit is

$$q'_{ult}=q_{ult}-\gamma D_f$$
$$=\frac{\gamma b}{2}N_\gamma+\gamma D_f N_q-\gamma D_f$$

(See equation 12.18)

or

$$q'_{ult}=\frac{\gamma b}{2}N_\gamma+(N_q-1).\gamma.\,D_f.$$

All symbols have the meanings already stated. One may generalize this equation and say that the net ultimate bearing capacity, q'_{ult} for a footing resting on a sand deposit is

$$q'_{ult}=a.\gamma.bN_\gamma+(N_q-1).\gamma.D_f \qquad \ldots(12.30\ a)$$

where every other symbol is the same as in equation (12.30) except a. a may be called the shape factor and a value of 0 5 for a continuous footing, 0.4 for a square footing with a side b and 0.3 for a circular footing with a diameter b.

Equation (12 30) shows that the net bearing capacity depends[2] upon four factors : (1) the unit weight of the soil, which depends upon the position of the water table (2) width of the footing (3) depth of the footing and (4) the relative density of the soil.

Factors N_γ and N_q in the above equation depend upon the angle of internal resistance ϕ of the soil. ϕ depends to a large extent on the relative density of the sand deposit. In practice, the relative density of a soil deposit is measured by a test called

---

1. For surface loading there is no difference between the gross and net bearing capacities, since the surcharge is nil.

2. These observations are true for any type of footing-square, circular etc. resting on a sand deposit or any granular material.

the *standard penetration test*, some details of which are also given in chapter 16. This test consists in counting the number of blows required to force a split-spoon sampler, 30 cms (12 inches[1]) into the soil under test. The blows are given with the help of a hammer weighing 65 kgs. (140 lbs) falling through a height of 75 cms (30 inches). The number of blows so counted measure the penetration resistance of the soil and are designated as N. For foundation design, the relative density D, related to the number of blows N and the angle of internal friction ϕ may be taken as indicated in table 12.2 below.

TABLE 12·2.
Relative Density D of Granular Soils Related to N and ϕ .

Relative Density D	0		15%		35%		65%		85%		100%
No. of belows N	0		4		10		30		50		
Angle of internal friction ϕ			28°		30°		36°		41°		
Relative Compactness	Very loose		Loose		Medium		Dense		Very Dense		

In table 12.2, ϕ may be increased by 5° for soils containing less than 5% fine sand and silt. Myerhof has given approximate relationships between ϕ and D. They are :

$$\phi = 25° + 0·15 \ D \qquad\qquad ...(12 \ 31)$$

for a granular soil containing more than 5% fine sand and silt and

$$\phi = 30° + 0·15 \ D \qquad\qquad ...(12 \ 32)$$

for a granular soil containing less than 5% sand and silt. In the above equations, D is expressed as percentage and not as a decimal.

The values of D given in table 12.2 can be plotted against the given ϕ values. Also it may be recalled that N_γ and N_q are also functions of ϕ and are shown in Fig, 12·15. So. for the general shear case, a consolidated plot of D. N_γ and N_q against ϕ can be drawn as shown in Fig. 12·17. Such a plot was first proposed by Peck, Hanson and Thornburn (15).

The penetration resistance for a granular soil at the site is determined by carrying out a number of tests, Value of N should be determined at intervals of 0·75 meteres in the vertical direction at a point going upto a depth approximately equal to width of the footing. An average value, of N for a particular point is then found out. A number of such points at the site should be probed. The smallest average value of N obtained in this manner should be used to find out ϕ, N_γ and N_q from Fig. 12·17 or table 12·2 and Fig. 12·15. While using the value of N in conjunction

---

1. The figures within the bracket were proposed by Terzaghi & Peck

with the above-mentioved figure of table, the following points may be kept in mind :

1. If the sub-soil consists of fine sand below water table, average number belows, from the test results may be too high. If N' designates the excessive number of blows then the following equation may be used to find out the equivalent N-value :

$$N = 15 + \tfrac{1}{2}(N' - 15) \qquad (...12\cdot33)$$

This equation is obviously applicable when $N' > 15$.

2. The co-relation between D and N given in table 12.2 'or in Fig. 12·17 is not reliable for gravels or for soils containing large

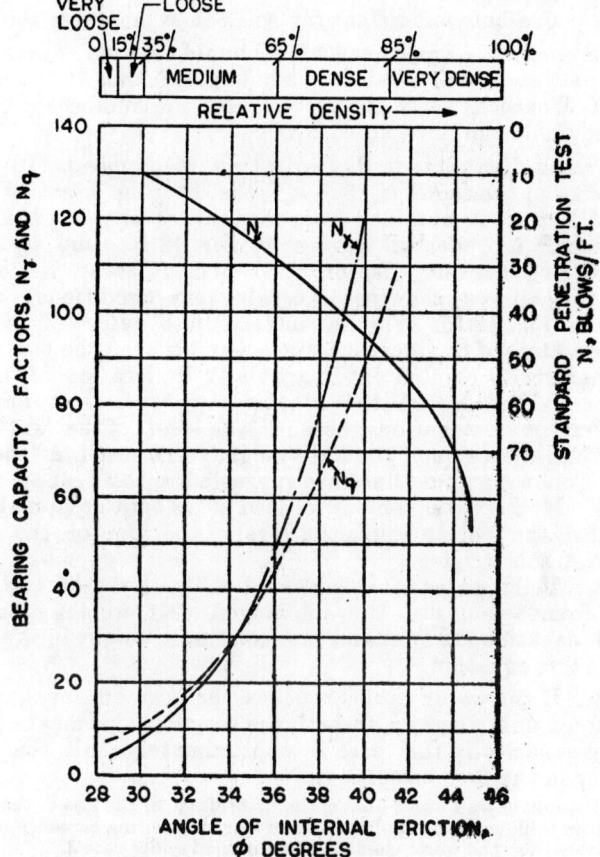

Correlation between N, N_γ N_q and ϕ for granular soils (After Peck, Hanson and Thornburn)

Fig. 12·17.

quantitis of graveles. In loose coarse gravel deposits, the split spoon sampler tends to slide into the large voids and hence given low values of N. Excessively large values of N may be expected if the sampler is blocked by a large stone.

3. If the test is made at shallow depths, the value of N may be too low, compared to the value that may be obtainable on the same soil at the same density but at a greater depth. The influence of the weight of the overburden on the value of N is given approximately by the equation ;—

$$N = N' \left(\frac{50}{p+10} \right) \qquad \qquad ...(12\cdot34)$$

where N = Adjusted value of standard penetration resistance.

N' = Standard penetration resistance as actually obtained.

p^1 = effective overburburden pressure, p.s.i not exceeding 40.
= weight of the soil above the level at which the test is made.

Use buoyant weight for soil below the water table.

Limitation. Equation(12·34) should be used conservatively as the data showing its validity is not substantial. It is suggested that, if N exceeds twice the value of N', the number of blows to be used in design must equal $N/2$.

The Indian Standard Institution recommends (10) that in the design of the footings, if $N \leqslant 5$, the bearing capacity factors N_γ and N_q may be obtained from the curves showing local shear conditions. *i.e.,* dashed curves in Fig. 12·15. and if $N \geqslant 30$, the bearing capacity factors N_γ and N_q may be obtained from the curves showing general shear conditions *i.e.,* solid curves in Fig. 12·15. For an intermediate value of N, N_γ N_c may be obtained by linear interpolation between the two curves.

As already pointed out in article 12·15, location of the water table has a bearing on the bearing capacity values obtainable from Terzaghi's equations, because the level of the water table determines the value of the unit weight to be used in the equation. Following guide lines are suggested in this respect :

1. If the water table is located at a depth greater than the width b of the footing below the level of bottom of the footing, use moist unit weight.

2. If the water table is located at the level of the bottom of the foundation, use the submerged unit weight γ_b in the N_γ term of equation (12·30a) and the moist unit weight in the $(N_c - 1)$ term of this equation.

3. If the water table touches the ground level, use the submerged unit weight γ_b in both the terms. This means that the bearing capacity in this case is approximately half the bearing capacity in the first case.

Even in cases 1 and 2 above, the possibility of the rise of the level of water table must be explored, otherwise the bearing capacity may be evaluated for the worst conditions mentioned under case 3.

---

1. Equation (12 34) is empirical. So p cannot be given in the c.g s. units unless the equation is modified. It is considered essential to keep the shape of the equation and mention p in p.s.i.

For an intermediate position of the water table, a linear interpolation may be made.

The design for footing on sands, in most of the cases, is governed by the settlement considerations. As such, it is essential that the settlement of a footing on a sand deposit is evaluated or the soil pressure chosen in such a manner that the settlement is within a reasonable limit.

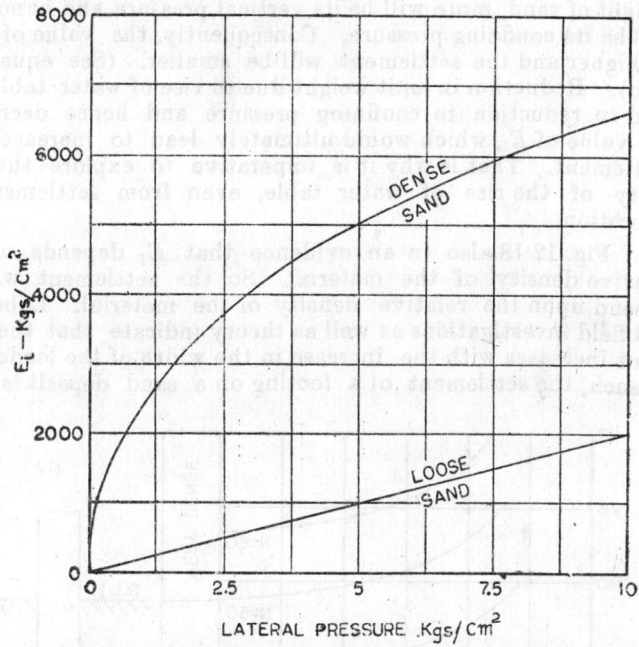

Variation of E_i with lateral pressure in case of sands.

Fig. 12·18

The settlement of a footing on a granular material depends upon the stress-strain characteristics of the material. A reference to Fig. 10·11 (a) will indicate that the stress-strain curve for a dense or a loose sand, at no stage, is a straight line. So the *modulus of elasticity* has no meaning for such soils. At best the ratio of the stress to strain for very low strains (initial parts of the two curves) could be represented by drawing tangents to the stress strain curves. A tangent to the curve for the dense sand is shown. The ratio of stress for this initial part of the curve and represented by such a straight line may be called the *initial tangent modulus* and be designed as E_i. It has been established, that the values of E_i both for dense as well as for loose sands depend upon the lateral pressure. Fig. 12·18 shows the general trend in the variation of E_i with the lateral confining pressure in triaxial testing, for dense as well as for loose sands. As a consequence, the rigidity of a sand measured by its E_i value, depends upon the confining pressure. Therefore, the settlement of a footing which depends upon the rigidity of the material. will depend upon the confining pressure. The confining pressure at a point in a sand deposit is roughly proportional to the vertical pressure which further depends upon its unit weight. More is the unit weight of sand, more will be its vertical pressure and hence more will be its confining pressure. Consequently, the value of E_i, will be higher and the settlement will be smaller. (See equation 9·1 also). Reduction in unit weight due to rise of water table would lead to reduction in confining pressure and hence decrease in the value of E_i, which would ultimately lead to increase in the settlement. That is why it is imperative to explore the possibility of the rise of water table, even from settlement consideration.

Fig. 12·18 also in an evidence that E_i depends upon the relative density of the material. So, the settlement will also depend upon the relative density of the material. Laboratory and field investigations as well as theory indicate that the settlement increases with the increase in the width of the loaded area. As such, the settlement of a footing on a sand deposit is a com-

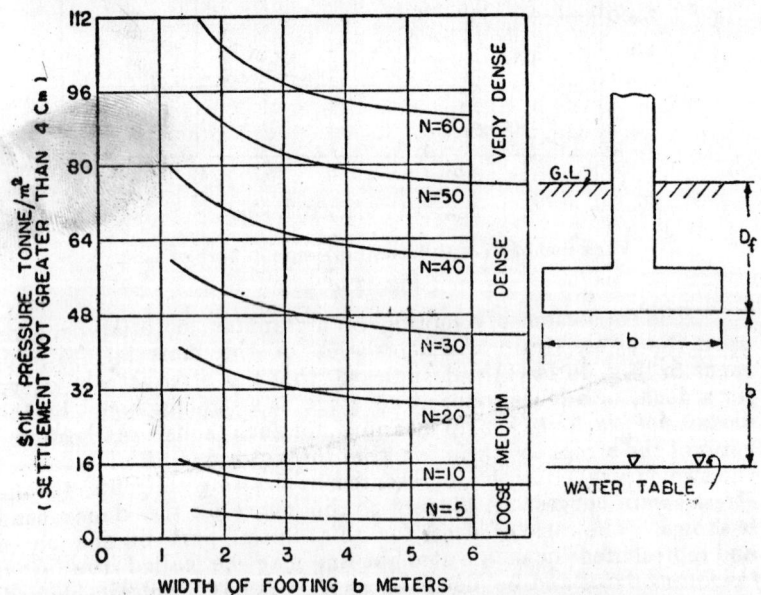

Soil pressure related to the width b of a footing and the penetration resistance N of the granular material for a settlement of 4 cms.
(After Peck, Hanson and Thornburn)
Fig. 12·19.

plex matter. The best solution to its determination lies in carrying out the plate loading test wherever applicable and economically feasible. The ultimate bearing pressure may be determined accordingly. In case it is not possible to carry out this test or if it is economically not feasible to do it, Fig. 12·19 may be used to find out the pressure which a given value of penetration resistance N and the given size b of a square footing, will cause the footing to settle 4 cms. If the pressure corresponding to some other settlement is required, it may be evaluated by assuming that the settlement is directly proportional to the soil pressure.

The figure is drawn on the assumption that the water table is at least a distance b below the base of the footing. If the water table is nearer or above the base of the footing, the pressure corresponding 4 cms. settlement may be taken as half of the value obtained from the figure. For intermediate positons of the water table, the appropriate pressure may be obtained by interpolation.

If the footing is not square, the least dimension be used in place of b.

Chart in Fig. 12·19 gives an insight into some very important points with respect to the design of footings. For a given soil pressure and a given relative density, there is only one size of the footing that will settle 4 cms. So, if a building has a number of footings. each of a different size, the settlement for each footing designed for the same soil pressure will be different, even if the soil density is assumed constant. Such a design may lead to serious differential settlements. However, if the soil pressure under the largest footing be used for the rest of the smaller footings, the settlement of all the footings will be less than 4 cms. Also, a fairly uniform deposit may have local variations and a footing located on the dense portion is likely to settle less than the one located on the loose portion. This can also lead to serious differential settlements. However, if the largest footing is located on the loosest portion of the deposit and the soil pressure for design of all footings is chosen on this basis, no footing will. settle more than 4 cms. and also the differeneial settlement between any two footings will not exceed 4 cms or so.

12·18. Footings on Clays

Equation (12·19) in article 12·15 gives the gross ultimate bearing capacity for a continuous footing resting on a cohesive soil. The equation reads as

$$q_{ult} = \gamma.D_f + 5.7c \qquad \qquad ...(12·35)$$

If the net ultimate bearing capacity is desired, it would be given by subtracting γD_f from both sides of equation (12·35).

$$q'_{ult} = 5·7c \qquad \qquad ...(12·35\ a)$$

If q_u represents the unconfined compressive strength of the soil, net ultimate bearing capacity is given by

$$q'_{ult} = 2.85 q_u \qquad \qquad ...(12.35\ b)$$

Corresponding equations for the circular and square footings would be

$$q'_{ult} = 1.3\ (5.7)c$$

(See equation (12.21) & (12.22). Substitute $N_\gamma = 0$, $N_q = 1$, $N_c = 5.7$ for $\phi = 0$ and subtract γD_f)

or $\qquad\qquad q'_{ult} = 7.4c \qquad\qquad ...(12\ 36)$

and $\qquad\qquad q'_{ult} = 3.7 q_u \qquad\qquad ...(12.36\ a)$

Net ultimate bearing capacity for rectangular footings may be calculated from Skempton's equation (12.28). Alternatively, following equation may be used for rectangular footings (15), on cohesive soils.

$$q'_{ult} = 2.85 q_u \left(1 + 0.3\ \frac{b}{L} \right) \qquad ...(12.37)$$

All these equations are only applicable if the clay is homogeneous in nature and does not show a network of hair cracks. If a clayey soil is preconsolidated, it will invariably show a network of cracks or slickensides and any procedure for bearing capacity determination based upon the unconfined compressive strength of the soil would be unsuitable. The strength of such a soil depends upon the nature and spacing of cracks. In such cases, it is always desirable to resort to the plateloading test.

Where the bearing capacity is based upon the unconfined compressive strength q_u of the soil, an average value of q_u must be used in the equation. An average value of q_u is determined by taking samples at 15 cms. intervals below the foundation, upto a depth equal to the width of the foundation and finding out the average of the values determined for all the samples.

Methods outlined in Chapter 9 are used to determine the settlements under these footings. If, however, the depth of the clay is sufficiently large, the plate-loading test would also give the settlement values.

12.19. Footings on Non-Uniform Deposits.

It is very seldom that one finds uniform soil deposits to build the structures on. Soil deposits are usually stratified and hence nonuniform. Just simply applying one or the other formulae for uniform soil deposits presented in the few previous articles, to actual field conditions may not be justified. Unless the engineer is thoroughly satisfied that the deposit is uniform, these formulae where applied have to be substantiated with checks and cross-checks.

If a soil profile[1] indicates that the deposit has a number of soft clay layers of different consistencies, it is essential to check the pressure at the top of each layer and compare it with the bearing capacity of that layer. Stress-distribution theories given

---

1. For definition see Chapter 16.

in Chapter 8 would be quite helpful in determining the pressure
at the top of each layer. Alternatively, the load may be considered to be dispersed at 2 : 1 slope (two vertical to one horizontal)
starting at the base of the footing and the pressure at the top

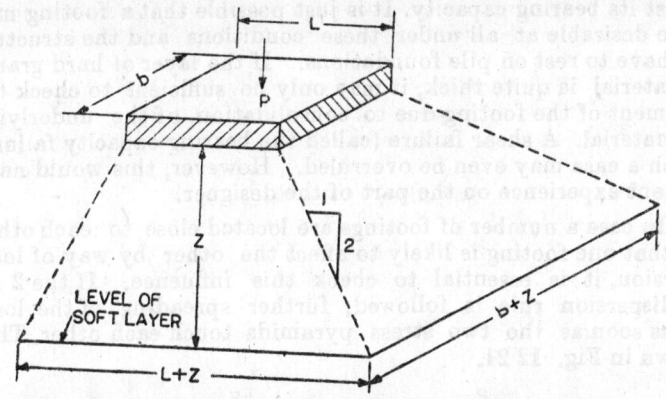

Approximate method of calculating stress increase.

Fig. 12·20.

of each layer may be calculated due to this dispersion. This is
indicated in Fig. 12·20. If z is the depth of the layer below the
base of the foundation, the above method assumes that the load
is supported by a flat-topped pyramid of soil whose sides slope
outward one metre for each two metres of depth, so that at the
level of the layer the area of the load is $(L+z)(b+z)$. The stress
increase $\triangle \sigma_s$ at this level due to a load P, would then be given
by,

$$\triangle \sigma_s = \frac{P}{(L+z)(b+z)} \qquad (12.38)$$

The method is equally applicable to a strip footing. If q is the
load intensity per unit area at the level of the foundation, then
the stress at a depth z will be.

$$\triangle \sigma_s = \frac{qb}{b+z} \qquad (12·39)$$

Instead of a pyramid, the stress dispersion is through a
trapezoidal wedge, in this case.

If a dense layer of soil is overlain by a thin layer of soft
clayey material, it may be more desirable to excavate the top soft
material and place the foundation on the dense material. In case
it is not economically advisable to remove the top soft layer, the
design may be worked out by assuming that the soft layer extends
to a large depth. Both the ultimate bearing capacity and the
settlement may be found on this basis.

If, however, a soft layer of material is overlain by a thin layer of hard granular material, it is very essential that the increase in stress at the level of the soft material is checked against its bearing capacity. It is just possible that a footing may not be desirable at all under these conditions and the structure may have to rest on pile foundations. If the layer of hard granular material is quite thick, it may only be sufficient to check the settlement of the footing due to consolidation of the underlying soft material. A shear failure (called the bearing capacity failure) in such a case may even be overruled. However, this would need sufficient experience on the part of the designer.

In case a number of footings are located close to each other such that one footing is likely to affect the other by way of load dispersion, it is essential to check this influence. If the 2 : 1 load dispersion rule is followed, further spreading of the load ends as soon as the two stress pyramids touch each other. This is shown in Fig. 12 21.

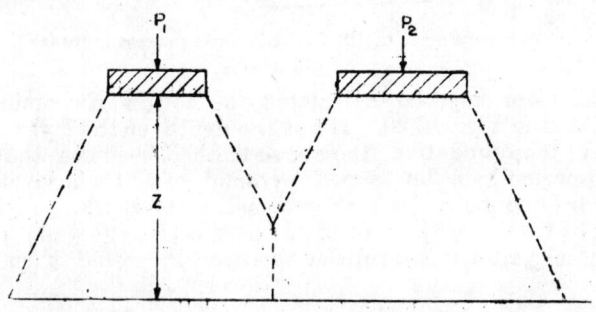

Fig. 12·21.

12·20. Design of Footings.

Following steps may be adopted for the design of footings.

Step 1. Calculate the loads that the footing has to support. For the bearing capacity analysis, the total load to be considered should include all the loads that the structure may have to carry anytime during its life. This would include the full dead weight including the estimated weight of the footing, plus the maximum live load including the wind and snow loads minus the weight of surcharge at the level of the footing. Vertical component due to any earth pressure or any earth-quake forces or the effect of bouyancy, if the water table is above the base of the footing, must be taken into account. In sandy soils, usually the settlement governs the design. In order to find out the settlements of such soil the load to be considered is the same as detailed above for the bearing capacity analysis. Unlike clays, settlement in sands takes place immediately on placing a load. Hence full values of the live loads have to be taken into account. In clays, the settlement is not affected by any loads that do not act for a long time. As such, only permanent loads should be taken into consideration. Permanent loads include the dead weight of the structure including the estimated weight of the footing and any permanent equipment such as furniture or machinery. Sometimes one-half of the live load is considered permanent. Wind and snow loads and any earthquake forces because of their transitory nature are not considered as permanent loads. The effect of bouyancy, in case the water table is above the base of the footing, has to be taken into account. The load that remains on the structure long enough to contribte to consolidation settlement is sometimes referred to as **service load.**

Numerical values of these loads are mentioned in the building code applicable in a particular area.

Step 2. Draw the soil profile showing the various strata underneath the surface on which the footing has to rest. Mark the level of water table on this profile. The level of the water table *must* be marked on the sketch *carefully.*

Step 3. Determine the depth of the footing. The footing must be carried below the top soil, artificial fill, debris or muck. In the case of a clayey soil, the footing must be placed below the zone that is affected by seasonal weather changes. In fine sands and silts, the footing must rest at a level below the depth of frost penetration. In heated buildings the interior footings are not affected by frost ; there-fore, they may be placed at a level determined by other considerations. The difference in the elevations of the footings should not be so much that the excavation for the lower footing disturbs the soil under the higher footing.

The Indian Standards Institution recommends a minimum depth of 80 cms. below the ground surface in sands and 90 cms. in clays. As a general guide, the maximum differecce between the elevations of two footings should not be more than one-half the clear distance between the footings.

Step 4. Determine the ultimate bearing capacity against shear failure, for the sub-soil. In order to find out the allowable bearing pressure, the ultimate bearing capacity is divided by a factor of safety. A factor of safety of three is recommended, as a general rule. In case the sub soil is very homogeneous or the superstructure is relatively light, a smaller factor of safety may be used. This must be based upon the experience of the engineer on the structures in the vicinity of the new building.

Step 5. Detemine the size of the footing based upon the allowable bearing capacity calculated in step 4. A settlement analys's as proposed in step 7 is an essential requirement. In case the footing size based upon the allowable pressure calculated in step 4 gives excessive settlement[1], the size is revised to bring the settlement within the permissible limits.

---

1. Some engineers prefer to proportion the footings in such a manner that the differential settlements are minimised See article 12·21.

Step 6.   Check the stresses at the top level of each stratum, in case the profile indicates a stratified deposit. This will be a check against any overstressing of the soil layer at greater depths.

Step 7.   Find out the total and differential settlements. In case they exceed the permissible limits, the footing size has to be revised, as already indicated in step 5.

Step 8.   Check the stability against horizontal forces.

Step 9.   Check for any uplift pressure on the footing.

Step 10.   Design the footing and make the provision for any foundation drains, water proofing or damp proofing.

12.21. Proportioning the Footings for Minimum Differential Settlement

Some engineers design the footings in such a manner that the settlement is minimised. This is done by choosing the same average pressure for all footings under the service load. This service load, for ordinary buildings, may be taken as the dead load plus one-half of the live load. In the case of cold-storages, ware-houses and other storage floors, larger percentage of live load should be used. The design procedure is as under :

Let P_{l+d} = Live load + dead load for the footing having the maximum live load/dead load ratio.

P_s = Service load for the above footing.

q_a = Allowable bearing capacity, taking into consideration the desired factor of safety.

q_d = Design bearing capacity for all the footings excepting the one having the maximum live load/dead load ratio.

A = Area of the footing with the maximum live load/dead load ratio.

Then, $A = \dfrac{P_{l+d}}{q_a}$ ...(12·40)

and $q_d = \dfrac{P_s}{A}$ ...(12·41)

so that area for other footings

$= \dfrac{\text{Corresponding Service Load}}{q_d}$ ...(12·42)

12·22.   Mat or Raft Foundation

Under the classification of foundations in article 12·6, a mat or a raft foundation was defined and Fig. 12·4 (d) gives the sketch of a mat foundation supporting 12 columns. Such a combined footing covering the entire area under a structure and supporting all the columns and walls becomes sometimes necessary. Following situations explain the typical conditions under which their use is called for.

1. Where the building loads are so heavy and the allowable bearing capacity so low that the small individual footings would cover more than half of the area under the building, the raft foundation usually works out economical.

2. In highly compressible soil deposits, settlements under individual footings are usually high. A raft foundation under such circumstances is the only solution. Sometimes the depth at which the raft is placed is so great that the unit load from the building equals the unit weight of the soil excavated. Such a type of raft is called a **floating foundation**.

The advantage of a mat foundation over the individual footings is that the unit pressure from the structure on the subsoil is reduced in the former case as compared to the latter. In addition, by combining all individual footings into a single mat, the bearing capacity is often increased. Structures like silos, chimneys and factories manufacturing heavy equipment are invariably founded on mat foundations.

The design of a mat foundation can be carried out by what are known as the *Rigid Method* and the *Method of finite differences*. The complete design procedures are outside the scope of this book. However, some important points relating to the bearing capacities of mat foundations resting on sand clay deposits are mentioned below.

Mats on Sands. The mat foundations are usually of a large size. The bearing capacity of sand increases with the width of the foundation. As a consequence, the bearing capacity failure of a mat foundation resting on a sand deposit may be ruled out. Large size of the mat also leads to high stresses in the underlying sand upto a considerable depth. If there are any loose pockets of sand at various depths scattered underneath the foundation, the effect of these loose pockets on differential settlement under the mat is relatively less compared to their effect on differential settlements in case individual footings were to rest on them (both types designed for the same soil pressure). For the same soil pressure, since the differential settlement is less for the mat foundation compared to the individual footings, some engineers recommend that allowable soil pressure in the case of a mat should be twice that allowed for individual footings (15). The allowable pressure for individual footings can be picked up from Fig. 12·19. The soil pressure for the mat foundation of equivalent size may be taken as twice the value obtained for an individual footing from this figure. If the water table is located at the base of the mat, the above value of allowable pressure may be reduced to 50 per cent. If the water table is at a depth intermediate between the base of the mat and a depth b below the base, an appropriate reduction ranging between 0 to 50 per cent may be made.

A standard penetration test is carried out for picking up the allowable pressure from Fig. 21·19. The observations relating to equation (12·33) and (12·34) must be kept in mind. In case the value of N is less than 10, either the sand must be compacted by some artificial means so that the sand attains density corresponding to $N=10$ or the structure must rest on deep foundation like piles.

Mats on Clays. The net ultimate bearing capacity of a mat foundation resting on a deep deposit of clay may be found from equation (12·37). This equation shows that the value of q'_{ult} is practically independent of the width[1] of the mat. For q'_{ult} to be of 1 kg./cm²., the unconfined compressive strength q_u should be of the order of 0·3 kg/cm²., assuming b/L as unity. This suggests that if an allowable pressure of 1 kg./cm². has to be provided for in the design, there is no factor of safety, if q_u equals approximately 0·3 kg./cm². To provide a factor of safety of 3, the unconfined compressive strength must equal about 1 kg./cm². The factor of safety for most of the mats provided on clay deposits is usually relatively small, since q_u may not equal 1 kg./cm²., q'_{ult} in the equation is the pressure in excess of the overburden pressure at the level of the foundation. So, by increasing the depth of excavation, the pressure that can safely be exerted by the building is correspondingly increased. This would increase the factor of safety. The recommended factor of safety for design of mat foundations is three.

The settlement calculations may be made by any one of the methods given in Chapter 9.

12.23. Bearing Capacity of Failure of a Cut in a Soft Clay

Refer to Fig. 22 which shows a cut made into a soft soil. The depth of the cut is D and its width is b. The cut is struted all along its length. If the surcharge is such that it equals the bearing capacity of the strips A_1B_1 or C_1E_1, the bearing capacity failure would take place. The failure wedge is shown in the figure. It starts at an angle of 45° at A_1 and beyond the point F, it continues as an arc of a circle upto the point C_1. There will be mirror reflection E_1FB_1.

Thus, $A_1F = E_1F = \dfrac{b}{\sqrt{2}}$;

since $A_1E_1 = b$

This means that $A_1B_1 = C_1E_1 = \dfrac{b}{\sqrt{2}}$

If p_v denotes the net intensity of vertical pressure at the level of A_1B_1 or C_1E_1, then taking unit length along the depth (perpendicular to the plane of the paper),

---

1. Because b appears in the ratio b/L.

$$p_o = \left[\gamma D \frac{b}{\sqrt{2}} - Dc \right] / \frac{b}{\sqrt{2}} = \gamma.D. - \frac{cD\sqrt{2}}{b} \quad ...(12.43)$$

where γ = Unit weight of the clay

c = Unit cohesion that acts vertically upwards as shown along C_1C or B_1B

b, D = Width and depth of the cut respectively.

The ultimate bearing capacity at the level of A_1B_1 or C_1E_1 may be taken as $q_{ult} = 5.7c$, where q_{ult} and c have their usual significance. Then the factor of safety against failure is given by,

$$F.O.S. = \frac{5.7c}{p_o} \quad ...(12.44)$$

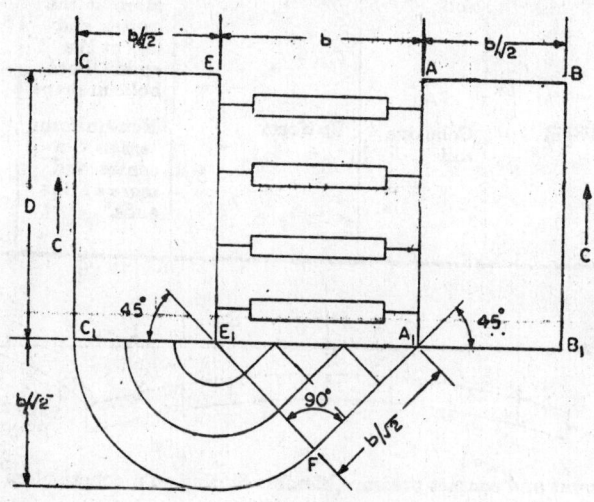

Fig. 12·22.

12·24. Contact pressures.

The pressure between the foundation and the subsoil is called *contact pressure*. The contact pressure distribution is used in the structural design of the foundation, since it determines the distribution of moments and shears. The distribution of contact pressure is a function of the type of the soil and the rigidity of the foundation. The foundations from the rigidity point of view may be classed as flexible—free to deflect and follow the theoretical settlement profile, and rigid—deflect uniformly. Table

12.3 gives the distribution of contact pressures and settlements of the two types of foundations-resting on cohesionless or cohesive material, in general terms.

TABLE 12·3

S. No.	Type of foundation	Type of Soil	Settlement	Contact pressure	Refer to figure.
1.	Flexible	Cohesionless soil	Non-uniform. Less at the centre, more at the ends	Uniform convex upwards	2.23
2.	Flexible	Cohesive soil	Non-uniform. More at the centre and less at the ends.	Uniform convex downwards	12.24
3.	Rigid	Cohesionless soil	Uniform	Non-uniform. More at the centre and less at the ends. Parabolic in shape	12.25
4.	Rigid	Cohesive soil	Uniform	Non-uniform, ess at the centre and more at the ends.	12.26

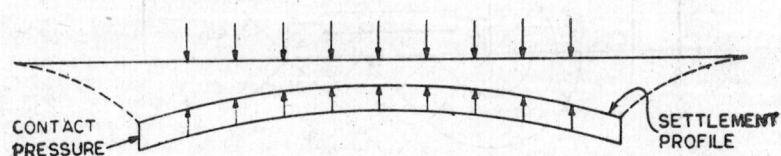

Settlement and contact pressure, flexible footing on a cohesionless soil
Fig. 12·23

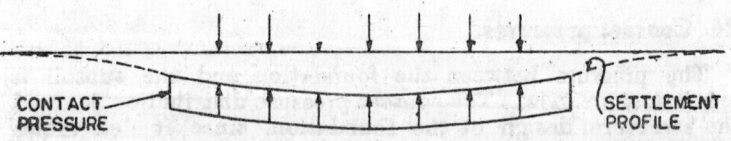

Settlement and contact pressure, flexible footing on a cohesionless soil
Fig. 12·24

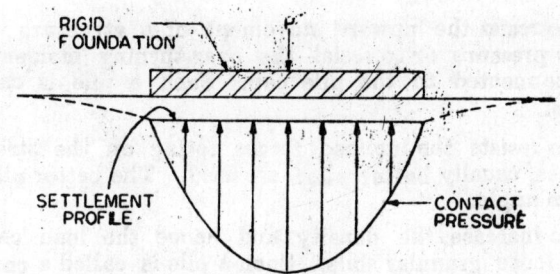

Settlement and contact pressure, rigid footing on a cohesive soil
Fig. 12·25

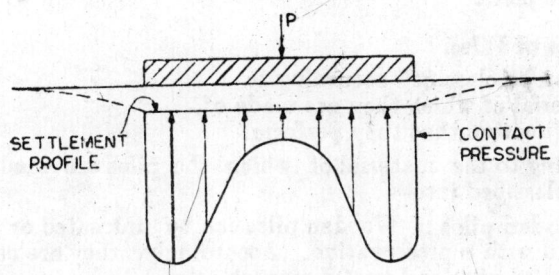

Settlement and contact pressure, rigid footing on a cohesive soil
Fig. 12·26

12.25. Pile Foundation.

When the sub-soil immediately underneath the foundation is not suitable to take up the loads from the structure, use is made of either the pile foundations or wells and caissons. A discussion on the wells and caissons is beyond the scope of this book. However, it is intended in this and following articles to give the salient points regarding pile foundations.

A pile, *by definition*, is a long but small diameter shaft which is used to transmit the loads to deeper soil layers capable of supporting the loads. Piles are used for the following purposes :

1. To transfer loads through soft soil or water to a stronger soil stratum. The pile in this case rests on the stronger stratum and transmits the load to it. Such a pile is called an **end-bearing pile.**

2. To transfer the loads to the weaker soil through *skin friction* between the surface of the pile and the soil through which it goes. Such a pile is called a **friction pile.**

3. To support loads applied at right angles to the axis of the piles. Such a pile is called **laterally loaded pile.** They are commonly used in the foundations of retaining structures and dolphins, etc.

4. To resist the upward movement of a structure due to hydrostatic pressure or to resist the over-turning moment of a structure connected to the pile cap. Such a pile is called a **tension pile.**

5. To resists the inclined forces acting on the structure. In such cases, usually **batter piles** are used. The better piles are driven at an angle.

6. To increase the density and hence the load carrying capacity of loose granular soils. Such a pile is called a **compaction pile.**

7. To carry the foundations through the depth of scour, in case of bridge piers.

12·26. Types of Piles.

Piles can be classified according to
a) material of which they are made of,
b) the function that they perform.

According to the material of which the piles are made of, piles can be classified into—

1. **Wooden piles :** Wooden piles can be untreated or they can be treated with a preservative. Accordingly, they are called *plain or untreated* piles and *treated* respectively.

2. **Concrete piles :** Concrete piles may either be precast or cast-in-place types. Many popular commercial varieties are available in this category.

3. **Steel piles :** Steel piles may be the H section or steel pipe piles. The sheet piles used as cut off walls under dams, coffer dams, and bulkheads, etc. may be considered under this class.

4. **Composite piles :** Composite piles made from two materials[1] are available.

From the functional point of view the piles may be called the *end-bearing piles, friction piles, laterally loaded piles, tension piles, batter piles or compaction piles.* Their use has already been detailed out in the previous article.

For a detailed *description* of the wooden, concrete, steel and composite piles, the reader may look up references 16 and 17, as such a description would be outside the scope of this book.

Factors that give the selection of one or the other type of piles are :

1. Size, weight and type of the structure to be supported.
2. Physical properties of the soil at the site.
3. Depth to which the pile has to be taken. Possibility of variation in depth to the supporting stratum has also to be taken into account.

---

1. Usually timber and concrete or steel and concrete.

4. Number of piles required.
5. Material available for the piles, near the site.
6. Driving equipment available.
7. Durability required.
8. Types of structures in the vicinity of the site of construction.
9. Depth and chemical composition of water through which the pile has to go.
10. Comparative costs in place.

The general requirements and the situations where a particular type of pile could be used are usually available in the building codes governing the design and construction requirements in an area.

12·27. Bearing Capacity of Piles in Sand

Piles in sands are used for the following purposes :

1. If the soil profile shows that the upper stratum is a soft or compressible material overlying a lower dense sand stratum, then piles are used to carry the load through the upper layer to the dense sand stratum.

2. If the soil profile shows that the material is loose sand, piles are used to compact the sand deposit in order to increase its density and hence its bearing capacity.

3. In river beds subjected to scour, piles are used to carry the load of the superstructure to dense sand layers below the maximum depth of scour.

The piles in sand derive their strength from both point bearing and skin friction. The bearing load cannot be determined from simple soil tests. There are two ways open for finding out the ultimate load carrying capacity of piles (1) the load test and (2) dynamic pile driving formulae. Some investigations are under way to find out the bearing capacity of piles based upon elastic theories but none of them has so far been put into practice.

As far as the load test on piles is concerned, there are two ways to carry out the load test : (1) the direct load test and (2) the cyclic load test. The direct load test constitutes of loading the pile directly with the help of a loading platform on which the load in the form of sand bags or cast iron bags or rails, etc. is placed. The loadsettlement curve is then drawn. A typical load-settlement curve (curve B) for such a test carried out in granular soils is shown in Fig. 12.27. The point of failure is evident in some load-settlement curves and only a factor of safety is required to determine the design load, while in others the settlement goes on increasing with the increase in load, so that picking

1. Bearing capacity fo mulae based on Terzaghi's and Ireland's works are also available. These formulae are at best approximations only. See reference 16, pages 212 & 213.

up the failure point becomes difficult. The latter situation is shown by curve B in Fig. 12·27. In such cases, limiting load is based upon some settlement value. Almost all building codes give some rules regarding the selection of the ultimate or allow-

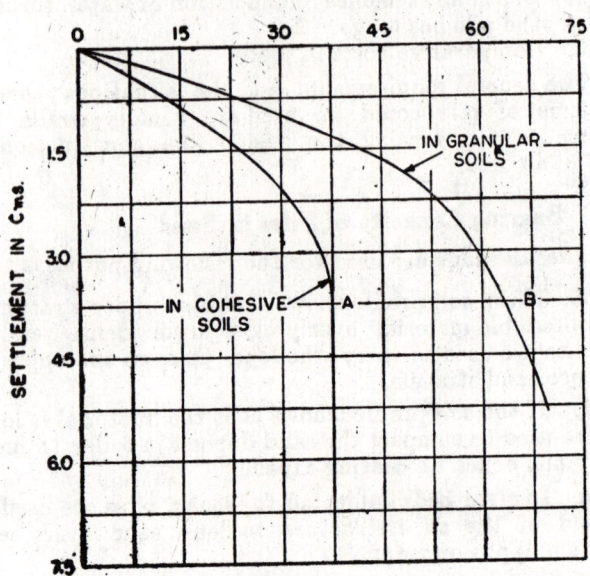

Load-settlement curves for a single pile driven in granular and cohesive soils.
Fig. 12·27

able load, based upon an optimum settlement value. A complete discussion of the various rules provided in different building codes is given on pages 668 to 670 of reference 18. As examples, only a few rules are mentioned here.

1. The allowable load is the one that corresponds to 10 cm. settlement.

2. Draw tangent lines to the upper and lower parts of a curve and let them intersect. Determine the load against the point of intersection and divide it by a factor of safety of 1·5 or 2.

3. If twice the design load is placed, on the pile, the pile should not settle more than 5 cms. Under the design load, it should not settle more than 2 cms.

4. The safe allowable load is 50 percent of the load which when applied for 48 hours produces a permanent settlement of 0·63 cms. (1/4 inch) measured at the top of the pile.

Each of these rules will give a different safe load. The engineer has to depend upon his judgment and the experience of other engineers in the area.

The cyclic load test involves the loading and unloading of the test pile. The use of a loading platform in such a case is cumbersome. Usually reaction loading procedures are adopted in such tests. The engineer must refer to an appropriate standard for conducting such a test and for the interpretation of the test data. Reference may be made to ASTM specifications Designation D–1143 for the standard test and the interpretation techniques.

The second method to determine the bearing capacity of pile foundations in granular materials is to find out the bearing value of a pile through a dynamic loading test. The pile is driven into the ground with the help of a drop hammer or a steam hammer, a number of varieties of which are commercially available. A number of corresponding pile driving formulae are available. The principle behind these formulae is that the energy supplied by the driving hammer to the pile is equal to the useful work done in penetrating the pile plus any losses.

Algebrically

Driving energy=Work done in penetrating the pile+losses.
The losses may be due to elastic compression of the pile, plastic behaviour of the soil, heat generation and vibrations, etc.

The above equation can mathematically be written as,

$$12E = RS + L \qquad \qquad \ldots(12 \cdot 45)$$

where
E=driving energy, in ft. lbs.
R=resistance of soil, lbs.
S=penetration of the pile per blow, in
L=loss in energy.

The loss in energy L, may be considered proportional to the resistance of the pile, so that

$$L = C.R,$$

where C is a constant and it equals $1 \cdot 0$ for drop hammers and $0 \cdot 1$ for steam hammers.

∴
$$12E = R.S + C.R$$

or
$$R = \frac{12E}{S+C} \qquad \qquad \ldots(12 \cdot 46)$$

If a factor of safety of six is applied, the above equation becomes

$$R = \frac{2E}{S+C} \qquad \qquad \ldots(12 \cdot 47)$$

The driving energy E equals the weight (in lbs.) of the hammer times the stroke of the hammer (in ft.), in case of drop or steam hammers.

Equation (12·47) is the famous Engineering-News formula[1]. Many other formulae are available and some of them are more rational than this formula. Whatever may be the form of a formula, its use has to be done very scrupulously since there are many shortcomings from which these formulae suffer. Some of the disadvantages are listed below :

1. A dynamic formula is based upon the assumption that the soil resistance during and after driving remains constant. This assumption may be alright for a granular soil, with usually high co-efficient of permeability and in which water can flow in and out very quickly so that the frictional resistance during driving is not affected. In clayey and silty soils, the permeability is low and since water cannot escape quickly, it acts as a lubricant and reduces driving resistance.

2. During driving, the soil gets disturbed and its shear strength gets reduced, specially in cohesive soils.

3. The pile driving formulae assume that impact between the hammer and the pile is that of two elastic bodies free to collide (Neutonian impact). Such an assumption is not justified since the pile because of its embedment cannot be treated as a free body.

The pile driving formulae are more used for piles driven into granular soils as the results obtainable in cohesive soils are not reliable.

In fact, the structure rests on a number of piles. The bearing capacity of a group of piles in granular deposits may be taken as the bearing value of a single pile multiplied by the number of piles in the group. A group of piles taken together will have a greater bearing capacity than the one calculated by the above procedure, since the effective width of piles as a group increases and hence the bearing capacity will increase as in sands the bearing capacity is a function of the width of the foundation. The usual spacing in a group ranges from 2 to 4 times the size (side of a square pile or diameter of a circular pile) of the pile.

12.28. Piles in Cohesive Soils

When it is not possible to have shallow footings or mat foundation in clays, the loads are carried to a deeper layer through piles foundations. Under such circumstances a pile develops friction all along its surface and acts as a friction pile.

A static load test is the best and the only reliable procedure to determine the safe bearing load that a pile driven in a clay deposit can carry. A load-settlement curve gives a well-defined failure point defined by a vertical asymptote. (See Fig. 12·27, curve A). A factor of safety of three is usually applied to this

---

1. Equation (12·47) has not been converted by the author into c. .s. units, just to maintain its identity.

ultimate load in order to find out the safe or allowable bearing load.

A friction pile in a clay deposit is supported by adhesion between the soil and the surface of the pile. An approximate method to find out the bearing capacity of such a pile is to multiply the embedded surface area with the unit adhesion between the soil and the pile surface. What should be the value of adhesion, is a debatable question. Soft clays come in direct contact with the surface of the pile due to high plasticity and the adhesion may, therefore, be taken equal to the cohesive strength of the clay. In stiff clays, usually a gap is created between the soil and the pile surface. As such, the adhesion cannot be taken equal to the cohesive strength of the clay. It is usually less. Unless found otherwise from load tests, values of unit adhesion given in table 12·4 may be taken for the purpose.

TABLE 12·4
Values of Adhesion for Different Clayey Soils
(After Tomlinson)

Pile Material	Cohesive lbs/sft.	Strength tonnes m²	Unit lbs./sft.	Adhesion tonnes m²
Concrete and timber	Soft. 0—750	0—3·65	0—700	0—3·41
	Medium 750—1500	3·65—7·30	700—900	3·41—4·38
	Stiff 1500—3000	7·30—14·6	900—1300	4·38—6·34
Steel	Soft 0—750	0—3·65	0—600	0—2·93
	Medium 750—1500	3·65—7·30	600—750	2·93—3·50
	Stiff 1500—3000	7·30—14·6	?	?

Piles are usually placed in groups under individual footings or under mats. The usual spacing is 2·5 to 5 times the size (side in the case of a square pile and diameter in the case of a circular pile). The bearing capacity of a group of piles may be calculated by multiplying the load carrying capacity of a single pile with the number of piles in the group. Since the bearing capacity in clayey soils does not depend upon the effective width of the foundation, it is just possible that the bearing capacity of the group may be less than the bearing capacity found by multiplying the individual capacity of a pile by the number of piles. As such, for the piles placed in clayey soils, bearing capacity of the group is found out by two methods.

1. Multiply the bearing capacity of an individual pile by the number of piles in the group.

2. Treat the group of piles as a composite block and find out its bearing capacity.

Bearing capacity in both the cases has to be found out both due to the bearing at the bottom as well as due to skin friction. Following methods may be employed.

1st Method. If the length of embedment of a pile is l and the diameter of each pile is D, then the bearing capacity of a single pile is,

$$q_d = \pi.D.l.c_d + \frac{\pi D^2}{4}(1.3)c.N_c \qquad \qquad ...(12.48)$$

where c_d = unit adhesion between the soil and the pile
    c = cohesive strength of the soil
    N_c = bearing capacity[1] factor due to Terzaghi, assuming $\phi = 0$

So, if the number of piles in the group is n, then the total bearing capacity of the group according to this method is

$$Q_d = nq_d = n\left[\pi.D.l.c_d + \frac{\pi D^2}{4}(1.3)c.N_c\right] \qquad ...(12.49)$$

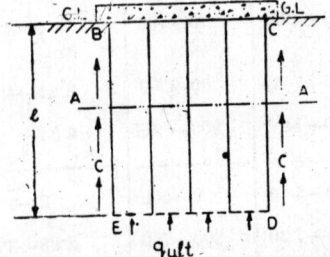 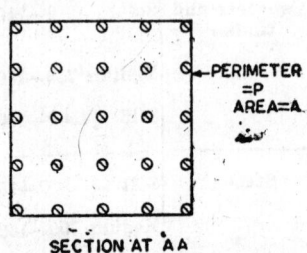

Fig. 12.28.

2nd Method. Refer to Fig. 12.28. It shows a group of n piles each embedded l units below the ground surface. The area occupied by the group as a whole is A as shown on the right hand side. If it is assumed that the block $BCDE$ acts as a single unit, then the bearing capacity of this block will be given by,

$$Q = p.l.c + (1.3)c.N_c.A \qquad \qquad ...(12.50)$$

    where p = periphery of the composite block
    c = cohesive strength of the soil

N_c = bearing capacity factor due to Terzaghi[1], assuming $\phi = 0$

Whichever of the above two methods gives the smaller value, it may be taken up for design.

12.29. Examples

Example 12 1. (a) Plot on one graph, the gross ultimate bearing pressure of a continuous 0.75 metres wide footing as a

---

1. Some authors suggest that N_c may be taken equal to 9 since the foundations are going to be deep.

function of depth of the footing. Let the depth D_f vary from 0 to 1·75 metres. Use Terzaghi's analysis. Soil properties are given below :

Soil properties	Soil A	Soil B
Unit weight	1·77 gms /c.c.	1·77 gms/c.c.
Cohesion c	0·44 kg./cm.²	0·0 kg./cm.²
Angle of internal friction ϕ	Zero	36°

(b) Assume that the footing is 0·75 m in width and it lies at the surface of the soil. Plot on one graph, the effect of increasing the width of the footing, from 0·75 metres to 2·0 metres in increments of 0·25 metres on the gross ultimate bearing capacity.

Assume in both the cases that the water table lies much below the zone of shear to have any effect on the bearing capacity.

Comment on the results obtained in the two cases.

Solution. (a) For a continuous footing, the gross ultimate bearing capacity is given by,

$$q_{ult} = cN_c + \gamma D_f N_q + \frac{\gamma b}{2} N_\gamma \qquad \text{...(eq. 12 16)}$$

with the usual significance of the various symbols.

Soil A. This soil is a cohesive material for which

$$\gamma = 1·77 \text{ gms./c.c.} \qquad c = 0·44 \text{ kg./cm}^2 \text{ and } \phi = 0$$
$$\gamma = 1·77 \text{ tonnes/m}^3 \qquad = 4·4 \text{ tonnes/m}^2$$

when $\phi = 0$, from Fig. 12·15 the bearing capacity factors are ∷

$$N_c = 5·7, \ N_\gamma = 0 \text{ and } N_q = 1$$

So, the equation for bearing capacity becomes

$$q_{ult} = 5·7c + \gamma D_f$$

Put $c = 4·4$ tonnes/m.² and $\gamma = 1·77$ tonnes/m.³

∴ $q_{ult} = 5·7 \times 4·4 + 1·77 D_f$

or $q_{ult} = 25·10 + 1·77 D_f.$ ...(12·51)

Soil B. The soil is a cohesionless material for which

$$\gamma = 1·77 \text{ gms./c.c.} \qquad c = 0 \text{ and } \phi = 36°$$
$$= 1·77 \text{ tonnes/m}^3$$

From table 12·2, an angle of internal friction of 36° corresponds to the borderline between medium and dense sand and further corresponds to a penetration value N of 30. The bearing capacity factors corresponding to general shear conditions may be picked up.

For $\phi = 36°$, $N_c = 60·0$; $N_\gamma = 50·0$ and $N_q = 43·0$

Substituting $c = 0$ and the above bearing capacity factors, the equation for bearing capacity becomes

$$q_{ult} = \gamma \; D_f \, (43{\cdot}0) + \frac{rb}{2}(50{\cdot}0)$$

or $q_{ult} = 43 \, \gamma.D_f + 25.\gamma.b.$ $(\cdots 12{\cdot}52)$

Put $\gamma = 1{\cdot}77$ tonnes/m.^{3} and $b = 0{\cdot}75$ metres.

$\therefore$ $q_{ult} = 43 \times 1{\cdot}77 \; D_f + 25 \times 1{\cdot}77 \times 0{\cdot}75$

or $q_{ult} = 36{\cdot}7 + 76{\cdot}7 \; D_f.$ ...(12·53)

Now vary D_f from zero to 1·75 metres and prepare a table giving the gross ultimate bearing capacity q_{ult} from equation (12·51) and (12·52), as a function of depth. Table 12·5 shows these values for the two soils.

TABLE 12.5
Values of q_{ult} as a Function of the Depth D_f

S. No.	Depth D_f in metres	Soil A $q_{ult} = 25{\cdot}10 + 1{\cdot}77 \, D_f$ tonnes/m²	Soil B $q_{ult} = 36{\cdot}7 + 76{\cdot}7 \, D_f$ tonnes/m²
1.	0	25·10	36·7
2.	0·25	25·54	55·75
3.	0·50	25·98	74·80
4.	0·75	26·42	93·84
5.	1·00	26·87	112·90
6.	1·25	27·31	131·95
7.	1·50	27·75	151·00
8.	1·75	28·20	170·05

q_{ult} versus D_f for both the soils is plotted in Fig. 12·29.

(b) If the footing lies at the surface, $D_f = 0$.

$\therefore$ For the cohesive soil, equation (12·51) gives

$q_{ult} = 25{\cdot}10$ tonnes/m².   ...(12·54)

For the cohesionless soil, equation (12·52) gives

$q_{ult} = 25\gamma b.$

If $\gamma = 1{\cdot}77$ tonnes/m.^{3}, then

$q_{ult} = 25 \times 1{\cdot}77 \; b = 44{\cdot}2 \; b$ tonnes/m³.   ...(12·55)

Equation (12·54) shows that the ultimate bearing capacity of a cohesive soil is independent of the width of the footing However, equation (12·55) indicates that the ultimate bearing capacity of cohesion soil depends upon the width of the footing.

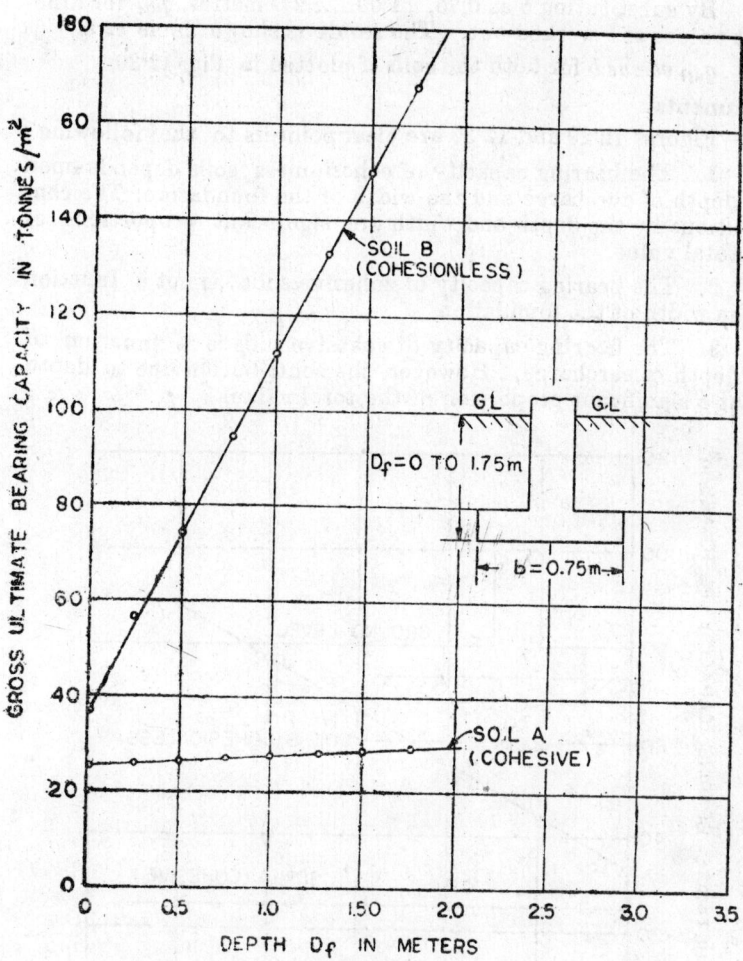

Fig. 12·29
TABLE 12·6
Values of q_{ult} as a Function of the Width b

S. No.	widtth in metres	Soil A q_{ult} constant in tonnes/m²	Soil B $q_{ult}=44·2b$ in tonnes/m²
1.	0·75	Constant	33·15
2.	1·00		44·20
3.	1·25	at 25·10	55·25
4.	1·50		66·30
5.	1·75	tonnes/m²	77·35
6.	2·00		88·40

By substituting b as 0·75, 1·00......2·0 metres. q_{ult} for different values of b worked out. The result is shown Table 12·6.

q_{ult} versus b for both the soils if plotted in Fig. 12·30.

Comments

Figures 12·29 and 12·30 are clear pointers to the following :

1. The bearing capacity of cohesionless soils depends upon the depth of surcharge and the width of the foundation. The contributions by the depth and width are significant proportions of the total value.

2. The bearing capacity of cohesive soils is not a function of the width of the foundation.

3. The bearing capacity of cohesive soils is a function of the depth of surcharge. However, the contribution due to depth is not a significant proportion of the total value.

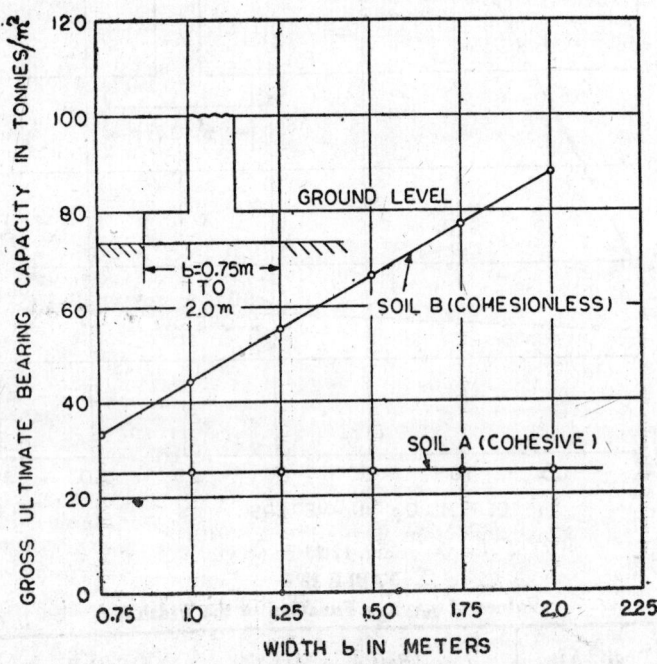

Fig. 12·30

4. The bearing capacity of cohesive soils is more a function of the cohesion value of the soil than any other thing.

5. As a consequence of 1 and 3 above the depth of surcharge is much more significant in the cohesionless soils than in the case of cohesive soils.

Example 12·2. (a) Calculate the gross ultimate bearing capacity of a (1·5 m × 1·5 m) square footing resting on the surface of a sand deposit with the following physical properties :

1. Unit weight 1·76 gms/c.c.
2. Angle of internal 36° friction
3. Cohesion=zero

(b) In case the footing is placed at a depth of 1·5 metres below the ground surface, find out its ultimate bearing capacity.

(c) In case the footing were to rest on clay whose unconfined compressive strength is 1·12 kg/cm²., indicate the bearing capacity with the conditions mentioned under (a) and (b) above.

(d) Comment on the results obtained.

Assume in all the cases that the water table lies at a distance greater than the width b below the footing i.e., it does not have any effect by way of reduction in bearing capacity.

Solution :   Size of the footing=(1·5 × 1·5) metres

$$\gamma = 1·76 \text{ gms/c.c.}$$
$$\phi = 36°$$

From table 12·2, an angle of internal friction of 36° corresponds to the borderline between medium and dense sand and it further corresponds to a penetration value of 30. The bearing capacity factors corresponding to the general shear conditions may be picked up.

Equation (12·22) will be applicable in this case.

(a) Since the footing rests on the surface of sand,
$$D_f = 0$$

Also $c = 0$

Substitution of these two values in

$$Q_{ult} = 1·3c \ N_c + \gamma . D_f . N_q + 0·4 \ \gamma b N_\gamma \qquad ...(12·56)$$

gives,

$$Q_{ult} = 0·4 \ \gamma . b . N_\gamma$$

For $\phi = 36°$, N_γ from fiure 12·15, is 50·0

$$\therefore \quad Q_{ult} = 0·4 \left(\frac{1·76}{10^0} \times 10^6 \right) \times 1.50 \times 50·0$$

$$= 52·8 \text{ tonnes/m.}^2 \text{ Ans.}$$

(b) When the footing is 1·5 meters below the ground surface :

$$D_f = 1·5 \text{ metres}$$

Also $c = 0$

Substitution of these values in the bearing capacity equation (12·56) gives

$$Q_{ult} = \gamma[D_f.N_q + 0.4.b.N_\gamma] \qquad ...(12.57)$$

For $\phi = 36°$, N_γ and N_q from figure 12·15 are 50·0 and 43·0 respectively.

$$Q_{ult} = \frac{1.76 \times 10^6}{10^3}\left[1.5 \times 43.0 + 0.4 \times 1.5 \times 50\right]$$

$$= 1.76 \times 10^3 [94.5]$$

$$= 166.3 \times 10^3 \text{ kg/m.}^2$$

$$= 166.3 \text{ tonnes/m.}^2 \text{ Ans.}$$

(c) For the cohesive soil, $q_u = 1.12$ kg./cm².

$\therefore$
$$c = \tfrac{1}{2}q_u = 0.56 \text{ kg./cm².}$$

$$= \frac{0.56}{1000} \times (100)^2 \text{ tonnes/m².}$$

or
$$c = 5.6 \text{ tonnes/m²}$$

when $\phi = 0$, from Fig. 12·15 ;

$$N_c = 5.7, N_\gamma = 0$$
and
$$N_q = 1.$$

For the footing resting on the surface, $D_f = 0$.

Under the above conditions, the bearing capacity is given by,

$$q_{ult} = 1.3 \times 5.7 \, c \qquad ...(12.58) \qquad \text{Equation (12.24)}$$

$\therefore$ $q_{ult} = 1.3 \times 5.7 \times 5.6 = 41.5$ tonnes/m². **Ans.**

For the footing resting 1·5 meters below the ground surface, $D_f = 1.5$.

N_γ term in equation (12·22) will vanish since $N_\gamma = 0$

$\therefore$ $q_{ult} = 1.3 \, c \, N_c + \gamma.D_f \, N_q.$ \qquad ...(12.59)

Substitutiug numerical values of various terms,

$$q_{ult} = 1.3(5.6)(5.7) + \frac{1.76 \times 10^6}{1000 \times 1000} 1.5 \times 1$$

or $q_{ult} = 41.5 + 2.64$

$$= 44.41 \text{ tonnes/m².} \text{ Ans.}$$

(d) Following comments are offered on the results obtained above :

(1) The ultimate bearing capacity of a cohesionless soil depends upon the width of the foundation as seen from the result in (a) above.

(2) In sandy soils, the effect of the depth of the base of the footing on the ultimate bearing capacity is considerable, as can be seen by comparing results in (a) and (b) above. As such, the depth of surcharge adds considerably to the bearing capacity of the soil.

(3) The ultimate bearing capacity of cohesive soils is a function of cohesion and the depth of surcharge. The cohesion is much more important than the depth of surcharge as can be seen in the last result under (c) above.

(4) The depth of surcharge is much more important in the case of sandy soils compared to the cohesive soils, since in the latter case when ϕ equals zero, N_q approaches unity.

Example 12.3. What will happen in part (b) of example 12.2. if

(a) the water table touches the base of the footing,

(b) the water table touches the ground surface ?

Solution. For the foundation resting at D_f metres below the ground surface, the bearing capacity in sandy soils is given by equation (12·57) i.e.

$$q_{ult}=\gamma D_f N_q+0\cdot4.b.\gamma N_\gamma \qquad\qquad ...(12.60)$$

(a) If the water table touches the base of the footing, γ. in the N_γ term in equation (12·60) is replaced by the submerged unit weight γ_b

γ for this soil is 1·76 gms/c. c.=1·76 tonnes/m³.

∴ γ_b for this soil will be 0·76 gms/ c.c.=0·76 tonnes/m.³

Since $\phi=36°$, $N\gamma$ and N_q from Fig. 12·15 are 50.0 and 43·0 respectively.

∴ $q_{ult}=1·76\times1·5\times43·0+0·4\times1·5\times0·76\times50$

$=113·7+22·7=136·4$ tonnes/m².

So, the gross ultimate bearing capacity is reduced from 166·2 tonnes/m² to 136·4 tonnes/m², when the water table rises to the bottom level of the fundation.

(b) If the water table touches the ground surface, γ in both the terms of equation (12.60) are replaced by the submerged unit weight γ_b.

∴ $q_{ult}=0·76[1·5\times43·0+0·4\times1·5\times50]$

$=0·76\times94·5=71·0$ tonnes/m².

So, the gross ultimate bearing capacity is reduced from 166·2 tonnes/m² to 71·0 tonnes/m² i.e. less than half the previous value, when the water table rises to the ground level.

Example 12·4 A load test was made with a 30 cm. square plate at a depth of 1 m. below the ground level in a highly cohesive soil with $\phi=0$. The water table was located at a depth of 5 m below the ground level. Failure occurred at a load of 4.500 kgs. What would be the ultimate bearing capacity per unit area for a 1 5 m wide continuous footing with its base located at the same depth in same soil ? (Take mass unit weight of soil as 1·9 gm/cm³ above water table.   Take $N_c=5.7$, $N_q=1$ and $N_\gamma=0$ with the soil with $\phi=0$).   (P.U. Nov. 1964)

Solution.   The gross bearing capacity of a square footing by Terzaghi's analysis is given by,

$$q_{ult}=1.3\ c\ N_c+\gamma.\ D_f\ N_q+0.4\ \gamma,\ b.\ N_\gamma \qquad ...(12.61)$$

For $\phi=0$, $N_c=5.7$, $N_q=1$ and $N\gamma=0$

$\therefore$ $q_{ult}=(1.3)(5.7)\ c+\gamma D_f$

or $q_{ult}=7.4\ c+\gamma D_f$ ...(12.62)

The water table has no effect since it lies at a distance b (b is the width of plate or the footing) below the level of placement of the plate or the foundation).   This can be seen in Fig. 12 31.

Here $\gamma=1.9$ gms./cm,$^3=1.9$ tonnes/m^3.

and $D_f=1$ m

$\therefore$ $q_{ult}=7.4\ c+1.9\times1=1.9+7.4\ c.$ ...(12.62 a)

When a square plate of 0.30 metre side is placed on the soil at a depth of 1 metre, the failure occurs under a load of 4500 kgs *i.e.* 4.5 tonnes.   So the q_{ult} equals $4.5/(0.3)^2=50$ tonnes/m^2.

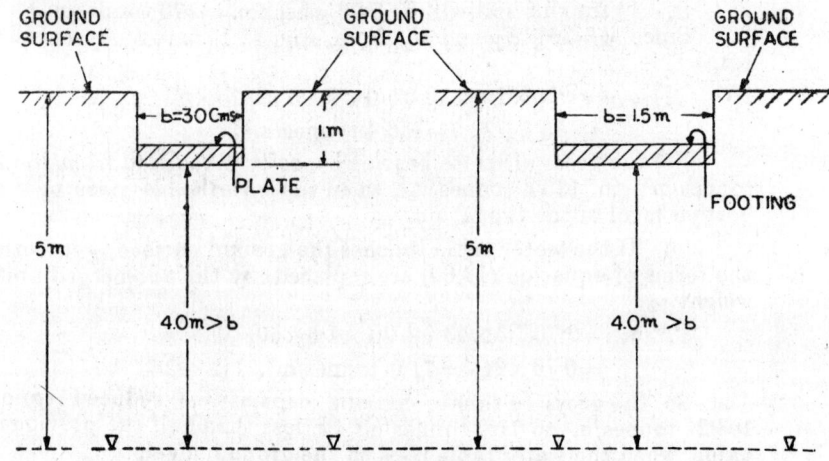

Fig. 12.31

Substituting this value in equation (12.62)

$50=7.4\ c+1.9$

$\therefore$ $c=6.5$ tonnes/m^2.

Now, the footing is continuous and the gross ultimate bearing capacity of a continuous footing is given by

$$q_{ult}=c\ N_c+\gamma.\ D_f\ N_q+\frac{b}{2}\ \gamma\ N_\gamma$$

For $\phi=0$, $N_c=5\cdot7$, $N_q=1$ and $N_\gamma=0$.

$\therefore$ $q_{ult}=5\cdot7c+\gamma . D_f$.

Here $c=6\cdot5$ tonnes/m^2

$\gamma=1\cdot9$ tonnes/m^3.

and $D_f=1$ metre

$\therefore$ $q_{ult}=5\cdot7\times6\cdot5+1\cdot9\times1=37\cdot01+1\cdot9$

$=38\cdot9$ tonnes/m^2 **Ans.**

CHAPTER—BIBLIOGRAPHY

1. Abbet, R.W., American Civil Engineering Practice, Vol. 1.

2. American Standard Building Code Requirements for Excavations and Foundations, ASCE Manuel on Engineering Practice, No. 32, 1953.

3. Bowls, E , "Foundation Analysis and Design," McGraw Hill Book Co., New York, N.Y., 1968.

4. Douglas, W.J., Foundations and Earthwork, American Civil Engineers Hand Book, New York, John Wiley and Sons. Inc., 1930.

5. Fellenlus, W., Erdstatische Berechnungen, N. Enist and Sohn, Berlin.

6. Housel, W.S., Load Tests on Flexible Surfaces, Proceeding of Highway Research Board, 1961.

7. Indian Standard Institution, Code of Practice for the Structural Safety of Building Foundations, IS-1904-61, New Delhi.

8. Indian Standard Institution, Method for Standard Penetration Test.

9. Johnson, S.M., and Kavanagh T.C., "The Design of Foundations for Building," McGraw Hill Book Co., New York, N.Y., 1968.

10. Lambe, T.W., and Whitman, R.V., Soil Mechanics, J. Wiley & Sons, Inc., N.Y.. 1969.

11. Mazanti, B., and Sawers, G.F., Unpublished Report on Load Test Research, Georgia Institute of Technology, Atlanta, Gai, 1948-55.

12. Meyerhof, G.G., Ultimate Bearing Capacity of Foundations, Geotechnique, Vol. 2.

13. National Building Code, National Board of Fire, Underwriters, New York, 1955.

14. Peck, R.B., Hanson, W.E., and Thornburn T.H., Foundation Engineering, New York, John Wiley & Sons, Inc., 1952.

15. Peririfoy, R.L., Construction Planning Equipment and
     Methods, New York, McGraw Hill Book Co., Inc. 1956.

16. Prandtl, L., Eindringungsfestigkeit and Festigkeit Von
     Schneiden, Zeitschrift fuer angewandte Mathematic und
     Mechanik, Vol. 1, No. 1, 1921.

17. Teng, W.C,, Foundation Design, New Jersey, Prentice
     Hall, Inc., 1962.

18, Terzaghi, K., Theoretical Soil Mechanics, New York,.
     J. Wiley & Sons, Inc.. 1943.

19. Terzaghi, K., and Peck, R.B., Soil Mechanics in En-
     gineering Practice, New York, J. Wiley & Sons, Inc.,
     1948.

20. Tomlinson, M.J., The Adhesion of Piles Driven in Clay
     Soils, Proceedings 4th International Conference on
     Soils Mechanics and Foundation Engineering, Vol. II,.
     London.

QUESTIONS

1. (a) What is pressure bulb ? Explain with sketches the
significance of pressure bulb for an elastic foundation.

(b) What do you understand by Bearing Capacity of a soil ?
How will you determine it in the field ? Describe the procedure and
mention the limitations.   (P.U. 1959)

2. Describe briefly with the help of a sketch, how you will
conduct a load test to ascertain the bearing capacity of a soil indicating
the limitations of the test ?

Explain how the bearing capacity of footings founded on clays
can be estimated.

What is the safe load in lbs. per sq. ft. you will permit on a
soft clay supporting a shallow continuous footing foundation, if the
unconfined compressive strength of an undisturbed sample of clay is
1000 lbs. per sq foot ? Assume a factor of safety of two.
                                                  (P.U. 1960)

3. (a) Give Terzaghi's general expression for the ultimate
bearing capacity of a continuous footing. State what are the factors
on which the ultimate bearing capacity of cohesionless and cohesive
soils depends and how.

(b) The footing of a column is 7 feet square and is founded
at a depth of 4 feet on a homogeneous cohesive soil of density 110 lbs./c.
ft. Calculate the safe load on this footing with a factor of safety of 3, if.

$c = 600$ lbs/sq. ft.
and $\phi = 0$.
For $\phi = 0$, $N_c = 5.7$, $N_\gamma = 0$ and $N_q = 1$ (P.U. 1961)

4. (a) Derive Terzaghi's formula for bearing capacity of
shallow footings. Show that the bearing capacity for a clay is independent of the size of the footing.

(b) What factor of safety would you use in the above
formula for evaluating the bearing capacity ? How far would you rely
on this value for the design of footings ?   (P.U. 1962)

5. What is a plate loading test ? How is it utilized to predict
the bearing capacity and probable settlement of footings ? Discuss the
limitations.   (P. U. 1963 Supp.)

6. (a) What are the assumptions made in the derivation of Terzaghi's bearing capacity formula ?

(b) A structure was built on a mat foundation 25 metres square. The mat rested at the ground surface on a stratum of uniform soft clay($\phi=0$), which extended to depth of 50 metres. If failure occurred at a uniformly distributed load of 30 tonnes per sq. metre. what was the average value of c ? (*P. U. April, 1964*)

7. Discuss the size effects of footing on the bearing capacity for cohesionless and cohesive soils ? (*P.U. Nov. 1964*)

8. (a) Derive Terzaghi's equation for bearing capacity of a continuous footing stating the assumptions made.

(b) Compute the allowable bearing capacity for a footing 3 m. square on dense sand ($\phi=36°$), if the depth of foundation is 1·5 m and the unit weight of the soil is 2·0 gms/c.c. Use a factor of safety of three. What will be the percentage increase in the value of the allowable bearing capacity if the depth is increased to 2·5 m ?
(*P. U. April, 1965*)

9. Draw the failure wedge, according to Terzaghi's theory, below a strip shallow footing resting on the ground surface. List all the forces acting fully for a rough base of the footing. (*P.U. April, 1966*)

10. (a) Explain the meaning and significance of the concept of "Pressure Bulb".

(a) A 20 by 20 metres of foundation supporting a ten-storey-edbuilding rests on a 40 metres deep clay layer. The average unconfined compressive strength of the clay was found to be 1·2 kg/cm². The unit pressure on the clay is 1·5 kg/cm². If the bottom of the mat is one metre below the ground surface, what is the safety factor against a shear failure of the clay ?

The unit weight of the clay may be assumed to be 1720 kg/cubic metre. (*P.U. Nov. 1966*)

11. Write Terzaghi's expression for ultimate bearing capacity of soils.

Discuss the factors on which the ultimate bearing capacity depends in (*i*) Sands (*ii*) clays. (*P.U. April, 1967*)

12. (a) Define safe bearing capacity and ultimate bearing capacity of soil.

(b) A footing carrying 10 tons per ft. rests at a depth of 3' from the existing ground surface. if the sub soils are pure clay and shear strength $s=c+\sigma \tan 0$ and unit weight is $\gamma m=120$ lbs/c ft. $C=1500$ lb/s ft. Calculate the width of footing required.
(*Mysore 1966*)

13. Discuss briefly on "the equations of bearing capacity of individual friction pile and point bearing piles. (*Baroda 1967*)

14. (a) Distinguish between shallow foundation and deep foundation. Give an example of each.

(b) Will the bearing capacity of soil remain the same for a square footing and a continuous strip footing ? If not why ?

(c) Calculate the ultimate unit bearing capacity for a square footing to be located at the depth of 3 feet from ground. $S=2000$ psf$+\sigma \tan 0°$. (*Raroda 1967*)

15. (a) Derive Terzaghi's expression for ultimate bearing capacity of long footings.

(b) The ultimate bearing capacity of shallow footings in sand is independent of width of footings. Comment on this statement.
(*J & K Univ*)

Stability of Slopes

13·1. Introduction.

Earthern embankments are invariably required for highways and railroads. Earthen dams and river training works constructed from earth are also often met with in engineering practice. A sensible design of the slopes of these earth structures is very essential, since a structural failure may lead to loss of human life in addition to the collosal economic loss, in the form of its own cost and damage to properties in the vicinity of the structure.

Collin (1846) a French Civil Engineer was the one who recognised the problems of slope stability while constructing some canals in France. He arrived at an important conclusion that the failure of an earthen embankment occurs along a surface which is a curve. He also advocated the importance of moisture content and desirability of shear tests on soils. His work did not receive much attention till the beginning of the nineteenth century. The difficulties in the construction of Panama and Kiel Canals and the railroads in Sweden invited the attention of the Engineering profession all over the United States and Europe. As a result, a number of methods have been evolved to check the stability of slopes.

Failure of embankments takes place mainly due to (1) the action of gravitational forces (2) the seepage forces within the soil. These forces (2) the seepage forces within the soil. These forces tend to cause shear stresses within the soil mass. The shear strength of the soil at all points should be sufficient to take care of these forces, if failure has to be prevented. The forces that tend to cause slippage are called the **disturbing** or **actuating forces** and the forces that tend to resist slippage are called the **restoring** or **resisting forces.**

Slope stability analysis presents many complex problems. One of the problems is the non-homogeniety of the soil of which the embankment is made of. Another problem can be the rough boundary conditions which define the flow net and thus determine the water pressures. As a consequence, some assumptions are invariably required for such analysis. The usual assumptions made are,

(1) The problem is two dimensional. An average cross section for the whole embankment is taken and it is assumed that there are no shearing stresses on the plane of the section.

(2)   The soil has the properties represented by the equation

$s = c' + \bar{\sigma} \tan \phi'$

s = shear strength of the soil

$\bar{\sigma}$ = effective normal stress.

c', ϕ' = cohesion and the angle of internal friction respectively, in terms of effective stresses.

In case the soil is homogeneous, c' and ϕ' will have only one value.   In case the soil constitutes of a number of different types, placed in well defined zones, then c' and ϕ' can have different values depending upon the zone being considered.

σ is the intergranular stress on the plane of failure and equals the total stress minus the water pressure, in case drainage is permitted and it is equal to the consolidation pressure σ_c in case the drainage is not permitted.

(3)   The flow net in case seepage is taking place can be drawn and the seepage forces could be evaluated.

The slope stability analysis could be split up into two categories :

(1)   The stability analysis of infinite slopes.

(2)   The stability analysis of finite slopes.

These two categories will be treated separately in the following articles.

13·2.   Stability Analysis of Infinite Slopes.

Refer to Fig. 13·1, which represents an infinite slope AB. The material is homogeneous in nature.   The slope is inclined at an angle i to the horizontal.   If this slope is subject to the action of gravity only, then the critical surface of failure would be a plane parallel to the slope AB. Let CD represent such a plane at a depth of z below the surface. Consider a prism of soil of inclined length b along the slope extending to this critical surface. The depth perpendicular to the plane of the paper may be considered unity.

The volume of this prism is $zb \cos i$, If γ represents the unit weight of the soil, then $\gamma zb \cos i$ represents the

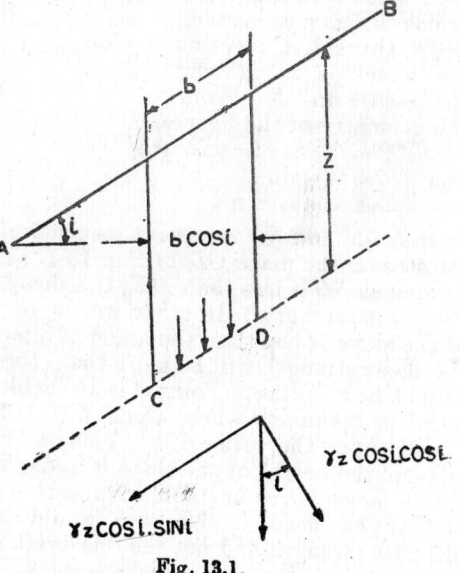

Fig. 13.1.

weight of the soil acting vertically downwards on CD Resolve this weight into two components one normal to CD and the other parallel to it. The component normal to CD is $\gamma zb \cos i \cos i$ and the component parallel to CD is $\gamma zb \cos i \sin i$. So, at this depth z, a plane parallel to the surface is acted upon by two stresses $i.e.$,

$$\sigma = \gamma z \cos^2 i \qquad \qquad ...(13\cdot4)$$
$$\tau = \gamma z \cos i \sin i \qquad \qquad ...(13.3)$$

since the area on which the forces act is $(b \times 1) = b$.

From here onwards the problem of stability can be spilt up into two cases.

Case I. When the soil is cohesionless in nature,
Case II. When the soil is cohesive in nature.
These two cases will be dealt with separately.

Case I. When the soil is cohesionless in nature.

If the soil is cohesionless, the strength envelope represented the by the Coulomb's equation $s = \bar{\sigma} \tan \phi'$ passes through the origin. Let OK represent the strength envelope as shown in Fig. 13·2. The envelop is inclined at an angle ϕ' to to the σ axis. Draw a line OP inclined at an angle i to σ—axis and take a point Q on it so that $OQ = \gamma z \cos i$. Draw a vertical MQN through Q meeting the envelope in M and the σ-axis in N From the geometry of the figures,

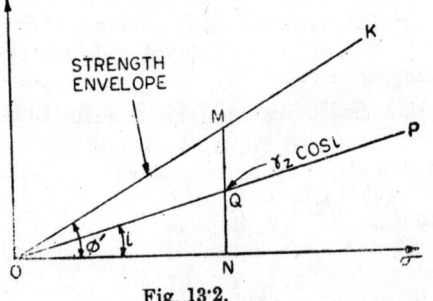

Fig. 13·2.

$$ON = OQ \cos i = \gamma z \cos^2 i$$
and $NQ = OQ \sin i$
$$= \gamma z \cos i. \sin i$$

so that ON and NQ represent respectively the normal and shear stresses on the plane CD of Fig. 13.1. For the normal stress ON, so long as NQ is less than MN, the shear stress would be less than the shear strength. In other words, so long as the inclination i of the slope is less than the angle of internal friction ϕ' of the soil, the shear strength will be more than the shear stress and there cannot be a failure. Point Q is an arbitrarily chosen point and could be taken any where along OP. Choice of Q will naturally depend upon the value of z. This obviously means that so long as $i < \phi'$ the height of the slope is immaterial, If, however, $i > \phi'$ the slope would be unstable irrespective of its height. The value of ϕ' to be used in this case would be the one derived from ultimate strength and not the one available at the peak strength.

Case II. When the soil is cohesive in nature.

The strength of a cohesive soil is represented by Coulomb's equation ; $s = c' + \sigma \tan \phi'$. This equation is represented by the straight line LK in Fig. 13·3. OL represents the cohesion c' in terms of effective stresses and the inclination of LK with the horizontal represents the angle of internal friction ϕ in terms of effective stresses. The stresses are represented by equations (13·2) and (13·3).

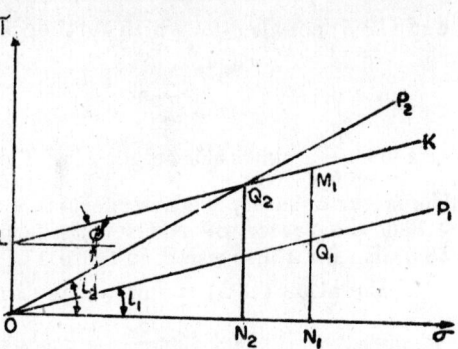

Fig· 13·3

Draw a line OP_1 at an angle $i_1 \leqslant \phi'$. Since the line can at the most be parallel to LK. It can never meet the strength envelop LK. So, if a point Q_1 is chosen on OP_1, and a vertical drawn through it, for the given normal stress ON_1 the strength will always be more than the stress i.e., N_1M_1 will always be more than N_1Q_1. Point Q_1 is an arbitrarily chosen point and could be taken anywhere along OP_1. Choice of Q_1 will naturally depend upon the value of z. This means that so long as $i_1 < \phi'$, the slope will be stable irrespective of the height of the slope. Again draw a line OP_2 at an angle $i_2 > \phi'$. This lines cuts the strength envelop LK in a point Q_2. Draw a vertical Q_2N_2 through this point Q_2. This means that for some depth limited by ON_2 the shear strength of the soil equals the shear stress. Let this depth be designated as z_{cr} so that $\gamma z_{cr} \cos i_2 = ON_2$. It follows that for a given angle of inclination of the slope i_2, there is a limiting height which may be called the critical height ($= z_{cr}$) upto which the slope can be constructed. The value of z_{cr} will vary with the variation in the angle of inclination i.e.,

Coulomb's law says that the shear strength is given by ;
$$s = c' + \overline{\sigma} \tan \phi'.$$

At the point Q_2 the strength s equals the stress $\tau = \gamma z \cos i \sin i$. $\overline{\sigma}$ in that case equals $\gamma z \cos^2 i$ (Assuming a general symbol i for the slope)

$$\therefore \quad \gamma z \cos i \sin i = c' + \gamma z \cos^2 i \tan \phi'$$
$$\therefore \quad \gamma z \cos^2 i (\tan i - \tan \phi') = c' \qquad \qquad \text{...(13.4)}$$

z in this equation is the critical value of height.

$$\therefore \qquad z_{cr} = \frac{c'}{\gamma \cos^2 i (\tan i - \tan \phi')} \qquad \text{...(13·5)}$$

One important conclusion from equation 13·4 is that the critical height z_{cr} is proportional to the cohesion c'.

If F_H represents some factor of safety, and if the developed cohesion $c_d = \dfrac{c'}{F_H}$ is used in equation 13·3 everything else remaining the same, the z given by

$$z = \frac{c_d}{\gamma \cos^2 i \; (\tan i - \tan \phi')} \qquad ...(13\cdot6)$$

is not the critical height. Ratio of z/z_{cr} from equations (13·4) and (13·5) equals c_d/c' which further is equal to $\dfrac{1}{F_H}$

i.e. $$\frac{z}{z_{cr}} = \frac{1}{F_H}$$

or z is $\dfrac{1}{F_H}$ times the height z_{cr}. This is the reason sometimes the factor of safety with respect to cohesion c' as taken above is called the factor of safety with respect to height. (See article 13·3 also for a discussion on factors of safety.)

Equation (13·5) can be written in a modified form as

$$\frac{c_d}{\gamma z} = \cos^2 i \; (\tan i - \tan \phi') \qquad ...(13\cdot7)$$

c_d, the developed cohesion can be taken in kg/cm² (say) γ can be taken in terms of kgs/cm³ and z in terms of cms. This makes the left hand side of equation (13·6) a dimensionless quantity. This dimensionless quantity is called the *stability number*. Equation (13·6) contains only three variables—the stability number the angle of inclination i and the angle of internal friction ϕ'. The use of a stability number reduces considerably the calcuation work and helps in the preparation of charts and tables for slope stability analysis. The charts and tables may not be as important for the cases of infinite slopes falling along parallel plane surfaces, but such simplified design data are quite useful in more complex cases of finite slopes.

13·3 Stability Analysis of Finite Slopes

There are a number of methods by which the finite slopes can be evaluated. The two important ones are ;

(1) The Swedish Circle Method or Method of Slices.

(2) The Friction Circle Method.

Brief details of the methods are as given below.

Swedish Circle Method or Method of Slices. The method assumes that the surface of sliding is an arc of a circle. This was established by studying the failure of embankments is Sweden. Firstly, we shall deal a case in which no seepage force acts.

Let AB be the arc of the circle with radius R, along which the slippage takes place (see Fig. 13·4). Divide the area above the arc into a number of slices by drawing verticals as shown. The horizontal distance between the verticals so drawn is usually equal but need not necessarily be equal. Whenever the soil stratum is non-homogeneous, the verticals should be drawn at.

each point where the arc passes from one material to another material. The length perpendicular to the plane of the paper is taken as unity.

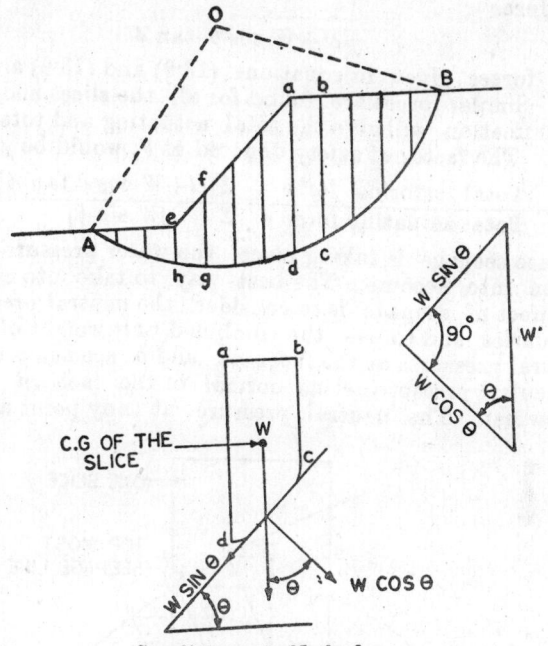

Swedish Circle Method
Fig. 13·4

Take one typical slice *abcd*. This slice is acted upon by the following forces :

1. Weight W of the slice acting in the downward direction.

2. The force *cl* acting along the face *dc* in the direction opposite to the direction of sliding. Here *c* is the cohesion of the soil and *l* is the length of the arc *i.e.* *l*= distance *dc*.

3. The soil reactions on the two vertical sides *ad* and *bc* of the slice. These reactions are neglected[1] in the analysis.

If θ is the angle made with horizontal by the tangent to *cd* at the point where the vertical through the *C.G.* of the slice passes, then W can be resolved into two components, one along the tang nt and the other normal to it. The tangential component of the weight forms the actuating force and the normal component contributes to the shear resistance.

So, the actuating force $= W \sin \theta$ ...(13·8)

---

1. The error involved in neglecting these reactions is very small.

If ϕ is the angle of internal friction of the soil, then the restoring force

$$= cl + W \cos \theta \tan \phi \qquad \ldots (13.9)$$

The forces given in equations (13·8) and (13·9) are for one slice *abcd*. Similar forces are found for all the slices and the respective summation will give the total actuating and total restoring forces. The factor of safety denoted as F_s would be given by,

$$F_s = \frac{\text{Total restoring force}}{\text{Total actuating force}} = \frac{\Sigma [cl + W \cos \theta \tan \phi]}{\Sigma \ [W \sin \theta]} \ldots (13.10)$$

In case seepage is taking place, the water pressure has also to be taken into account. The best way to take into consideration the effect of seepage is to consider the neutral pressures at the boundaries and to use the combined unit weight of the soil. The neutral pressures at the faces *ad* and *bc* are again neglected and the neutral pressure acting normal to the face *cd* is found from a flow net. The neutral pressure at any point along the

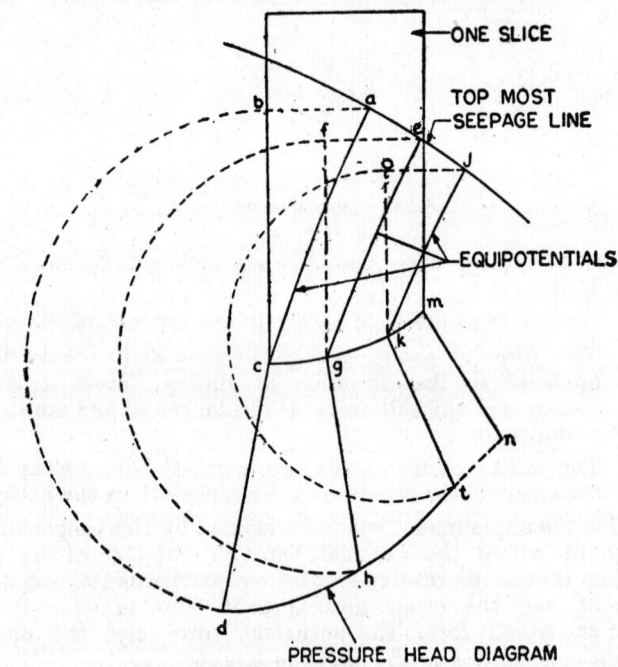

Fig. 13·5

surface AB can be found out from the equipotential lines. Take, for example a slice with a bottom surface *cm* as shown in Fig. 13·5. The todmost seepage line and three equipotentials touching *cm* are also shown. The pressure head at the point *c* is given by the distance *bc*. Point *d* is determined by drawing a normal to the are *cm* and cutting off *cd=bc*. Similarly, the points *h*, *l*, etc . are determined and the pressure head diagram *cdhlnm* can be drawn from which the neutral pressure normal to *cm* can be determined. This reduces the normal component to *cd* from $W \cos \theta$ to $(W \cos \theta - U)$ where U is the neutral force acting normal to the surface. So, the factor of safety in this case would be given by,

$$F_s = \frac{\text{Total restoring force}}{\text{Total actuating force}} = \frac{\Sigma \left[cl + (W \cos \theta - u) \tan\phi \right]}{\Sigma \left[W \cos \theta \right]} \quad \ldots(13 \cdot 11)$$

In actual computations, the area of the slice *abcd* is found by either using a graph paper or a plenimeter and its volume deduced by multiplying this area by unity. The weight W can then be found by using the unit weight of the soil. The resolution of W into the tangential and normal components is done by drawing a. force polygon as shown in the Fig. 13·4. It may be noted that for a slice of the type *efgh*, the tangential component may change its direction and become negative. So, this has to be evaluated carefully.

A set of calculations done according to the procedure laid down above is meant for one arc *AB* only. A number of such failure arcs must be tried and factors of safety arrived at, in order to find the lowest factor of safety corresponding to the critical arc. Solutions by Taylor and Fellenius given in article 13 5 help in obtaining the critical arc with the minimum number of trials.

Friction Circle Method. Refer to Fig. 13·6 (*a*) which shows a possible failure arc *AB*, for the embankment. Let the radius of

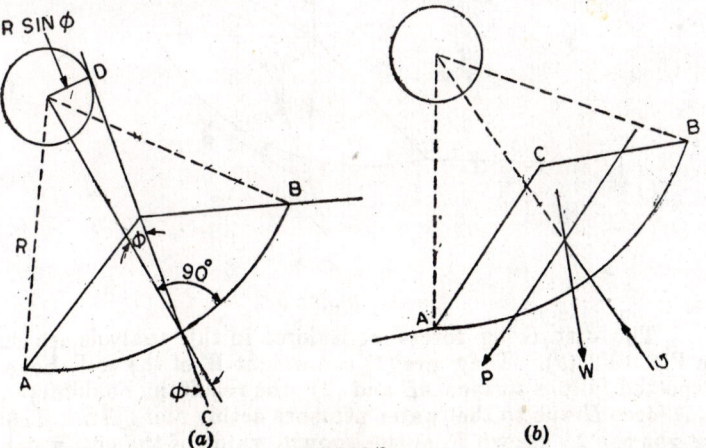

Fig. 13.6

this arc be R. A concentric circle with a radius $R \sin \phi$ is also shown. Any line *CD* tangential to the smaller circle must cut the arc AB at an obliquity φ. Conversely, any vector that represents the intergranular pressure at an obliquity ϕ to an element of the arc AB must be tangential to the small circle with a radius $R \sin \phi$. This small circle is called the *friction circle* or the *ϕ-circle*.

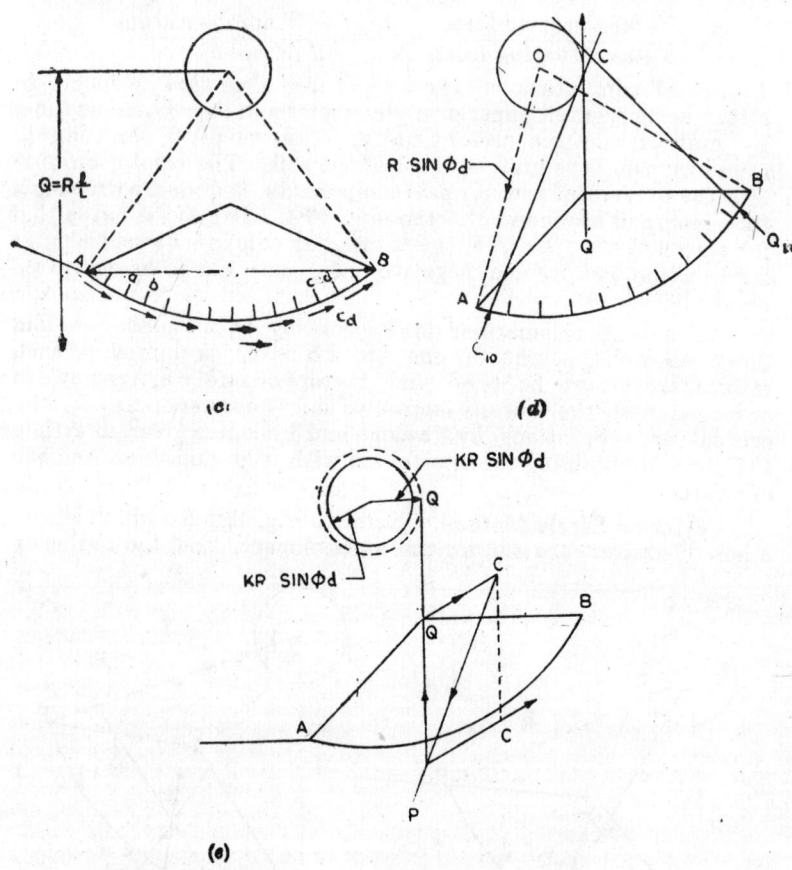

Fig. 13·6

The disturbing forces considered in this analysis are shown in Fig. 13·6 (b). They are (1) the weight W of the soil above the expected failure surface AB and (2) the resultant boundary neutral force U due to the water pressure acting along AB. The soil weight can be known from the known value of the unit weight of soil and the area of the section ABC. The line of action of the weight will pass through the centre of gravity of the area. The resultant boundary neutral force may be determined by drawing a pressure head diagram for the arc AB in the same manner as done for the arc cm as shown in Fig. 13·5. These two forces give a resultant disturbing force P.

Cohesive force acts all along the circular failure surface AB. Let the length of the arc AB be l and the length of the chord AB be L. Divided the arc into small elements, as shown in Fig. 13·6(c). The total cohesional force acting on the circular surface is c_d, l, where c_d is acting[1] unit cohesion. If, now, two small elements ab and cd are taken and the cohesion force along them is resolved into two components—one parallel to the chord and the other normal to it, then it would be found that the normal components cancel out. As a consequence, the total cohesional force on the surface equals c_d, L and it acts parallel to the chord AB. If the moments of $c_d l$ and $c_d L$ are taken about the respective centres, then

$$c_d.\ l.\ R = c_d.\ L.\ a. \qquad \qquad ...(13·6)$$

where a is the moment arm of the total cohesion force c_d. L. Put $c_d L = C$.

If the intergranular forces on the elements of the arc are considered, it can be seen that the lines of action of the two forces Q_1 and Q_{10} acting on two elements are tangential to the circle with radius $R \sin \phi_d$*, as seen in Fig. 13·6(d). The resultant of Q_1 and Q_{10} must pass through the point C, thus missing the tangency to the ϕ_d circle by a small margin. If the vector sum of all the forces on the small elements is taken, then this sum Q misses the tangency to the ϕ_d-circle by a small margin. Let the distance of this resultant from the centre of the ϕ_d circle be $KR \sin \phi_d$ where K is a coefficient whose value depends upon the distribution of the intergranular stress along the arc AB and the central angle AOB. The value of K can be picked up from Fig. 13·7, against the angle AOB. This curve is meant for a distribution having zero value at the ends of the arc and sinusoidal stress variation in between. This assumption of sinusoidal variation of intergranular stress is close to the actual distribution.

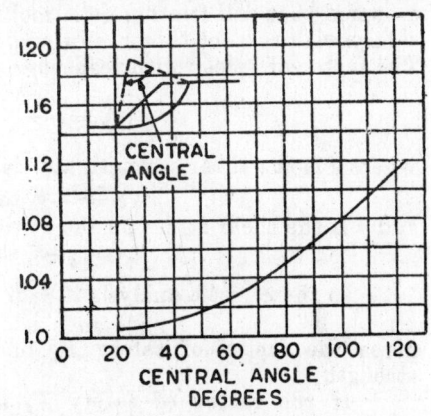

Fig. 13·7.

---

1. c_d is the working value of cohesion and is smaller than the ultimate value of c that the soil can develop.

*ϕ_d is the working value of the angle of internal friction and is less than ϕ.

If now all the three forces P, C and Q are considered, the element is in equilibrium under their action. The lines of action of all three forces must pass through a point. Lines of action of the forces P and C are known and the line of action of Q must pass through their intersection and at the same time be tangential to the circle with a radius $KR \sin \phi_d$. So, a force triangle as shown in Fig. 13·6 (e) can be drawn. From this triangle, value of C and Q can be found out. $\dfrac{C}{L}$ gives the value of c_d, the unit cohesion required to maintain stability.

The factors of safety in the analysis can be represented in three ways—the average factor of safety with respect to cohesion component of shear strength, the average factor of safety with respect to frictional component of shear strength and the factor of safety with respect to shear strength, designated as F_c, F_ϕ and F_s respectively. If c_d and ϕ_d are the working values of cohesion and angle of internal friction and c' and ϕ' are the effective shear strength components, then

$$F_c = \frac{c'}{c_d} \text{ and } F_\phi = \frac{\tan \phi'}{\tan \phi_d}. \qquad ...(13·7)$$

If F_ϕ is chosen as unity i.e. if ϕ_d equal ϕ' or in other words it is assumed that full friction gets mobilized, then F_c is the factor of safety with respect to cohesion under full mobilization of friction. The factor of safety with respect to shear is given by,

$$F_s = \frac{s}{\tau} \qquad ...(13·8)$$

where s is the shear strength and is given by
$$s = c' + \bar{\sigma} \tan \phi'$$
and τ is the shear stress on the critical surface and is given by,
$$\tau = c_d + \bar{\sigma} \tan \phi_d.$$

In the ϕ-circle analysis if ϕ_d is taken equal to ϕ', then $\dfrac{c'}{c_d}$ gives the factor of safety F_c under full mobilization of shear strength.

If the factor of safety F_s is needed, then ϕ_d is chosen in such a manner that the factors of safety F_ϕ and F_c turn out to be equal[1] and this value will represent the factor of safety F_s.

1. If $\dfrac{c'}{c_d} = \dfrac{\tan \phi'}{\tan \phi}$,

then $\dfrac{c'}{\tan \phi'} = \dfrac{c_d}{\tan \phi_d}$

∴ $\dfrac{c' + \tan \phi'}{\tan \phi'} = \dfrac{c_d + \tan \phi_d}{\tan \phi_d}$

or $\dfrac{c' + \tan \phi'}{c_d + \tan \phi_d} = \dfrac{\tan \phi'}{\tan \phi_d} = F.$

13·4 Modification for the case of Partial Submergence

In case one side of the slope is submerged under water due to a pond or a stream or if it is the slope of a dam, then in that case the boundry neutral forces will constitute of the water pressures on the submerged slope and on the failure arc AB. In such a case both the slices method as well as the friction circle method will need some modification.

Modification needed in the slices method is shown in Fig. 13·8. The failure mass includes the water contained in the arc ADE. So, the weight of water above AD must be taken into account in the slice 1, 2 and 3. Treatment of slices 2 and 3 does not need any modification. But along the base AE of slice 1, there is no shear strength. So, AE is excluded from the arc length l in the determination of cohesion component and the normal component of the weight of slice 1 also does not enter the friction determination. However the tangential component

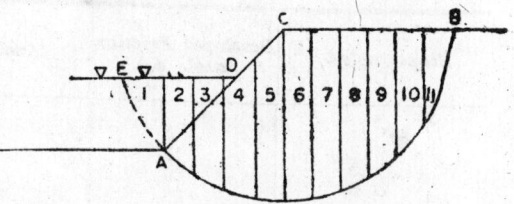

Partial Submergence Case.
Fig 13.8.

of the weight of slice 1 must be included in the sum of the weight components tending to disturb the slope. In fact the tangential component of the weight of slice 1 is negative relative to the tangential components of the weights of slices 6 to 11 and hence aids in resisting failure. In the friction circle method, the neutral force across AD acts normal to it at a point one-third of the distance from A to D, This force must be added vectorially to the neutral force across the arc AB to get the resultant boundary neutral force. This resultant force is added to the total weight to get the resultant disturbing force.

13.5. Location of Critical Arc

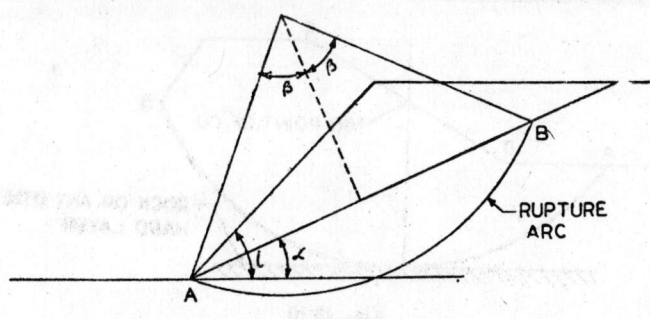

RUPTURE ARC

Fig. 13·9.

Taylor and Fellenius have found out solutions for simple finite slopes in which one slope is continuous and is bounded at the top and bottom by horizontal planes, the material of the embankment having been assumed homogeneous. From the data available from these solutions, the centre of the critical circle can be located. Reference in this connection may be made to Fig. 13·9 and table 13·1 given below.

It is only in the case of small values of ϕ that the critical arc passes below the toe. In all other cases the critical arc invariably passes through the toe.

If a hard soil layer exists at a shallow depth under a dam, the critical arc touches this hard layer tangentially as it can-

TABLE 13·1

Slope angle i	Developed Friction angle ϕ_d	Angle α	Angle β
60°	5	50	14
	15	56	13
	20	58	12
	25	60	11
40°	5	31·2	42·1
	15	36·1	37·2
	20	38	34·5
	25	40	31·0
30°	5*	23	48
	15	27	39
	20	28	31
	25	29	25
15°	5*	12·5	47
	10*	14·0	34

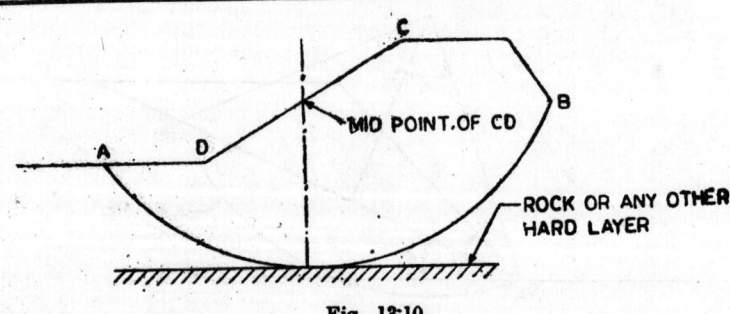

Fig. 13·10

* A more critical arc passes through the toe.

not cross this layer. If the inclination of the slope is less than 53° and ϕ is less than 15°, the critical arc passes below the toe. In such a case the centre of the critical arc lies on a vertical drawn through the middle point of the slope, as shown in Fig. 13·10.

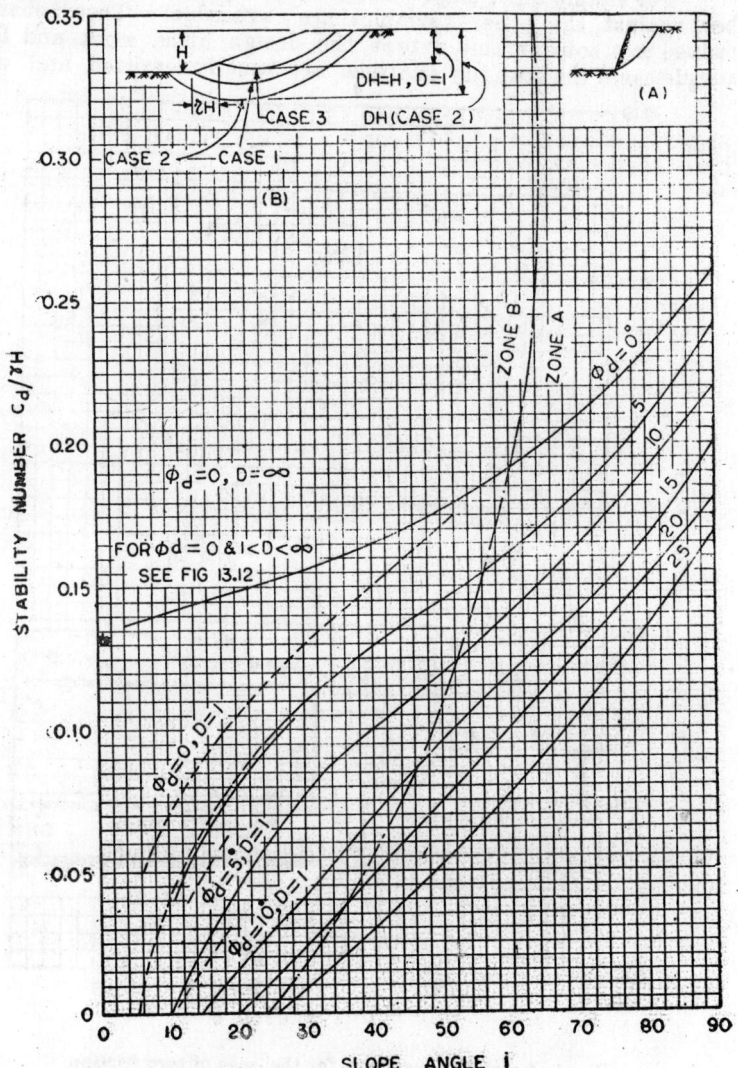

Chart of stability numbers (After Taylor)
Fig. 13·11.

The comments given above are only guiding factors, specially if the material is non-homogenous. However, they are helpful in reducing the number of trials.

13·6 Use of Charts

For simple finite slopes, charts showing the stability number against the other variables are available. These charts reduce to a considerable extent the design office work and for simple cases the stability analyses are usually omitted and use

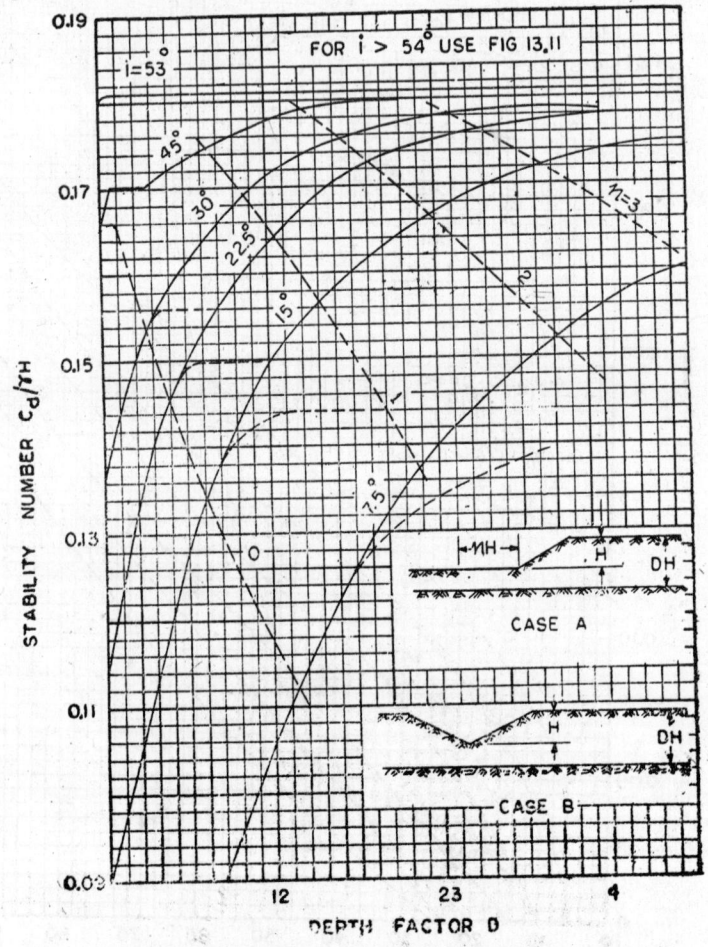

Chart of stability numbers for the case of zero friction
angle and limited depths (After Taylor)

Fig. 13·12.

is made of these charts alone. Two such charts are presented in figures 13 11 and 13·12

Fig. 13·11. shows a chart which gives the stability number $c_d/\gamma H$ against the slope angle i for various values of the angle of internal friction ϕ. It is assumed that for steep slopes, the critical circle passes through the toe of the slope with the lowest point of the arc at the toe as shown in the key sketch (A) This condition holds throughout the zone A of this figure. In zone B, the lowest point of arc is not at the toe of the slope. There are three cases shown in key sketch (B). For small friction angles or for small slope angles, the critical circle may pass below the toe rather than through it. This is shown as case 2 in the key sketch (B). For all ranges in which this holds, stability numbers are given in the chart by long dashed curves. Stabili y numbers for the most dangerous circles passing through the toe are given by solid lines in the chart both when there is and when there is not a more dangerous circle that passes below the toe. Where a solid line does not appear, the most dangerous circle passes below the toe.

Fig. 13·12 shows another case which is usually referred to as the $\phi=0$ case. Sometimes the shear strengths of soil samples from a given embankment do not show a consistent tendency towards increased strength at greater distances below ground surface. The shear strength in such cases is considered constant and such a case is represented in stability charts by zero friction angle. For this case the critical circle passes below the toe for slopes with inclinations less than 53 degrees. Theoretically, the critical arc in such cases extends to an infinite depth. But practically this is limited by hard material lying underneath. As such, the limiting depth governs the stability number. D in the figure under reference defines a ratio of the depth of failure to the height of the slope. The stability number is presented as a function of D and the slope angle i for $i<53°$ in this figure. For $i>54°$, Fig. 13·11 may be used. Full lines and short dashed lines in the chart of Fig. 13·12 represent the case A shown in the key sketch and the long dashed lines represent B also shown in the key sketch.

It may be noted that for $D=1$ and $\phi>0$, the solution for slope stability has been carried out only for 15 degree slopes as shown in the first chart.

The charts of Fig. 13·11 can be used in the following two ways.

(1) Given i and ϕ_d the stability number $c_d/\gamma H$ can be found. Knowing γ and H, the value of c_d can be determined. Then the factor of safety can be found out from,

$$F_H = \frac{c'}{c_d}$$

where c' is the effective cohesion.

(2) Given factor of safety, height and the unit weight γ, the stability number can be found out. Then from the chart, angle of inclination i can be determined against the permissible angle of internal friction. Chart of Fig. 13·12 can be used in a similar manner.

If the slope is submerged under water, the submerged unit weight γ_b should be used in place of γ. If there is a possibility of sudden draw down, then the saturated unit weight γ_{sat} may be used in place of γ. In this case, the angle of internal friction used is the weighted angle of internal friction. The weighted angle of internal friction is taken as

$$\left(\frac{\gamma_b}{\gamma_{sat}} \phi_d \right)$$

If the factor of safety with respect to cohesion is required, then ϕ_d equals ϕ' and in case the factor of safety with respect to shear strength is required, then $\tan \phi_d$ equals

$$\frac{\tan \phi'}{F_s}.$$

CHAPTER BIBLIOGRAPHY

1. Bowls, E., Foundation Analysis and Design, McGraw Hill Book Co., New York, N.Y., 1968.

2. Lambe, T.W., and Whitman, R.V., Soil Mechanics, J. Wiley & Sons, Inc., N.Y., 1969.

3. Skempton, A.W., A Slip in the West Bank of the Eau Brink Lut, Journal, Institution of Civil Engineering, 1945.

4. Taylor, D.W., Stability of Earth Slopes, Journal, Boston Society of Civil Engineers, 1937.

5. Taylor, D.W., Fundamentals of Soil Mechanics, New York, John Wiley & Sons, 1960.

6. Teng, W.C., Foundation Design, Prentice Hall, Inc., Englewood Cliffs. N.J., 1962.

7. Terzaghi, K., Theoretical Soil Mechanics, New York, John Wiley & Sons, 1943.

QUESTIONS

1. An embankment is inclined at an angle of 35° and its height is 10 metres. The angle of internal friction of the soil is 20° and its cohesive strength is 1·8 kg/cm². The unit weight of the soil is 1·9 gms /c.c. Find out the factor of safety with respect to cohesion, using Fig. 13.11.

2. An eight metre high road embankment has to be constructed from a soil whose angle of internal friction may be taken as zero. The value of c for the soil is 0·15 kg/cm². The unit weight of the soil is 1·80 gms/c.c. The soil profile shows a hard structure at 3 metres

below the ground level. Find out the permissible slope angle for the embankment.

Hint. Use Fig. 13·12

3. An embankment 10 metres high is inclined at an angle of 40° to the horizontal. A stability analysis by the method of slices gives the following forces per running metre.

Σ Shearing forces 45,000 kgs.

Σ Normal forces 87,320 kgs.

Σ Neutral forces 21,820 kgs.

The length of the failure arc is 22·0 metres, soil tests in the laboratory indicate the following results :

$$\phi' = 15 \text{ degrees}$$
$$c' = 0·20 \text{ kg/cm}^2.$$

Find out the factor of safety with respect to

(a) the shearing strength.

(b) the cohesion.

14

Compaction of Soils

14.1. Introduction

Compaction of a soil may be defined as the process of closely packing the soil particles together by reducing the air voids in the system. This is achieved by repetitive application of loads. Loads can be applied statically, dynamically, or through vibrations. These loads act only for a short duration. As a result of compaction, the dry density of the soil increases. Compaction should be distinguished from *consolidation*. In the consolidation process, a load acts for a long duration and expels the water from the pores of a saturated clay.

Compaction of a soil is an important process, as it helps it to achieve certain physical properties necessary for its proper behaviour under loads ; as *for example* proper compaction of a highway embankment, or an earthen dam reduces the chances of its settlement, increases the shear strength of the soil due to its increased density and reduces the permeability of the soil. Limits on the dry density that must be achieved by compacting the subgrades of highways or airports, form a part of the specifications laid down by many highway and airport agencies, for such construction jobs. Certain design procedures for pavements do require that a subgrade be compacted to a certain specified value of dry density before a pavement designed on the basis of these procedures is laid on it. The group index method[1] is an example of such design procedures. Strength of the soil in terms of the California Bearing Ratio can only be measured after the soil has been compacted under specified and controlled conditions. All this emphasises as to how important it is to have a scientific approach to soil compaction.

14.2. Factors Affecting Soil Compaction

Increase in the dry density of a soil due to compaction is affected by

1. The moisture content of the soil.
2. The mode and amount of compaction.

The additon of water to a dry soil sample helps in bringing the solid particles together. In fact, the water coats the solid soil

---

1 The student-engineer or the engineer interested in this design procedure may look up author's book "A text book on Highway Engineering and Airports" second edition. 1966.

particles. At low moisture contents the soil is stiff and it is diffi-cult to pack it together. As the water content increases, the water starts acting as a lubricant, the particles start coming closer due to increased workability and under a given amount of compaction, the soil-water-air mixture starts occupying lesser volume, thus affecting gradual increase in dry density. As more and more water is added and the given amount of compaction carried out, a stage is reached when the air-content of the soil attains a minimum volume under this given amount of compaction. The dry density, at this stage, for the given amount of compaction is maximum. The moisture content corresponding to this maximum dry density is called the **optimum moisture content.** Addition of water beyond the optimum moisture content, reduces the dry density because the extra water starts occupying the space which the soil could have occupied. If the moisture contents and the corresponding dry densities obtained under a given amount of compaction are plotted as shown in Fig. 14·1, the above observations can be clearly seen. A curve[1] (see curve 1) connecting the various points first rises, thus indi-

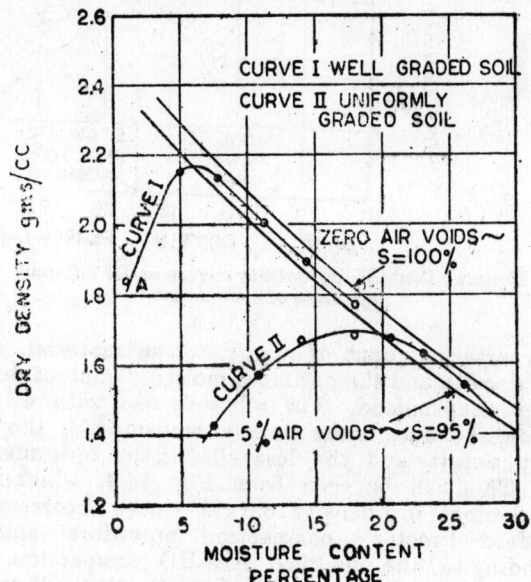

Moisture Content dry density curves
Fig. 14.1

cating an increase in dry density with the increase in moisture content, till it reaches a peak when the dry density is maximum and the moisture content has reached its optimum value and then it starts drooping down thus showing a decrease in dry density with the increase in moisture content. The curve also

---

1. A curve connecting the points obtained from moisture content-dry density readings is called the **moisture content dry density curve.**

shows the zero air voids lines and it can be seen that the peak corresponds to approximately 5 per cent air voids in the soil. The voids do not decrease appreciably by the addition of moisture content more than the optimum, although the dry density falls rapidly.

Compaction of a soil can be done by applying loads statically, dynamically or through vibrations. Each mode of compaction

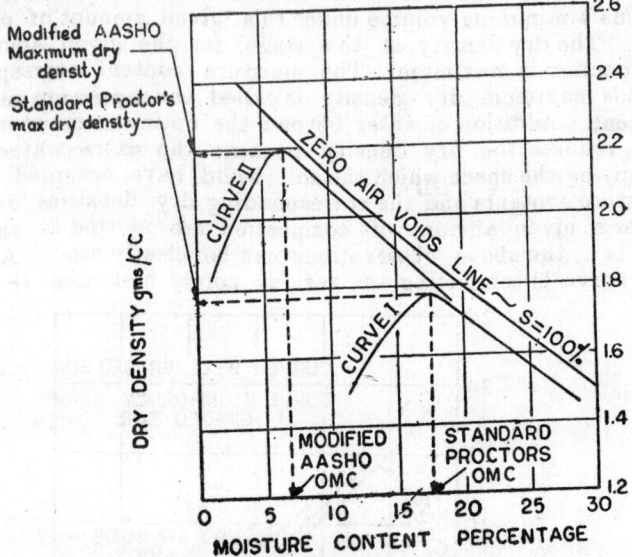

Moisture Content-dry density curves under different amounts of compaction

Fig. 14·2.

transmits certain amount of energy to the material. The maximum dry density and the optimum moisture content depend upon the energy transmitted. For all soils and with all modes of compaction, the more is the energy transmitted, the more will be the dry density and the less will be the optimum moisture content. This can be seen from Fig. 14·2, which shows two moisture-content dry density curves—curve I corresponding to the standard Proctor's compaction procedure[1] and curve II corresponding to the modified AASHO compaction procedure[1] showing the effect of the amount of compaction. Since the latter procedure employs heavier compaction, the amount of energy transmitted is more and, therefore, the dry density is more and the optimum moisture content is less, compared to the values obtainable from the former procedure.

The type of the soil is the third variable on which the maximum dry density and optimum moisture content depend.

---

1. See article 14·3 for these procedures

The maximum dry density at optimum moisture content may vary from 1·01 gms/c.c. to 2·29 gms/c.c. as shown in table 9 of chapter 4. In general, the coarser is the soil, the more is its maximum dry density and less is its optimum moisture content[1] and *vice-versa*. This trend is true irrespective of the mode of compaction. Gradation of the soil affects the shape of the moisture content and dry density curve. If a soil is well graded, this curve will have a pronounced peak and if it is uniformly graded, this curve will be flat. In general, the dry densty in the former case will be higher than in the latter case, with an opposite trend for the moisture contents. These trend may be seen in Fig. 14 1, in which two moisture content dry density curve—one for a well graded so 1 and the other for a uniformly graded soils— are shown. It has also been observed that the curves or clayey soils are relatively flat, whereas those of silts and granular materials are steep.

14.3 Laboratory Compaction Tests

The optimum moisture content and the maximum dry density for a soil can be measured in the laboratory. Then it is specified that the field density obtained by a field compaction procedure must correspond to a certain percentage of the laboratory value.

The types of compaction procedures are usually employed in the laboratory—one is called the *standard Proctor compaction test* after the name of its originator R R. Proctor (1933) and the other is called the *modified AASHO compaction test*. The first test was developed by Proctor for the construction of earthfill dams in the State of California (U.S.A.) which later on was adopted as the standard AASHO[2] laboratory procedure for compaction of soils (1) under AA8HO Designation T 99 38 and also by the ASTM[3] as their standard under ASTM designation D-558. The B.S. 1377 : 1949, Test No. 9 is also the description of this very procedure. With the coming in of heavier transport and military aircrafts, the Corps of Engineers, Department of Navy and the Federal Aviation Agency (all U.S. agencies) felt the need for a compaction procedure transmitting greater compactive effort to the soil, for the construction of air fields. So the standard Proctor test was modified and incorporated as the modified AASHO compaction test, in the AASHO testing procedures.

The Indian Standards Institution, New Delhi (5, 6) has also come up with their testing procedures.

The Standard Proctor's compaction test and the modified AASHO test procedure are described below. A brief comparison

1. Moisture content required to coat the soil particles depends upon the surface area of the particles in a given volume. The surface area of the coarser particles for a given volume is less than the surface area of fine-grained particles in the same volume.

2. American Association of State Highway Officials.

3. American Society of Testing Materials.

of these procedures with those suggested by the ISI is given towards the end of this article.

Proctor's Compaction Test. The apparatus used for the test consists of a cylindrical mould, with an internal diameter of 4 inches and an effective height of 4 6 inches, giving a volume of 1/30 cu ft. (see (Fig. 14 3) The mould is placed on a detachable square metallic plate and it carries a detachable collar of 2½ inches height on its top. The compaction of the soil in this mould is done with a hammer that weighs 5·5 lbs. and falls through a distance of 12 inches. The hammer is contained in an outer metallic sleeve that guides its movement. Air-dried soil sample passing a 3/4 inches ASTM (or B.S. or equivalent) sieve (3/16″ in AASHO test) is used for compaction. To start with, the dry soil is mixed thoroughly with a quantity of water and is com-

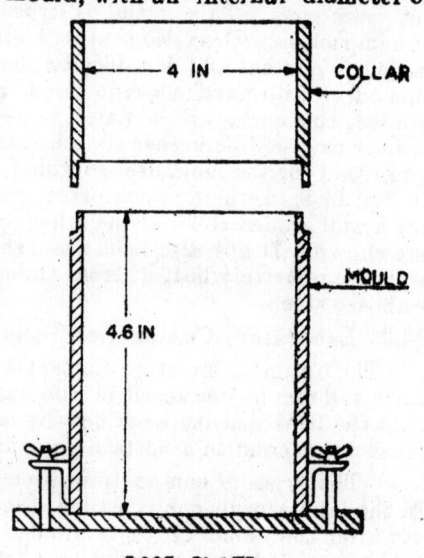

BASE PLATE

Compaction mould for the
Proctor's compaction Test
Fig. 14·3.

pacted in the mould in three equal layers giving a total height of 5 inches using 25 blows of the hammer falling through 12 inches during each below. Each compacted layer is raked with a spatula before placing the next layer. This ensures a good bond beeween the three layers. After compaction, the detachable collar is removed and the extra soil pr jecting above the top of the mould is trimmed off carefully, so as to level the soil surface with the top of the mould. The net weight of wet soil is determined by subtracting from the total weight of *wet soil, mould and the base plate,* the weight of *empty mould and the base. plate.* A representative soil sample is taken from this wet soil o determine its moisture content

Let,

W_1 = weight of the wet soil + mould + base plate

W_2 = weight of empty mould + base plate

V = volume of the mould (1/30 c. ft. in this case)

w = moisture content of the soil.

∴ $W_1 - W_2$ = weight of the wet soil,

Wet density or wet unit weight of the soil $= \dfrac{W_1 - W_2}{V}$

$$= \gamma(\text{say}) \quad ...(14\cdot1)$$

Dry density or dry unit weight, $\gamma_d = \dfrac{\gamma}{1+w} \qquad ...(14\cdot2)$

The moisture content against this dry density is already known. All this gives one point (e.g., point A on curve 1 in Fig. 14·1) on the moisture content dry density curve. The test is repeated five-six times by adding more water each time, in order to get five-six similar points through which the curve could be drawn. The value of weight W_1 gives an indication of whether the testing be stopped or carried on further. In the first few readings, W_1 measured after proper compaction and trimming, increases and then a stage reaches when W_1 decreases compared to the previous reading. This is the stage when the testing should be stopped, since the soil would have reached its maximum dry density.

From all this test data, the moisture content-dry density curve is drawn and the peak of this curve given the optimum moisture content and the corresponding maximum dry density.

Modified AASHO Compaction Test. This test is designed to give greater compactive energy to the soil. The testing equipment is similar to the one used in the standard Proctor's test except that the weight of the hammer is 10 lbs. and its fall is 18 inches. The testing procedure is also similar except that the soil is compacted in 5 equal layers giving a total height of 5 inches, using 25 blows of the hammer for each layer, the hammer falling through 18 inches during each blow. The calculations of dry density and moisture content are identical with the ones described in the preceding test. A curve of the type of curve II in Fig. 14·2 is obtained from the test data. The compactive energy transmitted by the modified AASHO test hammer is about 4·5 times the energy tranmitted by the Proctor's hammer.

ISI Standard on Compaction of Soils[1]. The Indian Standard Institution requires that the volume of the mould for compaction be either 1000 or 2250 c.c. The corresponding dimensions of diameter and height of the mould would be 100 m.m. and 127·3 m.m. in the first case and 150·0 m.m. and 127·3 m.m, in the second case, respectively.

For light compaction the weight of the hammer is specified as 2·6 kgs. and it drops through 310 m.m. For heavy compaction, the weight of the hammer is specified as 4·89 kgs. and it drops through 450 mm The soil under light compaction specifications has to be laid in three equal layers and under heavy compaction

1. For full details see IS : 2720 (Parts VII and VIII)—1965 Methods for test of soils—determination of moisture content-dry density relation using light and heavy compaction.

specifications it has to be laid in five equal layers. The number of blows to each layer will be 25 in case the 100 mm diameter mould is used and they will be 56 in case the 150 mm diameter mould is used.

Moisture content and dry-density calculations would be similar and from this data the moisture-content dry density curves can be drawn.

14·4. Dry Density-Moisture Current Relationships

The procedure to find out dry density against the moisture content of the soil has been explained in the previous article. Equation (14·2) gives the relationship between γ_d and w and typical moisture content dry density curves are shown in figures 14·1 and 14·2.

Equation (2·18 on page 20 gives the value of γ_d in terms of the specific gravity G, the void ratio e and the unit weight of water γ_w. Also equation (2·14) on page 19, gives a relationship between the ratio e the degree of saturation S, the moisture content w and the specific gravity G. From these two equations, one can express the dry density γ_d as a function of the degree of saturation, as under :

$$\gamma_d = \frac{G\gamma_w}{1+e} \qquad \text{...(equation 2·19)}$$

Also
$$S.e = w.G \qquad \text{...(equation 2·14)}$$

$$\therefore \quad e = \frac{w.G}{S}$$

$$\therefore \quad \gamma_d = \frac{G.\gamma_w}{1+\frac{w.G}{S}} \qquad \text{...(14·3)}$$

Now, if G and γ_w are known, one can draw the moisture content dry density curve for a given degree of saturation S. Two such curves for $S=100\%$ and $S=95\%$ are shown in Fig. 14·1 and one curve corresponding to $S=100\%$ is shown in Fig. 14·2. The $S=100\%$ curve means in other words the *zero air voids line*, since the air voids in this soil-water system will be zero. The specific gravity G in these cases has been taken as 2·70 and γ_w as 1 gm/c.c.

For the given specific gravity, the void ratio e at a particular dry density γ_d can also be found out, as under :

From equation (2·18)

$$\gamma_d = \frac{\gamma_w}{1+e}$$

$$1+e = \frac{G\gamma_w}{\gamma_d}$$

$$e = \frac{G.\gamma_w}{\gamma_d} - 1 \qquad \text{...(14·4)}$$

Sometimes γ_d is plotted as a function of e instead of the water content w.

Porosity η of the soil can also be evaluated from the known value of e, as given by equation (14·4),

$$\eta = \frac{e}{1+e} \qquad \text{(See page 22)}$$

$$\eta = \left(\frac{G.\gamma_w}{\gamma_d} - 1 \right) \frac{G.\gamma_w}{\gamma_d}$$

$$\eta = 1 - \gamma_d/G\gamma_w \qquad \qquad ...(14\cdot5)$$

Equation (14·5) gives the porosity η as a function of the dry density.

Example 14·1. The following data has been obtained on a soil, in a standard Proctor's compaction test.

Water content percentage   11·4   12·8   15·8   18·6   19·8

West density in gms/c.c.   1·90   1·96   2·07   3·06   2·03

The specific gravity of the solids is 2·75.

(*i*) Plot the dry density *versus* moisture content curve. Find out the optimum moisture content and the maximum dry density.

(*ii*) Draw the zero air voids line.

(Based upon P.U. 1963 Exam.)

Solution. (*i*) The wet density γ of the soil is given. The dry density can be found out from,

$$\gamma_d = \frac{\gamma}{1+w},$$

where w is the moisture content corresponding to the given wet density. Table 14·1 gives the values of w and γ_d.

TABLE 14·1

Moisture Content and Dry Density Values

Water Content, w	11·4%	12·8%	15·8%	18·6%	19·8%
Dry Density, γ_d	1·72	1·74	1·80	1·74	1·69

These values are shown plotted in Fig. 14·4. From the figure it is seen that the maximum dry density is 1·80 gm/c.c. and the corresponding optimum moisture content is 15·18%.

(*ii*) For drawing the zero air voids line take $S=1, \gamma_w=1$ gm/c.c. and $G=2·75$ in equation (14·3). Then give w the values 10%, 12% and so on to obtain the corresponding values of γ_d:

when $S=1, \gamma_w = 1$ gm/c.c.

and $G=2·75$

$$\gamma_d = \frac{2·75}{1+2·75w} \qquad \qquad ...(14\cdot6)$$

Tabulate the values of w and γ_d as in table 14·2 below :

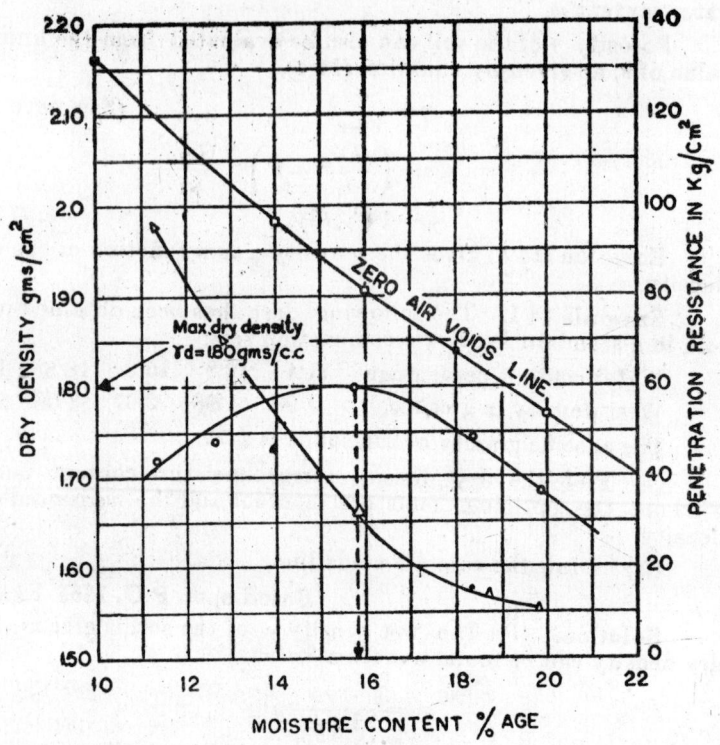

Fig. 14·4.

TABLE 14·2
Moisture Content and Dry Density Values
(From equation 14·6)

Water Content w	10%	12%	14%	16%	18%	20%	22%
Dry Density γ_d	2·16	2·07	1·98	1·91	1·84	1·77	1·71

Values in table 14·2 are plotted as shown in Fig. 14·4 to get the zero air voids line.

Comments. From the data plotted in Fig. 14·4, the reader would find that some of the data points lie away from the curve. This happens in almost all the experimental investigations. Scattering of points in an experimental work is inevitable.

14·5. Field Compaction of Soils

As stated earlier, many agencies have laid down specifications for compacting the embankments and subgrades of highway and airport pavements. These specifications are in the form of numerical limits on the dry density to be achieved in the field expressed as a percentage of the dry density obtainable in the laboratory. Table 14 3 is given as an example of such specifications. Requirements in this table have been reccmmended by the Highway Research Board Committee.

TABLE 14·3
Recommended Minimum Requirements for Compaction of Embankment[1]

Class of soil (PRA Classification page 106	Condition of Exposure					
	Condition 1 (not subject to inundation)			Condition 2[3] (subject to inundation)		
	Height of fill in metres[2]	Slope	Compaction (percentaje of AASHO maximum density)	Height of fill in metres	Slope	Compaction (percentage of AASHO maximum density
A—1	Not critical	1·5 to 1	95+	Not critical	2 to 1	95
A—3	Not critical	1·5 to 1	100+	Not critical	2 to 1	100
A—2—4 A—2—5	<15·25⎤ <15·25⎦	2 to 1	95+	<3.05 3·05 to 15·25 ⎤	3 to 1	95 ⎡ 95 to 100 ⎣
A—4 A—5	<15 25⎤ <15·25⎦	2 to 1	95+	<15·25	3 to 1	95 to 100
A—6 A—7⎦	<15·25	2 to 1	90—95	<15·25	3 to 1	95 to 100

Compaction of scils in the field is a complex process, as the soil for construction is obtained from the borrow pits and it may or may not have the moisture content at which it need be compacted for the embankment or the dam under construction. So, the first requirement is to achieve the moisture content at which

1. From H.R.B. Bulletin 58, 1952.

2. The original specifications are in feet. They have been converted in metres, with marginal adjustments of figures to the right of the decimal point in actual conversion.

3. Recommendations of condition 2 depend upon the height of the fill. If the height is of the order of 12 to 16 metres, the compaction should be 100 per cent of the laboraory AASHO value, at least for the part of the fill subject to periods of inundation.

they are to be laid on the construction job. This is done by first finding out the existing moisture content in the field and then adding more water, if needed. Addition of water to the soil is normally done either while excavating or during transport and seldom on the construction site. In case the soil is clayey, the water to be added must be added before evcavation and excavation carried out as soon as the moisture content has reached the desired value. This is essential because to bring about a change in moisture content after the clay has been excavated is not economically feasible. In case the soil excavated has more moisture than is required for compaction, the only way out is to air dry it after excavating and to compact it as soon as the desired moisture content has been attained.

A number of different types of equipment are available for compacting soils. To mention the more important ones, they are : *Smooth wheeled rollers, tamping or sheep's foot rollers, pneumatic rollers, impact rammers and vibrators.*

The conventional smooth three-wheeled roller imparts static compression to the soil. The soil is laid out in layers usually 10 to 15 cms. in thickness. The compaction is done layer by layer. Usually eight passes of the roller on each layer are sufficient to achieve approximately the standard Proctor dry density under optimum moisture content conditions. These rollers are suitable for silts, sand-silt and sand-clay mixtures, clayey silts, lean clays of low plasticity (P.I<10), well-graded sand-gravel mixtures having sufficient fines to act as binders and medium to heavy clayey soils. Many commercial designs are available in the weight range of 5·5 to 24 tonnes giving a pressure of roughly 30 kgs. to 80 kgs. per linear cm. of width of rear rolls. The size of the roller that would be most effective for a job will depend upon the type of the soil.

Pneumatic tyred roller is nothing but a box mounted on two axles. The front axles has one wheel less than the rear axles. The wheels of the front and the rear axles are staggared in their mounting arrangement so that when the roller passes over the soil, it covers all the area between the two extreme wheels of the rear axle. Actual number of wheels vary according to the make and the gross weight of the roller. The roller transmits the pressure to the soil through the types when the roller moves and as such gives a kneading action as well as compression to the soil underneath. The total weight of the roller, the contact pressure it transmits and the area of the contact with the soil are the important parameters for its performance. The equipment is suitable for moderately cohesive silty soils, clayey, gravelly and clean sands of close gradation. The compaction should not be done in layers exceeding 15 cms. Like smooth wheel rollers eight passes of the roller would be sufficient to achieve approximately the standard Proctor's dry density under the optimum moisture content conditions.

A sheep foot roller consists of a cylindrical drum, hollow from inside, which is provided with projecting *feet* all along its surface. The empty drum can be ballasted with wet sand or water to obtain the desired weight. The specifications for a typical roller are : length of drum 2 5 metres. diameter 1'5 metres, loaded weight 35 tonnes, length of the feet 15 to 22 cms. with a surface area of 45 to 80 sq. cms. and contact-pressure varying from 0'15 to 0 295 kg/cm.^{2} Some auxilliary construction unit is always required to tow a roller of this type. This type of equipment is more suitable for silty and clayey sands, clayey sands, clayey slits and medium to heavy clays. It is essential that the thickness of the layer to be compacted should not be more than the length of he feet plus five cms., in order to get better performance. These rollers are more effective where the modified AASHO compaction dry densty is required and this can be achieved by approximately 24 passes of the roller. Experience indicates that the use of sheep's foot rollers leads to higher void contents in the soils. As such, in construction jobs where the performance has to be based upon the minimum void content, this type f equipment is not suitable.

Impact rammers include the dropping weight tyre equipment including frog rammers and piling equipment. This type of equipment is suitable for compaction of small areas. The soils most receptive to compaction with the equipment are the cohesionless soils. Beds of drainage trenches and back fills of bridge abutments are usually compacted with such an equipment.

Vibrators are vibrating units of either the out-of-balance weignt type or the pulsating hydrolic type mounted on a roller or plate or a screen. Such units are highly effective for cohesionless soils and usually give densities higher than the ones obtainable in the laboratory Proctor compaction test, at optimum moisture contents much below the standard compaction values. For embankment construction, the thickness of the layer should not exceed 25 cms. However, behind the retaining walls where the soil is confined, the backfill much deeper in thickness may be compacted effectively with the help of this equipment.

All this discussion indicates that the choice of equipment will depend upon the type of the soil, the amount of compaction desired to be achieved and the moisture content. It is very essential that moisture content of the soil be controlled effectively in the field in order to get the desired dry density.

14.6 Compaction Control

The laboratory compaction test gives the maximum dry density and the optimum moisture content which would give this maximum dry density. In the field when the compaction is being carried out, it is essential to check the dry density and the moisture contents so as to affect proper control on the work.

The in-situ dry density of this soil can be checked by

1.   Core cutter method (B.S. 1377, test No. 10 C)
2.   Sand-replacement method (ASTM D—1556 and B.S. 1377 Test No. 10A)
3.   Volumenometer method (B.S. 1377, test No. 10 D)
4.   Rubber-Baloon method.

The first two methods are more commonly used. All these methods have their own limitations and must be properly understood before they are applied to a particular situation. Brief descriptions of the core-cutter and sand-replacement methods are given below :

CoreCutter Method: The core-cutter apparatus consists of a mild steel cutting ring and a dolly fits on its top. A typical cutting ring worm would be 12·5 cms. high 10 cms. internal diameter. The lower one cm. of the ring is sharpened and made into a cutting edge. The cutter is rammed into the soil with the dolly placed on its top, the ramming being done with a 14 cm. diameter metal rammer. When the top of the dolly is just about to touch the soil, the ramming is stopped and the cutter containing the-soil is dug out of the ground. The soil is trimmed level with the top and bottom of the cutter, so that the volume of the soil contained within it is equal to the internal volume of the cutter. The weight of the soil contained in the core-cutter and its moisture content are then determined. If W represent the weight of the soil, V the volume of the cutter and w the moisture content of the soil, then the density γ_d is given by,

$$\gamma_d = \frac{W}{V(1+w)} \qquad \qquad ...(14·7)$$

This apparatus is more suitable for soft cohesive soils but cannot be applied to stiff clays, sandy soils and soils containing too many stones damage the cutting edge.

Sand Replacement Method. The apparatus consists of a cylinder which contains sand graded between No. 25 and No. 52 B.S. sieves. A hole about 4 inches (about 10 cms.) in diameter is cut into the soil layer whose density is required. The cylinder with the sand is first weighed and then it is placed over this hole and the sand poured into the hole, so as to bring the sand level flush with the edge of the hole. The cylinder of sand is again weighed. The difference of weights before and after pouring the sand gives the weight of the sand in the hole. The volume of the sand is then calculated using the bulk density of the sand. This gives the volume of the hole and hence the volume of the soil dug out of the hole. The soil dug out of the hole is carefully preserved and its weight and moisture content are determined. Knowing the weight, volume and the moisture content of the soil, its dry density can be found out.

This method can be used for any type of soil.

There are many methods to find out the moisture content of the soil. In the first place a soil sample can be brought from the field and its moisture content determined by the usual laboratory procedure given in article 3.4 on page 39. This procedure is not, however helpful in affecting proper control on the moisture content at which the soil is to be compacted in the field since the sample has to be left in the oven for twenty-four hours for drying and by the time the result would be known, the layer of soil would have been compacted in the field. What is required in the field is a ready reckoner from which the moisture content could be read on the spot. This can be made available with the help of soil penetrometers. A **soil penetrometer** is a device which measures the penetration resistance of the soil at the given moisture content and for the given compactive effort. It may be mentioned here that the penetration resistance of a particular soil under the given compactive effort depends only upon the moisture content. So, when the laboratory compaction test is being conducted, the resistance to penetration against the moisture content added to the soil is measured, after the compaction at that moisture content is over. This would give one point on the peneteration resistance *versus* moisture content graph. If the test is repeated five-six times by adding more and more water as is done in the usual laboratory compaction test, five-six points on the penteration resistance *versus* moisture content graph would be available. Through these points a smooth curve can be drawn. One such curve is shown in Fig. 14·4 just as an *example*. This curve can be taken to the field. The same penetrometer when used in the field will give the penetration resistance of the soil and from this graph the moisture content can be read. If any moisture alteration is needed in the field, this can be readily done.

Among the more popular penetrometers are the TVA and the Proctor penetrometers. The proctor penetrometer has been accepted as a standard device by the ASTM under their Designation D. 1558. This penetrometer consists of a special calibrated spring dynamometer with a pressure scale on the stem of the handle. The pressure scale is calibrated to 130 lbs in one pound sub-divisions. Maximum load obtained is given by a sliding ring. Interchangeable needles of 1, 3/4, 1/2, 1/3, 1/5, 1/10, 1/20 square inches[1] end areas are supplied with the equipment. The needles are called plasticity needles. The size to be chosen for a particular soil should be such that the readings could be accurately read *i e.* the load read on the stem is neither too small nor too large.

---

1. Since this is a standard equipment the author feels it essential to give the scale and sizes in F.P.S. units and not to convert them to C.G.S. units as the manufacturers have not come out with units that are calibrated in C.G.S. system.

Electrical gadgets like the Bouyoucos moisture meters are also available for measuring the field moisture content. Nuclear devices have also been devised for the purpose. However, the Proctor's penetrometer is the simplest device for the purpose.

CHAPTER BIBLIOGRAPHY

1. American Association of State High Ways Officials Standard Laboratory Method of Test for Compaction and Density of Soils, Standard Specification for Highway Materials and Testing, Wash. D.C., 1942.

2. Bowls, E., Foundation Analysis and Design, McGraw Hill Book Co., New York, N.Y., 1968.

3. Department of Scientific and Industrial Research, Road Research Laboratory, Soil Mechanics for Road Engineers, London, Her Majesty's Stationary Office, 1952.

4. Hews, L.I. and Oglesby, C.H., Highway Engineering, Bombay, Asia Publishing Hou e, 1962.

5. Indian Standard Institution, Methods of Test for Soils, Part VIII, Determination of Moisture Content-dry Density Relation, using light Compaction, IS 2720 (Part VII)-1965, New Delhi.

6. Indian Standards Institution, Methods of Test for Soils, Part VIII, Determination of Moisture Content-dry Density Relation, using heavy compaction, IS : 2720 (Part VIII)-1965, New Delhi.

7. Leonard, G.A., Foundation Engineering, McGraw Hill Book Co,, Inc., New York, N.Y., 1962.

8. Scott, R.F.. Principles of Soil Mechanics, Addison-Wiley & Sons, Inc., New York, N.Y,, 1951.

9. Tomlinson, M.J., Foundation Design and Construction, Wiley Interscience John Wiley & Sons, Inc., New York, N.Y., 1969.

10 Woods, K.B., at el, Compaction of Earth Embankment, proceedings Highway Research Board, Wash., 1938.

QUESTIONS

1. Why are soils for earth dams, highway fills and other embankments compacted ? Discuss the factors affecting compaction and describe the Proctor's test. Show roughly the lines of 80% and 100% saturation on a compaction curve. (P.U. 1959 Supp.)

2. Write a short note on "Standard Proctor's test."

(P.U. 1960 Supp.)

3. (a) What is the difference between "Compaction" and "Consolidation" ?

(b) What are the factors that influence compaction of soils ? Briefly discuss their effect on compaction.

(c) The following data have been obtained in a standard Proctor compaction test on a soil.

Water content percentage	19·9	22·8	25·7	27·4
Wt. of container and compacted soil in lbs	9·53	9·69	9·63	9·59

The specific gravity of solids is 2·77. The container is 1/30 c.ft. in volume and its weight is 5·20 lbs.

(i) Plot the compaction curve and find the optimum moisture content and maximum dry density.

(ii) Draw 100% saturation curve. (P.U. 1961)

4. The following data were recorded in a standard Proctor test on a soil sample from a highway embankment.

Water content %	11·0	15·7	19·9	22·8	27·4
Wt. of container plus compacted soil in lbs.	9·05	9·27	9·53	9·66	9·59

Weight of empty container = 5·20 lbs.

(a) Plot the compaction curve. Indicate the maximum dry unit weight and O.M.C. of the soil.

(b) If the contractor is required to secure 98% of compaction in the field, what is the range of water content that would be allowed ?

(b) During construction, a sample was taken from the fill, which gave the following results :

Dry weight   160·0 lbs/c.ft.

Water content   15%.

What directions would you give to the contractor to improve the compaction of subsequent layers ? (P.U. 1962)

5. (a) Describe the various methods of compacting a sandy soil ?

(b) The following data has been taken from a standard compaction test :

Compaction water content %	11·4	12·8	15·8	18·6	19·8
Moist density lbs/c.ft.	118·4	122·0	128·8	128·4	127·3

The specific gravity of solids is 2·75

(i) Plot the dry density vs water content.

(ii) Plot saturated density vs saturated water content. (P.U. 1963)

6. (a) A core-cutter, 5 inches in height and 4 inches in diameter and weighing 1077 gms. when empty, is used to determine the in situ density of an embankment. The weight of the core-cutter full of soil is 2968 gms. If the water content of the soil is 7%, what is the in-situ dry unit weight and porosity ?

If the embankment gets fully saturated due to heavy rains what will be the increase in its water content and the bulk density, assuming no volume charge on saturation ? The specific gravity of the soil particles is 2.68

(b) Give the methods of field compaction and their suitablity for different types of soils. (*P.U. 1963 Supp.*)

7. The following data was obtained from a "Proctor's Compaction Test"

Water content %	5·90	7·60	9·61	11·60	13·81
Wt. of wet sample in lbs.	4.00	4.30	4·46	4.67	4.42

Specific Gravity of grains 2·70

Volume of mould 1/30 cu. ft.

(a) Plot the moisture dry density curve and find out the O.M.C. and the maximum dry density.

(b) What is Proctor's penetrometer ? Sketch a penetration resistance graph on the above graph paper. How is this curve useful in compaction control ? (*PU April 1966*)

8. Describe Standard Proctor Compaction Test. How are the results of such a test used in practice ? (*P.U. April 1966*)

9. (a) Sketch a typical compaction curve and zero air void curve obtained from B.S. Compaction test on a uniform sand. Also show on the same graph, the disposition of the compaction curve when the same soil is compacted under a modified A.A.S H.O. standards. Also show on the same graph the compaction curve that will be obtained from a test. if the soil were well graded instead of a uniform sand.

(b) Explain the methods of controlling compaction in the field.

10. What are the advantages of compacting soil mass ? Also describe briefly the standard Proctor Test. (*Baroda 1966*)

11. (a) A dry soil compacted to a bulk-density of 1.84 gms/c.c ; after adding 20% water to it, Calculate the percentage of air-voids. If soil is compacted under a moisture content of 25% and bulk density remains same, how are the dry density and the percentage of air-voids altered ? Assume the specific gravity of soil solids as 2·74.

(b) Write general formula for the evaluation of bearing capacity of soil as stated by Dr. Terzaghi. What are the assumptions made in this formula ? (*Baroda 1967*)

Note : For part (b), refer to chapter on "Bearing capacity"

12. (a) Explain the mechanism of compaction. State and sketch the nature of relation between density and O.M.C., maximum density and compaction energy and O.M.C. and compaction energy.

(b) State for what purpose it is necessary to take undisturbed soil samples. Briefly outline the methods to obtain undisturbed soil simples in the following textured soils (i) clay soil (ii) cohesionless soil (iii) gravelly soil. (*A.P. 1968*)

Note : For part (b), refer to chapter on "Soil Exploration."

13. (a) Distinguish between field compaction and laboratory compaction.

(b) Density of rolled earth fill was checked in the field by excavating a hole and weighing the excavated material the volume of the hole was determined by measuring the weight of loosely deposited standard Ennor sand required to fill the hole. The unit weight of similarly deposited sand was found by calibration tests to be 1 48 gm/c.c. The weight of sand required to fill the hole was 6·2 kg.

The weight of excavated soil at its field moisture was 8.45 kg. and its own dry weight was 6.78 kg. The specific gravity of soilds was 2.74. What were the field density, water content and degree of saturation of the fill ?

(*Regional Engg. College A.P. 1968*)

Soil Stabilization

15.1. Introduction

As pointed out in chapters 1 and 4, soils can be classified into two distinct catagories—cohesionless and cohesive soils. It has been observed that regions that are predominently clayey do not usually have sandy materials. As subgrades of highways either of the two types alone can not take the traffic independently since sands lack cohesion and spread laterally under vertical loads and clays when wet lose all strength. Combination of the two in certain specific proportions and thorough compaction with or without the use of additives like lime, cement, bitumen, etc., may result in a stable subgrade. A surface which has been prepared with the coarser and finer varieties of the soils both available at or near the site with or without the use of admixtures and compacted so as not to allow it to lose its strength under traffic is called a **stabilized surface**. The process of preparing such a surface may be termed as **soil stabilization**.

In order to cut down the cost of the stabilized surface, it is necessary to know exactly the locations alongside the road or in the near vicinity wherefrom one or the other type of material shall be available. A thorough soil-survey is required to be conducted for the said purpose before the making the final alignment of the road and designing it. All other useful available data such as geological maps indicating the rock out-crops, the agricultural maps showing the soil-type and the photographic maps giving the overall pictorial view of the entire area under survey etc, should also be collected. Collection of this data together with the actual soil-survey shall also help determine the character of the subgrade soil and the behaviour of the borrowpits installed for the collection of the materials. Samples of various materials should then be subjected to laboratory testing such as Grain size Distribution, and Aterberg Limits. Determination and establishing moisture-density relationship. etc.

Based on the above testing, the materials may now be divided under the two heads, v z., the aggregates (or granular fraction) and the binder soil (or clay fraction). Broadly speaking; the fraction passing No. 20 ASTM sieve or its equivalent and retained on No. 80 ASTM silver or its equivalent may be termed as granular fraction and may include sands, crushed stone, gravel, slag or such other material of which the main characteristics are internal friction, srength and hardness. And, the fraction passing No. 200 ASTM sieve or its equivalent may be termed as clay and silt fraction which provides cohesion and better moisture-hold, capable of binding the particles together. The agreegates possess high unit weight, excellent drainage properties, good mechanical stability, low capillarity and do not undergo appreciable change in volume when wet. The binder soils, on the other hand, have low unit weight poor drainage properties, very low supporting power when wet, high capillarity, and excessive shrinkage and swelling properties.

15.2. Various Types of Admixtures Used in Soil Stabilization
Various types of admixtures used in soil stabilization can been catagorised under the tollowing heads depending upon the properties imparted to the soil:

(i) **Cementing Materials.** They increase the strength of the soil by cementing action. Portland cement imparts strength by hydration and by modifying the clay-minerals to some extent. Lime reduces the thickness of water film surrounding individual soil particles and imparts strength as a result of flocculation and aggregation. Lime and fly-ash which speed up the pozzolonic action used together, modify the clayminerals and impart strength to the soil. Addition of sodium-silicate results in agel which gains in strength after setting.

(ii) **Modifiers.** They improve plasticity characteristics of the soil but may or may not improve strength. Cement and lime will change the water film on the soil particles, shall modify the clay-minerals to some extent and shall decrease the plasticity index. Bitumen used in small quantities with low-grade aggregates retards moisture/sorption of the clay fraction in the soil-aggregate mixture.

(iii) **Water Proofing Agents.** Bituminous materials coat the soil and retard or completely stop sorption of moisture. Plastic membranes can also help retard or stop moisture-movement into the soil. These materials are used when the soil is of such a type that the natural moisture in it is adequate and any more of it if sorbed shall produce unstable conditions.

(iv) **Water retaining Agents.** Some moisture is essential for stabilization but certain types of soils such as sands do not sorp moisture as such. Calcium chloride and sodium chloride help increase rate of water sorption. These materials, in addition, lower the vapour pressure of soil water and also lower its freezing point.

(v) **Water retarding Agents.** Materials such as calcium arcylate, resins and certain other organic compounds if used, help retard the ingress of moisture into the soil and thus act as water repellent agents.

(vi) **Miscellaneous Agents.** Lignin and molasses can be used as cheap additives for binding the soil particles together. Lignin derivatives can be used as excellent dispersing agents for clays.

The method of application of these admixture and the effect of the same on the soil when stabilized has been discussed in the articles that follow.

15·3. Soil Stabilization Types

The following are the principal types of soil stabilization which can be used in highway construction ;

1. Mechanical stabilization.
2. Chemical stabilization.
3. Cement stabilization.
4. Bituminous stabilization.
5. Other methods of stabilization such as electrical and thermal stabilization, complex stabilization. etc.

Brief description of each of the above-mentioned types is given below.

15·4 Mechanical Stabilization.

It consists in processing the aggregate and the binder soil in a suitable gradation[1] so that the mixture duly compacted, depends on the soil alone for its stability. No admixture is used in this type of soil-stabilization. and emphasis is laid on the exclusive use of the local soil. If the existing soil naturally contains all the particle sizes in the appropriate proportions, the problem becomes very simple. When such a naturally-graded soil is not locally available it may be necessary to import soils from nearby sources and blend them in suitable proportions, so that requisite amount of internal friction and cohesion are ensured.

The process of mechanical stabilization is used both for base-courses as well as for surface-courses. A higher proportion of finer sizes is required for the surface than for the base courses, to retain moisture for cohesion, to reduce moisture-penetration from top, and to avoid dusting during dry weather. Particle-grading is done in such a manner that the compacted base has adequate strength and the compacted top surface provides the necessary abrasive resistance.

The factors affecting mechanical stability are :

(i) Mechanical strength of the aggregates, weak aggregates being better.

---

1. Grading involves mixing coarse aggregate, sand, silt and clay *i.e.* all the particle-sizes from the maximum to the minimum in proper proportions so that the voids in the coarse aggregate are filled up by sand and those in sand are filled up by silt and clays and the resulting mix when compacted is the densest.

(ii) Mineral composition of the materials, so as to be resistant to weathering :

(iii) Particle-size distribution of the granular and the binder soils ;

(iv) Characteristics of the soil-mortar, as determined by the plasticity properties of the fraction passing No. 40 ASTM sieve or equivalent ; and

(v) Compaction in the field, being such as to aim at about 95% of the laboratory-value of the maximum density.

A series of ten sieves (sieves 2″. 1½″, 1″, 3/4″, Nos. 4, 10, 20, 40 and 200 according to ASTM specifications or 3″, 1½″, 3/4″, 3/8 3/16″. Nos 7,14,25,52 and 200 according to B.S. Specifications or their equivalents of the ISI specifications) are used in order to obtain approximate gradation for the proposed mixture. Some specified proportion of the designated mixture must be retained on each of the sieves of the series for producing maximum density. It has generally been accepted that a granular mass attains a very high density after compaction when the distribution of particle size is such that the following law due to Fuller is followed.

Percentage of particles passing any sieve

$$= 100 \sqrt{\frac{D}{d}}$$

where

D is the aperature size of a particular sieve, and d is the size of the largest particle passing.

Table 15·1 gives the American tentative limits of particles size distribution for base and surface-courses. Of the gradations A, B and C available, any one may be employed depending upon the maximum size of the particles available. The soil mixtures if produced with these proportions, agree considerably with the Fuller's law.

The gradation given in table 15·1 are used with certain reservations and limitations the discussion of which is beyond the scope of this book. It is, however, necessary to mix the native soil with the appropriate quantities of one or more imported soils in such proportions that the mixture gives the percentages of various fractions or particle-sizes so as to conform to one of the gradations A, B or C. There are a number of methods available for such a mix-design. Some of these are given below :

(a) *Method of Trials*. It consists in preparing a number of mixtures of the native and the imported soil/soils in various proportions based on experience. After being tested, the mixture which meets the gradation requirements and plasticity requirements is finally selected.

TABLE 15·1

Tentative Limits of Particle-size Distribution for Base and Surface-Courses—ASTM
(Mechanical Stablization)

ASTM Sieve Designation	Gradation A (Sand-clay Base or surfacing)	Gradation B (Gravel sand-clay) Base		Surfacing	Gradation C Base or surfacing
		2″-Maxm. size	1″-Maxm size		
2 in.	—	100	—	—	
1½ in.	—	70—100	—	—	
1 in	—	55—85	100	100	
¾ in.	—	50—80	70—100	85—100	100
3/8 in.	—	40—70	50—80	65—100	
No. 4	—	30—60	35—65	55—85	70—100
No. 10	100	20—50	25—50	40—70	35—80
No. 20	55—90	—	—		
No. 40	35—70	10—30	15—30	25—45	25—50
No. 200	8—15	5—15	5—25	10—25	8—25

(b) *Combining on the basis of Sieve Analysis.* According to this method, the materials are proportioned either by making use of some mathematical formula or by using graphical method such that the grading of the mixture will meet the desired specifications. After having so determined the proportions, a sample of the mix is taken and tested for liquid-limit and plasticity index which should be in accordance with the actual requirements of the mix.

(c) *Combining on the basis of plasticity Index.* This method consists in proportioning the materials with the help of a formula or using tables such that the $P.I.$ of the mixture falls within the required range. The trial mix after being tested for $P.I.$ is also tested for appropriateness of gradation.

(d) *Graphical Method.* Of the graphical methods available for mix design. the one due to Rothfuchs is the most suitable. It is reasonably quick and simple and can be used for mix design of any number of components. The method consists in plotting comulative curve of the required particle-size distribution by choosing a scale of sievc size in such a way that the particle-size distribution plots as a straight line. On ths same scale are then plotted particle-size distribution curves of the aggregates to be mixed. The straight lines that most nearly approximate to the particle-distribution curves of single aggregates are then drawn by selecting a straight llne for each curve such that the areas enclosed between it and the curve are a minimum and are balanced about the straight line. Then if the opposite ends of these straight lines are joined together, the proportions for mixing can be read off from the points where these joining lines cross the straight line representing the required mixture. (For details see reference 1).

Whatever the method of mix design, the trial mix is actually prepared and checked for gradation, Plasticity characteristics are also determined, for the fraction passing No. 40 ASTM (or No, 36 BS) sieve or equivalent sieve, For bases, the L.L. should not be more than 25 and P.I. not more than 6. For surfaces, the maximum allowable value of L L is 35 and PI between 4 and 9.

15.5 Mehra's Method of Mechanical Stabilization

This method simplifies the procedure by reducing the number of sieves for mechanical analysis to only three (being Nos. 10, 40 and 200 ASTM or their equivalents) and also the number of laboratory tests to a minimum. He recommends a compacted thickness of 3″ for the base-course having a sand content of a minimum of 50% (fraction passing No. 40 and retained on 200) P.I. for the mixture between 5 and 7. The surface-course according to him should also have a compacted thickness of 3″, prepared from soil mixture and brick aggregates in the ratio of 2 : 1 respectively, with a minimum sand content of 33% in the mix, the mix having P,I. between 9.5 and 12·5 when the road is not to be surface-treated and between 8 and 10 if surfacing is desired. It is clear that the gradation in this case is rather poor since some of the fractions may be altogether absent so that maximum density and hence maximum strength may not be obtained. This has been done to reduce the cost since the alternative would have been to pay for increased cartage to bring the deficient material from elsewhere. In India, since the traffic in the rural areas particularly, is not heavy, comparatively low specifications would serve the propose at least for the time being.

Mechanical stabilization is, thus, by far, the cheapest and probably the simplest of all the types, and so has been widely accepted inasmuch as future road construction in India is con-

cerned.   Quite a number of roads in the conntry have made use
of the process of mechanical stabilization as moderated by Prof.
S.R. Mehra to suit Indian conditions and almost all the roads so
far constructed with or without surface-treatment have behaved
in a very satisfactory manner under mixed traffic (including the
iron-tyred bullock carts).

15·6. Chemical Stabilization

   As has been mentioned in the above paragraphs, if both the
types of the constituent materials are available locally, mechanical stabilization may be the best suited, Bus areas predominant
in one type of soil are generally deiicient in the other type.   In
order to stabilize mechanically, highly clayey soils, as much as
70 to 80% by weight of sandy soil is needed, the cost of transportation of which, if not locall available, shall be tremendous.
In such cases, some type of additive which when added to the
soil shall react chemically with it producing stable particles by
changing some of the basic properties of the soil, can be made
use of advantageously.   The materials which have been found
to combine chemically with the soils are : lime, calcium chloride,
sodium chloride, resin, sodium silicate, lignin and molasses. Of
all these, lime is probably the best suited for highly clayey soils
(such as the black cotton soils) which are found in abundance
in India.

   Addition of 2 to 10% by weight of lime has been found
to be very effective in case of clayey soils. Hydrated-lime, which
is easy to work with may be preferred to quick lime which though
cheaper produces burning effect on the skin.   Addition of lime
effectively reduces the plastscity index and lineal shrinkage since
there is material decrease in the thickness of the moisture-film
surrounding the clay particles.   There is also some base-exchange
 effect producing coagulation and flocculaticn of the soil particles
resulting in some improvement in grain-size.   There is decrease
in capillary action of clay and its water-holding properties. Lime
is also presumed to act with silica and alumina of clay and as a
result of curing, strength is developed in due course, so that there
is an effective increase in C.B.R. (California Bearing Ratio) and
an increase in durability.   The detrimental effects of moisture
are considerably reduced. Swelling is also effectively arrested,
reducing the overall swell by about 80%.   Indian black cotton
soils which with the absorption of water are responsible for very
high volume-changes, becoming highly plastic in wet weather and
developing deep and and wide cracks with the evaporation of
water in dry wether, can be stabilized by the addtion of lime.
This chemical is also active with clayey gravels but not effective
with silty and loamy soils. nor is it appreciably reactive with
sandy soils.

   The properties known as *Deliquescence* and *Hygroscopicity*
whereby a material absorbs moisture from the atmosphere have

been made use of in the case of **calcium chloride** for use as an additive for soil-stabilization. Its addition decreases the rate of evaporation by lowering the vapour pressure of water and by increasing the surface tension. It also facilitates compaction so that the desired density is obtained with comparatively less compactive effort. Further it lowers the freezing point and thus considerably reduces the detrimental effect due to frost-heave. A soil which is predominantly sandy can be made to improve its water-retaining properties by the addition of this chemical to the extent required for providing necessary cohesion. The quantity needed may be up to $\frac{1}{4}$ kilogram per square metre per 2·5 cms, depth of the crust. The main use of this chemical in road construction before being recognised as an admixture for stabilization was as a dust palliative, inducing dust-repelling properties in the surface-crust which has been treated with it. **Sodium Chloride** is quite similar in action to calcium chloride but is slightly less water-retentive. In addition, it produces crystallisation within the surface, when after some evaporation of moisture, the solution becomes concentrated. After this crystallization, further evaporation is considerably arrested. The rate of application is a litte more than that in the case of calcium chloride. **Resins** are water-repellant materials so that a soil which is predominantly clayer and ordinarily holds excessive water can be profitably treated either by some artificially-prepared resin or natural resin e.g. vinsol resin. The addition of this chemical to the extent of about 1 to 2% by weight can provide natural stability both during dry as well as wet weather. It prevents the entry of excessive amount of water and reduces susceptibility to damage by frost-saction.

In desert areas, it may rather be difficult to procure finer materials i.e. clay so that in general, cement or bituminous stabilization is recommended in such cases, both the methods being expensive due to the cost of the material and its cartage to the site. It has been found that the addition of very small quantity of **sodium silicate** (2-3% compared to 6-16% in the case of cement) imparts greater strength to the mixture compared to cement and increases the abrasive resistance of the treated compacted mass. The material i.e. sodium silicate, though costlier than cement, can be locally manufactured from sand and only a very small quantity is necessary.

Lignin which is a waste bi-product from paper industry has been found to improve the compressive strength and density of some soils considerably. Similarly, molasses, a bi-product from sugar industry, can be made use of to improve the internal cohesion of the soil.

15.7 Cement Stabilization or Soil-Cement Stabilization.

Cement is suitable for almost all types of soils, especially the coarse grained soils. For fine-grained soils, if the clay-fraction is up to 30% stabilization with cement may be economical, since when the finer fraction is excessive, comparatively large quantity of cement—slurry is needed to coat the particles, This additive is very popular in the United States and some other countries both for the highway as well as for the air-field construction but in India the cost of cement which is needed to the extent of 6 to 16% is prohibitive, so that not many roads have been constructed here by this method of construction.

The exact quantity of cement required to produce a satisfactory soil-cement mix under a certain set of conditions is found out by performing actual laboratory tests such as compressive strength test, wetting and drying test of the compacted mix, and freezing and thawing test etc. The mix requires an effective miosture content in the case of sandy soils and a tittle more moisture content in the case of silty and clayey soils. Thorough compaction aimed at maximum density followed by 7 to 14 days of damp curing is necesary to attain the maximum possible strength for which time, adequacy and method of mixing are important factors.

As has been stated above, almost every soil which can be pulverised can be stabilized with cement. There is, however. that group of soils containing excessive quantity of organic mate-and certain sulphates, which render the soil unfit for action with cement. Soil-cement material is more suited for base-courses in India due to the predominance of bullock-cart traffic ; the base being finally treated at top with bituminous surfacing.

15.8 Bituminous Stabilization

The type of stabilization is suitable for sandy soils and soils containing a minimum of silt and clay-size particles, but again, in this case also, the high cost of bitumen prohibits its use in general in India.

Bitumen adds Cohesion so that clays are unsuitable for treatment with it, since its addition would make the soils fluidy. So, it is the optimum fluid content which matters here. The other purpose served by this binder is the proofing of the mix whereby it preserves the bonding action of moisture-films between the particles and reduces the detrimental effect of water which may enter the surface at a later stage.

The quantity of bituminous material required to be mixed depends upon the intended purpose, the climatic conditions and the moisture content of the soil and may generally vary from 4 to 7%. If tar is available, the varieties R T-5 or RT-6 may be useful for coarser soils and heavier varieties i.e. RT-3 and RT-4 for

---

[1]Soil cement ie the term used for the mixture of soil and cement prepared after laboratory investigation and observing field conditions. Such a mixture, when compacted and cured, is much stronger and more durable than the soil and possesses its own characteristic properties.

comparatively fine soils. On the other hand, out of asphaltic products, cut-back of the medium curing type or a slow-breaking emulsion may be found to be useful. In some cases a small quantity of cement is also added with the emulsion. Excess quantity of water of the emulsion is utilized by cement which hydrates, imparts strength to this mix and increases its stability. This process is specially useful in arid regions where there is scarcity of water and treatment with emulsion brings down considerably the quantity of water required. Addition of small quantity of lime may render excessively wet sand also suitable for bituminous stabilization.

The relative stability of soil-bitumen mixture is generally evaluated by water absorption tests and stability tests such as Modified Hubbard Field Test, Cone-penetration Test, Florida Bearing Test and Compressive-strength Test. The strength durability and stability shall depend upon the thoroughness of mixing, the moisture content and compaction.

15·9. Other Methods of Soil Stabilization

Quite a large number of other methods such as *Electrical Stabilization, Thermal Stabilization, Chemical Solidification, Complex Stabilization, etc,*. are also available for the treatment of soils, depending upon the actual field conditions. In certain special cases, where the treatment of soils with cement, bitumen or chemicals alone does not ensure the desired results, we may resort to Complex Stabilization, wherein the additive may be of two or more materials, of which at least one must be a binder. Thus, for water-logged areas, lime alone in most cases may not be as effective as a mixture of lime and a small quantity of bitumen.

CHAPTER—BIBLIOGRAPHY

1. Catton, M.D., Laboratory Investigation of soil-cement Mixtures for subgrade Treatment in Kansas, Proceedings H.R.B., Wash., Vol. 17.

2. Her Majesty's Stationary Office, soil Mechanics for Road Engineers, 1955.

3. McKesson, C.L., and Mohr, A.W., Soil-emulsified Asphalt and Pavement, Proceedings H.R.B., wash., Vol. 21.

4. Portland Cement Association, Soil Cement Roads, Construction Hand Book, Chicago, Illinois, 1943.

5. Rothfuchs, G., Particle Size Distribution of Concrete Aggregates to Obtain Greatest Density, Zement, 1935.

6. Rothfuchs, G., Graphical Determination of the Proportioning of the Various Aggregates Required to Produce a Mix of a Given Grading. Betonstrasse, 1939.

QUESTIONS

1. Define soil stabilization and state the different methods used to stabilize soil. Mention briefly the conditins under wihich each of these methods is applicable. Describe in detail the method of Soil Cement stabilization as practised in India. *(P.U. 1959)*

2. Write a brief note on "Bituminious stabilization."
(P.U. 1959 Supp.)

3. Write a short note on "Soil-Cement" stabilization.
(P.U. 1960 Supp.)

4. Write a short note on Soil Stabilization of bituminous mixtures.
(P.U. 1961)

5. Writa a short note on the various aspects of Soil Cement for road construction. *(P.U. 1962)*

6. Describe the utility and mode of application of hygroscopic and water retentive chemicals for Soil Stabilization ?
(P.U. 1963 Supp.)

7. Suggest a suitable method for stabilizing black cotton soils, for road purposes. *(P.U. 1963 Supp.)*

8. What do you understand by "mechanical and chemical" stabilization of the soil ? Discuss giving the stability with respect to the type of the soil. *(P.U. 1964)*

9. Write a short note on the different methods of soil stabilization.
(P.U. 1965)

10. Write a short note on "Soil-Cement Stabilition."
(P.U. Nov., 1966)

16

Soil Exploration

16.1. Introduction

A reasonably accurate assessment of the physical properties of the soils is very essential for the foundation design of new structures. The selection of the type and depth of the foundations, the bearing capacity and settlement analysis and the determination of the earth pressures, all depend upon soil properties. The field and laboratory investigations required to obtain the necessary soil data for design are called **soil exploration.**

Soil exploration is not only needed for the design and construction of new structures but is also needed, sometimes, for remedial measures if a structure starts cracking due to settlement. The alignment (both horizontal and vertical) of highway and airfields must also be based upon proper soil data. Location of local sources of materials for construction of these facilities and any subgrade treatment if needed, also call for proper soil exploration programmes.

The *objectives* of an exploration programme are :

1. Determination of the extent in the horizontal direction and the thickness in the vertical direction of the each type of soil upto the required depth.

2. Determination of the depth of ground water table below the ground surface.

3. Obtaining of samples of ground water, soil and rock of a size and condition adequate for positive identification of the material and for the laboratory tests required to be conducted for determining the physical properties.

4. Field observation and testing for evaluating the in-situ soil properties to substantiate the laboratory determinations.

5. In case the survey is meant for exploring the sources of materials, it is essential to collect data regarding the cubic contents of each deposit.

16.2. Methods of Sub-Soil Exploration

There are many methods of sub surface exploration. All these methods can be classified into three categories :

1. Semi-direct methods.
2. Direct methods.

3.   Indirect methods.

The *semi-direct methods* include all operations that involve drilling of bore-holes into the ground for purpose of taking out the samples.   Auger borings, displacement borings, wash borings, percussion drilling and continuous samplings all fall under the category of semi-direct methods.   It must be understood at this stage itself that the *"advance of bore-holes"* and *"taking of the samples"* are two separate, though interconnected processes.   The semi-direct methods enumerated above are meant for advancing the bore-holes.   Sampling techniques will be described in Article 16.6.   The *direct methods* include the digging of trenches and trial pits.   Accessible borings are also included in this category.   Geophysical methods and soundings or probings fall into the third category *i.e. indirect methods.*

16.3. Phases in Sub-Soil Exploration

The sub-soil exploration involves four distinct phases.

1.   Fact finding and geological survey.

2.   Reconnaissance exploration.

3.   Detailed explorations.

4.   Special explorations.

Before the actual explorations at site are started, it is always advisible and beneficial to know the geological history of the area from the geological reports and maps.   Soil survey maps of the area are also sometimes available and they can help in providing information about the soils that the engineer may encounter in the field. A visit to the area would add to the knowledge gained from the above documents.   Case histories of some of the adjoining structure, if available, are of great help in planning the actual exploration programme.   General information about the soil in the area can also be gathered from the inhabitants of the area.

The second stage of the whole programme is the reconnaissance exploration.   The main objective of this phase of exploration is to obtain rough idea about the soil type in the area.   A rough soil profile and representative sampling of the major soil strata may be aimed during the operations for this phase. Ground water table must also be located during this programme. Geophysical methods are helpful in the preliminary exploration of large and medium sized projects, as these methods are quick and cheap.   With proper interpretation, data from these methods can be directly used in design.   However, it is always a good practice to supplement the data with trial borings.   Auger borings, displacement borings and wash borings are commonly used during this phase of investigation. Data from reconnaissance exploration helps in further planning of detailed and special explorations.

The purpose of the detailed explorations is to obtain comprehensive soil profiles and representative soil samples of the important deposits. This helps in locating deposits of good construction materials. Some times undisturbed samples of the soil strata are also taken during this phase and sent to the laboratory for testing. Test pits and trenches are usually dug for shallow explorations. Anyone of the semi-direct methods can be used for these explorations.

Special explorations are carried out to obtain undisturbed soil samples from the critical soil layers. The samples are usually 10 cms in diameter. Accessible borings for visual inspection are drilled during this programme. Field tests, if any, required to be conducted, are carried out during this phase of investigation. Power driven augers, wash borings, percussion and rotary drilling methods are employed to advance the bore holes. Large drive samplers and core-barrels are used to obtain the samples. Pore-water pressure are determined in the materials of low permeability and piezometer or hydrostatic pressure cells are employed for this purpose.

Consolidation characteristics of a clayey layer, unconfined or triaxial compression strength of a soil and direct or torsion testing of a soil is done on the samples obtained during this phase of investigation.

16·4 Semi-Direct Methods of Sub-surface Exploration (Boring Methods).

As already stated in article 16·2, the semi direct methods are the auger borings, displacement borings, wash borings, percussion drilling and continuous sampling. Due to their greater importance in sub-surface exploration, a brief description of each of these methods is given first in this article. Other methods will be described subsequently.

1. Auger Borings. *Soil Auger* is a device that helps in advancing a bore-hole into the ground. Many types of soil augers, both hand-operated and power operated, are available in the market. Two types of hand operated augers are shown in Fig. 16·1. Fig. 16·1 (b) shows a small helical auger and Fig 16·1(a) is a post hole auger. These augers[1] are pressed into the ground and rotated with the handle at top. A simultaneous pressing and rotation drives the auger into the soil As soon as the auger gets filled with the soil, it is taken out and the soil is brushed off the blades. The soil sample so obtained is a *disturbed sample* The auger can be used where the sides of the bore-hole can stand unsupported. Below the water

---

1. Many other types of augers such as large helical or worm type augers, spoon augers and continuous flight augers have not been shown for want of space. All these augers are available in a variety of sizes.

table it is difficult to work with an auger since the soil usually has a tendency to cave in. In such situations, usually a pipe casing is used to support the soil. This usually reduces the progress of the work since each time that the casing has to be driven, the auger is required to be taken out. Wash borings described subsequently is the best solution under such circumstances.

Hand-driven augers are usually suitable for soft soils and upto a depth of 5 6 metres For greater depths and for harder materials, power driven augers are more suitable This method of exploration is quite suitable for reconnaissance and detailed explorations to shallow depths.

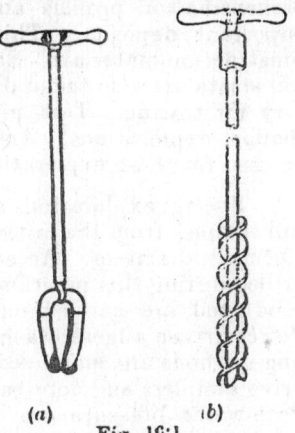

(a)   (b)

Fig. 16·1

2. **Displacement Boring.** This is the simplest method of boring. Use is made of the slit, cup or piston samplers. The operation involves (1) driving of a closed sampler into the soil until the desired depth is reached (2) rotation of the sampler or release or withdrawal of the piston and (3) taking the sample. After the withdrawal of the sampler and removal of the sample, the sampler is again inserted into the bored hole and is forced down to the depth where a new sample is desired. This method is suitable for loose to medium cohesionless soils and soft to stiff cohesive soils. It is used both for reconnaissance as well as for detailed boring programme.

3. **Wash Boring.** A wash boring is advanced partly by lightly chopping and twisting action of a light bit attached at the end of a drilling rod and partly by jetting water strongly through the hollow drilling rod. Cuttings are removed from the hole by circulating water. The complete set up is shown in Fig. 16·2. The equipment consists of a set of pipes 1·5 metres long and 6 to 10 cms. internal diameter. They can be coupled with one another. This set constitutes the casing. To start with a piece of the casing is driven into the ground with the help of the driving weight. The inside of the casing is, then cleared, with a chopping bit attatched by the lower end of the hollow drilling rod. The drilling rod ranges from 2·5 cms. to 5 cms. in internal diameter. Water under pressure is circulated through the hollow drilling rod (also called wash pipe). The water emerges at a high velocity through the water ports in the chopping bit. The water and the eroded soil form a slurry and this slurry starts moving upwards through the annular space between the casing and the drilling rod and through the horizontal pipe. A falls into

the settling tank *B*, After the first pipe has been cleaned, another pipe is coupled to it and the assembly is driven deeper

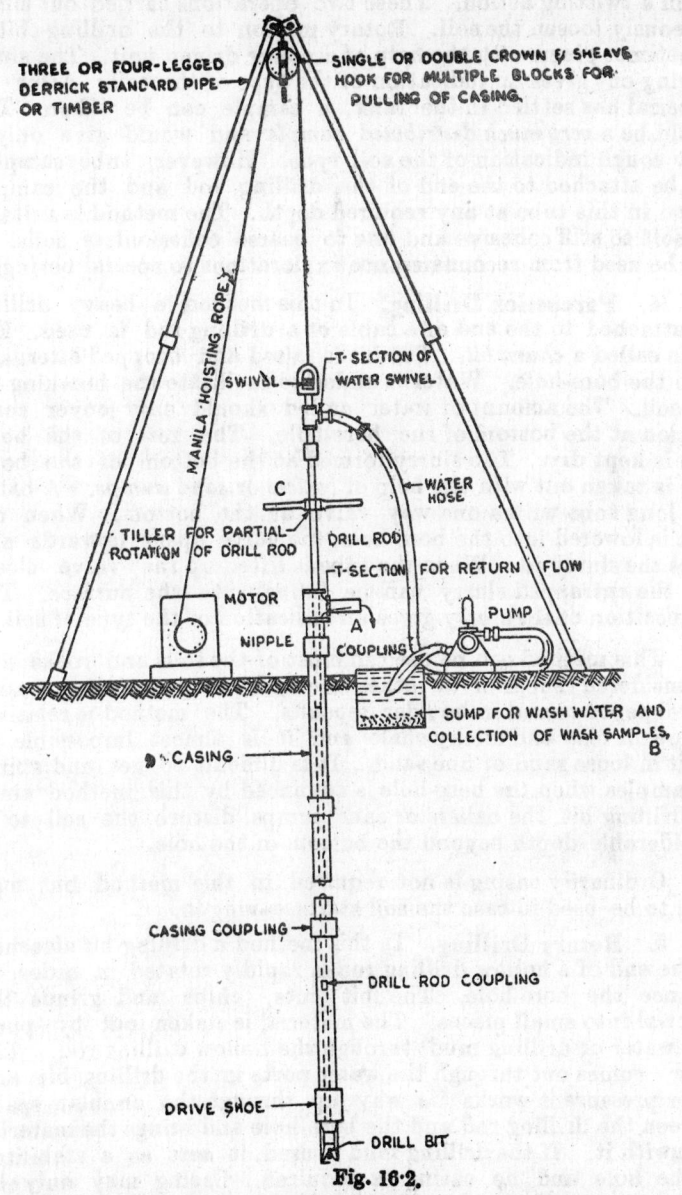

THREE OR FOUR-LEGGED DERRICK STANDARD PIPE OR TIMBER

SINGLE OR DOUBLE CROWN & SHEAVE HOOK FOR MULTIPLE BLOCKS FOR PULLING OF CASING.

MANILA HOISTING ROPE

SWIVEL

T-SECTION OF WATER SWIVEL

TILLER FOR PARTIAL ROTATION OF DRILL ROD

WATER HOSE

DRILL ROD

T-SECTION FOR RETURN FLOW

MOTOR

PUMP

NIPPLE

A

COUPLINGS

CASING

SUMP FOR WASH WATER AND COLLECTION OF WASH SAMPLES, *B*

CASING COUPLING

DRILL ROD COUPLING

DRIVE SHOE

DRILL BIT

Fig. 16·2.

into the ground surface. The process of washing is repeated. Simultaneously, with the help of the tiller marked *C*, the bit is

given a twisting action. These two operations carried out simultaneously loosen the soil. Rotary motion to the drilling bit is
sometimes given with the help of a power driven unit. The slurry
flowing out gives an indication of the type of the soil. After the
material has settled in the tank, a sample can be taken. This
would be a *very much distributed sample* and would give only a
very rough indication of the soil type. However, tube samplers
can be attached to the end of the drilling rod and the samples
taken in this tube at any required depth. The method is suitable
for soft to stiff cohesive and fine to coarse cohesionless soils. It
can be used from reconnaissance explorations to special borings.

4. **Percussion Drilling.** In this method a heavy drilling
bit attached to the end of a cable or a drilling rod is used. The
bit is called a *churn bit.* The bit is raised and dropped alternately in the bore-hole. Water is added to facilitate the breaking of
the soil. The amount of water added should only cover some
portion at the bottom of the bore-hole. The rest of the bore-
hole is kept dry. The slurry formed at the bottom of the bore-
hole is taken out with the help of *bailers* or *sand pumps.* A bailer
is a long tube with a one-way valve at the bottom. When the
tube is lowered into the bore-hole, the valve opens upwards and
takes the slurry in. When the tube is lifted up the valve closes
and the entrapped slurry can be brought to the surface. The
composition of tht slurry gives an indication of the type of soil.

This method can be used in most of the soils and rocks and
is considered superior to other methods in penetrating coarse
gravel deposits and in boulder deposits. The method is relatively slow in clay and sticky shale and it is almost impossible to
use it in loose sand or fine sand. It is difficult to get undisturbed samples when the bore-hole is advanced by this method since
the drilling bit, the bailers or sand pumps disturb the soil to a
considerable depth beyond the bottom of the hole.

Ordinarily casing is not required in this method but mav
have to be used in case the soil starts caving in,

5. **Rotary Drilling.** In this method a drilling bit attached
to the end of a hollow drilling rod is rapidly rotated in order to
advance the bore-hole. The bit cuts, chips and grinds the
material into small pieces. The material is staken out by pumping water or drilling mud[1] through the hollow drilling rod. The
water comes out through the water ports in the drilling bit and
under pressure it works its way up through the annular space
between the drilling rod and the bore-hole and brings the material
alongwith it. If the drilling mud is used, it acts as a stabilizer
for the hole and no casing is required. Casing may only be
needed to start the work. Most of the depth of the hole remains

---

1. Drilling mud is a slurry of bentonite in water. Bentonite is a
montmorrillonitic clay having thixotropic properties.

uncased. At the end of the day's work, this mud becomes a gel and holds the chippings in suspension, thus preventing these chipping from collecting near the drilling bit.

The method can be used in most of the soils but is very effective in highly resistant rock formations. In coarse gravels, boulders and very porous deposits, this method is not normally employed, since this would involve a large consumption of the drilling fled. The bore-holes in rotary drilling range from 10 cms. to 75 cms.

6. **Continuous Sampling.** In this technique, the boring is accomplished entirely by sampling. In exploration of rocks, coredrilling is done for obtaining the rock samples continuously. In the soils too, when core barrels and piston samplers are used continuously to obtain the soil samples, the boring proceeds side by side.

16·5. Stabilization of bore-holes

The problem of preventing caving of the sides and bottom of a hole and to avoid disturbance of the soil to be sampled is common to all boring methods.

Uncased, dry bore-holes are generally stable when they are *shallow* and above the *ground water table,* but danger of caving increases rapidly with depth and the presence of free ground water. In firm cohesive soils the hole may remain open for a limited length of time. Bore-hole without any provisions for stabilization and extending below ground water level are often used in displacement boring and continuous sampling with piston samplers.

There are a number of methods for stabilization the bore-holes. A brief description of these methods given below :

1. **Stabilization with water.** Suitable in rock and often in stiff cohesive soils. Water alone is not suitable or preventing caving of borings in soft or cohesionless soils.

2. **Stabilization with drilling fluid.** An uncased bore-hole can be stabilized by filling it with a proportioned drilling fluid or 'mud', which when circulated also serves to remove the churne up soil from the bottom of the hole.

Drilling fluid. A satisfactory drilling fluid can be obtained by mixing locally available clays with water but it is often advantageous and necessary to add commercially prepared products such as vol-clay or aquagel. These commercial clay products consist of highly collidal, gel forming, thixotropic clays primarily bentonite with certain chemicals added to con rol dispersion, thixotropy, gel strength and viscosity.

The stabilizing effect of drilling fluid is caused partly by its higher specific gravity compared to water and partly by the formation of relatively imperv ious lining or *mud cake* on the sides of a bore hole.

3. Stabilization with Casing. Casing or the lining of the bore hole with steel pipe provides the safest though relatively expensive method of stabilization of the bore-hole. Different types of casing are available.

As already stated in article 16 4, casing is used extensively with wash boring, percussion drilling and eep auger borings. It is generally required for all borings in very soft soils where large open drive samplers are used and where accurate observations of the ground water levels are made.

4. Stabilization by Freezing. A bore-hole may be stabilized by freezing the soil around it as the boring progresses. This may be accomplished by replacing water with kerosene or brine cooled by means of "dry ice". This method cannot be used when the ground is dry or nearly so and when there is a strong ground water flcw. So far it has been used for experimental purposes.

5. Stabilization by Grouting. A bore-hole passing through zones in rock cavities, faults and broken rock etc. may be stabilized by filling the lower part of the hole with cement grout and then re-boring the concrete plug.

16·6. Common Techniques of Sampling.

Soil samples are of two types : (1) disturbed ssmples (2) undisturbed samples. The disturbed soil samples may further be classified into (a) representative samples and (b) non-representative samples.

As pointed out in Chapter 1, the soil in the natural state of deposition has some structure. The properties of the soil depend to a large extent on the structural configuration of the soil particles in the natural state. This is specially true in the case of cohesive soils. If the structure of a sample during the process of sampling gets disturbed, the sample is designated as a disturbed sample. Samples obtained from auger borings, wash borings and percussion drilling are all disturbed. In case soil sample brought to the surface from some depth retains all the constituents present in the soil in its natural state at that particular depth from which it is taken, the sample is designated as a representative sample. In case some constituents of the soil are lost during the process of sampling or if some constituents of other layers get mixed up with the sample from a particular depth, the sample may be designated as non-representative. The *representative* samples of a soil are suitable for classification and chemical tests, but the *non-representative* samples only give an indication of the major stratifications of the soil. A sample of soil which retains the natural soil structure after sampling is called an undisturbed sample. With the sampling techniques at the disposal of the soil engineer, it is not possible to obtain a truely undisturbed soil sample. All that can be done is to reduce the amount of disturbance, through better sampling procedures. This type of sample is used for more precise tests like the shear strength or the consolidation test.

The type of the equipment used in sampling depends upon the type of the soils and the type of the sample required. Generally properties of granular soils are determined from the disturbed samples, since it is very difficult to get an undisturbed sample from such soil deposits. The standard penetration tests measures the in-situ relative density of such soils and substantiates the results from the disturbed sample. Penetration tests are made at frequent intervals along the depth of the bore-hole. A *split spoon sampler* is used in the standard penetration test to obtain the soil sample from the depth at which the test is conducted. Properties of cohesive soils are obtained from the undisturbed samples and for sampling thin-walled tubes popularly known as *shelby tubes* are used. Brief description of the split spoon sampler and the thin-walled tube sampler are given below.

Split Spoon Sampler. A split spoon sampler consists of a thick walled steel tube split lengthwise. It carries a cutting edge called a cutting shoe at the low end and a head at the upper end. The makes of the sampler currently available in the market have an outer diameter ranging from 2^2 inches to $4\frac{1}{2}$ inches and the internal diameter ranging from $1\frac{1}{2}''$ to $4''$. The overall length in all cases is 24 inches. The Indian Standards Institution recommends (7) a size, outer diameter 50.8 mm. and inner diameter 35 mm. The sampler is lowered into the bottom of the bore-hole with the help of drilling rods and is driven into the soil with hammer blows. After penetration to full depth, it is pulled out. The cutting shoe and the head are removed and the soil is examined visually. Then. it can be bottled and sealed for shipment to the laboratory for further examination.

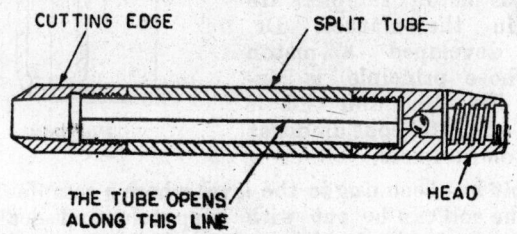

CUTTING EDGE   SPLIT TUBE

THE TUBE OPENS ALONG THIS LINE   HEAD

Fig. 16.3 Split Spoon Sampler

1. There are many other types of penetration tests. But the standard penetration test is the one that is most widely used.

2. A model with an outer diameter of 2 inches and an internal diameter of $1\frac{1}{2}$ inches is considered standard penetration equipment.

Split tube samplers with ring liners are also available. In case it is not required to test the sample in the field, the liner can be inserted in the split tube. After taking the sample, the liner is taken out and sealed with wax on both sides for shipment to the laboratory. Fig. 16·3 shows the main parts of a split tube sampler.

Thin walled tube sampling. A thin walled tube sampler is a seamless tube with very thin walls. The Indian Standards Institution recommends that the tubes shall be 40 mm. to 125 mm. in outside diameter and should be made of steel, brass, or aluminium. The sizes and wall thicknes s as recommended by ISI of these tubes are given in Table 16·1.

TABLE 16·1
Sampling Tube Dimensions

Outside diameter in mm.	40	50	55	65	75	125
Wall thickness in mm.	125	125	125	17	17	315

The length of tubes shall be 10 times the diameter for sampling in sandy soils and it shall be 10—15 times the diameter for sampling in clays, subject to a maximum of 60 cms.

The lower end of the tube is sharpened to form a cutting edge. One typical tube is shown in Fig. 16·4.

The tube can be fitted to a head attached to a drilling rod. Static force is applied to push the tube in the bottom of the borehole dug to the required depth. Sometimes it happens that the soil tends to drop out from the sampler while being withdrawn. In such cases piston samplers are available in the market. Dr. Osterberg developed a piston sampler whose principle is explained in Fig. 16·5 and this is the most advantageous amongst all the piston samplers.

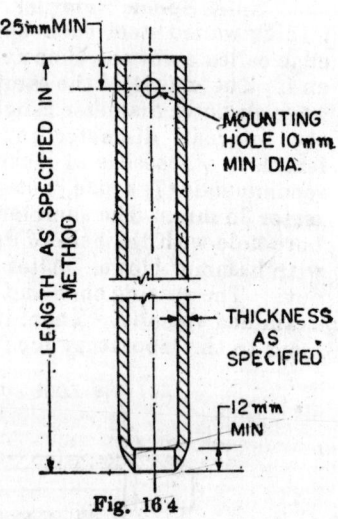

Fig. 16·4

If a pit has been dug to the level where a sample is needed, a cake of the soil can be cut with a spatula and sealed on all sides. This probably is the best method of sampling but is expensive.

There are many other devices for sampling but for want of space, it is not possible to describe all of them.

Sample Disturbance. Some sample disturbance by any one of the sampling techniques, however sophisticated they may

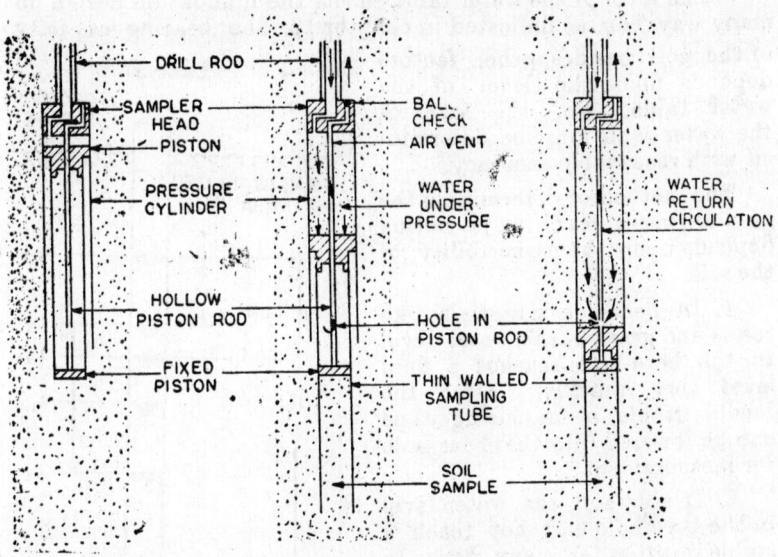

16·5 Osterberg's Piston Samplers

be, usually occurs. As far as the sampling by the use of the tube samplers is concerned the **area ratio** is an index to the sample disturbance. Area ratio is defined as,

$$A_r = \frac{D_e^2 - D_i^2}{D_i^2} \times 100$$

where D_e = external diameter of the sampling tube

D_i = internal diameter of the sampling tube.

and it represents approximately the ratio between the volume of displaced soil and the volume of the sample. If this ratio is less than 10 per cent, the disturbance is considered negligible. The area ratio of the split spoon sampler is approximately 30% and hence a sample obtained is a disturbed sample. The area ratios for the sampling tubes mentioned in table 16·1 are less than 10% and hence a sample obtained with these tubes is an undisturbed sample.

---

1.   Reproduced by permission of the ISI.
2.   Reproduced by permission of the ISI.

16·7. Measurement of Ground Water

The level of the water table effects the foundation design in many ways, *e.g.* as indicated in chapter 12, the bearing capacity of the soil besides other factors depends upon the level of the water table. So, the level of the water table must be determined with reasonable accuracy.

The method of determining the ground water level in a boring depends upon the permeability of the soil.

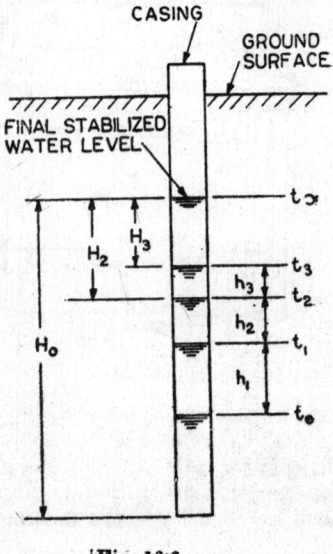

[Fig. 16·6

1. In pervious materials like sands and gravels, the water level in the bore-hole reaches a final level very quickly. After the level is stable, a measuring tape can be lowered into the bore-hole for measurement.

2. In silty soil, the water level in the bore-hole may not reach a stable position for many days. In such cases, following procedure may be adopted.

Refer to Fig. 16·6

(*a*) Determine rise or fall of water level in two or more equal time intervals t, *i.e.* $t = t_1 - t_0 = t_2 - t_1 = \ldots\ldots\ldots$

(*b*) The final water level is given by,

$$H_0 = \frac{h_1{}^2}{(h_1 - h_2)}$$

or

$$H_2 = \frac{h_2{}^2}{h_1 - h_2}$$

or

$$H_3 = \frac{h_3{}^2}{h_2 - h_3}$$

3. In impervious clayey soils special pressure measuring devices are employed for finding out the final water level in the bore-hole.

In most of the projects the water table level is measured after 24 hours of boring. However, such a measurement could be misleading in silty and clayey soils in which it takes days and sometimes weeks for the water level to come to a stable position.

16·8 Planning an Exploration Programme

Planning an exploration programme is not an easy task. The planner must have a clear concept of the objectives of the programme, must know the latest methods of soil exploration and must keep in mind the relative cost of the soil exploration versus the cost of the foundations.

There are many questions that the pose the engineer. They are :

1. What would be spacing of the bore-holes.
2 How many minimum bore holes should be planed for the site.
3. What should be the depth upto which each-boring should be taken.
4. What equipment should be used for advancing the bore holes and for taking the samples.

The last question has partly been replied briefly in the previous article. The first three questions are very important and need careful attention on part of the engineer incharge of planning the exploration programme.

The nature and the size of the job govern to some extent the planning of a sub-soil exploration programme. Larger is the job, naturally more elaborate will be the exploration programme. However, number of borings, their location and depth is not solely governed by the size and nature of the project, but to a large degree they depend upon the sub-surface conditions. Reconnaissance exploration comes handy in deciding these issues. In fact, the exploration is often a series of progressive approximations in which each step is determined by the results of the already completed part of the exploration. It is, therefore, difficult to give hard and fast rules as to how many borings should be taken, what should be the depth of each bore hole and what should be the spacing between the borings. As a general rule, one may say that more uniform a soil is in its horizontal stratification, the more will be the spacing between two bore-holes and hence less will the number of bore holes required to cover the area under investigation. Table 16·2 may be taken as a general guide for spacings and minimum number of borings required at a construction site.

The IS : 1892-1962 recommends that for a compact building site covering an area of 0.4 hectare (=4000 sq. metres), one boring at each of the corners and one at the centre should be quite adequate. For smaller buildings; even one bore hole should be enough. For large industrial and residential areas, cone penetration tests may be performed at every 100 metres by dividing the area into a grid.

The depth of a bore hole depends to some extent on the type and size of the structure, design considerations like the safety against foundation failure, permissible total and differential settlements, seepage and earth pressure, etc·, but to a

TABLE 16·2

Suggested spacing of borings and minimum number of borings required

S. No.	Type of Job	Type of soil in its horizontal stratification			Minimum number of borings at each job.
		Uniform	Average	Erratic	
		Distance between borings (metres)			
1.	1 or 2 storeyed structure	60	30	15	3
2.	Multi-storeyed structure	45	30	15	4
3.	Bridge piers, abutments and transmission towers	—	30	1	1—2 for each foundation unit
4.	Highways and airports	300	150	100	—
5.	Borrow pits for embankments	300—150	150—160	30—15	—

large extent it depends upon the character and sequence of subsurface strata in the vertical direction. It is sometimes advisible, therefore, to carry out the trial borings to the full depth of the sub-soil until the rock is met with. As a general rule, borings should extend to the soil stratum of adequate bearing capacity and must penetrate all deposits which are unsuitable for foundation purposes. Such deposits include highly compressible materials like peat, soft clay, organic soil and unconsolidated fill. Even if a soft layer is sandwiched between soil layers of better bearing capacities it must be penetrated, since the settlement of a foundation may depend upon this soft layer. If a foundation has to rest on a compressible clay deposit of large thickness, the boring must penetrate a depth where the stress increase beyond that level will not cause appreciable settlement of the foundation. Unless loads are very heavy, the borings are stopped as soon as the rock is reached. Table 16·3 gives the rules of the thumb, for depth to which the trial borings should be taken. These rules are subject to the general considerations given above. This table has been prepared according to the recommendations of the committee on sampling and testing of the ASCE (1).

TABLE 16·3

Suggested Depth of Penetration of Trial Boring

S. No.	Type of Structure	Design consideration and criterian for depth of exploration	Depth of penetration of trial borings	Remarks
1.	Building Foundations	Excessive total and differential settlement	Minimum D-10 metres or upto the depth where the increase in stress due to the structure is 10 per cent of original vertical stress at that depth	D is the depth of exploration below loaded rectangular area. See Fig. 16·7.
2.	Retaining and quay walls	Bearing capacity failure primary consideration. Settlements, total or differential secondary consideration	D=0.5 to 2H	D is the depth of exploration below the bottom of the wall and H is the net height of the wall See Fig. 16·8.
3.	Terraces & Fills	-do-	D=1.25 L for terraces and 0.5 L for traingular fills.	D is the depth of exploration below the original ground surface and L is the average horizontal length of the slope. See Fig. 16·9.
4.	Deep cuts	Stability of slopes	D=0·75B to 1·0 B	Depth of the probale surface of sliding is limited by the bottom width B of the cut D as in Fig. 16·10.
5.	Earthen Dams and Levees	Bearing capacity failure and see—page	D=L	D is the depth of exploration below the original ground surface and 2 L is the bottom width See Fig. 16·11.
6.	Concrete Dams	-do-	D=1·5 to 2·0 H	D as under 4 above and H is the net height of the dam. See Fig. 16·11.

TABLE 16·3 (*Continued*)

7.	Highways, railroads and airfields	General stability and drainage conditions	D=1 to 2 metres for light loads and 2 to 3·5 metres for heavy loads	D is the depth of exploration as shown in Fig. 16·12. All unsuitable strata to be fully explored *See* Fig. 16·12
8.	Tunnels	Stability of materials and the pressure exerted against the tunnel lining.	D=B	D is the depth of exploration below the invert elevation and B is the gross width of the tunnel. *See* Fig. 16·13.

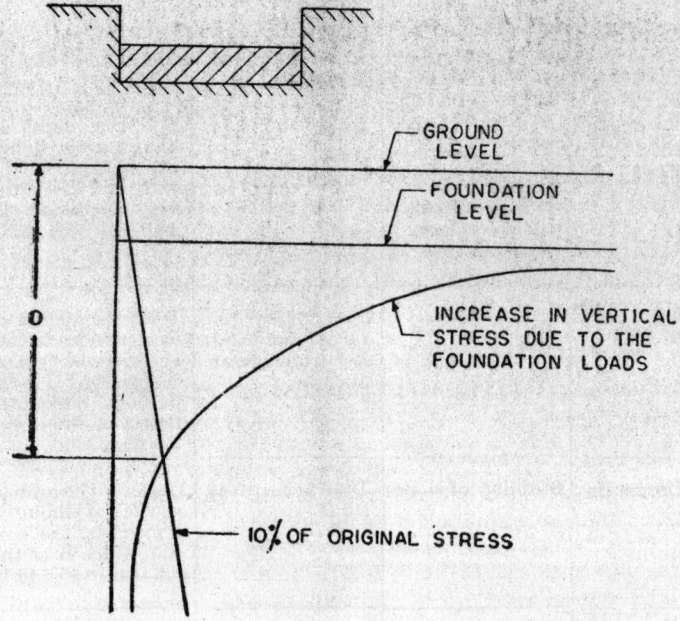

Fig. 16·7.

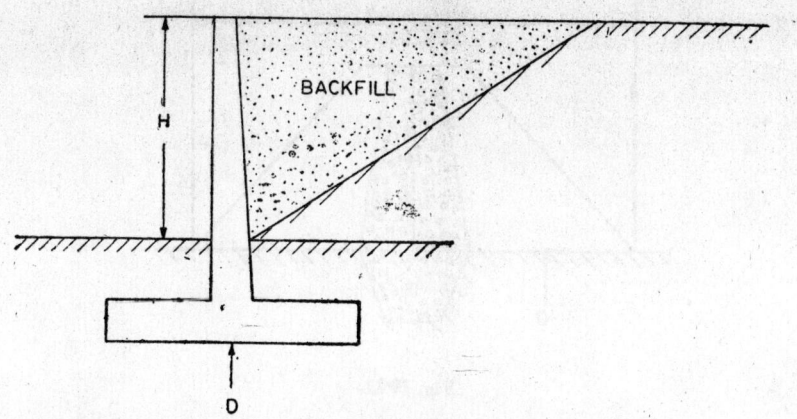

Fig. 16·8.

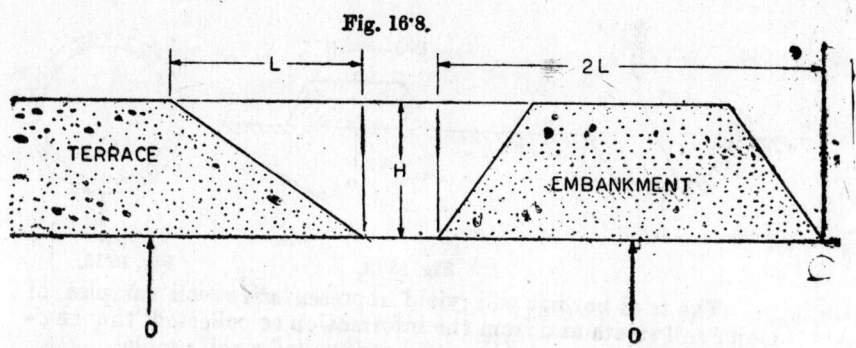

Fig. 16.9.

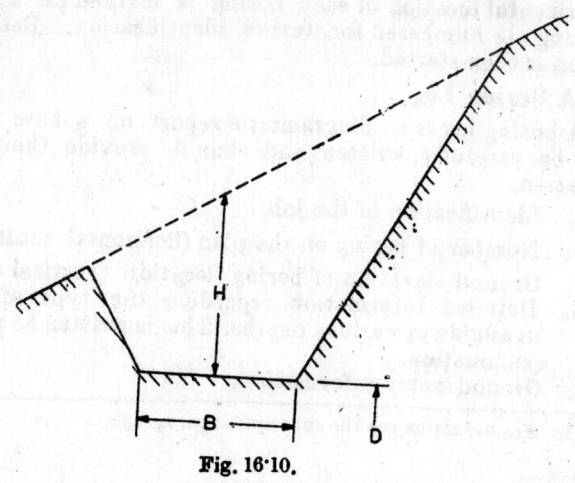

Fig. 16·10.

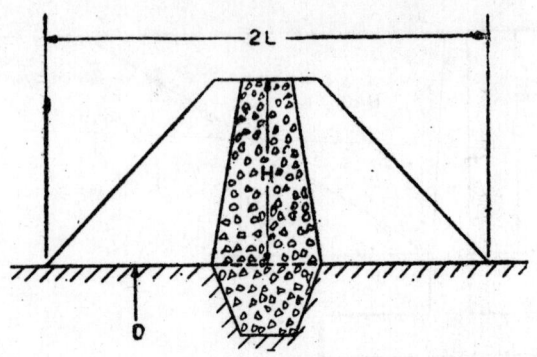

Fig. 16·11.

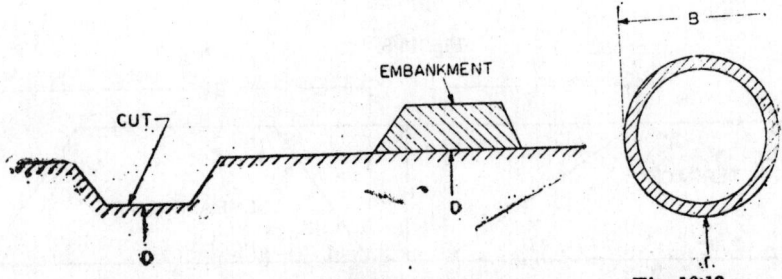

Fig. 16·12.   Fig. 16·13.

The trial borings will yield representative soil samples of major soil strata and from the information so collected the engineer can have a rough idea of what is called a soil profile[1].

Then a careful planning of the rest of the borings is done. The horizontal location of each boring is marked on a site plan. All boring are numbered for future identification. Detailed exploration is then started.

16·9. A Boring Log

A boring log is a diagrammtic report on a bore hole. It should be carefully written and should provide the following information.

1. Identification of the job.
2. Number of boring on the plan (horizontal control).
3. Ground elevation of boring location (vertical control).
4. Detailed infoimation regarding the type of the soil available at various depths. This is written as per visual examination.
5. Ground water information.

---

1. For definition see the subsequent paragraph.

6. Strength test values, if any tests are conducted in the field. These values must be shown on the boring log at the relevant elevations at which the tests are conducted or from which the samples are taken for the tests (The tests usually conducted in the field are the unconfined compression test, the vane shear test and the standard penetration test).

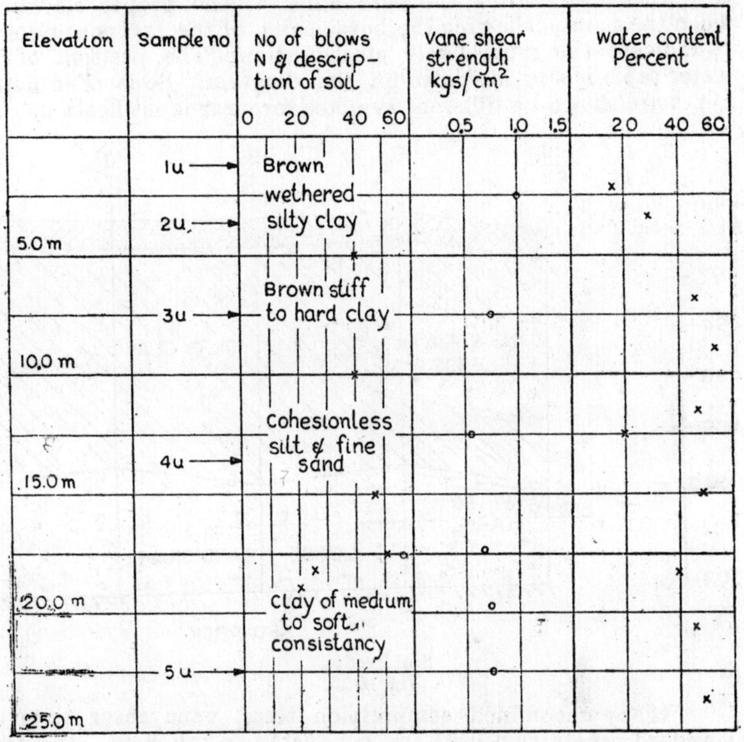

Elevation	Sample no	No of blows N & description of soil				Vane shear strength kgs/cm²			Water content Percent		
		0	20	40	60	0.5	1.0	1.5	20	40	60
	1u →	Brown									
		wethered								x	
5.0 m	2u →	silty clay									x
		Brown stiff									x
	3u →	to hard clay				o					x
10.0 m											x
		cohesionless				o			x		x
15.0 m	4u →	silt & fine sand									x
					x						x
						o	o				x
20.0 m		clay of medium	x							x	
		to soft consistancy				o					x
	5u →						o				
25.0 m											x

u denotes unconfined compression sample

Data field in a boring log.
Fig. 16·14.

Fig. 16·14 is an example of the data field in a boring log.

The boring logs of the various bore-holes form a part of the boring report which is kept as a permanent record.

Soil samples taken in the field are properly identified and identification recorded in the boring report.

16·10. A Soil Profile

A soil profile is a graphical representation of the sub-soil conditions along a given line on the ground. A detailed soil profile can only be prepared after the detailed exploration along a line has been done, since the detailed exploration will yield the soil data at reasonably spaced points on the line along which the profile is to be drawn. Fig.16·15 gives an example of a detailed soil profile. The borings along the profile line are plotted as vertical lines drawn to a suitable horizontal scale. Soil layers encountered in each bore hole are then plotted along these vertical lines representing the bore holes at the proper elevations. Then the points showing the boundaries of the layers are joined to indicate the most likely stratification. The position of the water table is also indicated on the diagram. Soils of different types are shown by different symbols for clear identification.

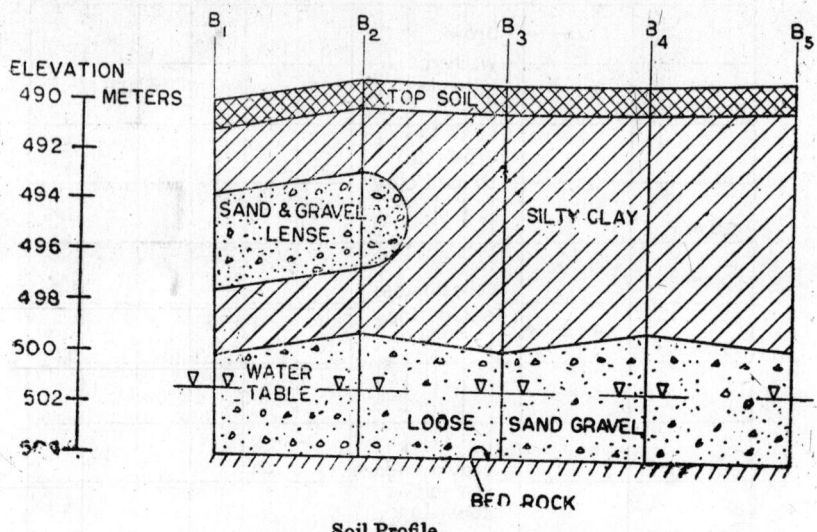

Soil Profile.
Fig. 16·15.

If any unconfined compression tests, vane shear tests or standard penetration tests are conducted in the field, the test results are also shown on the soil profile against the elevations, at which they are conducted or from which sampler are taken to conduct the tests. (These tests values are not shown in the Fig. 16·15).

16.11. Direct and Indirect Methods

The direct and indirect methods of sub-surface exploration were referred to in article 16.2.

The direct accessible borings are expensive. Information of a specialized nature can be obtained from these methods, although it has always to be supplemented with information from additional borings, so as to have a complete picture of the whole area. Specialized information can be : (1) obtaining of special samples (2) study of ground water conditions and the anticipated construction difficulties for moderate jobs (3) an appraisal of the soil conditions for the foundations of heavy structures like dams. Test pits, trenches or accessible borings are only dug if the reconnaissance exploration indicates any need for them. Sometimes the pits are dug to carry out load-tests.

The indirect methods like the electrical resistivity method the gravitational and magnetic methods, the siesmic method, and the soundings are only employed to know the major changes in sub-soil conditions as no specific information can be obtained from these methods. These methods are not very common in India and are therefore not described here for want of space. The interested student may look up references 1 and 3.

CHAPTER—BIBLIOGRAPHY

1. AASHO. Standard specifications for Highway Materials and Methods of Sampling and Testing, Part 1 & 2, 9th edition, published by the Association, Wash, D.C. 1966.

2. ASSHO, Manual on Foundation Investigation, ASSHO Washington D.C., 1967.

3. Bowls, E., Foundation Analysis and Design, McGraw Hill Book Co., New York, N.Y., 1968.

4. Ivorslev, M.J., Subsurface Exploration and Sampling of Soil for Civil Engineering Purposes, report on a research project of the Committee on Sampling and Testing. S.M. & F. Division, American Society of Civil Engineers, 1962.

5. Indian Standards Institution, Code of Practice, for the site investigation for Foundations, IS ; 1892-1962, New Delhi.

6. Indian Standards Institution, Code of Practice for Thin Wall Tube Sampling for Soils, IS : 2132 1963, New Delhi.

7. Indian Standards Institution, Methods of Standard Penetration Test for Soils, IS : 2131-1963, New Delhi.

8. Lambe, T.W., and Whitman, R.V., Soil Mechanics, J. Wiley & Sons, Inc., N.Y., 1969.

9. Mohr, H.A., Exploration of Soil Conditions and Sampling Operation, a report from H.A. Mohr, Consulting Engineer, 250 Stuart Street, Boston, 1962.

10. Teng, W.C., Foundation Design, New Jersey, Prentice Hall, Inc., 1962.

11. Terzaghi, K., and Peck, R.B., Soil Mechanics in Engineering Practice, New York, J. Willey & Sons, 1948.

12, Tomlinson, M.J., Foundation Design and Construction, Wiley & Sons, Inc., New York, N.Y., 1969.

QUESTIONS

1. What are the important features which need attention during soil exploration ? Describe the different methods of soil exploration. *(P.U. 1959 Supp)*

2. What is the purpose of soil surveys and soundings? Describe the methods of taking disturbed and undisturbed soil samples.

If free water is encounted in a bore-hole made in connection with a soil survey, what information concerning it should be recorded and explain how this information is obtained *(P.U, 1960 Supp)*

3. Write a short note on "Soil Exploratory programme for a Highway". *(P.U. 1961)*

4. (a) Discuss the various geophysical methods available for soil exploration. Mention their relative advantages and disadvantages.

   (b) How are undisturbed samples obtained and preserved prior to testing ? Sketch the e--log p curves from consolidation tests on an undisturbed sample, on a partly disturbed sample and on a remoulded sample of a normally consolidated clay. *(P.U. 1962)*

Note : For the latter part of question 4 (b). see Chapter 7 of this book.

5. Describe the various methods of "Soil Exploration" and discuss their merits and demerits. *(P. U. 1963)*

6. What is the purpose of a soil survey for new highways ? How is it conducted ? What type of samples and in what quantity would you like to collect during the soil survey ? Enumerate the field and laboratory tests you would recommend on the soil, which may be helpful for the design and construction of a proposed highway.
(P.U. 1963 Supp)

7. What do you understand by (i) representative sample (ii) undisturbed sample (iii) non-representative sample of soil ? Discuss briefly the methods of taking an undisturbed sample in cohesionless soil. *(P.U. No. 1964)*

8. Write short notes on :
   (a) Auger and Wash borings.
   (b) Standard split spoon sampler.
   (c) Standard penetration value. *(P.U. Nov. 1945)*

9. Describe the procedure to be adopted for sub-soil exploration for a large ware-house with a raft foundation. *(P.U. Nov. 1966)*

10. (a) What is the purpose of Soil exploration programme ?
    (b) Decribe a displacement boring. How is a sample taken using thi method. What kind of sample do you expect and why ?
    (c) What is meant by accessible exploration ? Name these types.
    (d) You are to prepare a subsurface exploration programme for a 12 storeyed building 300 feet by 200 feet in area. The area upon which this is to be built is a fairly recent till material of about 15 to 20 feet in depth underlain by a compressible clay layer approximately 10 to 15 feet in depth. Below the soft clay, we expect medium stiff clay down to bed rock at about 80 to 100 feet. below ground level. Outline your exploration programme considering number and depths of holes, types of exploration tools to be used and type and location of samples to be taken. What tests will you require for the purpose of design.
    (Mysore 1968)

APPENDIX

APPENDIX

TABLE I

Influence Values for Solution of Boussinesq Equation (Point Load)

(After G. Gilboy)

r/z	N_B	r/z	N_B	r/z	N_B	r/z	N_B
0·00	0·4775	0·40	0·3294	0 80	0·1386	1·20	0·0513
1	0·4773	1	0·3238	1	0·1353	1	0·0501
2	0·4770	2	0·3181	2	0·1320	2	0·0489
3	0·4764	3	0·3124	3	0·1288	3	0·0477
4	0·4756	4	0·3068	5	0·1257	4	0·0466
5	0·4745	5	0·3011	5	0·1226	5	0·0454
6	0·4732	6	0·2955	6	0·1196	6	0·0443
7	0·4717	7	0·2899	7	0·1166	7	0·0433
8	0·4699	8	0·2843	8	0·1138	8	0·0422
9	0·4679	9	0·2788	9	0·1110	9	0·0412
0·10	0·4657	0·50	0·2733	0·90	0·1083	1·30	0·0402
1	0·4633	1	0·2679	1	0·1057	1	0·0393
2	0·4607	2	0·2625	2	0·1051	2	0·0384
3	0·4579	3	0·2571	3	0·1005	3	0 0374
4	0·4548	4	0·2518	4	0·0981	4	0·0365
5	0·4516	5	0·2466	5	0·0956	5	0·0357
6	0·4482	6	0·2414	6	0·0933	6	0·0348
7	0·4446	7	0·2363	7	0·0910	7	0·0340
8	0·4409	8	0·2313	8	0·0887	8	0·0332
9	0·4370	9	0·2263	9	0·0865	9	0·0324
0·20	0·4329	0·60	0·2214	1·00	0·0844	1·40	0·0317
1	0·4286	0	0·2165	1	0·0823	1	0·0309
3	0·4242	2	0·2117	2	0 0803	2	0·0302
3	0 4197	3	0·2070	3	0 0783	3	0·0295
4	0·4151	4	0·2024	4	0·0764	4	0 0288
5	0·4103	5	0·1978	5	0·0744	5	0 0282
6	0·4054	6	0·1934	6	0·0727	6	0·0275
7	0·4004	7	0·1889	7	0·0709	7	0·0269
8	0 3954	8	0·1846	8	0·0681	8	0 0263
9	0·3902	9	0·1804	9	0·0674	9	0·0257
0·30	0·3849	0·70	0·1762	1·10	0·0658	1·50	0·0251
1	0·3796	1	0·1721	1	0·0641	1	0·0245
2	0·3742	2	0·1681	2	0·0626	2	0·0240
3	0·3687	3	0·1641	3	0·0610	3	0·0234
4	0·3632	4	0·1603	4	0 0595	4	0·0229
5	0·3577	5	0·1565	5	0·0581	5	0·0224
6	0 3521	6	0·1527	6	0·0567	6	0·0219
7	0·3465	7	0·1491	7	0 0553	7	0·0214
8	0·3408	8	0·1455	8	0·0539	8	0·0209
9	0·3351	9	0·1420	9	0 0526	9	0·0204

TABLE I—(Continued)

r/z	N_B	r/z	N_B	r/z	N_B	r/z	N_B
1·60	0·0200	2·10	0·0070	2·60	0·0029	3·10	0·0013
1	0·0195	1	0·0069	1	0·0028	1	0·0013
2	0·0191	2	0·0068	2	0·0028	2	0·0013
3	0·0187	3	0·0066	3	0·0027	3	0·0012
4	0·0183	4	0·0065	4	0·0027	4	0·0012
5	0·0179	5	0·0064	5	0·0026	5	0·0012
6	0·0175	6	0 0063	6	0·0026	6	0·0012
7	0·0171	7	0·0062	7	0·0025	7	0·0012
8	0·0167	8	0·0060	8	0·0025	8	0·0012
9	0·0163	9	0·0059	9	0·0025	9	0·0011
1·70	0·0160	2·20	0·0058	2·70	0·0024	3·20	0·0011
1	0·0157	1	0·0057	1	0·0024	1	0·0011
2	0·0153	2	0·0056	2	0·0023	2	0·0011
3	0·0150	3	0·0055	3	0·0023	3	0·0011
4	0·0147	4	0·0054	4	0·0023	4	0·0011
5	0·0144	5	0·0053	5	0·0022	5	0·0011
6	0·0141	6	0·0052	6	0·0022	6	0·0010
7	0·0138	7	0·0051	7	0·0022	7	0·0010
8	0·0135	8	0·0050	8	0·0021	8	0·0010
9	0·0132	8	0·0049	9	0·0021	9	0·0010
1·80	0·0129	2·30	0·0048	2·80	0·0021	3·30	0 0010
1	0·0126	1	0·0047	1	0·0020	1	0·0009
2	0·0124	2	0·0047	2	0·0020	2	0 0009
3	0·0121	3	0·0046	3	0·0020	3	0·0009
4	0 0119	4	0·0045	4	0·0019	4	0·0009
5	0·0116	5	0·0044	5	0·0019	5	0·0009
6	0·0114	6	0·0043	6	0·0019	6	0·0009
7	0·0112	7	0·0043	7	0·0019	7	0·0009
8	0 0109	8	0·0042	8	0·0018	8	0·0009
9	0·0107	9	0·0041	9	0·0018	9	0·0009
1·90	0·0105	2·40	0·0040	2·90	0·0018	3·40	0·0009
1	0·0103	1	0·0040	1	0·0017	3·41	
2	0·0101	2	0·0039	2	0·0017	to	0·0008
3	0·0099	3	0·0038	3	0·0017	3·49	
4	0·0097	4	0·0038	4	0·0017	3·50	
5	0·0095	5	0·0037	5	0·0016	to	0·0007
6	0·0093	6	0·0036	6	0·0016	3·61	
7	0·0091	7	0·0036	7	0·0016	3·62	
8	0·0089	8	0·0035	8	0·0016	to	0·0006
9	0·0087	9	0·0034	9	0·0015	3·74	
						3·75	
						to	0·0005
						3·90	
2·00	0·0085	2·50	0·0034	3·00	0 0015	3·91	
1	0·0084	1	0·0033	1	0·0015	to	0·0004
2	0·0082	2	0·0033	2	0·0015	4·12	
3	0·0081	3	0·0032	3	0·0014	4·13	
4	0·0079	4	0·0032	5	0·0014	to	0·0003
5	0·0078	5	0·0031	5	0·0014	4·43	
6	0·0076	6	0·0031	6	0·0014	4·44	
7	0·0075	7	0·0030	7	0·0014	to	0·0002
8	0·0073	8	0·0030	8	0·0013	4·90	
9	0·0072	9	0·0029	9	0·0013	4·91	
						to	0·0001
						6·15	

TABLE II

Influence Values for Solution of Boussinesq Equation (Rectangular load)

n

m	0·1	0·2	0·3	0·4	0·5	0·6	0·7	0·8	0·9	1·0	1·2	1·4
0·1	0·00470	0·00917	0·01323	0·01678	0·01978	0·02223	0·02420	0·02576	0·02698	0·02794	0·02926	0·03007
0·2	0·00917	0·01790	0·02585	0·03280	0·03866	0·04348	0·04735	0·05042	0·05283	0·05471	0·05733	0·05894
0·3	0·01323	0·02585	0·03735	0·04742	0·05593	0·06294	0·06858	0·07308	0·07661	0·07938	0·08323	0·08561
0·4	0·01678	0·03280	0·04742	0·06024	0·07111	0·08009	0·08734	0·09314	0·09770	0·10129	0·10631	0·10941
0·5	0·01978	0·03866	0·05593	0·07111	0·08403	0·09473	0·10340	0·11035	0·11584	0·12018	0·12626	0·13003
0·6	0·02223	0·0438	0·06294	0·08009	0·09473	0·10688	0·11679	0·12474	0·13105	0·13605	0·14309	0·14749
0·7	0·02420	0·04735	0·06858	0·08734	0·10340	0·11679	0·12772	0·13653	0·14356	0·14914	0·15703	0·16199
0·8	0·02576	0·05042	0·07308	0·09314	0·11035	0·12474	0·13653	0·14607	0·15371	0·15978	0·16843	0·17389
0·9	0·02698	0·05283	0·07661	0·091770	0·11584	0·13105	0·14356	0·15371	0·16185	0·16835	0·17766	0·18357
1·0	0·02794	0·05471	0·07938	0·10129	0·12018	0·13605	0·14914	0·15978	0·16835	0·17522	0·18508	0·19139
1·2	0·02926	0·05733	0·08323	0·10631	0·12626	0·14309	0·15703	0·16843	0·17766	0·18508	0·19584	0·20278
1·4	0·03007	0·05894	0·08561	0·10941	0·13003	0·14749	0·16199	0·17389	0·18357	0·19139	0·20278	0·21020
1·6	0·03058	0·05994	0·08709	0·11135	0·13241	0·15028	0·16515	0·17739	0·18737	0·19546	0·20731	0·21510
1·8	0·03090	0·06058	0·08804	0·11260	0·13395	0·15207	0·16720	0·17967	0·18986	0·19814	0·21032	0·21836
2·0	0·03111	0·06100	0·08867	0·11342	0·13496	0·15326	0·16856	0·18119	0·19152	0·19994	0·21235	0·22058
2·5	0·03138	0·06155	0·08948	0·11450	0·13628	0·15483	0·17036	0·18321	0·19375	0·20236	0·21512	0·22364
3·0	0·03150	0·06178	0·08982	0·11495	0·13684	0·15550	0·17113	0·18407	0·19470	0·20341	0·21633	0·22499
4·0	0·03158	0·06194	0·09007	0·11527	0·13724	0·15598	0·17168	0·18469	0·19540	0·20417	0·21722	0·22600
5·0	0·03160	0·06199	0·09014	0·11537	0·13737	0·15612	0·17185	0·18488	0·19561	0·20440	0·21749	0·22632
6·0	0·03161	0·06201	0·09017	0·11541	0·13741	0·15617	0·17191	0·18496	0·19569	0·20449	0·21760	0·22644
8·0	0·03162	0·06202	0·09019	0·11543	0·13744	0·15621	0·17195	0·18500	0·19574	0·20455	0·21767	0·22652
10·0	0·03162	0·06202	0·09019	0·11544	0·13745	0·15622	0·17196	0·18502	0·19576	0·20457	0·21769	0·22654
∞	0·03162	0·06202	0·09019	0·11544	0·13745	0·15623	0·17197	0·18502	0·19577	0·20458	0·21770	0·22656

TABLE II—Continued

n

m	1·6	1·8	2·0	2·5	3·0	4·0	5·0	6·0	8·0	10·0	∞
0·1	0·03058	0·03090	0·03111	0·03138	0·03150	0·03158	0·03160	0·03161	0·03162	0·03162	0·03162
0·2	0·05994	0·06058	0·06100	0·06155	0·06178	0·06194	0·06199	0·06201	0·06202	0·06202	0·06202
0·3	0·08709	0·08804	0·08867	0·08948	0·08982	0·09007	0·09014	0·09017	0·09018	0·09019	0·09019
0·4	0·11135	0·11260	0·11342	0·11450	0·11495	0·11527	0·11537	0·11541	0·11543	0·11544	0·11544
0·5	0·13241	0·13395	0·13496	0·13628	0·13684	0·13724	0·13737	0·13741	0·13744	0·13745	0·13745
0·6	0·15028	0·15207	0·15326	0·15483	0·15550	0·15598	0·15612	0·15617	0·15621	0·15622	0·15623
0·7	0·16515	0·16720	0·16856	0·17036	0·17113	0·17168	0·17185	0·17191	0·17195	0·17196	0·17197
0·8	0·17739	0·17967	0·18119	0·18321	0·18407	0·18469	0·18488	0·18496	0·18500	0·18502	0·18502
0·9	0·18737	0·18986	0·19152	0·19375	0·19470	0·19540	0·19561	0·19569	0·19574	0·19576	0·19577
1·0	0·19546	0·19814	0·19994	0·20236	0·20341	0·20417	0·20440	0·20449	0·20455	0·20457	0·20458
1·2	0·20731	0·21032	0·21235	0·21512	0·21633	0·21722	0·21749	0·21760	0·21767	0·21769	0·21770
1·4	0·21510	0·21836	0·22058	0·22364	0·22499	0·22600	0·22632	0·22644	0·22652	0·22654	0·22656
1·6	0·22025	0·22372	0·22610	0·22940	0·23088	0·23200	0·23236	0·23249	0·23258	0·23261	0·23263
1·8	0·22372	0·22736	0·22986	0·23334	0·23495	0·23698	0·23735	0·23671	0·23681	0·23684	0·23686
2·0	0·22610	0·22986	0·23247	0·23614	0·23782	0·23912	0·23954	0·23970	0·23981	0·23985	0·23987
2·5	0·22940	0·23334	0·23614	0·24010	0·24196	0·24344	0·24392	0·24412	0·24425	0·24429	0·24432
3·0	0·23088	0·23495	0·23782	0·24196	0·24394	0·24554	0·24608	0·24630	0·24646	0·24650	0·24654
4·0	0·23200	0·23698	0·23912	0·24344	0·24554	0·24729	0·24791	0·24817	0·24836	0·24842	0·24846
5·0	0·23236	0·23735	0·23954	0·24392	0·24608	0·24791	0·24857	0·24885	0·24907	0·24914	0·24919
6·0	0·23249	0·23671	0·23970	0·24412	0·24630	0·24817	0·24885	0·24916	0·24939	0·24946	0·24952
8·0	0·23258	0·23681	0·23981	0·24425	0·24646	0·24836	0·24907	0·24939	0·24964	0·24973	0·24980
10·0	0·23261	0·23684	0·23985	0·24429	0·24650	0·24842	0·24914	0·24946	0·24973	0·24981	0·24989
∞	0·23263	0·23686	0·23987	0·24432	0·24654	0·24846	0·24919	0·24952	0·24980	0·24989	0·25000

TABLE III[1]

Maximum Safe Bearing Capacity

(*Courtesy Indian Standards Institution*)

S. No.	Types of Rocks and Soils	Maximum Safe Bearing Capacity tonne/m²	Remarks
	I Rocks		
1.	Rocks-head without laminations and defect for example granite, trap and diorite	330	
2.	Laminate rocks (for example sand stone and lime-stone) in sound condition	165	
3.	Residual deposits of shattered and broken bed rock and hard shale, cemented material	90	
4.	Soft Rock	45	
	II Non-Cohesive Soils		
5.	Gravel, sand and gravel, compact and offering high resistance to penetration when excavated by tools	45	See note 2.
6.	Coarse sand, compact and dry	45	Dry means that the ground water level is at a depth not less than the width of the foundation below the base of the foundations.
7.	Medium sand, compact and dry	25	
8.	Fine sand, silt (dry lumps easily pulverised by the fingers)	15	
9.	Loose gravel or sand gravel mixture; loose coarse to medium sand, dry	25	See note 2.
10.	Fine sand, loose and day	10	
	III Cohesive Soils		
11.	Soft shale, hard or stiff clay in deep bed, dry	45	This group is susceptible to long term consolidation settlement.
12.	Medium clay, readily indented with a thumb nail	25	

1. Reproduced from IS: 1904—1961, by permission of the Indian Standards Institution, New Delhi.

13.   Moist clay and sand clay mixture   15
        which can be indented with strong
        thumb pressure

14.   Soft clay indented with moderate   10
        thumb pressure

15.   Very soft clay which can be penetrated   5
        several inches with the thumb

16.   Black cotton soil or other shrinkable   15
        or expansive clay in dry condition
        (50% saturation)

IV Peat

17.   —Peat   —   See note 3. To be
                                   determined after
                                   investigation.

V Made-up Ground

18.   Fills or made-up ground   —   See note 3. To be
                                                     determined after
                                                     investigation.

Note 1. Increase or decrease the allowable bearing values as follows :

(a) The allowable bearing values may be increased by an amount equal to the weight of the material removed from above the bearing level, that is, the base of the foundation.

(b) For non-cohesion soil, the allowable bearing value shall be reduced by 50 percent if the water table is above or near the bearing surface of the soil. If the water table is below the bearing surface of the soil at a distance at least equal to the width of the foundation, no such reduction shall apply. For intermediate depths of water table, proportional reduction of the allowable value may be made.

Note 2. Compactness or looseness of non-cohesive materials may be determined by driving a wooden picket of dimensions $5 \times 5 \times 70$ cm with a sharp point. The picket shall be pushed vertically into the soil by the full weight of a person weighing 70 kg. If the penetration of the picket exceeds 20 cms, the loose state shall be assumed to exist.

Note 3. No generalized values for safe bearing pressures can be given for these types of soils. In such. cases, adequate site investigation (see IS : 1892—Code of Practice for Site Investigation for Foundations) shall be carried out and expert advice shall be sought.

Peat may occur in a very soft spongy condition or may be quite firm and compact. While ultimate bearing capacity may be high in the compact cases, very large consolidation settlements occur even under small pressures and movements continue for decade.

The strength of made-up ground depends on the nature of the material, its depth and age and methods used for consolidating it (if any).

INDEX

A

Accuracy of settlements, 299
Active earth of pressure, 373
—effect of inclination of ground surface, 396
Active Rankine state, 382
Actuating forces, 492
Absorbed moisture, 118
Aeration, zone of, 119
Angle of contact, 121
Angle of internal friction, 322
—effective, 324
—true, 327, 330
Arching in ideal soils, 415
Area correction, 340
Atterberg limits, 46, 67
Auger, soil, 541
—boring, 541
—helical, 541
—post-hole, 541

B

Base parabola, 195
Bed rock, 1, 426
Bearing capacity, 426
—failure, 433
—gross, 427
—net, 427
—safe, 426
—tables, 432
—ultimate, 426
Bearing capacity of,
—a cut in soft clay, 470
—isolated footings, 455
—piles in sand, 475
Bentonite, 88
Black cotton soil, 89
Brownian movement, 11
Boulders, 86
Bound water, 118
Boundary conditions, 177
—limiting, 177
Boussinesq's case, 265
Bulk density, 17

C

Capillary,
—forces, 120
—fringe, 119
—head, 127, 164
—head test, 163
—moisture, 118

Clay, 87
—lean, 88
—varved, 88
Coefficient of,
—compressibility, 226
—consolidation, 226, 236
—permeability, 117, 138
—permeability in the horizontal direction, 152
—permeability in the vertical direction, 150, 151
—uniformity, 66
—volume compressibility, 227
—earth pressure at rest, 373, 379
Cohesion, 322
—effective, 324
—factor, 329
—true, 327, 330
Cohesive water, 118
Colloidal suspension, 10
Compaction,
—equipment for, 520
—factors affecting soil, 543
—field, 519
—ISI standard on, 515
—modified AASHO test, 513, 515
—Proctor's test, 513, 514
Compression index, 222
Compressibility of soils, 205
Compressibility characteristics, measurement of, 208
Confined acquifer, 157
Conjugate planes, stresses, 398
Consolidation, of soils, 207
—ratio, 226
Contact,
—moisture, 119
—pressure, 471
Consolidated quick test, 331
Consolidated undrained test, 331
Continuity equation, 168, 171
Continuous phase, 10
Critical arc location of, 503
Critical hydraulic gradient, 39, 181, 184, 200
Culmann's graphical solution, 408

D

Darcy's
—coefficient of permeability, 138
—law, 137, 169
Deep foundations, 430
Degree of, consolidation, 232, 239
—average consolidation, 240